Antiviral Secrets of Medicinal Plants

Editors:
Deepu Pandita
Government Department of School Education
Jammu and Kashmir, India

Anu Pandita
Vatsalya Clinic, New Delhi, India

CRC Press
Taylor & Francis Group
Boca Raton London New York

CRC Press is an imprint of the
Taylor & Francis Group, an **informa** business
A SCIENCE PUBLISHERS BOOK

First edition published 2026
by CRC Press
2385 NW Executive Center Drive, Suite 320, Boca Raton FL 33431

and by CRC Press
4 Park Square, Milton Park, Abingdon, Oxon, OX14 4RN

CRC Press is an imprint of Taylor & Francis Group, LLC

Library of Congress Cataloging-in-Publication Data (applied for)

ISBN: 978-1-032-56052-6 (hbk)
ISBN: 978-1-032-56053-3 (pbk)
ISBN: 978-1-003-43358-3 (ebk)

DOI: 10.1201/9781003433583

Typeset in Times New Roman
by Prime Publishing Services

Dedication

This book is lovingly dedicated to our mother Mrs. Raj Dulari, who has been our pillar of strength and fountain of wisdom. Your unwavering support, inspiration, and boundless love have been our guiding light.

Preface

In an era where viral infections pose significant health challenges globally, the search for effective antiviral treatments has never been more critical. Humanity has continually sought ways to combat these invisible adversaries. According to the World Health Organization (WHO) statistics, three-quarters (80%) of biosphere inhabitants depend primarily on plant-based medicines to improve health. Plant-based therapies are in tune with nature and the golden truth is that these have no side effects or minor side effects. Of the estimated 300,000 plant species, two-thirds of global floral species have therapeutic worth. It is estimated that plant-based natural product drugs make up 25% of the total drugs in developed countries like the United States, and 80% in developing countries, for instance, India and China which supply two-thirds of the plants used in the Western medicinal system.

Severe viral infections such as Acquired immunodeficiency syndrome (AIDS) by causal agent of Human Immunodeficiency Virus (HIV-1, HIV-2), Ebola Virus Disease (EVD) caused by Ebola virus, Herpes simplex by Herpes simplex virus (HSV-1, HSV-2), Influenza by influenza A virus, Hepatitis by hepatitis virus *A*, *B*, *C*, *D*, *E*. and devastating pandemics of human coronavirus (COVID-19) due to Severe Acute Respiratory Syndrome Coronavirus 2 (SARS-CoV-2) represent some significant global problems. In the present emerging urgency pandemics like COVID-19 have no successful treatment for viral load of SARS-CoV-2 and potential antiviral medicines are deficient. Immune system boosting and rejuvenating plants are needed to fight against unhealthy environments and viruses without any side effects as people with strong immunity show a higher recovery rate against *e.g.*, COVID-19.

WHO has focused on herbal and modern systems of medicine. There are several medicinal plants with a repository and treasure of antiviral metabolites and other phytochemicals that have proven their effectiveness in killing the invaded viruses and can be designed into novel antiviral drugs. The following chapters compile cutting-edge research on the antiviral properties of various plants, discussing their chemical components and bioactive constituents. This book seeks to bridge the gap between traditional knowledge and contemporary scientific inquiry by delving into the mechanisms through which these plants exert their antiviral effects. The gingerols and shogaols from fresh ginger have antiviral activity on the human respiratory syncytial virus (HSRV). Glycyrrhizin, 18β-glycyrrhetinic acid (18β-GA), and 18β-GA of *Glycyrrhiza glabra* inhibit human respiratory syncytial virus (HRSV). Curcumin of *Curcuma longa* has antiviral activity against parainfluenza virus type 3 (PIV-3), dengue virus type 2, vesicular stomatitis virus (VSV), feline infectious peritonitis virus (FIPV), enterovirus 71, Ebola virus, Ebstein-Barr virus, Herpes Simplex Virus (HSV), chikungunya virus, Respiratory Syncytial Virus (RSV), Hepatitis C Virus, Human Immunodeficiency Virus type 1(HIV), human papillomavirus, influenza virus A, Japanese encephalitis virus, murine norovirus, Zika virus, SARS-CoV-2 and Flock House virus (FHV) etc. *Tinospora cordifolia* extracts show efficacy against HSV and SARS-CoV-2. *Acorus calamus* β-asarone inhibits HSV-1 and HSV-2 and tatanan A is anti-Dengue virus-2 (anti-DENV2) by decreasing DENV replication and infection. Calendula flower extract containing dichloromethane inhibits HIV1- Reverse transcriptase and suppresses HIV-mediated fusion. A *Camellia sinensis* (tea) catechin epigallocatechin gallate (EGCG) reduces dengue viral growth reducing viral infection by affecting internalization or by modulating the endosomal pH as well as influenza A and B viruses, HIV, SARS-COV-2 and HSV. *Tamarindus* chitinase (chi)-like lectin (TCLL) exhibited a notable reduction in chikungunya viral infection. The vinblastine, vincristine, catharanthine, and vindoline of *Catharanthus roseus* have demonstrated potent antiviral activity against polio, HSV, dengue, influenza, and SARS-COV-2 viruses. *Centella asiatica* asiaticoside is effective in treating COVID-19. *Gymnema sylvestre* gymnemic acid A and B demonstrate antiviral activity against the Asian influenza virus, HSV, SARS-CoV-2, and moderately inhibited feline calicivirus whereas gymnemagenol inhibited the proliferation of Grouper Nervous Necrosis Virus (GNNV), and Noda virus. Hypericin from *Hypericum perforatum* can inhibit the replication of HIV and HSV. As we navigate the complexities of viral diseases, the potential of medicinal plants as antiviral agents offers a beacon of hope. This book will mainly focus on the green viral arsenals, viral defenders, and plant fighter metabolites and their modes of action. This book is an essential resource for researchers, healthcare professionals, and anyone interested in the intersection of botany and virology. We hope this compilation will inspire further research and innovation in the quest for natural antiviral solutions.

Poetic Musings on Antiviral Plants

In nature's garden green of meadows to the ancient trees, where healing blooms and weaves,
Antiviral wonders across the land and whispers through petals, roots, and leaves.
With roots deep in the earth and branches high,
These plants stand guard 'neath the azure sky.
Boosting immunity, a warrior's might,
Against viral foes, it takes its flight.
Catharanthus roseus, a bloom so divine,

Nature's pharmacy, a sacred sign.
Centella asiatica, leaves that entwine,
In the tapestry of health, a thread so fine.
Azadirachta indica, Neem's embrace,
Guardian of wellness in a verdant space.
Hypericum perforatum, a sunlit bloom,
Chasing shadows of illness, dispelling gloom.
Xanthoceras sorbifoli, a blossom rare,
Antiviral whispers lingering in the air.
Ginger (*Zingiber officinale*) root, with a fiery sting,
A remedy from the soil's spring.
Antiviral warmth, in teas that brew,
Nature's healer, tried and true.
So, let us wander in nature's embrace,
Embrace the healing, with humble grace.
In every plant, a story unfolds,
A poem of wellness, in green it molds.
Antiviral whispers, a harmonious refrain,
In nature's embrace, healing shall reign.

Editors
Deepu Pandita
Anu Pandita

Contents

List of Contributors

Abhijit Kumar
University Institute of Biotechnology, Chandigarh University, Gharuan, Mohali, Punjab, India

Abrar Ahmad
Department of Eastern Medicine and Surgery, Government College University Faisalabad, Pakistan

Anu Pandita
Vatsalya Clinic, Krishna Nagar, New Delhi 110051, India

Ashraf Dar
Department of Biochemistry, University of Kashmir, Srinagar 190006, India

Bushra Hafeez Kiani
Department of Biological Sciences (Female Campus), International Islamic University, Islamabad 44000, Pakistan

Chanin Sillapachaiyaporn
Department of Clinical Chemistry, Faculty of Allied Health Sciences, Chulalongkorn University, Bangkok 10330, Thailand
Center of Excellence on Natural Products for Neuroprotection and Anti-ageing, Chulalongkorn University, Bangkok 10330, Thailand

Deepu Pandita
Government Department of School Education, Jammu, Jammu and Kashmir, India

Dhaneswar Prusty
Department of Biochemistry, School of Life Sciences, Central University of Rajasthan, NH-8, Bandarsindri, Kishangarh, Ajmer, Rajasthan 305817 India

Dhanush S Sadhu
Kittur Rani Channamma College of Horticulture, Arabhavi 591 218, University of Horticultural Sciences, Bagalkot, Karnataka, India

Ehsan Asgarian
Department of Horticultural Sciences, Faculty of Agricultural Science and Engineering, University of Tehran, Iran

Ghousia Liaquat
Institute of Agro-Industry and Environment, The Islamia University of Bahawalpur, Pakistan

Ghulam Hassan Abbasi
Institute of Agro-Industry and Environment, The Islamia University of Bahawalpur, Pakistan

Ghulam Murtaza
Department of Food Science, Government College University, Faisalabad, Pakistan

Hafiza Salma
Institute of Agro-Industry and Environment, The Islamia University of Bahawalpur, Pakistan

Hamaad Raza Ahmad
Institute of Soil and Environmental Sciences, University of Agriculture Faisalabad 38040, Pakistan

Hanzla Siddique
Department of Eastern Medicine and Surgery, Government College University Faisalabad, Pakistan

Himanshu Dawda
Department of Botany, HVPS Ramniranjan Jhunjhunwala College (Empowered Autonomous), Ghatkopar (W), Mumbai 400 086, India

Ishita Pandey
University Institute of Biotechnology, Chandigarh University, Punjab, India

Issa Sulaiman Al-Amri
Department of Biological Sciences and Chemistry, College of Arts and Sciences, University of Nizwa, Birkat Al-Mouz, Nizwa, Sultanate of Oman

Karishma Rajbhar Khan
Department of Botany, HVPS Ramniranjan Jhunjhunwala College (Empowered Autonomous), Ghatkopar (W), Mumbai 400 086, India

K. Venkata Ratnam
Department of Botany, Rayalaseema University, Kurnool, Andhra Pradesh 518 007, India

Lepakshi Md. Bhakshu
Department of Botany, Dr. YSR Government Degree College, Vedurukuppam, Chittoor District 517 569, Andhra Pradesh, India.

Mahantesha Naika B.N.
Centre for Biotechnology Research, Department of Biotechnology and Crop Improvement, College of Horticulture, Bagalkot 587104, University of Horticultural Sciences, Bagalkot, Karnataka, India

Mani Iyer Prasanth
Department of Clinical Chemistry, Faculty of Allied Health Sciences, Chulalongkorn University, Bangkok 10330, Thailand
Center of Excellence on Natural Products for Neuroprotection and Anti-ageing, Chulalongkorn University, Bangkok 10330, Thailand

Manisha Sharma
Amity Institute of Biotechnology, Amity University Madhya Pradesh, India,

Masarat Bashir
COTS, Sheri-Kashmir University of Agricultural Science and Technology, Srinagar, India.

Mithraa T
Division of Plant Genetic Resources, ICAR-Indian Agricultural Research Institute, New Delhi 110012.

Mitra Aelaei
Department of Horticulture Sciences, Faculty of Agriculture, University of Zanjan, Zanjan, Iran

Mubeen Fatima
Centre for Applied Molecular Biology (CAMB), University of the Punjab, Lahore, Pakistan

Muhammad Ali
Institute of Agro-Industry and Environment, The Islamia University of Bahawalpur, Pakistan

Muhammad Ashar Ayub
Institute of Agro-Industry and Environment, The Islamia University of Bahawalpur, Pakistan

Muhammad Imran
Institute of Agro-Industry and Environment, The Islamia University of Bahawalpur, Pakistan

Muhammad Nadeem
Institute of Agro-Industry and Environment, The Islamia University of Bahawalpur, Pakistan

Nagesh Nilawar
Department of Quality Assurance, SVKM's Dr. Bhanuben Nanavati College of Pharmacy, Mumbai, Maharashtra, India

Neha Sharma
Amity Institute of Biotechnology, Amity University Madhya Pradesh, India

Noman Ahmad
Institute of Soil and Environmental Sciences, University of Agriculture Faisalabad 38040, Pakistan

Seyed Alireza Salami
Department of Horticultural Sciences, Faculty of Agricultural Science and Engineering, University of Tehran, Iran

Sayed Sartaj Sohrab
Special Infectious Agents Unit, King Fahd Medical Research Center, King Abdulaziz University, Jeddah, Saudi Arabia

Tewin Tencomnao
Department of Clinical Chemistry, Faculty of Allied Health Sciences, Chulalongkorn University, Bangkok 10330, Thailand
Center of Excellence on Natural Products for Neuroprotection and Anti-ageing, Chulalongkorn University, Bangkok 10330, Thailand

Shashidhar M. Dodamani
Department of Plantation, Spices, Medicinal and Aromatic crops, KRCCH, Arabhavi 591218, University of Horticultural Sciences, Bagalkot, Karnataka, India

Siriporn Chuchawankul
Department of Transfusion Medicine and Clinical Microbiology, Faculty of Allied Health Sciences, Chulalongkorn University, Bangkok 10330, Thailand
Immunomodulation of Natural Products Research Unit, Faculty of Allied Health Sciences, Chulalongkorn University, Bangkok 10330, Thailand

Parthasarathy Seethapathy
Amrita School of Agricultural Sciences, Amrita Vishwa Vidyapeetham, Coimbatore 642109

Pooja Bhadrecha
University Institute of Biotechnology, Chandigarh University, Punjab, India

Pooja Bhadrecha
University Institute of Biotechnology, Chandigarh University, Gharuan, Mohali, Punjab

Priya Godara
Department of Biochemistry, School of Life Sciences, Central University of Rajasthan, NH-8, Bandarsindri, Kishangarh, Ajmer, Rajasthan 305817, India

Pulala Raghuveer Yadav
Department of Biotechnology, Indian Institute of Technology Hyderabad, Kandi, 502 285, Telangana State, India

Rushikesh R Mali
Department of Quality Assurance, SVKM's Dr. Bhanuben Nanavati College of Pharmacy, Mumbai, India

Shuchi Kaushik
State Forensic Science Laboratory, Home (Police) Department, Government of Madhya Pradesh, India

Sudip Prasad Jena
Department of Biochemistry, School of Life Sciences, Central University of Rajasthan, NH-8, Bandarsindri, Kishangarh, Ajmer, Rajasthan 305817 India

Tabassum Khan
Department of Pharmaceutical Chemistry & Quality Assurance, SVKM's Dr. Bhanuben Nanavati College of Pharmacy, Mumbai, India

Quazi Mohammad Imranul Haq
Biological Sciences and Chemistry-Biology Section, University of Nizwa, Sultanate of Oman, Visiting Research Scientist at University of Arizona, USA

Rimsha Hassan
Department of Eastern Medicine and Surgery, Government College University Faisalabad, Pakistan

Rukhsana Akhter
Department of Biochemistry, University of Kashmir, Srinagar 190006, India
Department of Higher Education, Jammu & Kashmir, India.

Safdar Hussain
Centre for Applied Molecular Biology (CAMB), University of the Punjab, Lahore, Pakistan

Saurabh Gupta
Department of Microbiology, Mata Gujri College, Fatehgarh Sahib 140406, Punjab, India

Sengamuthu Kumaresan
Amrita School of Agricultural Sciences, Amrita Vishwa Vidyapeetham, Coimbatore 642109.

Subbiah Sankaralingam
PG and Research Department of Botany, Saraswathi Narayanan College, Madurai, 625022, India

Swaati Sharma
University Institute of Biotechnology, Chandigarh University, Punjab, India

Tabasum Ashraf
Department of Biochemistry, University of Kashmir, Srinagar 190006, India

Tabassum Khan
Department of Pharmaceutical Chemistry & Quality Assurance, SVKM's Dr. Bhanuben Nanavati College of Pharmacy, Mumbai, Maharashtra, India

Tahira Jameel
Institute of Agro-Industry and Environment, The Islamia University of Bahawalpur, Pakistan

Usha Mukundan
Department of Botany, HVPS Ramniranjan Jhunjhunwala College (Empowered Autonomous), Ghatkopar (W), Mumbai 400 086, India

1 Artemisia-Mother of Herbs: Unlocking the Antiviral Potential of Nature's Arsenal

Seyed Alireza Salami[1*], Anu Pandita[2], Mitra Aelaei[3], Ehsan Asgarian[1] and Deepu Pandita[4]

1. Introduction

Medicinal plants, many of which remain overlooked, play a significant role in modern medicine because of the wide array of bioactive compounds that nature has developed over millions of years of evolution (Kshirsagar and Rao, 2021). Approximately 40% of modern medicines originate from plants and nearly 70– 80% of the global population rely on alternative medicine, mainly from herbal sources, for their primary healthcare needs (Barkat et al., 2021; Chan, 2003). Artemisia L. is a large genus of small herbs and shrubs in the sunflower family (Asteraceae). It is commonly known as "mugwort", "common wormwood", "wild wormwood", "sailor's tobacco", "felon herb", "old Uncle Henry", "chrysanthemum weed", "old man or St. John's plant" and "naughty man" (Abiri et al., 2018). There are over 500 species found worldwide, commonly in North America, Asia, and Europe with the highest diversity in Asia (Bora and Sharma 2011; Nakamura and Hattori, 2011). Some species of the *Artemisia* genus are used as herbs for flavoring, and for ornamental purposes, and as medicine with pharmacological applications in both traditional and modern medicine systems (Bora and Sharma, 2011; Taleghani et al., 2020; Apaza Ticona et al., 2020), such as anti-malarial (Woerdenbag et al., 1990), anticancer (Rezazadeh et al., 2023), anti-pyretic (Goud et al., 2015), antioxidant (Brisibe et al., 2009), antibacterial (Juteau et al., 2002), antifungal (Soylu et al., 2005), anti-spasmodic (Koul et al., 2017), anti-rheumatic (Hadi et al., 2014), anti-tumor (Wang et al., 2016), anti-inflammatory (Ghlissi et al., 2016), anticoagulant (Kshirsagar and Rao, 2021), anti-diabetic (Nofal et al., 2009), cytotoxic (Zheng, 1994), hepatoprotective (Gilani et al., 2005), treat detrimental neuron function and memory loss (Bora and Sharma, 2010), worm infestation (Abiri et al., 2018), cardiac stimulant (Kshirsagar and Rao, 2021), treatment of

[1] Department of Horticultural Science, Faculty of Agriculture, University of Tehran, Karaj, Iran.
[2] Vatsalya Clinic, Krishna Nagar, New Delhi 110051, India
[3] Department of Horticulture Sciences, Faculty of Agriculture, University of Zanjan, Zanjan, Iran
[4] Government Department of School Education, Jammu, Jammu and Kashmir, India
All authors have equal contributions
* Corresponding author: asalami@ut.ac.ir[1]; deepupandita@gmail.com[4]

DOI: 10.1201/9781003433583-1

systemic lupus erythematosus (Fuzimoto, 2021), osteoinductive activities (Ahuja et al., 2011) and antiviral properties (Kshirsagar and Rao, 2021) are among those reported for Artemisia (Hussain, 2022).

Artemisia annua L. or *Artemisia chamomilla* C. Winkl., commonly known as the sweet wormwood or annual mugwort, is a medicinal Artemisia species native to Asia and broadly distributed all over temperate zones and used in traditional Chinese medicine to treat fever and malaria (Tu, 1999; Ho et al., 2014; Cheong et al., 2020; Uzun and Toptas, 2020). *A. annua* possesses sesquiterpene lactones mainly polycyclic artemisinin, monoterpenes, triterpenoids, steroids, and flavonoid bioactive compounds (Septembre-Malaterre et al., 2020). The artemisinin (quinghaosu in China) derivatives called artemisinins consist of artesunate (artesunic acid), dihydro-artemisinin ($C_{15}H_{24}O_5$), artemether (artemetherum), arteannuin B ($C_{15}H_{20}O_3$) and artemisone (artemifone), and others having significant pharmacological potential in therapy of Rheumatoid Arthritis (RA), malaria, Systemic Lupus Erythematosus (SLE), allergic contact dermatitis, and cancer (Fuzimoto, 2021; Slezakova and Ruda-Kucerova, 2017). The World Health Organization (WHO) has recommended Artemisinin-based Combination Therapy (ACT) as a preliminary prescription for the uncomplicated multi-drug-resistant Plasmodium falciparum malaria (Efferth, 2018; Gendrot et al., 2020). The Nobel Prize in Medicine (2015) was awarded for the discovery of artemisinin, which drew the attention of researchers to this plant to explore its other medicinal properties (Efferth et al., 2015). The antiviral broad-spectrum potential of artemisinin has been shown in various studies against a wide variety of DNA and RNA viruses, including Polyomavirus BK (BKV) and Zika Virus, Human Immunodeficiency Virus (HIV), Human Herpes Simplex Virus (HSV), Human CytoMegaloVirus (HCMV), Hepatitis B Virus (HBV) and Hepatitis C Virus (HCV) (Efferth et al., 2008; Flobinus et al., 2014). *In vitro* studies on the infection of Severe Acute Respiratory Coronavirus 2 (SARS-CoV-2) on permissive cells verified the efficacy of artemisinin and its derivatives in various cell models (Han et al., 2019; Zhou et al., 2021). Activities of nine artemisinin-associated compounds in *in vitro* pharmacokinetic model predicted the therapeutic potential linked to anti-SARS-CoV-2 (Cao et al., 2020). In silico molecular simulations/docking studies investigated the artemisinin-based inhibition of the binding interaction of coronavirus spike protein to host cell surface receptor proteins and so preventing both the virus endocytosis and Nuclear factor-kappa B (NF-κB) signaling pathway activation (Cao et al., 2020; Rolta et al., 2021; Sehailia and Chemat, 2021; Uckun et al., 2021). *Artemisia dracunculus* a bushy aromatic and medicinal herb commonly known as wild tarragon is antidiabetic and anticoagulant and used for culinary purposes for pungent anise-like flavor and aroma of leaves and flowering tops (https://en.wikipedia.org/wiki/Tarragon#cite_note-3) (Swanston-Flatt et al., 1991; Shahriyary and Yazdanparast, 2007). *Artemisia campestris* commonly known as the field wormwood finds various pharmacological uses as an antiseptic and anti-helmintic, in the treatment of pleurisy, common colds due to coronavirus or rhinovirus, respiratory, cutaneous, and digestive disorders (Palmer, 1975; Breverton, 2013; Al-Snafi, 2015; Apaza Ticona et al., 2020).

2. Taxonomy and Scientific Classification

Asteraceae, commonly known as the Sunflower Family, represents the largest family of angiosperms, encompassing over 25,000 species. This family includes numerous species of medicinal and economic importance, with Artemisia recognized as one of its prominent genera. (Funk et al., 2009) (Table 1). The classification of the Artemisia genus was initially established by Linnaeus in 1753 (Jiao et al., 2023). The name of this particular genus is derived from the Greek goddess of chastity, Artemis, as a tribute to her (Ahuja et al., 2011). Another etymology refers to the Persian/Greek king's wife, Queen Artemisia, who was known as a botanist and medical researcher (died *c.* 350 BC) (Abiri et al., 2018).

Table 1 The Taxonomic Classification of the genus Artemisia

Kingdom	*Plantae*
Division	**Tracheophyta**
Class	**Magnoliopsida**
Order	**Asterales**
Family	**Asteraceae**
Tribe	**Anthemideae**
Genus	**Artemisia**

Synonyms

The genus Artemisia has several synonyms, which are alternative scientific names used to refer to the genus or its species in the past. Some of the synonyms of the genus Artemisia (*https://en.wikipedia.org/wiki/Artemisia_(plant)*) include:

Absinthium Mill.
Chamartemisia Rydb.
Oligosporus Cass.
Artemisiastrum Rydb.
Artanacetum (Rzazade) Rzazade.
Abrotanum Mill.
Draconia Heist. ex Fabr.
Artemisia subg. Seriphidium Less.
Hydrophytum Eschw.
Seriphidium (Besser ex Less.) Fourr.
Dracunculus Ruppr. ex Ledeb, 1845.

The genus Artemisia is extensively distributed and includes over 500 species, commonly known as mugwort or wormwood (Bora and Sharma, 2011). The species are found in various habitats, ranging from arid deserts to humid forests, and are adapted to different climatic conditions. The scientific names of some well-known species of Artemisia include:

(a) *A. annua* (Sweet wormwood or annual mugwort)
(b) *A. dracunculus* (Tarragon)
(c) *Artemisia absinthium* (Absinthe wormwood)
(d) *Artemisia vulgaris* (Common mugwort)
(e) *Artemisia tridentata* (Big sagebrush)

The classification of Artemisia is still a matter of debate among taxonomists due to the morphological complexities, ploidy, and complex genetics within its species (Haghighi et al., 2014; Hussain, 2020; Ahadi et al., 2015). However, recent molecular studies have provided new insights into the relationships among the species within the genus Artemisia and have helped to clarify the taxonomy of Artemisia.

3. Botanical Description of the Plant

The following is a general botanical description of the genus Artemisia:

1. Habit: Artemisia species are mostly herbaceous perennials, biennials, and annual herbs, but some are shrubs or subshrubs (Iranshahi et al., 2007; Watson et al., 2002). Some species are tall and have a bushy appearance, while others are more compact and have a dense growth habit.
2. Leaves: The leaves of Artemisia species are typically alternately arranged on the stem, and are simple or pinnately divided. The shape and size of the leaves can vary greatly among species and can be used to help identify the plant. The leaves are usually grayish-green or silver in color and are covered with tiny hairs, giving them a distinctive appearance.
3. Flowers: The flowers of Artemisia species are small and arranged in clusters or heads. The flowers are typically yellow or white and have both male and female parts. The flowers are not showy and are not commonly grown for ornamental purposes.
4. Fruit: The fruit of the Artemisia species is achene, which is a small, dry, one-seeded fruit that does not open at maturity.
5. Aroma: Many Artemisia species have a strong, aromatic aroma due to the existence of volatile essential oils in their leaves and stems. The aroma can be used to help identify the plant and is also utilized in traditional medicine and aromatherapy.

Among the different species of Artemisia, *A. annua* is a shrub that is usually single-stemmed and reaches a height of about 2 m, with alternate branches. The plant is extremely vigorous and free of diseases and pests (Jelodar et al., 2014).

4. Geographical Distribution

Artemisia is a genus of plants that is widely distributed around the world. The exact distribution of Artemisia species depends on the species in question, but the genus is found in a range of habitats, from arid deserts to humid forests, and is adapted to different climatic conditions. Some of the most well-known species of

Fig. 1 Illustration of Artemisia vulgaris (1897) (https://en.wikipedia.org/wiki/Mugwort).

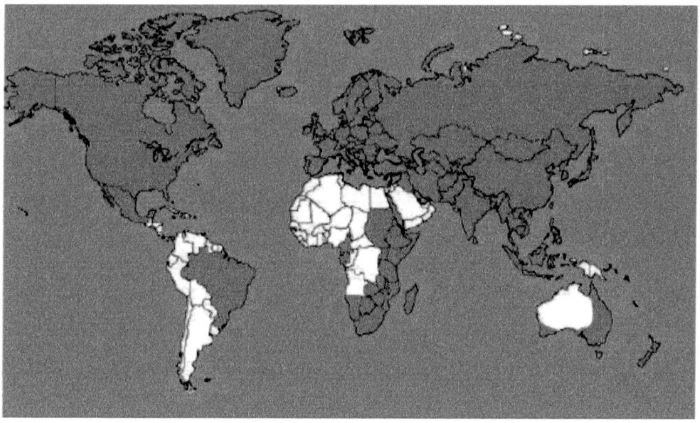

Fig. 2 Distribution of Artemisia on all continents except Antarctica (Koul et al., 2017).

Artemisia are found in the Northern Hemisphere, particularly in Asia and North America. For example, *A. annua* is native to Asia, while *A. dracunculus* is native to Central and Western Asia.

Artemisia species are also found in additional regions of the globe, including Europe, Africa, and South America. Some species, such as *A. absinthium*, have been introduced to other parts of the world and have become naturalized in new environments.

In summary, the geographical distribution of Artemisia species is quite broad, and the genus is found in a range of habitats and regions around the world (Fig. 2).

Artemisia annua has been successfully cultivated in various soil types and geographical locations, indicating its ability to adapt to different environments. While certain A. annua varieties may not tolerate acidic soils with a pH of 5.0– 5.5, most soil types can be utilized for cultivation. At the same time, growth and production of secondary metabolites are strongly affected by abiotic environmental conditions (*e.g.*, temperature, light, water, and nutrient levels) (Zhang et al., 2023).

5. Phytochemical Constituents and Metabolite Profiling

Artemisia species contain a variety of phytochemical compounds, including essential oils, flavonoids, terpenoids, phenols, and others. The phytochemical profile of Artemisia species is influenced by different factors, including the species, part of the plant being analyzed, growth conditions, and extraction methods.

Metabolite profiling of Artemisia species involves the primary and secondary metabolite analysis of the plant. The primary metabolites are involved in the normal functioning of the plant, while the secondary metabolites have various biological activities, such as providing defense against herbivores, pathogens, and environmental stress.

Some of the most commonly detected phytochemical constituents in Artemisia species include:

(i) Essential oils: The plant essential oils are volatile compounds produced by the plant and accord the characteristic aroma to the Artemisia species. The essential oils of Artemisia species have various biological actions, including anti-inflammatory, anti-spasmodic, and anti-bacterial properties.

(ii) Flavonoids: The flavonoids phyto-compounds have a wide distribution in the kingdom plantae. Artemisia species contain a variety of flavonoids, including quercetin, kaempferol, and rutin.

(iii) Terpenoids: Terpenoids are a large and diverse group of compounds found in many plant species. Artemisia species contain a variety of terpenoids, including sesquiterpenes, monoterpenes, and diterpenes.

(iv) Phenols: Phenols or phenolic acids are a class of aromatic compounds that are widely distributed in the kingdom plantae. Artemisia species contain various phenols, including caffeic acid and rosmarinic acid.

(v) In summary, Artemisia species contain a variety of phytochemical compounds, including essential oils, flavonoids, terpenoids, and phenols, and metabolite profiling of the plants can help identify the specific compounds present. The specific phytochemical profile of Artemisia species can diverge based on various factors, necessitating further research to completely understand the phytochemistry of this genus.

The compounds present in different *Artemisia* species are terpenoids (isoprenoids), flavonoids (bioflavonoids), coumarins (cumarin), sterols, acetylenes, caffeoylquinic acids (chlorogenic acid), and volatile aromatic oils (Koul et al., 2017). Artemisia species are known for their aromatic compounds, many of which have been found to have insecticidal and medicinal properties. However, some species are also toxic and can cause digestive issues if consumed in large amounts. Artemisinin and its derivatives (ARTs) are among the notable phytochemicals. Chloroquine (CQ) and hydroxychloroquine (HCQ) are antimalarials that are used for conditions like rheumatoid arthritis RA and systemic lupus erythematosus SLE (Cheong et al., 2020).

Among the significant phytochemicals are artemisinin and its derivatives (ARTs). Chloroquine (CQ) and hydroxychloroquine (HCQ), both known for their antimalarial properties, are also employed in the treatment of autoimmune disorders such as rheumatoid arthritis (RA) and systemic lupus erythematosus (SLE) (Cheong et al., 2020). Artemisia has also exhibited promising antiviral activity against viruses like HBV, HIV, SARS-CoV-2, HSV, and Human and Avian Influenza Viruses (Orosco, 2023). The structures of some bioactive metabolites of Artemisia are given in Fig. 3.

6. Medicinal/Therapeutic Uses

Artemisia species have a long history of usage in traditional medicine, and many species are used for various medicinal purposes. Some of the most commonly reported medicinal uses of Artemisia species include:

(i) Anti-inflammatory and analgesic effects: Some species of Artemisia, such as *A. annua* and *A. capillaris* are used to treat inflammatory and painful disorders, for instance, rheumatoid arthritis and menstrual cramps.

(ii) Anti-malarial: *Artemisia annua* is used in traditional medicine to cure malaria, and its anti-malarial properties have been confirmed in laboratory and animal studies. The World Health Organization (WHO) recognizes the potential of *A. annua* as an anti-malarial agent, but more research is needed to determine its efficacy and safety in humans.

(iii) Anti-cancer: Some studies have shown that certain compounds found in Artemisia species, such as artemisinin, have anti-cancer properties and may be useful in treating various types of cancer, including liver cancer, breast cancer, and lung cancer.

(iv) Antioxidant and neuroprotective effects: Some studies have shown that Artemisia species have antioxidant properties, and may find use in the treatment of various neurodegenerative diseases, for instance, Parkinson's and Alzheimer's disease.

(v) Anti-inflammatory and anti-viral effects: Some species of Artemisia, such as *A. dracunculus* (tarragon), are used in traditional medicine systems to treat viral infections, for instance, the common cold and influenza, and show anti-inflammatory and anti-viral effects in the laboratory studies.

In conclusion, Artemisia species have a long history of use in the traditional medicinal system, and many species are used for various medicinal purposes, including anti-inflammatory, anti-malarial, anti-cancer, antioxidant, and anti-viral effects. However, more investigation is required to determine the efficacy and

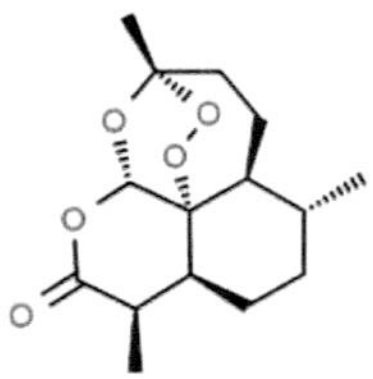

Artemisinin

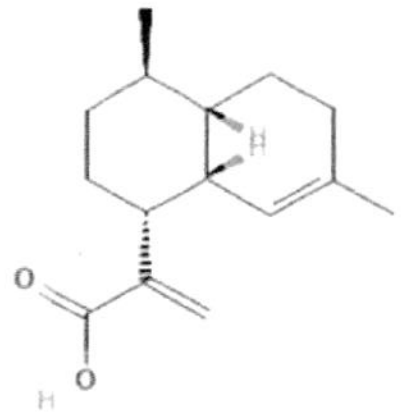

Artemisinic acid

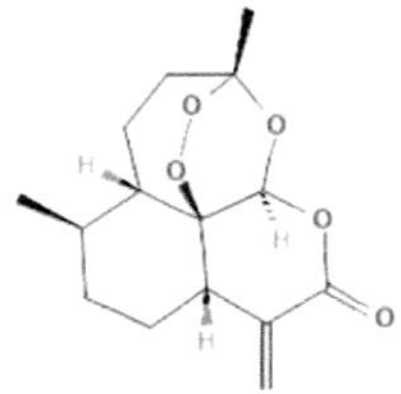

Artemisitene

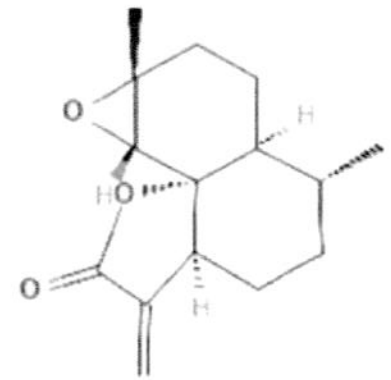

Arteannuin B

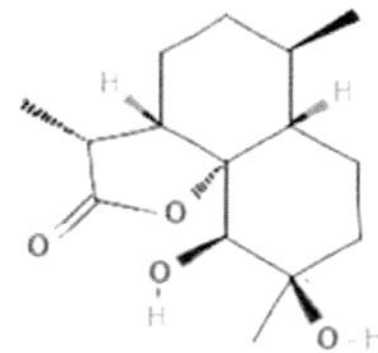

Arteannuin M

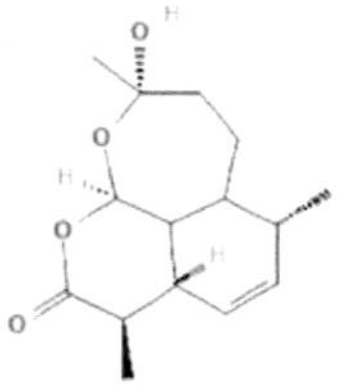

Deoxyartemsinin

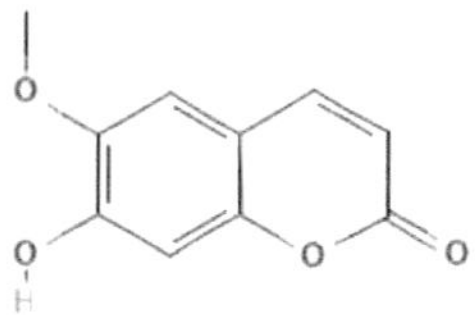

Scopoletin

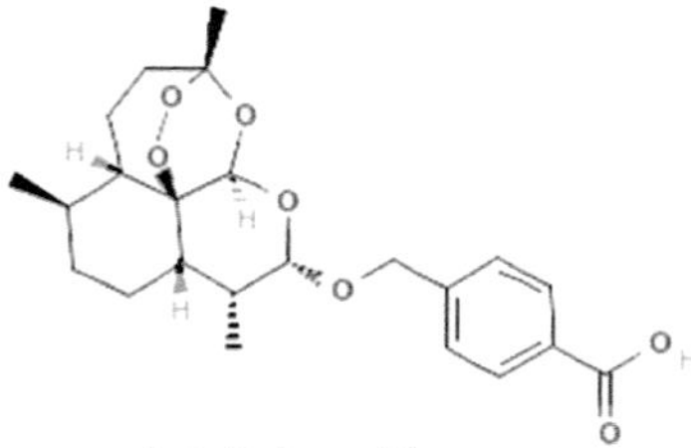

Artelinic acid

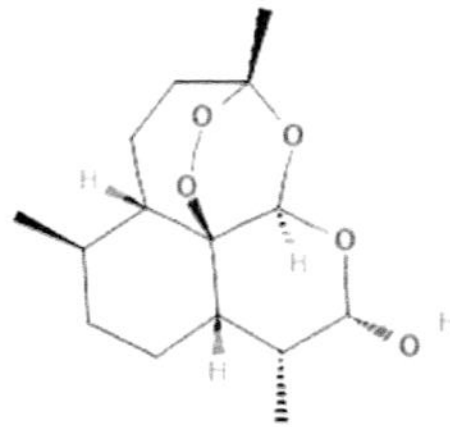

Dihydroartemisinin

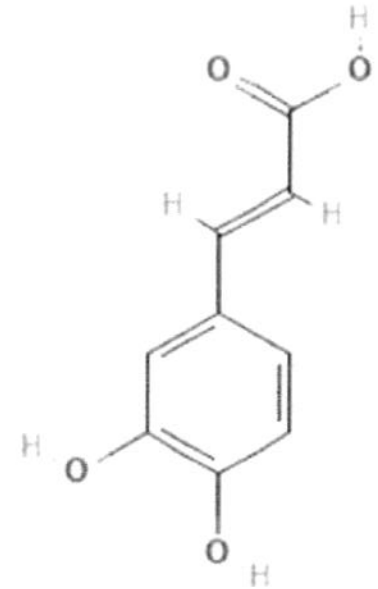

Caffeic Acid

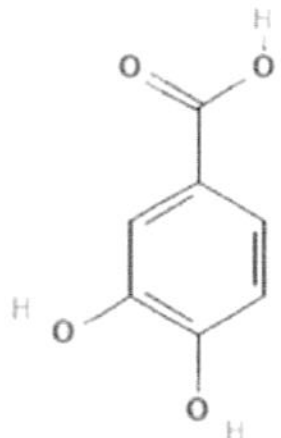

Protocatechuic acid

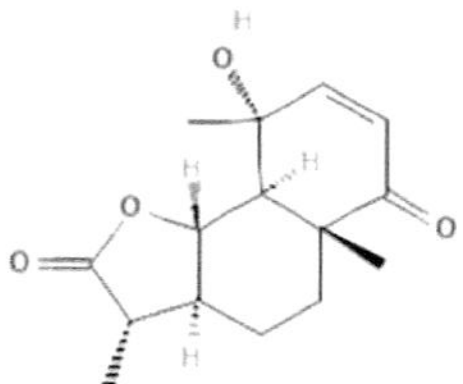

Vulgarin

Camphor

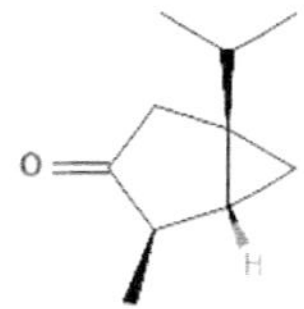

α-thujone

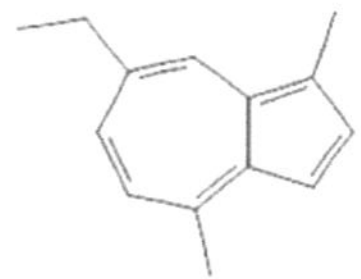

Chamazulene

Contd.

Contd.

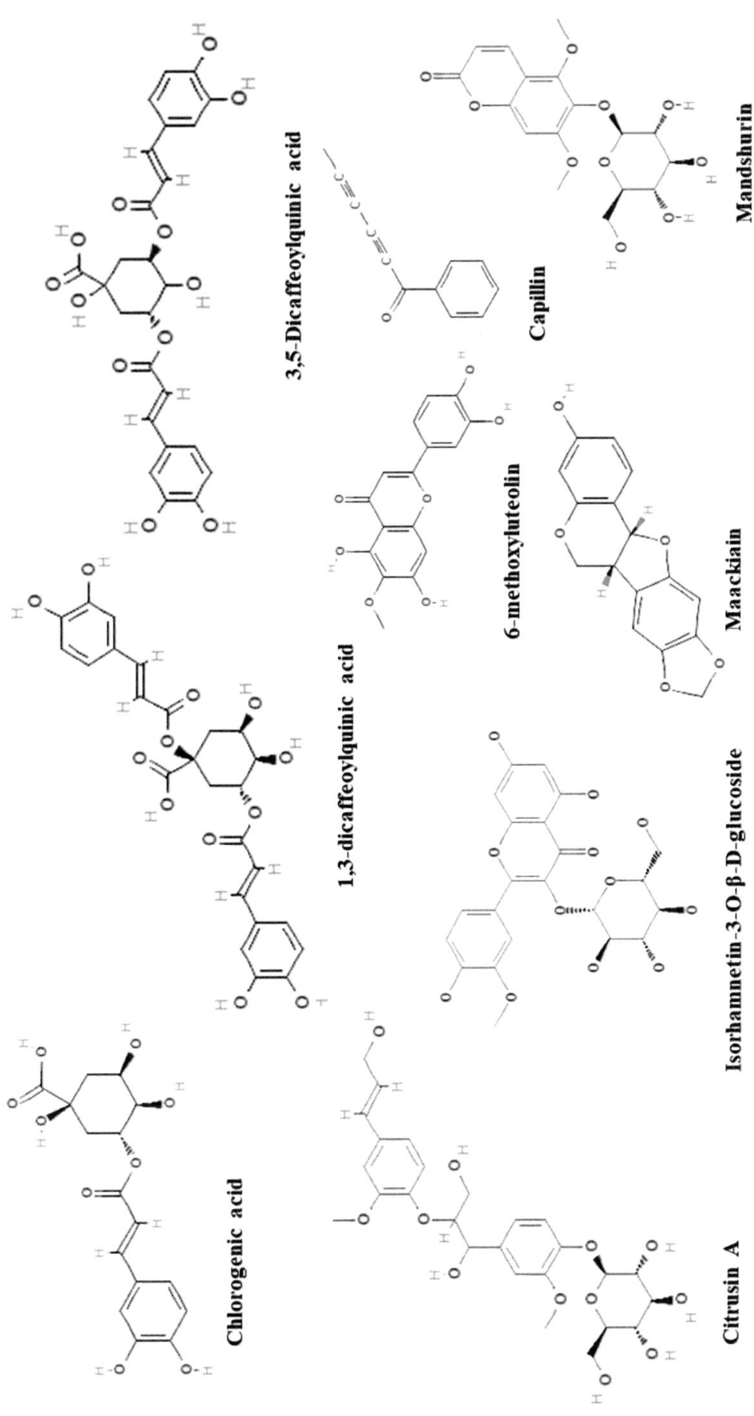

Chlorogenic acid
Citrusin A
1,3-dicaffeoylquinic acid
Isorhamnetin-3-O-β-D-glucoside
6-methoxyluteolin
Maackiain
3,5-Dicaffeoylquinic acid
Capillin
Mandshurin

Contd.

Contd.

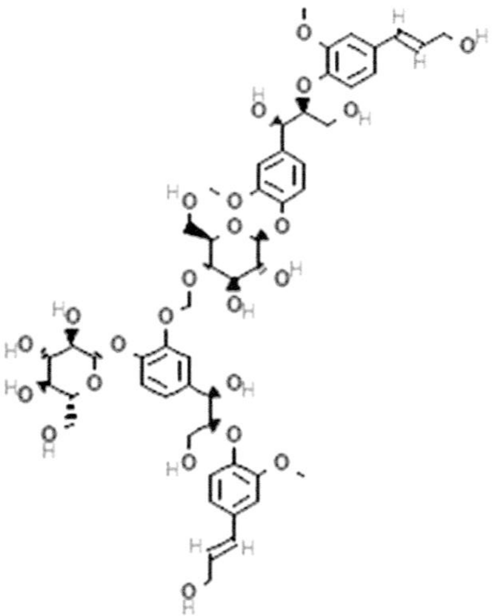

Alaschanioside A

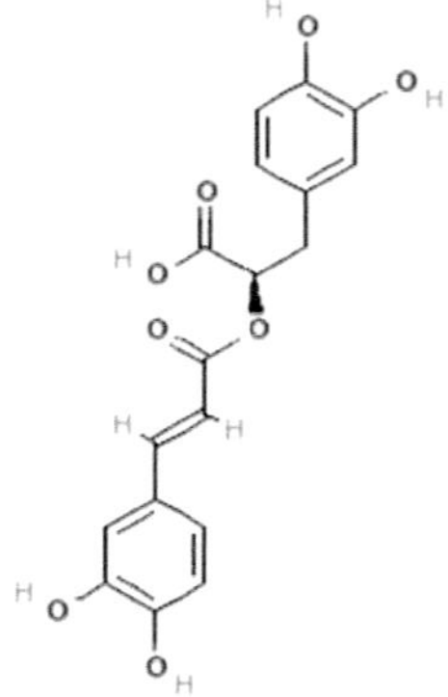

Rosmarinic acid

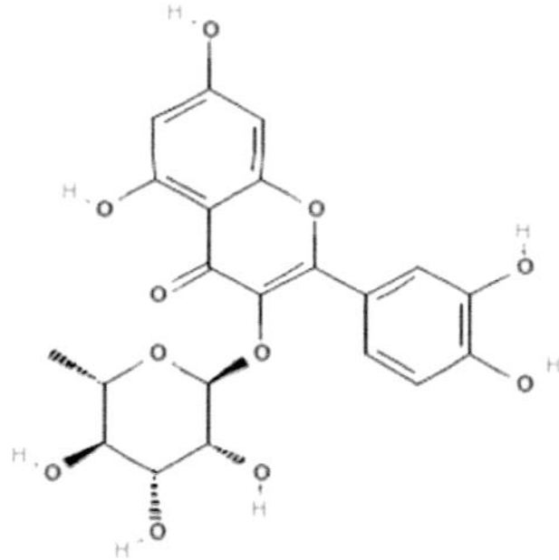

Quercitrin

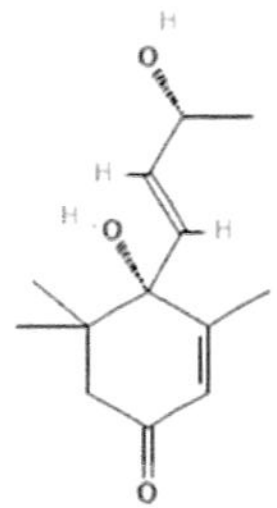

Blumenol A

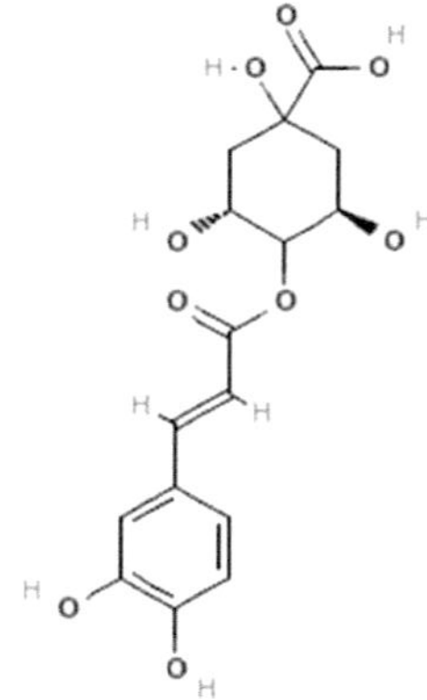

Cryptochlorogenic acid

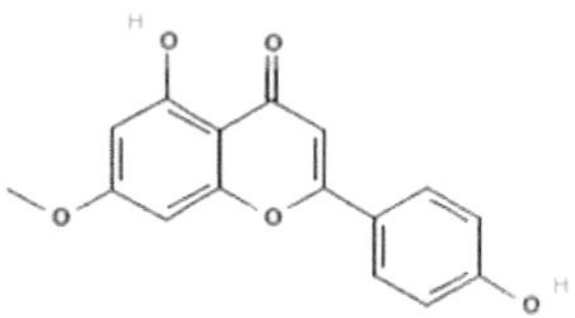

Genkwanin

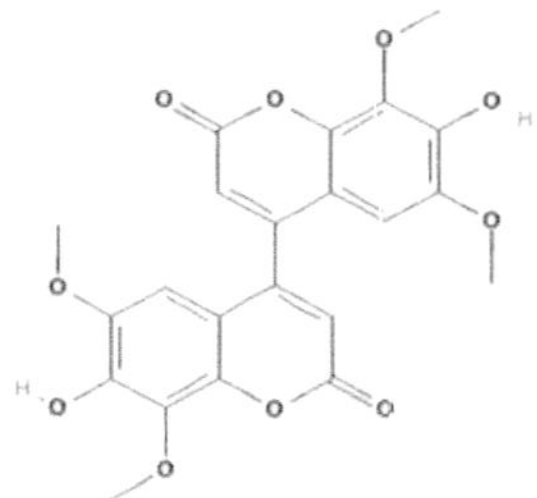

Biisofraxidin

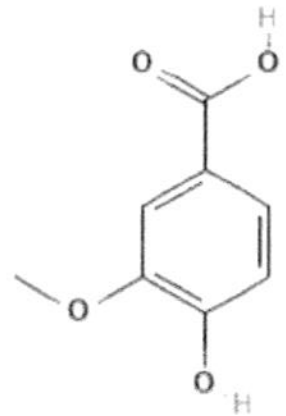

Vanillic acid

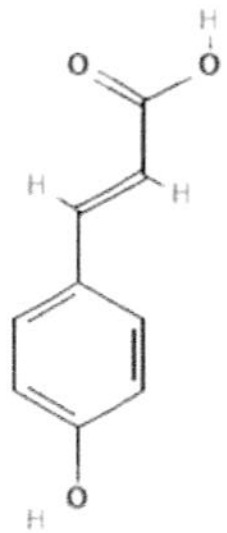

p-coumaric acid

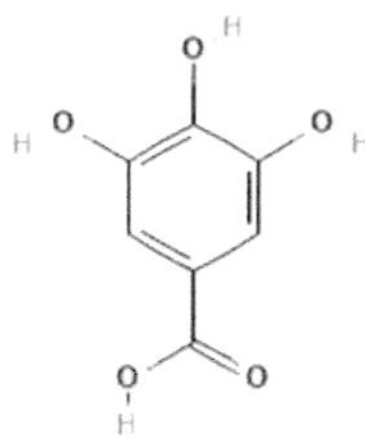

Gallic acid

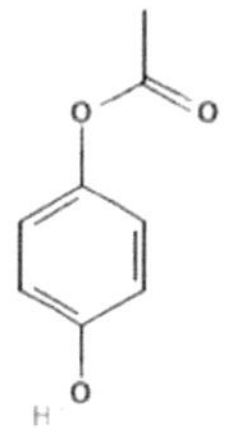

4-hydroxyphenyl acetate

Contd.

Contd.

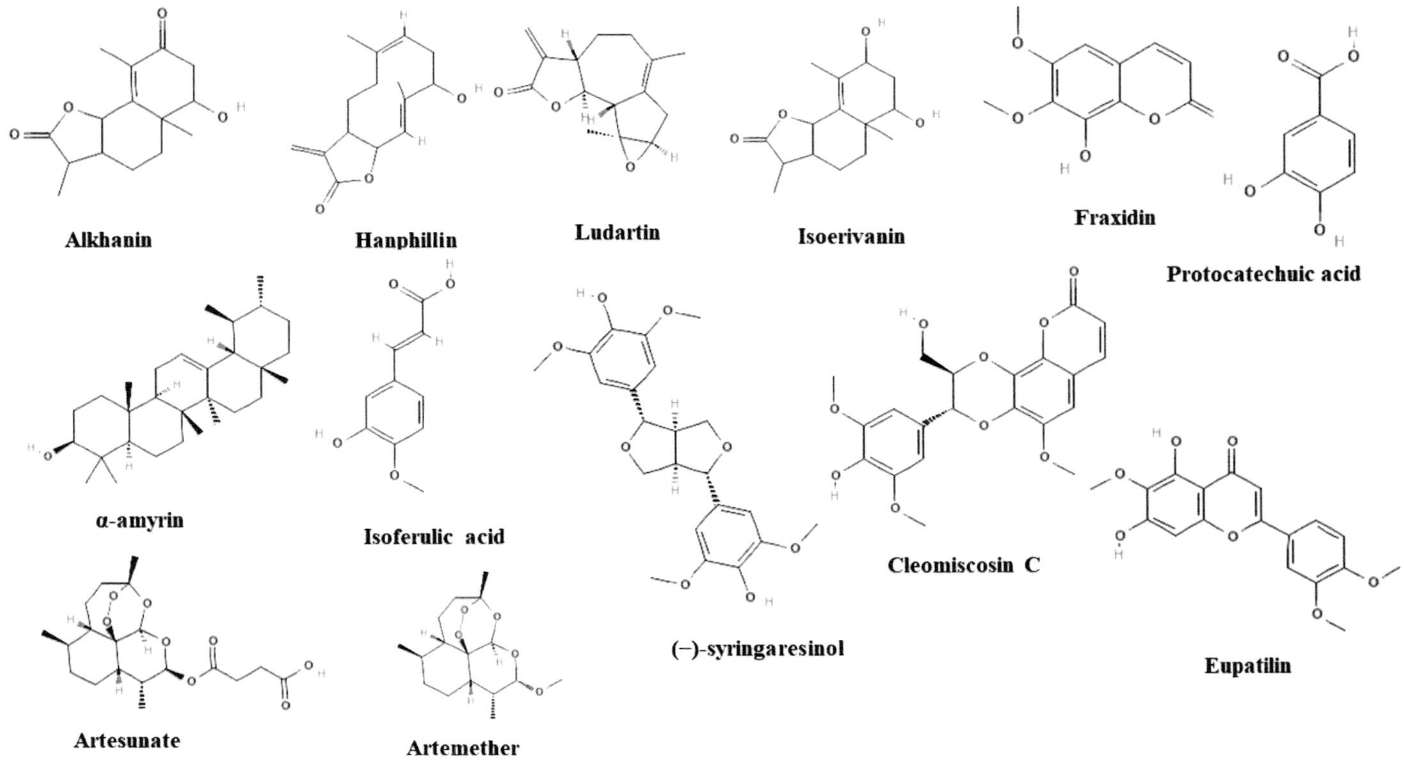

Fig. 3 The structure of phytochemicals in Artemisia was accessed from PubChem open chemistry database (https://pubchem.ncbi.nlm.nih.gov/).

Table 2 Pharmacological Properties of Artemisia species.

Scientific Name of Artemisia Species	*Common Name*	*Isolated Bioactive Compounds*	*Therapeutic Roles*	*Reference/s*
Artemisia absinthium (L.)	Common wormwood	(Z)-thujone, (E)-thujone myrcene, linalool, monoterpene sesquiterpene lactones, polyphenolic compounds, lignins, trans-sabinyl acetate hydrocarbons, flavonoids, tannins	Anthelmintic, cardiac and memory stimulant, restores decline in mental functions and liver inflammation	Wake et al., 2000; Guarrera 2005; Kordali et al., 2005; Daise et al., 2008; Orosco, 2023b
Artemisia abrotanum (L.)	Southernwood, lad's love, southern wormwood	Flavonols, tannins, coumarins 1,8-cineole, Linalol (linalool), davanone C, thujyl alcohol, 3,4-Dihydroxycinnamic acid (caffeic acid)	Insecticide, liver conditions	Kranen-Fiedler, 1956; Tunon et al., 2006; Bora and Sharma, 2010; Orosco, 2023b
Artemisia annua (L.)	Sweet wormwood, sweet annie, sweet sagewort	Eucalyptol (cineole), α-pinene, camphor, 2,2-Dimethyl-3 methylenenorbornane (camphene), borneol, Isogermacrene D, hydroxyl amino isoartemisia ketone, flavones, flavonols, endoperoxide sesquiterpene lactone (artemisinin), coumarins, phenolic acids, artemisinic acid, arteanniun B, artesunate, trioxanes, artemether artemisinin G and arteether	Fever, malaria, fibrosis	Bhakuni et al., 1990; Wei et al., 1992; Woerdenbag et al., 1994; Mueller 2000; Han et al., 2007; Han et al., 2008; Miller and Xinzhuan, 2011; Talman et al., 2019; Liu et al., 2019
Artemisia asiatica (Nakai)/***Artemisia indica***	Indian wormwood	Volatile oils, flavones, alkaloids, artemisolide	Cancer, inflammation, infections, ulcerogenic diseases	Reddy et al., 2006; Lim et al., 2008; Orosco, 2023b

Contd.

Table 2 *Contd.*

Scientific Name of Artemisia Species	*Common Name*	*Isolated Bioactive Compounds*	*Therapeutic Roles*	*Reference/s*
Artemisia arborescens (L.)	Tree wormwood	Terpenes thujone, borneol, thujol, flavone, fatty acids	Anti-inflammatory, antihistaminic, blood decongestant	Bora and Sharma, 2010; Orosco, 2023b
Artemisia douglasiana (Bess.)	California mugwort, Douglas's sagewort, dream plant	Monoterpenes cineole, 2-Camphanone, linalol, beta-Thujone, thujone, sesquiterpene lactones Tauremisin and Cordilin	Premenstrual syndrome, dysmenorrhea, joint pain, headaches, abrasions, rashes	Garcia and Adams 2005; Orosco, 2023b
Artemisia dracunculus (L.)	Estragon, wild tarragon	Menthol, limonene, anethole, anisol, α-phellandrene, anisic acid, anisaldehyde, d-sabinene, estragole, myrcene, ocimene, coumarins, β-pinene, glucoside, polyphenolic compounds, 1-methoxy-4-(2-Propenyl)-benzene, 1R-α-pinene	Antidiabetic, anticoagulant	Bayrak et al., 1986; Zhang et al., 2005; Shahriyary and Yazdanparast, 2007
Artemisia maritima (L.)	Sea wormwood, old woman	beta-thujone, alpha-thujone, alpha-pinene, sabinen, dolcymene, sabinol cuminaldehyde, isobutanoic acid, delphinate, sesquiterpene peroxysemiketal, sesquiterpene lactones, fatty acids, polyphenolic compounds, 1,8-cineole, camphor, borneol, chrysanthenone	Anthelmintic, liver function, GI issues	Ishibashi et al., 1965; Jaitak et al., 2008; Bora and Sharma, 2010; Orosco, 2023b

Contd.

Table 2 *Contd.*

Scientific Name of Artemisia Species	*Common Name*	*Isolated Bioactive Compounds*	*Therapeutic Roles*	*Reference/s*
Artemisia scoparia Waldst. and Kit.	Capillary wormwood, virgate wormwood, redstem wormwood	alpha-pinene, geranyl acetate, methylheptenone ethylene ketal, 1,8-cineole, karvone, 1-thujone, α-thujadicarboxylic acid, 1-thujyl alcohol, eugenic acid, cinnamaldehyde, (Z)-beta-ocimene, gamma-terpinene, coumarins, pyrogallol tannins, cholagogic components, flavonoids, flavones, beta-myrcene, p-cymene and dl-limonene	Antibacterial, antiseptic, antipyretic	Singh et al., 2008; Singh et al., 2009; Bora and Sharma, 2010; Orosco, 2023b
Artemisia tripartita (Rydb.)	Threetip sagebrush	Guaianolides cumambrin B, biologically-active polysaccharides	Cold, sore throats, tonsillitis, headaches, wounds	Xie et al., 2008; Bora and Sharma, 2010; Orosco, 2023b
Artemisia verlotiorum (Lamotte)	Chinese mugwort	Valeric acid, palmitic acid, comphene, eucalyptol (cineole), fenchone, alpha-phellandrene, thujyl alcohol, beta-thujone, cadinene	Hypertension	Covello, 1941; Bora and Sharma, 2010; Orosco, 2023b
Artemisia vestita (Wall. ex-Bess.)	Russian Wormwood	Flavonoids, germacrene D, 1,8- eucalyptol (cineole), α-, β-, γ-himachalene, himachalol, santolina alcohol, (–)-trans-caryophyllene, thujones, thujanols	Inflammatory diseases	Weyerstahl et al., 1987; Ye et al., 2008; Orosco, 2023b
Artemisia vulgaris (L.)	Common mugwort	Coumarins, sterols, p-cymene, fenchone, α-thujone, beta-thujone, cineole, comphor, beta-pinene, 4-terpinenol, isoborneol, alpha-terpineol, lemonol (geraniol), (–)-trans-caryophyllene, caffeoylquinic acids	Analgesic, anti-inflammatory, antispasmodic, liver disease	Carnat et al., 2000; Ragasa et al., 2008; Govindaraj et al., 2008; Temraz and El-Tantawy 2008; Orosco, 2023b

safety of using Artemisia species for medicinal purposes, and it is important to consult with a healthcare professional before using Artemisia for medicinal purposes. The Artemisia species have various pharmacological arsenals in driving biological activity to fight against innumerable dreaded diseases (Table 2).

7. Antiviral Activities of Artemisia

Artemisia with a repository of bioactive metabolites has antiviral pharmacotherapeutic potential. This suggests that Artemisia is indispensable as an antiviral agent and supports the fight against various virus variants (Table 3).

Recently, SARS-CoV-2 caused the global pandemic COVID-19 leading to symptoms like dyspnea, headache, cough, myalgia, fever, and fatigue in humans (Wu et al., 2020). A lot of scientific data is available on the antiviral action of *A. annua* and artemisinin derivatives against SARS-CoV-2 (Alsaffar et al., 2020; Cao et al., 2020; Das et al., 2021; Li et al., 2021; Peele et al., 2020; Sehailia and Chemat, 2021; Sharma and Deep, 2022). The gelseminic acid (scopoletin), arteannuin B and (+)-artemisinic acid (artemisic acid) of *A. annua* show antiviral effects by interacting with 3-chymotrypsin-like protease (3CLpro) and Spike (S) glycoprotein (Spike proteins) of SARS-CoV-2. The 3CLpro and Spike proteins play roles in the insertion of the virus in the host cell and replication of the virus inside the host cell. Thus, gelseminic acid (scopoletin), arteannuin B, and (+)-artemisinic acid (artemisic acid) interfere with the virus insertion into the host cell and the replication pathways of SARS-CoV-2 (Baggieri et al., 2023). Artemisinin and its derivatives from diverse Artemisia species potentially inhibit the SARS-CoV-2 binding with Angiotensin Converting Enzyme-2 (ACE-2) receptor by interacting with the SARS-CoV-2 3-chymotrypsin-like proteases (3CL pro also known as the main protease (MPro)), spike ectodomain structural trimer, spike receptor-binding domain (RBD), and spike (S) glycoprotein (Baggieri et al., 2023).

Other mechanisms of action include artemisinin-based compounds (ARTs) based suppression of Transforming Growth Factor beta (TGF-β) expression and TGF-β-mediated inflammatory damage which influences various metabolic pathways (Fig. 4) (Gendrot et al., 2020; Trieu et al., 2021; Orosco, 2023b). Consequently, ARTs can prove useful in antiviral drug design.

A. scoparia exhibits inhibitory effects on influenza A viruses of Orthomyxoviridae family by disrupting the replication of the influenza virus, reducing the surface glycoprotein hemagglutinin (HA) and intriguing transmembrane protein Matrix-2 (M2) proteins, and the decrease of the activity of surface glycoprotein neuraminidase (NA) (Fig. 5) (Ponticelli et al., 2023; Orosco, 2023b).

A. annua inhibits DNA replication of HIV, while *A. campestris subsp. glutinosa* (Besser) Batt works by blocking the entry of HIV viruses and inhibiting viral transcription by interference with Sp1 transcription factor and nuclear factor kappa B (NF-κB) (Fig. 6) (Oguariri et al., 2010; Apaza Ticona et al., 2020; Orosco, 2023b). *A. campestris* subsp. glutinosa extracts inhibit the activity of HIV in *in vitro*

Table 3 Antiviral activities of Artemisia species.

Target Viruses	*Artemisia Species and/ or Compounds*	*Mode of Action and/or Consequences*	*Reference/s*
Severe Acute Respiratory Coronavirus 1 (SARS-CoV-1)	***A. annua*** L	Virus-induced Cytopathic Effect (CPE)	Li et al., 2005
	A. herba-alba Asso.	High content of oxygenated monoterpenes characterizes its essential oil	Asdadi et al., 2020
SARS-CoV-2	Artemisinin and Artemisinin-derivatives	Inhibits binding with Angiotensin-Converting Enzyme 2 receptor (ACE-2)	Rolta et al., 2021
		Transforming Growth Factor-beta (TGF-β) suppression	Trieu et al., 2021
	Artemisinin-based compounds (ARTs)	Influences several indispensable metabolic pathways	Gendrot et al., 2020
	Artemisinin-piperaquine blend	Not determined	Li et al., 2021
	A. annua L	Susceptibility of omicron variants to hot water extracts of ***A. annua***	Nair et al., 2023
	A. afra Jacq. ex	Plaque reduction and cell viable assay	Nie et al., 2021
Influenza Viruses (IVs)	***A. scoparia***	Disrupts viral replication, reduces hemagglutinin antigenic glycoprotein and Matrix-2 (M2) proteins, and attenuates the activity of receptor-destroying enzyme (sialidase/neuraminidase)	Ponticelli et al., 2023
	A. annua L.	***A. annua*** aqueous extract at 0.976 mg/ml concentration showed an inhibitory effect on cytopathy caused by influenza A virus FM1 (H1N1)	Tao et al., 2020
		Activation of interleukin 6 (IL-6) and tumor necrosis factor-α (TNF-α) expression	Ou et al., 2020
	A. cina (Santonica flowers)	Anti-H5N1 efficacy	Hegazy et al., 2022

Contd.

Table 3 *Contd.*

Target Viruses	*Artemisia Species and/ or Compounds*	*Mode of Action and/or Consequences*	*Reference/s*
	A. rupestris L.	Not determined. Anti-influenza to A3 and B strains	Yong and Aisa, 2011
	A. caruifolia Roxb.	Reduced pulmonary edema and suppressed IL -6 and IL-10 expression in BALFs of mice infected with influenza virus BJ501 strain. But no decrease in virus titer or prevent mortality from severe influenza	Yuexia et al., 2016
	A. scoparia Waldst. and Kit.	***A. scoparia*** flavonoids have inhibitory effect against several influenza virus strains. The cirsimaritin (CST) flavonoid inhibited replication of influenza A virus by downregulating NF-κB signal transduction pathway	Wang et al., 2017
	A. glabella Kar. et Kir.	It was experimentally found that EOAG causes 100% inhibition of influenza virus in the entire concentration range studied (5 – 100 M)	Seidakhmetova et al., 2002
Human Immunodeficiency Virus (HIV)	***A. annua*** L.	Replication inhibition	Oguariri et al., 2010
	A. absinthium L.	The caffeoyl and dicaffeoylquinic acids (diCQAs) polyphenolics from plant aqueous extract inhibit HIV-1	Ali et al., 2016
	A. afra Jacq. ex ***A. abyssinica*** Schtz. Bip	Anti-HIV cytotoxic assay	Asres et al., 2001
	A. annua L	Interaction between HIV-1 envelope glycoprotein (Env), *i.e.*, gp120/gp41 complex and cluster of differentiation 4 (CD4) membrane protein of T lymphocytes	Chang and Woo, 2003
	A. campestris subsp. glutinosa (Besser) Batt	Terpenes and flavonoids of plant ethanolic extract block the entry of HIV and inhibit transcription and replication	Apaza Ticona et al., 2020

Contd.

Table 3 *Contd.*

Target Viruses	*Artemisia Species and/ or Compounds*	*Mode of Action and/or Consequences*	*Reference/s*
	A. caruifolia Roxb.	***A. caruifolia*** methanolic extract containing N1, N5, N10-tri-p-coumaroylspermidine showed inhibitory activity on HIV-1 protease (38.5% at 100mg/ml).	Nakamura and Hattori, 2001
	A. capillaris Thunb.	Five constituents, artemisidin A (flavonoid), and four coumarins, artemicapins A, B, C and D, and 70 already known compounds, have been isolated from ***A. capillaris***. However, three compounds confirmed action against HIV replication in H9 (T-lymphoid cells).	Wu et al., 2001
Hepatitis B virus (HBV)	***A. annua*** L	Artemisinin and its semisynthetic derivative artesunate inhibit the HBV virus.	Efferth et al., 2008
	A. morrisonensis	The compound p-hydroxyacetophenone (PHAP) isolated from ***A. morrisonensis*** was found to have potential anti-HBV effects in HepG2 2.2.15 cells.	Huang et al., 2014
	A. absinthium L.	The study found that ***S. lappa*** and ***A. absinthium*** have antiviral effect against HBV in CHB besides being quite safe for a treatment duration of 12 weeks.	Shabnam et al., 2018
	A. capillaris Thunb.	Without affecting cell viability, KCT-01 significantly lowered antigen secretion, virion production, and pgRNA synthesis in HepG2.2.15 cells.	Kim et al., 2018
	A. argyi essential oil (AAEO)	Inhibition of membrane lipids (sphingolipids), NF-κB, and tumor necrosis factor (TNF) signaling pathways.	Zhang et al., 2023
Hepatitis C	***A. annua*** L	Artemisinin and its semisynthetic derivative artesunate inhibit the hepatitis C virus.	Efferth et al., 2008
Herpes Simplex Virus (HSV)	***A. vulgaris*** L.	Interacts directly with HSV virus.	Xiao et al., 2023

Contd.

Table 3 *Contd.*

Target Viruses	*Artemisia Species and/ or Compounds*	*Mode of Action and/or Consequences*	*Reference/s*
	A. argyi (AEE)	Disruption of membranes of HSV-1, thus hindering attachment and penetration of virus.	Liu et al., 2023
(HSV)1	***A. absinthium*** L. A. ***herba-alba*** Asso.	Evaluated against herpes simplex, Sindbis, and polio viruses.	Mouhajir et al., 2001
	fragrans Willd. ***incana*** (L.) Druce. ***A. persica*** Boiss. ***A. vulgaris*** L.	Not defined.	Karamoddini et al., 2011
	A. kermanensis Podl.	Increased oil concentrations inhibit the formation of virus plaque.	Gavanji et al., 2015
	A. mendozana D.C.	The essential oils showed inhibitory virucidal action against HSV-1.	Duschatzky et al., 2005
	A. douglasiana Bess.	***A. douglasiana***, ***A. gratissima***, ***Eupatorium patens*** and ***T. absinthioides*** essential oils inactivated HSV-1 at 65–125ppm.	García et al., 2003
	A. campestris L. ***A. chamaemelifolia*** Vill. ***A. caruifolia*** Roxb.	Not defined.	Karamoddini et al., 2011
	A. annua L.	Artemisinin and its semisynthetic derivative artesunate inhibit the HSV1 virus.	Efferth et al., 2008
	A. arborescens L.	The essential oil of ***A. arborescens*** in Phospholipon 90H (P90H) liposome vesicles increased antiherpetic activity ***in vitro.***	Sinico et al., 2005
	A. aucheri.	Spring extract of 50 µg/ml, 75µg/ml, 100 µg/ml and autumn extract of 50 µg /ml and 75 µg/ml concentrations declined the viral infection.	Zamanian et al., 2021.

Contd.

Table 3 *Contd.*

Target Viruses	*Artemisia Species and/ or Compounds*	*Mode of Action and/or Consequences*	*Reference/s*
HSV-2	***A. arborescens*** L.	***A. arborescens*** essential oil inhibits the lateral diffusion of HSV-1 and HSV-2.	Saddi et al., 2007
	A. chamaemelifolia Vill.	***A. chamaemelifolia*** aqueous extract affected cell monolayer ~70% in two strains BA and DD and inhibited virus yield by ~80% for strain BA and ~70% for strain DD via yield reduction assay.	Angelova et al., 2019
Flaviviruses	***A. vulgaris*** L.	***A. vulgaris***, ***Lippia alba***, ***Lippia origanoides***, and ***Origanum vulgare*** essential oils from Colombia inhibited the replication of Yellow fever virus (YFV) in ***in vitro*** conditions.	Meneses et al., 2009
	A. douglasiana Bess.	***A. douglasiana*** and ***Eupatorium patens***, inhibited Dengue serotype 2 (DEN-2) infection with VC50 values of 60 and 150 ppm, respectively.	García et al., 2003
Bovine Viral Diarrhea Virus (BVDV)	***A. annua*** L	Artemisinin and its semisynthetic derivative artesunate inhibit the BVDV virus.	(Efferth et al., 2008)
Human papillomavirus (HPV)	***A. argyi***	Inhibition of binding between human papillomavirus (HPV) E7 oncoprotein and the retinoblastoma (RB) tumor suppressor (Rb1) protein.	Lee et al., 2005
Newcastle disease virus (NDVs)	***A. glabella*** Kar. et Kir.	It was experimentally found that EOAG produced 40% inhibition of Newcastle disease virus over the studied concentration range (5 – 100 M). The antiviral effect was independent of the concentration.	Seidakhmetova et al., 2002
Dengue virus type 2 (DENV-2)	***A. mendozana*** D.C.	***A. mendozana*** oils showed variable inhibitory activity against DENV2.	Duschatzky et al., 2005

Contd.

Table 3 *Contd.*

Target Viruses	*Artemisia Species and/ or Compounds*	*Mode of Action and/or Consequences*	*Reference/s*
Junin virus ***(JUNV)***	***A. douglasiana*** Bess.	The essential oils extracted from seven plants showed a variable effect on viruses.	García et al., 2003
	A. mendozana D.C.	The essential oils produced variable action on viruses.	Duschatzky et al., 2005
Feline immunodeficiency virus (FIV)	***A. verlotiorum*** Lamotte.	***A. verlotorum*** aqueous extract exhibited strong activity against feline immunodeficiency virus (FIV).	Calderone et al., 1998
Murine Noroviruses (MNV1)	***princeps*** var. orientalis.	The alpha-thujone compound present in ***A. princeps*** essential oil strongly inhibited MNV-1 virus.	Chung, 2017
Human gamma herpes virus type 4 (Epstein-Barr virus) or HHV-4 (EBV)	***A. capillaris*** Thunb.	Diacetylenic spiroketal enol ether epoxide AL-1 (45) from ***A. capillaris*** strongly inhibited tumor promoter-induced Epstein-Barr virus activation in Raji cells (IC50 0.5 μM).	Dembitsky and Levitsky, 2006
Sindbis virus (SV)	***A. absinthium*** L.	***A. absinthium*** was evaluated against herpes simplex, Sindbis and polio viruses.	Mouhajir et al., 2001
Feline Calicivirus FCV-F9	***princeps*** var. orientalis.	The alpha-thujone compound present in ***A. princeps*** essential oil strongly inhibited FCV-F9 virus.	Chung, 2017
Polio virus (PV)	***A. absinthium*** L.	***A. absinthium*** was evaluated against herpes simplex, Sindbis, and polio viruses.	Mouhajir et al., 2001
	A. herba-alba Asso.	***Pinus halepensis*** and ***Punica granatum*** extracts were most active against poliovirus and inhibited the virus at a minimum concentration of 6.5mg/ml. The activities of ***Artemisia herba-alba***, ***Quercus faginea***, ***Rumex bucephalophorus***, ***Pinus halepensis and Punica granatum*** were light independent.	Mouhajir et al., 2001

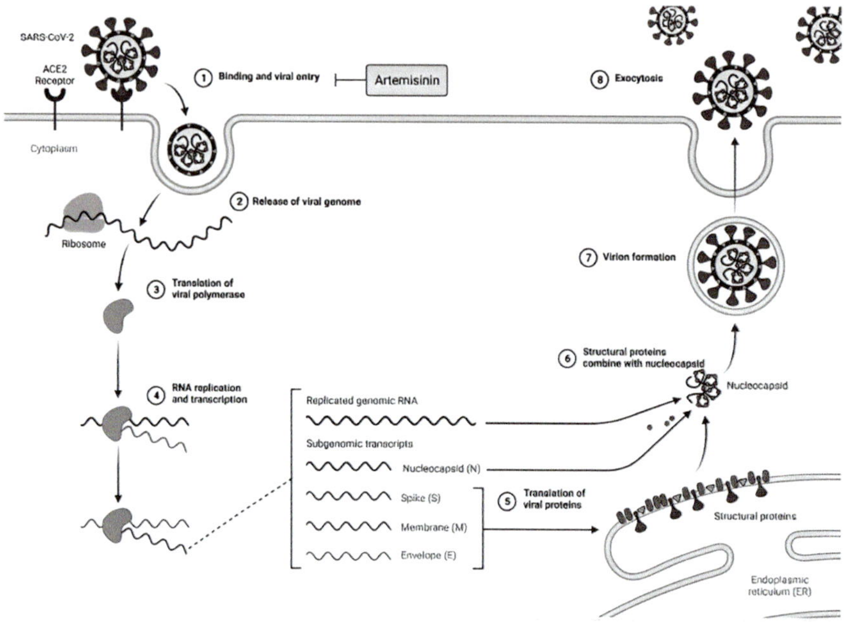

Fig. 4 Artemisinin and its derivatives from *Artemisia* show antiviral action against SARS-CoV-2. (*Source:* Orosco, 2023)

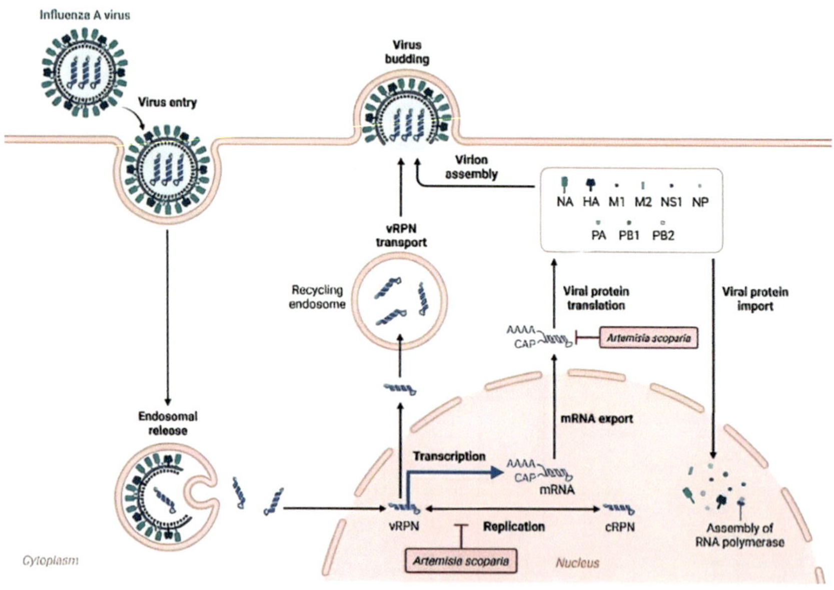

Fig. 5 Antiviral action of *Artemisia scoparia* against the influenza A virus. (*Source:* Orosco, 2023)

conditions. The anti–HIV mode of action of the extracts of *A. campestris* is due to two terpenes *i.e.*, damsin and canrenone and four flavonoids *i.e.*, 5,7-dihydroxy-2-(3-hydroxy-4-methoxyphenyl)-6,8-dimethoxychromen-4-one (acerosin), 6,2',4'–trimethoxyflavone, cardamomin (cardamonin/alpinetin chalcone) and 4',5-dihydroxy-6,7,8-trimethoxyflavone (xanthomicrol). These terpene and flavonoid compounds produce an inhibitory effect on the two dissimilar cellular targets of HIV replication, *i.e.*, by blockage of viral entrance and/or transcription in *in vitro* (wild type HIV–1 and Vesicular Stomatitis Virus (VSV) pseudotyped HIV–1) (Apaza Ticona et al., 2020).

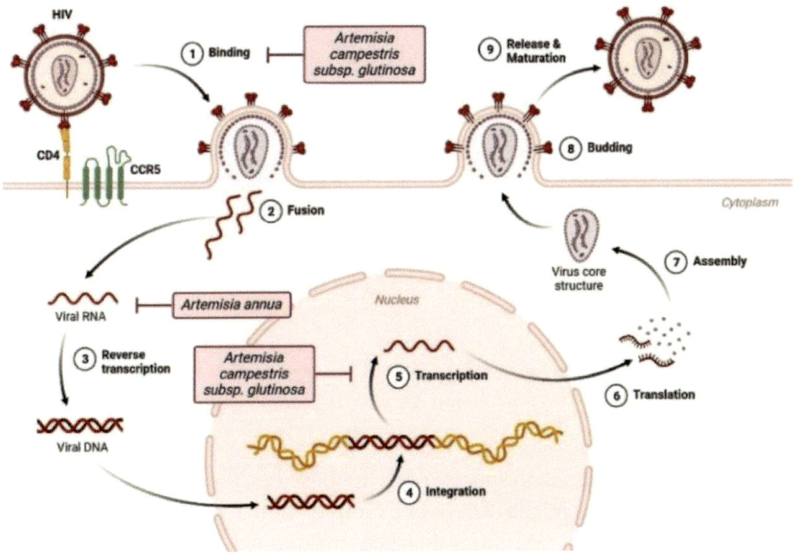

Fig. 6 The anti-HIV potential of *Artemisia annua* and *A. campestris subsp. glutinosa.* (*Source:* Orosco, 2023b)

Hepatitis B viruses (HBVs) lead to chronic liver diseases. *A. argyi* essential oil (AAEO) shows anti-HBV potential by inhibiting DNA replication of HBV and secretion of the HBV antigens. Another potential mode of action by nanostructured lipid carriers (NLCs) loaded with AAEO includes inhibition of immune response-related signaling pathways (Fig. 7) (Gong et al., 2017; Zhang et al., 2023; Orosco, 2023b).

8. Toxicity and Safety

Artemisia species contain various phytochemical compounds that can be toxic or have medicinal properties, depending on the dose and the specific compound in question. Some Artemisia species, such as *A. absinthium* (wormwood), are toxic in high doses, while others, such as *A. annua* (sweet wormwood), are used as traditional medicines for various ailments.

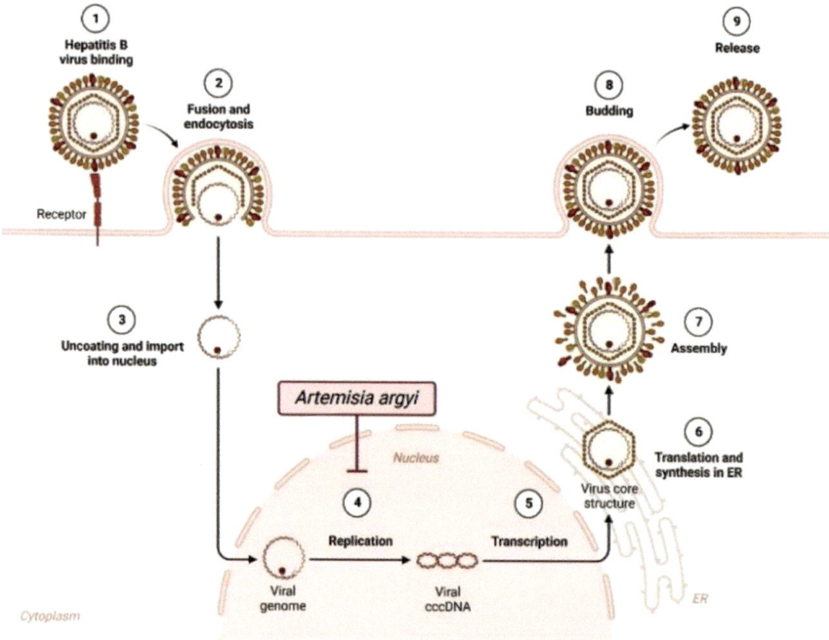

Fig. 7 Antiviral activity of *Artemisia argyi* against the HBV. (*Source:* Orosco, 2023)

The toxicity of Artemisia species can vary based on the species, the part of the plant being consumed, and the dose. For example, high doses of *A. absinthium* can cause symptoms such as vomiting, diarrhea, muscle weakness, and convulsions, while lower doses can have a milder, bitter taste.

A. annua, on the other hand, is used in traditional medicine systems to treat malaria, and the World Health Organization (WHO) has recognized its potential as an anti-malarial agent. However, the safety of using Artemisia annua as a medicine has not been fully established, and more research is needed to determine its efficacy and safety in humans.

In general, it is important to exercise caution when consuming Artemisia species, as some species can be toxic, and the safety of others has not been fully established. If you are considering using Artemisia species for medicinal purposes, it is important to have a proper consultation with a healthcare professional before using it.

In conclusion, the toxicity and safety of Artemisia species varies based on the species, the part of the plant being consumed, and the dose, and it is important to exercise caution when consuming Artemisia species or using them for medicinal purposes.

References

Abiri, R., Silva, A.L.M., de Mesquita, L.S.S., de Mesquita, J.W.C., Atabaki, N., de Almeida, E.B., Shaharuddin, N.A. and Malik, S. (2018). Towards a better understanding of Artemisia vulgaris: Botany, phytochemistry, pharmacological and biotechnological potential. *Food Research International*, *109*, 403–15. https://doi.org/https://doi.org/10.1016/j.foodres.2018.03.072.

Ahadi Dolatsara, E., Salami' A., Shekarpoor M., Nagavi, M.R. and Soreni, A. (2015). Evaluation of Chloroplast DNA diversity and phylogenetic relationship among 28 Iranian Artemisia species. *Iranian Journal of Horticultural Sciences*, 45(4), 401–415.

Ahuja, J., Suresh, J., Paramakrishnan, N., Mruthunjaya, K. and Naganandhini, M. (2011). An ethnomedical, phytochemical and pharmacological profile of Artemisia parviflora Roxb. *Journal of essential oil bearing plants*, *14*(6), 647–657.

Al-Snafi, A.E. (2015). The pharmacological importance of Artemisia campestris-A review. *Asian Journal of Pharmaceutical Research*, *5*(2), 88–92.

Ali, M., Abbasi, B.H., Ahmad, N., Ali, S.S., Ali, S. and Ali, G.S. (2016). Sucrose-enhanced biosynthesis of medicinally important antioxidant secondary metabolites in cell suspension cultures of Artemisia absinthium L. *Bioprocess and Biosystems Engineering*, *39*(12), 1945–1954. https://doi.org/10.1007/s00449-016-1668-8.

Alsaffar, D.F., Yaseen, A. and Jabal, G. (2020). In silico molecular docking studies of medicinal arabic plant-based bioactive compounds as a promising drug candidate against COVID-19. *Int. J. Innov. Sci. Res. Technol*, *5*, 876–896.

Angelova, P., Hinkov, A., Tsvetkov, V., Todorov, D., Shishkova, K., Dragolova, D., Shishkov, S. and Kapchina-Toteva, V. (2019). Antiherpes virus activity of extracts from Artemisia chamaemelifolia vill. *Comptes rendus de l'Académie bulgare des Sciences*, *72*(11).

Apaza Ticona, L., Bermejo, P., Guerra, J.A., Abad, M.J., Beltrán, M., Martín Lázaro, R., Alcamí, J. and Bedoya, L.M. (2020). Ethanolic extract of Artemisia campestris subsp. glutinosa (Besser) Batt. inhibits HIV–1 replication *in vitro* through the activity of terpenes and flavonoids on viral entry and NF–κB pathway. *Journal of Ethnopharmacology*, *263*, 113163. https://doi.org/https://doi.org/10.1016/j.jep.2020.113163.

Asdadi, A., Hamdouch, A., Gharby, S. and Hassani, L.M.I. (2020). Chemical characterization of essential oil of Artemisia herba-alba asso and his possible potential against covid-19. *Journal of Analytical Sciences and Applied Biotechnology*, *2*(2), Anal. Sci. Appl. Biotechnol.-72.

Asres, K., Bucar, F., Kartnig, T., Witvrouw, M., Pannecouque, C. and De Clercq, E. (2001). Antiviral activity against human immunodeficiency virus type 1 (HIV-1) and type 2 (HIV-2) of ethnobotanically selected Ethiopian medicinal plants. *Phytotherapy Research*, *15*(1), 62–69. https://doi.org/https://doi.org/10.1002/1099-1573(200102)15:1<62::AID-PTR956>3.0.CO;2-X

Baggieri, M., Gioacchini, S., Borgonovo, G., Catinella, G., Marchi, A., Picone, P., Vasto, S., Fioravanti. R., Bucci, P., Kojouri, M., Giuseppetti, R., D'Ugo, E., Ubaldi, F., Dallavalle, S., Nuzzo, D., Pinto, A. and Magurano, F. (2023). Antiviral, virucidal and antioxidant properties of Artemisia annua against SARS-CoV-2. *Biomedicine and Pharmacotherapy*, *168*, 115682. https://doi.org/https://doi.org/10.1016/j.biopha.2023.115682.

Barkat, M.A., Goyal, A., Barkat, H.A., Salauddin, M., Pottoo, F.H. and Anwer, E.T. (2021). Herbal medicine: Clinical perspective and regulatory status. *Combinatorial chemistry and high throughput screening*, *24*(10), 1573–1582.

Bora, K.S. and A. Sharma (2010). "Neuroprotective effect of Artemisia absinthium L. on focal ischemia and reperfusion-induced cerebral injury." *Journal of Ethnopharmacology 129*(3): 403–409.

Bora, K.S. and A. Sharma (2011). "The Genus Artemisia: A Comprehensive Review." *Pharmaceutical Biology 49*(1): 101–109.

Breverton, T. (2013). *Breverton's complete herbal*. Quercus.

Brisibe, E.A., et al. (2009). "Nutritional characterisation and antioxidant capacity of different tissues of Artemisia annua L." *Food Chemistry 115*(4): 1240–1246.

Cao, R., Hu, H., Li, Y., Wang, X., Xu, M., Liu, J., Zhang, H., Yan, Y., Zhao, L., Li, W., Zhang, T., Xiao, D., Guo, X., Li, Y., Yang, J., Hu, Z., Wang, M. and Zhong, W. (2020). Anti-SARS-CoV-2 Potential of Artemisinins *In Vitro*. *ACS Infectious Diseases*, *6*(9), 2524–2531. https://doi.org/10.1021/acsinfecdis.0c00522.

Chan, K. (2003). Some aspects of toxic contaminants in herbal medicines. *Chemosphere*, *52*(9), 1361–71. https://doi.org/https://doi.org/10.1016/S0045-6535(03)00471-5.

Chang, Y.-S. and Woo, E.-R. (2003). Korean medicinal plants inhibiting to Human Immunodeficiency Virus type 1 (HIV-1) fusion. *Phytotherapy Research*, *17*(4), 426–429. https://doi.org/https://doi.org/10.1002/ptr.1155.

Cheong, D.H., Tan, D.W., Wong, F.W. and Tran, T. (2020). Anti-malarial drug, artemisinin and its derivatives for the treatment of respiratory diseases. *Pharmacological Research*, *158*, 104901.

Chung, M.S. (2017). Antiviral activities of Artemisia princeps var. orientalis essential oil and its α-thujone against norovirus surrogates. *Food Science and Biotechnology*, *26*(5), 1457–1461. https://doi.org/10.1007/s10068-017-0158-3.

Das, S., Sarmah, S., Lyndem, S. and Singha Roy, A. (2021). An investigation into the identification of potential inhibitors of SARS-CoV-2 main protease using molecular docking study. *Journal of Biomolecular Structure and Dynamics*, *39*(9), 3347–3357. https://doi.org/10.1080/07391102.2020.1763201.

Duschatzky, C.B., Possetto, M.L., Talarico, L.B., García, C.C., Michis, F., Almeida, N.V., de Lampasona, M.P., Schuff, C. and Damonte, E.B. (2005). Evaluation of Chemical and Antiviral Properties of Essential Oils from South American Plants. *Antiviral Chemistry and Chemotherapy*, *16*(4), 247–251. https://doi.org/10.1177/095632020501600404.

Efferth, T. (2018). Beyond malaria: The inhibition of viruses by artemisinin-type compounds. *Biotechnology Advances*, *36*(6), 1730–1737. https://doi.org/https://doi.org/10.1016/j.biotechadv.2018.01.001.

Efferth, T., et al. (2015). "Nobel Prize for artemisinin brings phytotherapy into the spotlight." *Phytomedicine 22*(13): A1–A3.

Efferth, T., Romero, M.R., Wolf, D.G., Stamminger, T., Marin, J.J.G. and Marschall, M. (2008). The Antiviral Activities of Artemisinin and Artesunate. *Clinical Infectious Diseases*, *47*(6), 804–811. https://doi.org/10.1086/591195.

Flobinus, A., Taudon, N., Desbordes, M., Labrosse, B., Simon, F., Mazeron, M.-C. and Schnepf, N. (2014). Stability and antiviral activity against human cytomegalovirus of artemisinin derivatives. *Journal of Antimicrobial Chemotherapy*, *69*(1), 34–40. https://doi.org/10.1093/jac/dkt346.

Funk, V.A., Stuessy, T. and Bayer, R. (2009). *Systematics, evolution, and biogeography of Compositae*. International Association for Plant Taxonomy Vienna.

Fuzimoto, A.D. (2021). An overview of the anti-SARS-CoV-2 properties of Artemisia annua, its antiviral action, protein-associated mechanisms, and repurposing for COVID-19 treatment. *Journal of Integrative Medicine*, *19*(5), 375–388. https://doi.org/https://doi.org/10.1016/j.joim.2021.07.003

García, C.C., Talarico, L., Almeida, N., Colombres, S., Duschatzky, C. and Damonte, E.B. (2003). Virucidal activity of essential oils from aromatic plants of San Luis, Argentina. *Phytotherapy Research*, *17*(9), 1073–1075. https://doi.org/https://doi.org/10.1002/ptr.1305.

Gavanji, S., Sayedipour, S.S., Larki, B. and Bakhtari, A. (2015). Antiviral activity of some plant oils against herpes simplex virus type 1 *in Vero* cell culture. *Journal of Acute Medicine*, *5*(3), 62–68. https://doi.org/https://doi.org/10.1016/j.jacme.2015.07.001.

Gendrot, M., et al. (2020). "Antimalarial artemisinin-based combination therapies (ACT) and COVID-19 in Africa: In vitro inhibition of SARS-CoV-2 replication by mefloquine-artesunate." *International Journal of Infectious Diseases 99*: 437–440.

Ghlissi, Z., et al. (2016). "Antioxidant, antibacterial, anti-inflammatory and wound healing effects of Artemisia campestris aqueous extract in rat." *Biomedicine & Pharmacotherapy 84*: 115–122.

Gilani, A.H., et al. (2005). "Hepatoprotective activity of aqueous–methanol extract of Artemisia vulgaris." *Phytotherapy Research 19*(2): 170–172.

Gong, P.-Y., et al. (2017). "Anti-HBV Activities of Three Compounds Extracted and Purified from Herpetospermum Seeds." *Molecules 22*(1): 14.

Goud, B.J., et al. (2015). "A review on history, controversy, traditional use, ethnobotany, phytochemistry and pharmacology of Artemisia absinthium Linn." *International Journal of Advanced Research in Engineering and Applied Sciences 4*(5): 77–107.

Hadi, A., et al. (2014). "Anti-inflammatory and analgesic activities of Artemisia absinthium and chemical composition of its essential oil." *Int J Pharm Sci Rev Res 38*: 237–244.

Haghighi, A.R., Belduz, A.O., Vahed, M.M., Coskuncelebi, K. and Terzioglu, S. (2014). Phylogenetic relationships among Artemisia species based on nuclear ITS and chloroplast psbA-trnH DNA markers. *Biologia*, *69*, 834–839.

Han, Y., et al. (2019). "Antimalarial drugs and their metabolites are potent Zika virus inhibitors." *Journal of Medical Virology 91*(7): 1182–1190.

Ho, W.E., et al. (2014). "Artemisinins: Pharmacological actions beyond anti-malarial." *Pharmacology & Therapeutics 142*(1): 126–139.

Huang, T.-J., Liu, S.-H., Kuo, Y.-C., Chen, C.-W. and Chou, S.-C. (2014). Antiviral activity of chemical compound isolated from Artemisia morrisonensis against hepatitis B virus *in vitro*. *Antiviral Research*, *101*, 97–104. https://doi.org/https://doi.org/10.1016/j.antiviral.2013.11.007.

Hussain, A. (2020). The Genus Artemisia (Asteraceae): A Review on its Ethnomedicinal Prominence and Taxonomy with Emphasis on Foliar Anatomy, Morphology and Molecular Phylogeny: Ethnomedicinal Prominence and Taxonomy of Artemisia. *Proceedings of the Pakistan Academy of Sciences: B. Life and Environmental Sciences*, *57*(1), 1–28.

Iranshahi, M., Emami, S.A. and Mahmoud, S.M. (2007). Detection of sesquiterpene lactones in ten Artemisia species population of Khorasan provinces.

Jelodar, N.B., Bhatt, A., Mohamed, K. and Keng, C.L. (2014). New cultivation approaches of Artemisia annua L. for a sustainable production of the antimalarial drug artemisinin. *J. Med. Plants Res*, *8*(10), 441–447.

Jiao, B., et al. (2023). "Phylogenomics and morphological evolution of the mega-diverse genus Artemisia (Asteraceae: Anthemideae): implications for its circumscription and infrageneric taxonomy." *Annals of Botany 131*(5): 867–883.

Juteau, F., et al. (2002). "Antibacterial and antioxidant activities of Artemisia annua essential oil." *Fitoterapia 73*(6): 532–535.

Karamoddini, M.K., Emami, S.A., Ghannad, M.S., Sani, E.A. and Sahebkar, A. (2011). Antiviral activities of aerial subsets of Artemisia species against Herpes Simplex virus type 1 (HSV1) in vitro. *Asian Biomedicine*, *5*(1), 63–68.

Kim, H., Jang, E., Kim, S.-Y., Choi, J.Y., Lee, N.R., Kim, D.S., Lee, K.T., Inn, K.S., Kim, B.J. and Lee, J.H. (2018). Preclinical Evaluation of *In Vitro* and *In Vivo* Antiviral Activities of KCT-01, a New Herbal Formula against Hepatitis B Virus. *Evidence-Based Complementary and Alternative Medicine*, *2018*(1), 1073509.

Kshirsagar, S.G. and R.V. Rao (2021). "Antiviral and Immunomodulation Effects of Artemisia." Medicina 57(3): 217.

Koul, B., Taak, P., Kumar, A., Khatri, T. and Sanyal, I. (2017). The Artemisia genus: A review on traditional uses, phytochemical constituents, pharmacological properties and germplasm conservation. *J. Glycomics Lipidomics*, *7*(1), 142.

Kshirsagar, S.G. and R.V. Rao (2021). "Antiviral and Immunomodulation Effects of Artemisia." Medicina *57*(3): 217.

Lee, H.G., Yu, K.A., Oh, W.-K., Baeg, T.W., Oh, H.C., Ahn, J.S., Jang, W.C., Kim, J.W., Lim, J.S., Choe, Y.K. and Yoon, D.Y. (2005). Inhibitory effect of jaceosidin isolated from Artemisia argyi on the function of E6 and E7 oncoproteins of HPV 16. *Journal of Ethnopharmacology*, *98*(3), 339–43. https://doi.org/https://doi.org/10.1016/j.jep.2005.01.054.

Li, G., Yuan, M., Li, H., Deng, C., Wang, Q., Tang, Y., Zhang, H., Yu, W., Xu, Q., Zou, Y., Yuan, Y., Guo, J., Jin, C., Guan, X., Xie, F. and Song, J. (2021). Safety and efficacy of artemisinin-piperaquine for treatment of COVID-19: an open-label, non-randomised and controlled trial. *International Journal of Antimicrobial Agents*, *57*(1), 106216. https://doi.org/https://doi.org/10.1016/j.ijantimicag.2020.106216.

Li, S.Y., Chen, C., Zhang, H.Q., Guo, H.Y., Wang, H., Wang, L., Zhang, X., Hua, S.N., Yu, J., Xiao, P.G., Li, R.S. and Tan, X. (2005). Identification of natural compounds with antiviral activities against SARS-associated coronavirus. *Antiviral Research*, *67*(1), 18–23. https://doi.org/https://doi.org/10.1016/j.antiviral.2005.02.007.

Mandel, J.R., Barker, M.S., Bayer, R.J., Dikow, R.B., Gao, G., Jones, K.E., Keeley, S., Kilian, N., Ma, H., Siniscalchi, C.M., Susanna, A., Thapa, R., Watson, L. and Funk, V.A. (2017). The Compositae Tree of Life in the age of phylogenomics. *Journal of Systematics and Evolution*, 55(4), 405–410. https://doi.org/10.1111/jse.12265.

Meneses, R., Ocazionez, R.E., Martínez, J.R. and Stashenko, E.E. (2009). Inhibitory effect of essential oils obtained from plants grown in Colombia on yellow fever virus replication in vitro. *Annals of Clinical Microbiology and Antimicrobials*, *8*(1), 8. https://doi.org/10.1186/1476-0711-8-8.

Mouhajir, F., Hudson, J., Rejdali, M. and Towers, G. (2001). Multiple antiviral activities of endemic medicinal plants used by Berber peoples of Morocco. *Pharmaceutical Biology*, *39*(5), 364–374.

Nakamura, N. and Hattori, M. (2001). Inhibitory effects on HIV-1 protease of tri-p-coumaroylspermidine from Artemisia caruifolia and related amides. *Chemical and pharmaceutical bulletin*, *49*(7), 915–917.

Nofal, S.M., et al. (2009). "Anti-diabetic effect of Artemisia judaica extracts." *Research Journal of Medicine and Medical Sciences 4*(1): 42–48.

Oguariri, R.M., et al. (2010). "Evaluation of the effect of pyrimethamine, an anti-malarial drug, on HIV-1 replication." *Virus Research 153*(2): 269–276.

Orosco, F.L. (2023). "Current Advances in Antiviral Potential of Artemisia Against Major Viral Infections." *Journal of Bacteriology and Virology 53*(2): 61–73.

Palmer, G. (1975). Shuswap Indian ethnobotany. *Syesis*, *8*, 29–81.

Peele, K.A., Potla Durthi, C., Srihansa, T., Krupanidhi, S., Ayyagari, V.S., Babu, D.J., Indira, M., Reddy, A.R. and Venkateswarulu, T. C. (2020). Molecular docking and dynamic simulations for antiviral compounds against SARS-CoV-2: A computational study. *Informatics in Medicine Unlocked*, *19*, 100345. https://doi.org/https://doi.org/10.1016/j.imu.2020.100345.

Rezazadeh, N., et al. (2023). "Anticancer effect of Artemisia extract and cisplatin on induction of apoptosis and inhibition of proliferation in A2780 human ovarian cancer." *Nova Biologica Reperta 9*(4): 246–256.

Rolta, R., Salaria, D., Sharma, P., Sharma, B., Kumar, V., Rathi, B., Verma, M., Sourirajan, A., Baumler, D.J. and Dev, K. (2021). Phytocompounds of Rheum emodi, Thymus serpyllum, and Artemisia annua Inhibit Spike Protein of SARS-CoV-2 Binding to ACE2 Receptor: In Silico Approach. *Current Pharmacology Reports*, *7*(4), 135–149. https://doi.org/10.1007/s40495-021-00259-4.

Saddi, M., et al. (2007). "Antiherpevirus activity of Artemisia arborescens essential oil and inhibition of lateral diffusion in Vero cells." *Annals of Clinical Microbiology and Antimicrobials 6*(1): 10.

Sehailia, M. and Chemat, S. (2021). Antimalarial-agent artemisinin and derivatives portray more potent binding to Lys353 and Lys31-binding hotspots of SARS-CoV-2 spike protein than hydroxychloroquine: potential repurposing of artenimol for COVID-19. *Journal of Biomolecular Structure and Dynamics*, *39*(16), 6184–6194. https://doi.org/10.1080/07391102.2020.1796809.

Seidakhmetova, R., Beisenbaeva, A., Atazhanova, G., Suleimenov, E., Pak, R., Kulyyasov, A. and Adekenov, S. (2002). Chemical composition and biological activity of the essential oil from Artemisia glabella.

Septembre-Malaterre, A., et al. (2020). "Artemisia annua, a Traditional Plant Brought to Light." *International Journal of Molecular Sciences 21*(14): 4986.

Shabnam, A., Mohammad Akhtar, S., Sharad, M. and Mohammad, M. (2018). Antiviral Efficacy of Qust (Saussurea lappa) and Afsanteen (Artemisia absinthium) for Chronic Hepatitis B: A Prospective Single-Arm Pilot Clinical Trial. *Pharmacognosy Research, 10*(3).

Shahriyary, L. and Yazdanparast, R. (2007). Inhibition of blood platelet adhesion, aggregation and secretion by Artemisia dracunculus leaves extracts. *Journal of Ethnopharmacology, 114*(2), 194–198. https://doi.org/https://doi.org/10.1016/j.jep.2007.07.029.

Sharma, S. and Deep, S. (2022). In-silico drug repurposing for targeting SARS-CoV-2 main protease (Mpro). *Journal of Biomolecular Structure and Dynamics, 40*(7), 3003–3010.

Slezakova, S. and Ruda-Kucerova, J. (2017). Anticancer Activity of Artemisinin and its Derivatives. *Anticancer Research, 37*(11), 5995. http://ar.iiarjournals.org/content/37/11/5995.abstract.

Swanston-Flatt, S.K., Flatt, P.R., Day, C. and Bailey, C.J. (1991). Traditional dietary adjuncts for the treatment of diabetes mellitus. *Proceedings of the Nutrition Society, 50*(3), 641–51. https://doi.org/10.1079/PNS19910077.

Taleghani, A., et al. (2020). "Artemisia: a promising plant for the treatment of cancer." *Bioorganic & Medicinal Chemistry 28*(1): 115180.

Trieu, V., et al. (2021). "Targeting TGF-β pathway with COVID-19 drug candidate ARTIVeda/PulmoHeal accelerates recovery from mild-moderate COVID-19." MedRxiv: 2001. 2024.21250418.

Tu, Y.-y. (1997). "New antimalarial drug—Qinghaosu and Dihydro-Qinghaosu." *Chinese Journal of Integrated Traditional and Western Medicine 3*(4): 312–312.

Uckun, F.M., Saund, S., Windlass, H. and Trieu, V. (2021). Repurposing Anti-Malaria Phytomedicine Artemisinin as a COVID-19 Drug [Mini Review]. *Frontiers in Pharmacology, 12*. https://doi.org/10.3389/fphar.2021.649532.

Uzun, T. and O. Toptas (2020). "Artesunate: could be an alternative drug to chloroquine in COVID-19 treatment?" *Chinese Medicine 15*(1): 54.

Wang, L., Si, L., Li, Y., Wang, H., Xu, F., Bian, H., Shi, Y. and Huang, H. (2017). Study on mechanism and active ingredient of Artemisia scoparia extracts against influenza virus. *Lat. Am. J. Pharm, 36*(7), 1355–1360.

Watson, L.E., Bates, P.L., Evans, T.M., Unwin, M.M. and Estes, J.R. (2002). Molecular phylogeny of Subtribe Artemisiinae (Asteraceae), including Artemisia and its allied and segregate genera. *BMC Evolutionary Biology, 2*(1), 17. https://doi.org/10.1186/1471-2148-2-17.

Woerdenbag, H.J., et al. (1990). "Artemisia annua L.: a source of novel antimalarial drugs." *Pharmaceutisch Weekblad 12*(5): 169–181.

Wu, C., Chen, X., Cai, Y., Xia, J. A., Zhou, X., Xu, S., Huang, H., Zhang, L., Zhou, X., Du, C., Zhang, Y., Song, J., Wang, S., Chao, Y., Yang, Z., Xu, J., Zhou, X., Chen, D., Xiong, W.,…Song, Y. (2020). Risk Factors Associated With Acute Respiratory Distress Syndrome and Death in Patients With Coronavirus Disease 2019 Pneumonia in Wuhan, China. *JAMA Internal Medicine, 180*(7), 934–943. https://doi.org/10.1001/jamainternmed.2020.0994.

Wu, T.S., Tsang, Z.J., Wu, P.L., Lin, F.W., Li, C.Y., Teng, C.M. and Lee, K.H. (2001). New constituents and antiplatelet aggregation and anti-HIV principles of Artemisia capillaris. *Bioorganic and Medicinal Chemistry, 9*(1), 77–83. https://doi.org/https://doi.org/10.1016/S0968-0896(00)00225-X.

Yuexia, M., Wei, Z., Zhongpeng, Z., Min, L., Jian, L. and Yuguang, W. (2016). Combination of ribavirin and reduning protects mice against severe pneumonia induced by H1N1 influenza a virus. *Journal of Traditional Chinese Medicine, 36*(2), 181–186. https://doi.org/https://doi.org/10.1016/S0254-6272(16)30025-5.

Zhang, N., Yang, H., Han, T., Kim, H.S. and Marcelis, L.F. (2023). Towards greenhouse cultivation of Artemisia annua: The application of LEDs in regulating plant growth and secondary metabolism. *Frontiers in Plant Science, 13*, 1099713.

Zheng, G.-Q. (1994). "Cytotoxic terpenoids and flavonoids from Artemisia annua." *Planta medica 60*(01): 54–57.

Zhou, Y., et al. (2021). "In vitro efficacy of artemisinin-based treatments against SARS-CoV-2." *Scientific Reports 11*(1): 14571.

2 Antiviral Effects of *Curcuma longa* L. and its Phytochemical Constituents

Chanin Sillapachaiyaporn[1,2], Mani Iyer Prasanth[1,2], Tewin Tencomnao[1,2] and Siriporn Chuchawankul[3,4*]

1. Introduction

Curcuma longa L. (*C. longa*, syn. *Curcuma domestica*), commonly known as turmeric, belongs to the genus *Curcuma* and the family Zingiberaceae. It is distributed throughout tropical and subtropical areas, especially in Asian countries (Nisar et al., 2015; Ooko et al., 2017). *C. longa* is a perennial rhizomatous plant with pointed leaves, 3–5 ft height, and funnel-shaped yellow flowers (Mathai et al., 2018). The rhizomes are distinguished by their orange-yellow color. (Sharma et al., 2021). *C. longa* rhizome powder is widely used as a spice in Asian cuisine (Dasgupta, 2019); moreover, it has been reported to have a broad range of pharmaceutical properties such as antioxidant (Braga et al., 2003), anti-inflammation (Lee et al., 2020), antitumor (Wang et al., 2017), and antimicrobial (Gupta et al., 2015).

2. Taxonomy and Scientific Classification

Kingdom : Plantae
Phylum : Tracheophyta
Class : Liliopsida
Order : Zingiberales
Family : Zingiberaceae
Genus : *Curcuma*
Species : *Curcuma longa*

[1] Department of Clinical Chemistry, Faculty of Allied Health Sciences, Chulalongkorn University, Bangkok, Thailand.
[2] Center of Excellence on Natural Products for Neuroprotection and Anti-ageing, Chulalongkorn University, Bangkok, Thailand.
[3] Department of Transfusion Medicine and Clinical Microbiology, Faculty of Allied Health Sciences, Chulalongkorn University, Bangkok, Thailand.
[4] Immunomodulation of Natural Products Research Unit, Faculty of Allied Health Sciences, Chulalongkorn University, Bangkok, Thailand.
* Corresponding author: siriporn.ch@chula.ac.th

DOI: 10.1201/9781003433583-2

3. Phytochemical Constituents and Metabolite Profiling

C. longa is composed of sesquiterpenoids, flavonoids, phenols, diarylheptanoids, tannins, monoterpenoids, and steroids, as tabulated in **Table 1**. The chemical constituents of *C. longa* depend on extraction methods and solvents. Hexane (non-polar) extract mainly comprises sesquiterpenoids, such as curcuminol and methenolone (El-Hawaz et al., 2018). Ethanol, a polar solvent extract is enriched with phenolic compounds such as curcumin, demethoxycurcumin, curcumin-O-glucuronide, *isorhamnetin*, and valoneic acid bilactone (Sabir et al., 2020). Comparative studies demonstrated the different chemical compositions between fresh and dry rhizhomes of *C. longa* (Singh et al., 2010; Salem et al., 2022).

The major active components of *C. longa* are curcuminoids and zedoary turmeric oils. Curcuminoids are polyphenolic non-volatile compounds, including curcumin, demethoxycurcumin, and bisdemethoxycurcumin, with dark yellow powder characteristics (Praveen et al., 2021; Wu and Tong, 2022). Zedoary turmeric oils are volatile oils mainly composed of terpenoids, including curcumol, germacrone, and curdione (Wu and Tong, 2022). The chemical structures of these compounds found in *C. longa* are illustrated in **Fig. 1**.

4. Medicinal/Therapeutic Uses

4.1 Uses Described in Pharmacopoeias and in Traditional Systems of Medicine

C. longa has been used in pharmacopeias in several countries, including China (Liu et al., 2022), India (Prakash et al., 2017), Morocco (Babili et al., 2022), Thailand (Pantharos et al., 2022), and Vietnam (Van Dau et al., 1998). From ancient history, humans have utilized *C. longa* in the Ayurvedic medicine system to treat several diseases such as jaundice, skin infections, wound healing, flatulence, sprains, arthritis, and stomach disturbances (Hussain et al., 2022). In Chinese pharmacopoeia, *C. longa* has been used to promote blood circulation; relieve itching, soreness, swelling, and pain; and eliminate acne (Liu et al., 2022).

4.2 Traditional /Folklore Uses Described in Folk Medicine

In traditional medicine, accumulating evidence suggests that *C. longa* has been widely used to cure various diseases in Asian countries. In Indonesia, *C. longa* is one of the most traditional herbal medicines, used to attenuate gastric disorders (Permatananda et al., 2021). In Hawaii, *C. longa* has been traditionally used to treat sinus and ear infections, and gastrointestinal ulcers (Kumar and Sakhya, 2013). Indian women apply a *C. longa* rhizome on the skin to prevent hair growth (Kumar et al., 2011). Moreover, in Pakistan, they orally or locally use boiled *C. longa* rhizomes to kill internal or external parasites (Aziz et al., 2018).

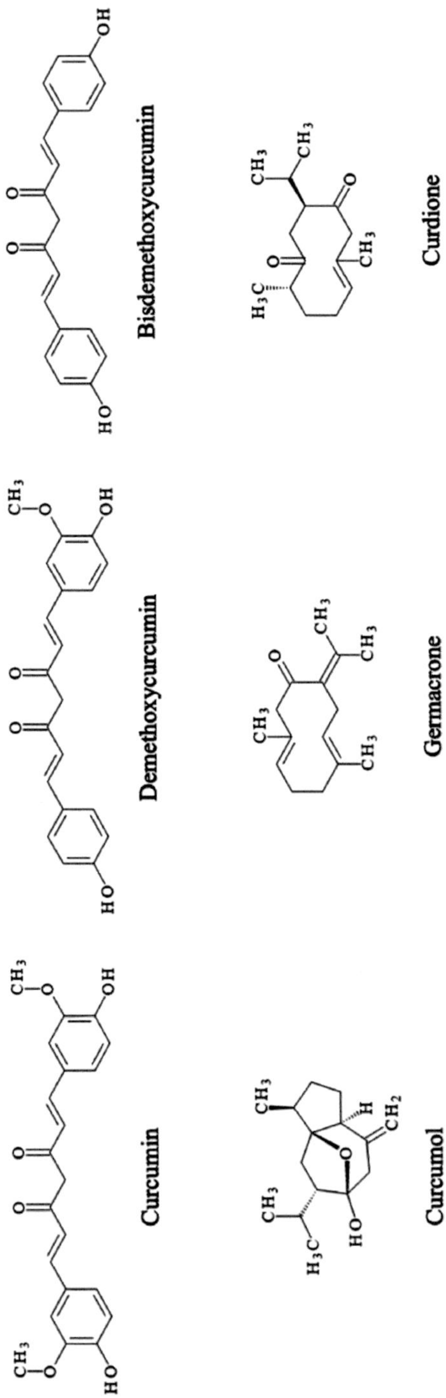

Fig. 1 Chemical structures of major chemical constituents found in *Curcuma longa*.

Table 1 Reported phytochemical constituents of *Curcuma longa*

Compound	*Molecular Formula*	*Molecular mass*	*Classification*	*References*
alpha-Santalene	$C_{15}H_{24}$	204.35	Sesquiterpene	(Singh et al., 2010)
alpha-Zingiberene	$C_{15}H_{24}$	204.35	Sesquiterpene	(Singh et al., 2010)
Apigenin	$C_{15}H_{10}O_5$	270.24	Flavonoid	(Mohsin et al., 2022)
aromatic-Curcumene	$C_{15}H_{22}$	202.33	Sesquiterpene	(Singh et al., 2010)
aromatic-Turmerone	$C_{15}H_{22}O$	218.33	Sesquiterpene	(Singh et al., 2010)
beta-Bisabolene	$C_{15}H_{24}$	204.35	Sesquiterpene	(Singh et al., 2010)
beta-Bisabolol	$C_{15}H_{26}O$	222.37	Sesquiterpene	(Singh et al., 2010)
beta-Caryophyllene	$C_{15}H_{24}$	204.35	Sesquiterpene	(Singh et al., 2010)
beta-Elemene	$C_{15}H_{24}$	204.35	Sesquiterpene	(El-Hawaz et al., 2018)
beta-Elemenone	$C_{15}H_{22}O$	218.33	Sesquiterpene	(El-Hawaz et al., 2018)
beta-Sesquiphellandrene	$C_{15}H_{24}$	204.35	Sesquiterpene	(Singh et al., 2010)
beta-Turmerone	$C_{15}H_{22}O$	218.33	Sesquiterpene	(Singh et al., 2010)
Bisabolone (6S, 7R)	$C_{15}H_{24}O$	220.35	Phenol	(Singh et al., 2010)
Bisdemethoxycurcumin	$C_{19}H_{16}O_4$	308.3	Diarylheptanoid	(Praveen et al., 2021; Sabir et al., 2020)
Caffeic acid	$C_9H_8O_4$	180.16	Phenol	(Sabir et al., 2020)
Caffeic acid hexoside	$C_{15}H_{18}O_9$	342.3	Phenol	(Sabir et al., 2020)
Casuarinin	$C_{41}H_{28}O_{26}$	936.6	Tannin	(Sabir et al., 2020)
Catechin	$C_{15}H_{14}O_6$	290.27	Phenol	(Mohsin et al., 2022)
Coumaric acid	$C_{15}H_{18}O_8$	326.3	Phenol	(Mohsin et al., 2022; Sabir et al., 2020)
Curcumenol	$C_{15}H_{22}O_2$	234.33	Sesquiterpene	(El-Hawaz et al., 2018)
Curcumenone	$C_{15}H_{22}O_2$	234.33	Sesquiterpene	(El-Hawaz et al., 2018)
Curcumin	$C_{21}H_{20}O_6$	368.4	Diarylheptanoid	(Praveen et al., 2021; Sabir et al., 2020; Taha et al., 2015)
Curcumin-*O*-glucuronide	$C_{27}H_{28}O_{12}$	544.5	Phenol	(Sabir et al., 2020)
Curcumol	$C_{15}H_{24}O_2$	236.35	Sesquiterpene	(Wu and Tong, 2022)
Curdione	$C_{15}H_{24}O_2$	236.35	Sesquiterpene	(Sabir et al., 2020; Wu and Tong, 2022)
Demethoxycurcumin	$C_{20}H_{18}O_5$	338.4	Diarylheptanoid	(Praveen et al., 2021; Sabir et al., 2020)
Digalloyl-hexoside	$C_{20}H_{20}O_{14}$	483.08	Phenol	(Sabir et al., 2020)

Contd.

Table 1 *Contd.*

Compound	*Molecular Formula*	*Molecular mass*	*Classification*	*References*
Emodin	$C_{15}H_{10}O_5$	270.24	Phenol	(Mohsin et al., 2022)
epi-beta-Santalene	$C_{15}H_{24}$	204.35	Sesquiterpene	(Singh et al., 2010)
Ferulic acid	$C_{10}H_{10}O_4$	194.18	Phenol	(Mohsin et al., 2022)
Germacrene A	$C_{15}H_{24}$	204.35	Sesquiterpene	(El-Hawaz et al., 2018)
Germacrene B	$C_{15}H_{24}$	204.35	Sesquiterpene	(El-Hawaz et al., 2018)
Germacrene D	$C_{15}H_{24}$	204.35	Sesquiterpene	(El-Hawaz et al., 2018)
Germacrone	$C_{15}H_{24}$	204.35	Sesquiterpene	(El-Hawaz et al., 2018; Singh et al., 2010; Wu and Tong, 2022)
Isocurcumenol	$C_{15}H_{22}O_2$	234.33	Sesquiterpene	(El-Hawaz et al., 2018)
Isorhamnetin	$C_{16}H_{12}O_7$	316.26	Flavonoid	(Sabir et al., 2020)
Kaempferol	$C_{15}H_{10}O_6$	286.24	Flavonoid	(Mohsin et al., 2022)
Methenolone	$C_{20}H_{30}O_2$	302.5	Steroid	(El-Hawaz et al., 2018)
Myricetin	$C_{15}H_{10}O_8$	318.23	Flavonoid	(Mohsin et al., 2022)
Neocurdione	$C_{15}H_{24}O_2$	236.35	Sesquiterpene	(El-Hawaz et al., 2018)
Plumbagin	$C_{11}H_8O_3$	188.18	Phenol	(Mohsin et al., 2022)
Quercetin	$C_{15}H_{10}O_7$	302.23	Flavonoid	(Mohsin et al., 2022)
Quercetin-3-D-galactoside	$C_{21}H_{20}O_{12}$	464.4	Flavonoid	(Sabir et al., 2020)
Rutin	$C_{27}H_{30}O_{16}$	610.5	Flavonoid	(Mohsin et al., 2022)
Santalenone	$C_{15}H_{22}O$	218.33	Sesquiterpene	(Singh et al., 2010)
Sesquisabinene	$C_{15}H_{24}$	204.35	Sesquiterpene	(Singh et al., 2010)
Sinapic acid	$C_{11}H_{12}O_5$	224.21	Phenol	(Sabir et al., 2020)
Syringic acid	$C_9H_{10}O_5$	198.17	Phenol	(Mohsin et al., 2022)
Terpinolene	$C_{10}H_{16}$	136.23	Monoterpene	(Singh et al., 2010)
Thymoquinone	$C_{10}H_{12}O_2$	164.2	Phenol	(Mohsin et al., 2022)
trans-alpha-Atlantone	$C_{15}H_{22}O$	218.33	Sesquiterpene	(Singh et al., 2010)
trans-alpha-Bergamotene	$C_{15}H_{24}$	204.35	Sesquiterpene	(Singh et al., 2010)
trans-beta-Farnesene	$C_{15}H_{24}$	204.35	Sesquiterpene	(Singh et al., 2010)
Valoneic acid bilactone	$C_{21}H_{10}O_{13}$	470.3	Tannin	(Sabir et al., 2020)
Vanillic acid	$C_8H_8O_4$	168.15	Phenol	(Mohsin et al., 2022)

4.3 Applications Supported by Clinical Data

Hundreds of completed clinical trials of curcumin, a responsible compound of *C. longa*, have been reported. The studies suggest that curcumin has therapeutic and pharmacological benefits for various populations of patients (Sivani et al., 2022). Adjuvant therapy of *C. longa* curcumin showed beneficial effects on patients infected with severe acute respiratory syndrome coronavirus 2 (SARS-CoV-2) (Pawar et al., 2021; Sankhe et al., 2022).

4.4 Pharmacological Activities Mainly Antiviral

Bioactive compounds obtained from *C. longa* have been documented to exhibit various pharmacological properties such as antioxidant (Braga et al., 2003), anti-inflammatory (Lee et al., 2020), antitumor (Wang et al., 2017), antibacterial (Gupta et al., 2015), and antiviral (Prasansuklab et al., 2021; Srivastava et al., 2022) activities. Kim et al. reported that *C. longa* aqueous extract had anti-hepatitis B virus (HBV) in both *in vitro* HepG2.2.15 cell line and *in vivo* HBV X protein (HBx) transgenic mice models (Kim et al., 2009; Kim et al., 2011). Besides, *C. longa* supplement capsule could inhibit SARS-CoV-2 replication in Vero E6 cells with the half-maximal effective dose (EC_{50}) at 7.4 μg/mL (Bormann et al., 2021). Accumulating data demonstrates that curcumin, the most abundant active ingredient of *C. longa* shows antiviral activity against various types of viruses, including bovine herpesvirus type 1 (BHV-1) (Zhu et al., 2015), Chikungunya virus (Mounce et al., 2017), dengue virus type 2 (DV2) (Chen et al., 2013), Ebola virus (Setlur et al., 2017), Ebstein-Barr virus (EBV) (Hergenhahn et al., 2002), enterovirus 71 (EV71) (Qin et al., 2014), HBV (Wei et al., 2017), hepatitis C virus (HCV) (Anggakusuma et al., 2014) (Chen et al., 2012), herpes simplex virus type 2 (HSV-1) (Ferreira et al., 2015), human adenovirus (hADV) (Jennings and Parks, 2020), human cytomegalovirus (HCMV) (Lv, An, et al., 2014; Lv et al., 2015; Lv, Lei, et al., 2014), human immunodeficiency virus type 1 (HIV-1) (Ali and Banerjea, 2016; Ferreira et al., 2015; Kumari et al., 2015; Vajragupta et al., 2005), human papillomavirus (HPV) (Prusty and Das, 2005), influenza virus A (IVA) (Ou et al., 2013), Japanese encephalitis virus (JEV) (Chen et al., 2013), murine norovirus (MNV) (Yang et al., 2016), Newcastle disease virus (NDV), pseudorabies viruses (PRV) (Chen et al., 2013), respiratory syncytial virus (RSV) (Obata et al., 2013), Rift Valley fever virus (RVFV) (Narayanan et al., 2012), SARS-CoV-2 (Bormann et al., 2021; Zupin et al., 2022), vaccinia virus (Chen et al., 2013), viral hemorrhagic septicemia virus (VHSV) (Jeong et al., 2015), and Zika virus (Mounce et al., 2017). Moreover, other curcuminoids: bisdemethoxycurcumin and demethoxycurcumin also provide antiviral effects against Chikungunya virus (Mounce et al., 2017), HCV (Anggakusuma et al., 2014), SARS-CoV-2 (Mulu et al., 2021), and Zika virus (Mounce et al., 2017) **(Table 2)**.

Table 2 Antiviral activities and mode of actions of active compounds from *C. longa*

Virus	*Active compounds*	*Effective dose*	*Mode of action/ Target*	*Experimental model*	*References*
Bovine herpesvirus type 1 (BHV-1)	Curcumin	5-20 μM	Inhibit viral entry and viral replication	BHV-1-infected Madin-Darby bovine kidney (MDBK) cells	(Zhu et al., 2015)
Chikungunya virus (CHIKV)	Bisdemethoxycurcumin	IC_{50} = 4.84 μM	Inhibit viral entry	Vral envelope inhibition	(Mounce et al., 2017)
	Curcumin	IC_{50} = 3.89 μM	Inhibit viral entry	CHIKV-infected Hela cells, viral envelope inhibition	(Mounce et al., 2017)
	Demethoxycurcumin	IC_{50} = 0.89 μM	Inhibit viral entry	Vral envelope inhibition	(Mounce et al., 2017)
Dengue virus type 2 (DV2)	Curcumin	10 μM	Inhibit viral envelope	DV2-infected Vero cells	(Chen et al., 2013)
Ebola virus	Curcumin	N/A	Inhibit viral matrix protein 30 (VP30)	Molecular docking	(Setlur et al., 2017)
Ebstein-Barr virus (EBV)	Curcumin	15 μM	inhibit BZLF-1 gene transcription	Raji DR-CAT cell	(Hergenhahn et al., 2002)
Enterovirus 71 (EV71)	Curcumin	40 μM	Inhibit viral replication	EV71-infected Vero cells	(Qin et al., 2014)
Hepatitis B virus (HBV)	*C. longa* aqueous extract	200 and 500 μg/mL	Inhibit viral replication	HepG2.2.15 cells (HBV genome-containing hepatocytes)	(Kim et al., 2009)
		50 mg/kg/d	Decrease HBV X gene expression	HBV X protein (HBx) transgenic mice	(Kim et al., 2011)
	Curcumin	20 μM	Inhibit viral replication	HepG2.2.15 cells (HBV genome-containing hepatocytes)	(Wei et al., 2017)

Contd.

Table 2 *Contd.*

Virus	*Active compounds*	*Effective dose*	*Mode of action/ Target*	*Experimental model*	*References*
Hepatitis C virus (HCV)	Bisdemethoxycurcumin	IC_{50} = 8.46 µM	Inhibit viral entry	HCV-infected hepatocytes	(Anggakusuma et al., 2014)
	Curcumin	5-20 µM	Inhibit viral entry	HCV-infected hepatocytes	(Anggakusuma et al., 2014)
		10-20 µM	Inhibit viral replication and induction of HO-1/ Akt signaling of host cells	HCV-infected hepatocytes	(Chen et al., 2012)
	Demethoxycurcumin	5-20 µM	Inhibit viral entry	HCV-infected hepatocytes	(Anggakusuma et al., 2014)
Herpes simplex virus type 2 (HSV-1)	Curcumin	5 and 50 µM	Inhibit viral replication	HSV-2-infected human genital epithelial cells	(Ferreira et al., 2015)
Human adenovirus (hADV)	Curcumin	50-100 µM	Inhibit viral replication	hADV-infected A549 human lung adenocarcinoma	(Jennings and Parks, 2020)
Human cytomegalovirus (HCMV)	Curcumin	0.2-0.8 µg/mL	Inhibit immediate early antigen (IEA) and UL83A expressions of viruses Down-regulate heat shock protein 90 (Hsp90) of host cells	HCMV infected human embryonic lung fibroblast (HELF) cells and molecular docking	(Lv, An, et al., 2014; Lv et al., 2015)
		25-50 mg/kg	Inhibit viral replication	HCMV-infected Balb/c mice	(Lv, Lei, et al., 2014)

Contd.

Table 2 *Contd.*

Virus	*Active compounds*	*Effective dose*	*Mode of action/ Target*	*Experimental model*	*References*
Human immunodeficiency virus type 1 (HIV-1)	Curcumin	N/A	Inhibit HIV-1 intergrase	In silico molecular docking	(Vajragupta et al., 2005)
		N/A	Inhibit HIV-1 protease	In silico molecular docking	(Vajragupta et al., 2005)
		IC_{50} = 0.7 µM (infected T-cells) IC_{50} = 12 µM (infected PBMCs)	Inhibit HIV-1 reverse transcriptase	HIV-1-infected CEM-T-cells/ PBMCs	(Kumari et al., 2015)
		5 and 50 µM	Inhibit viral replication	HIV-1-infected T-cells	(Ferreira et al., 2015)
		80 µM	Promoting Tat protein degradation	Tat-transfected HEK-293T cells	(Ali and Banerjea, 2016)
Human papilloma virus (HPV)	Curcumin	100 µM	Inhibit viral replication	HPV18-positive cervical carcinoma cells (HeLa cells)	(Prusty and Das, 2005)
Influenza virus A (IVA)	Curcumin	5-30 µM	Inhibit viral entry (interfering hemagglutination) and viral replication	IVA-infected Madin-Darby canine kidney (MDCK) cells and molecular docking	(Ou et al., 2013)
Japanese encephalitis virus (JEV)	Curcumin	10 µM	Inhibit viral envelope	JEV-infected Vero cells	(Chen et al., 2013)
Murine norovirus (MNV)	Curcumin	0.25-2 mg/mL	Inhibit viral entry	MNV-infected RAW264.7 macrophages	(M. Yang et al., 2016)

Contd.

Table 2 *Contd.*

Virus	*Active compounds*	*Effective dose*	*Mode of action/ Target*	*Experimental model*	*References*
Newcastle Disease Virus (NDV)	Curcumin	≥ 31.2 μM	Inhibit hemagglutination	Hemagglutination inhibition	(Chen et al., 2013)
Pseudorabies viruses (PRV)	Curcumin	30 μM	Inhibit viral envelope	PRV-infected PK-15 cells	(Chen et al., 2013)
Respiratory syncytial virus (RSV)	Curcumin	5 and 10 μg/mL	Inhibit viral replication	RSV-infected human nasal epithelial cells	(Obata et al., 2013)
Rift Valley fever virus (RVFV)	Curcumin	10 μM	Inhibit viral replication	RVFV-infected A549 lung epithelial cells	(Narayanan et al., 2012)
Severe acute respiratory syndrome coronavirus 2 (SARS-CoV-2)	Bisdemethoxycurcumin	N/A	Inhibit main protease (Mpro)	Molecular docking	(Mulu et al., 2021)
	C. longa supplement capsule	EC_{50} = 7.4 μg/mL	Inhibit viral replication	SARS-CoV-2-infected Vero E6 cells	(Bormann et al., 2021)
	Curcumin	N/A	Inhibit main protease (Mpro)	Molecular docking	(Ibrahim et al., 2020)
		N/A	Inhibit non-structural protein 12 (nsp 12) (RNA-dependent RNA polymerase)	Molecular docking	(Singh et al., 2021)
		N/A	Inhibit non-structural protein 15 (nsp 15) (EndoRNAse)	Molecular docking	(S. Kumar et al., 2021)

Contd.

Table 2 *Contd.*

Virus	*Active compounds*	*Effective dose*	*Mode of action/ Target*	*Experimental model*	*References*
		N/A	Inhibit non-structural protein 9 (nsp 9) (RNA-binding viral protein)	Molecular docking	(M. Kumar et al., 2021)
		N/A	Inhibit S protein	Molecular docking	(Jena et al., 2021; Nag et al., 2021; Verma et al., 2021)
		EC_{50} = 7.9 µg/ mL	Inhibit viral replication	SARS-CoV-2-infected Vero E6 cells	(Bormann et al., 2021)
		10 µM	Inhibit viral replication	SARS-CoV-2-infected Vero E6 cells	(Zupin et al., 2022)
		N/A	Interact with ACE2 of host cells	Molecular docking	(Jena et al., 2021; Nag et al., 2021; Verma et al., 2021)
		N/A	Interact with cathepsin K of host cells	Molecular docking	(Oso et al., 2022)
		N/A	Interact with GRP78 of host cells	Molecular docking	(Allam et al., 2020)
		N/A	Interact with TMPRSS2 of host cells	Molecular docking	(Umadevi et al., 2022)
	Demethoxycurcumin	N/A	Inhibit main protease (Mpro)	Molecular docking	(Mulu et al., 2021)

Contd.

Table 2 *Contd.*

Virus	*Active compounds*	*Effective dose*	*Mode of action/ Target*	*Experimental model*	*References*
	Diacetylcurcumin	N/A	Inhibit non-structural protein 12 (nsp 12) (RNA-dependent RNA polymerase)	Molecular docking	(Singh et al., 2021)
	O-Demethyl demethoxycurcumin	N/A	Inhibit S protein	Molecular docking	(Umashankar et al., 2021)
Vaccinia virus	Curcumin	60 μM (30% inhibition)	Inhibit viral envelope	Vaccinia virus-infected BSC-1 cells	(Chen et al., 2013)
Viral hemorrhagic septicemia virus (VHSV)	Curcumin	120 μM	down-regulate heat shock cognate 71 (HSC71) of host cells	VHSV-infected Fathead minnow (FHM) cells	(Jeong et al., 2015)
Zika virus (ZIKV)	Bisdemethoxycurcumin	IC_{50} = 3.61 μM	Inhibit viral entry	Vral envelope inhibition	(Mounce et al., 2017)
	Curcumin	IC_{50} = 1.9 μM	Inhibit viral entry	ZIKV-infected Hela cells, viral envelope inhibition	(Mounce et al., 2017)
	Demethoxycurcumin	IC_{50} = 5.91 μM	Inhibit viral entry	Vral envelope inhibition	(Mounce et al., 2017)

4.5 Mode of Action of Antiviral Compounds

Antiviral agents are mostly designed to target the processes of the viral life cycle. *C. longa* extract and its chemical constituents have been reported as antiviral agents by interfering with viral attachment, entry, genome replication and transcription, and post-transcriptional modification. Moreover, it could affect the host cellular proteins mediated in viral infection.

Inhibition of Viral Attachment and Entry

The initial step of the viral life cycle is a viral entry, in which viruses attach to the host cell membrane and penetrate the cells. Pre-treatment of envelope viruses, including IVA, DV2, JEV, PRV, vaccinia virus, and NDV with curcumin could reduce viral infectivity by inhibiting the hemagglutination activity (Chen et al., 2013; Ou et al., 2013). Moreover, curcuminoids: curcumin, bisdemethoxycurcumin, and demethoxycurcumin potentially inhibit Chikungunya virus, Zika virus, and HCV infections by blocking the viral entry (Anggakusuma et al., 2014; Mounce et al., 2017). Zhu L *et al* reported that curcumin (5-20 μM) exhibited anti-BHV-1 inhibitory activity in Madin-Darby bovine kidney (MDBK) cells via altering lipid bilayer plasma membrane of the host cells and inhibiting viral entry (Zhu et al., 2015). Besides, pre-incubation of MNV with curcumin (0.25-2 mg/mL) could decrease viral infectivity against RAW264.7 macrophages (M. Yang et al., 2016). Based on computational analysis, curcumin and O-demethyl demethoxycurcumin potentially serve as anti-SARS-CoV-2 compounds with strong binding affinity to a SARS-CoV-2 spike protein (S protein), which mediated viral attachment (Jena et al., 2021; Nag et al., 2021; Umashankar et al., 2021; Verma et al., 2021).

Inhibition of Viral Genome Transcription

Curcumin was demonstrated to suppress HIV-1 reverse transcription by inhibiting HIV-1 reverse transcriptase activity, an enzyme used to transcribe viral RNA to complementary DNA (cDNA). Kumari *et al* found that curcumin reduced HIV-1 infection in both lymphoblastoid T cells (IC_{50} = 0.7 μM) and primary peripheral blood mononuclear cells (PBMCs) (IC_{50} = 12 μM) through HIV-1 RT inhibition (Kumari et al., 2015). Increasing the curcumin concentration up to 80 μM could inhibit HIV-1 transcription by promoting Tat protein degradation in HEK293 cells (Ali and Banerjea, 2016). Moreover, molecular docking studies showed that curcumin may have the ability to inhibit several kinds of viral enzymes that are important for viral genome replication, including HIV-1 integrase (Vajragupta et al., 2005), SARS-CoV-2 non-structural proteins (nsp 9, nsp 12, and nsp 15) (M. Kumar et al., 2021; S. Kumar et al., 2021; Singh et al., 2021), and viral matrix protein 30 (VP30) of Ebola virus (Setlur et al., 2017).

Inhibition of Viral Post-translational Modification

After the translation of viral gene expression, the immature proteins need to be modified to make functional proteins. In the HIV-1 cycle, HIV-1 protease

(HIV-1 PR) plays a crucial role in viral maturation by cleaving polyproteins into functional proteins necessary for viral replication (Sillapachaiyaporn et al., 2019). Computation docking performed by Vajragupta O. *et al* demonstrated that curcumin could fit well to the catalytic site of HIV-1 PR and may provide HIV-1 PR inhibitory activity (Vajragupta et al., 2005). Like HIV-1 replication, SARS-CoV-2 requires the main protease (Mpro or 3CLpro) to cleave its polyproteins into mature proteins. Now, several natural products have been reported to inhibit SARS-CoV-2 by blocking the Mpro activity. Curcuminoids: curcumin, bisdemethoxycurcumin, and demethoxycurcumin isolated from *C. longa* exhibited a possible inhibitory activity against Mpro by blocking the catalytic domain of this enzyme (Ibrahim et al., 2020; Mulu et al., 2021).

Interference Against Host Cells

Apart from inhibiting the viral proteins, targeting the host cellular proteins involved in viral infection also prevents viral infectivity. Furthermore, direct inhibitions of host cellular proteins will advantage fast mutation viruses, including SARS-CoV-2. According to *in silico* studies, curcumin probably prevents SARS-CoV-2 infection by blocking host proteins such as angiotensin-converting enzyme-2 (ACE2) (Jena et al., 2021; Nag et al., 2021; Verma et al., 2021), glucose-regulating protein 78 (GRP78) (Allam et al., 2020), transmembrane serine proteinase (TMPRSS2) (Umadevi et al., 2022), and cathepsin K (Oso et al., 2022), which are human receptors or enzymes that participate viral entry. Besides, Jeong EH *et al* demonstrated that pre-treatment of curcumin (120 μM) on fathead minnow (FHM) epithelial cells protected VHSV infection through down-regulation of heat shock cognate 71 (HSC71), a protein-mediated viral entry, expression of the host cells (Jeong et al., 2015). Moreover, curcumin (0.2-0.8 μg/mL) significantly downregulated the expression of heat shock protein 90 (Hsp90), a cellular protein that promotes viral infectivity, in HCMV-infected lung fibroblasts (Lv et al., 2015).

4.6 Future of Research and Drug Delivery

Although *C. longa* shows a broad range of pharmaceutical activities with low side effects, the bioavailability of the active compounds of *C. longa*, especially curcumin is limited (Mathew and Hsu, 2018). Based on the physicochemical characteristics of curcumin, it has a hydrophobic nature, photo-degradation, rapid metabolism, chemical instability, and a short half-life (Hussain et al., 2022). To challenge these limitations, the modification of curcumin with nanoparticle construction has been developed. The nanotechnology-based delivery system has been utilized to improve curcumin's delivery and biological function by encapsulating curcumin with carriers such as liposomes, nanostructured lipids, solid lipids, nanoemulsions, nanocomposite systems, polymeric micelles, polymeric nanoparticles, and hydrogels (Hussain et al., 2022). Accumulating evidence suggests that modifying curcumin into nanoparticles (NPs) could reduce toxicity and enhance the antiviral activity of curcumin. Researchers developed curcumin-modified silver nanoparticles (Cur-AgNPs) against RSV and HIV-1 infections. Interestingly, the Cur-AgNPs showed high solubility and stability in the cell culture medium; moreover, Cur-AgNPs provided high antiviral activity with low cytotoxicity (Sharma et al., 2017; X. X. Yang et al., 2016). Curcumin carbon quantum dots (Cur-CQDs) were synthesized to improve the solubility and reduce the toxicity of curcumin; furthermore, the Cur-CQDs demonstrated high antiviral activity against JEV by directly inhibiting the viral E protein, which facilitates viral entry (Chen et al., 2022). The formulation of curcumin in poly lactic-co-glycolic acid nanoparticles (Cur-PLGA NPs) was developed to enhance the anti-Zika virus. The Cur-PLGA NPs provided high effectiveness on Zika virus inhibition, compared to the native curcumin (Pacho et al., 2021). However, these developments in nanotechnology to enhance drug delivery were investigated *in vitro* systems, and further research of clinical trials is required to confirm the antiviral activity of both curcumin and curcumin-nanoparticles.

5. Toxicity and Safety

C. longa aqueous extract was tested for cytotoxicity against human liver cell lines (Chang and HepG 2.2.15 cell lines) by a group of Kim HJ. The results suggest that the *C. longa* extract at concentrations of 200 and 500 mg/L, did not cause any toxicity against the human liver cell lines (Kim et al., 2009). Moreover, studies suggest that the administration of *C. longa* or curcumin is safe for humans at high dosages. Oral taking of curcumin at 6 g/day for 4–7 weeks showed no adverse effect in humans (Ryan et al., 2013); however, it may be caused by the low intestinal absorption of curcumin (Soleimani et al., 2018). According to the Food and Drug Administration (FDA) report, generally recognized as safe (GRAS), the dosage for healthy consumption without any adverse effects in humans, of curcumin is up to 12 kg/day (Butnariu et al., 2022).

Conclusion

C. longa rhizome is widely used as a spice in the cuisines of Asian countries for its antioxidant, anti-inflammatory, antitumor, and antimicrobial properties among many more. It has been used in Ayurvedic and other traditional medicinal systems to treat various diseases including jaundice, skin infections, wound healing, flatulence, sprains, arthritis, and stomach disturbances, which has also been proven in recent research studies. Interestingly, the plant was also reported to exhibit antiviral activity against a range of viruses including SARS-CoV-2, among many more, as the different phytoconstituents of *C. longa* could interfere with viral attachment, entry, genome replication and transcription, and post-transcriptional modification, which has been explained in detail in Table 2. Curcumin was observed to be non-toxic in higher doses as well. However, this could be attributed to the low bioavailability of the phytoconstituents in the circulatory system. This can be overcome by incorporating nanotechnology into curcumin-based research to improve curcumin's delivery and biological function by encapsulating curcumin with different nano-carriers which has been successfully utilized in enhancing the antiviral activity. More research into this field to analyze the other health benefits of curcumin can also be considered. The advent of nanotechnology in the efficient delivery of curcumin will improve the overall benefits of curcumin, which can bring vast benefits to human healthcare.

References

Ali, A. and Banerjea, A.C. (2016). Curcumin inhibits HIV-1 by promoting Tat protein degradation. *Scientific Reports*, *6*(1), 27539.

Allam, L., Ghrifi, F., Mohammed, H., El Hafidi, N., El Jaoudi, R., El Harti, J., Lmimouni, B., Belyamani, L. and Ibrahimi, A. (2020). Targeting the GRP78-dependant SARS-CoV-2 cell entry by peptides and small molecules. *Bioinformatics and biology insights, 14*, 1–11

Anggakusuma, Colpitts, C.C., Schang, L.M., Rachmawati, H., Frentzen, A., Pfaender, S., Behrendt, P., Brown, R.J.P., Bankwitz, D., Steinmann, J., Ott, M., Meuleman, P., Rice, C.M., Ploss, A., Pietschmann, T. and Steinmann, E. (2014). Turmeric curcumin inhibits entry of all hepatitis C virus genotypes into human liver cells. *Gut*, *63*(7), 1137–1149. https://doi.org/10.1136/gutjnl-2012-304299.

Aziz, M.A., Khan, A.H., Adnan, M. and Ullah, H. (2018). Traditional uses of medicinal plants used by Indigenous communities for veterinary practices at Bajaur Agency, Pakistan. *Journal of ethnobiology and ethnomedicine*, *14*(1), 1–18.

Babili, F.E., Nigon, C., Lacaze, L., Millé, J., Masiala, A., Simm, J., Lamade, V.M. and El Haj, A.A. (2022). A New Colorimetric DPPH Radical Scavenging Activity Method: Comparison with Spectrophotometric Assay in Some Medicinal Plants Used in Moroccan Pharmacopoeia. *Pharmaceutical Fronts*, *4*(02), e89–e102.

Bormann, M., Alt, M., Schipper, L., van de Sand, L., Le-Trilling, V.T.K., Rink, L., Heinen, N., Madel, R.J., Otte, M. and Wuensch, K. (2021). Turmeric root and its bioactive ingredient curcumin effectively neutralize SARS-CoV-2 in vitro. *Viruses*, *13*(10), 1914.

Braga, M.E., Leal, P.F., Carvalho, J.E. and Meireles, M.A.A. (2003). Comparison of yield, composition, and antioxidant activity of turmeric (Curcuma longa L.) extracts obtained using various techniques. *Journal of Agricultural and Food Chemistry*, *51*(22), 6604–6611.

Butnariu, M., Quispe, C., Koirala, N., Khadka, S., Salgado-Castillo, C.M., Akram, M., Anum, R., Yeskaliyeva, B., Cruz-Martins, N. and Martorell, M. (2022). Bioactive Effects of Curcumin in Human Immunodeficiency Virus Infection Along with the Most Effective Isolation Techniques and Type of Nanoformulations. *International Journal of Nanomedicine*, 3619–3632.

Chen, H.-H., Lin, C.-J., Anand, A., Lin, H.-J., Lin, H.-Y., Mao, J.-Y., Wang, P.-H., Tseng, Y.J., Tzou, W.-S. and Huang, C.-C. (2022). Development of antiviral carbon quantum dots that target the Japanese encephalitis virus envelope protein. *Journal of Biological Chemistry*, *298*(6).

Chen, M.-H., Lee, M.-Y., Chuang, J.-J., Li, Y.-Z., Ning, S.-T., Chen, J.-C. and Liu, Y.-W. (2012). Curcumin inhibits HCV replication by induction of heme oxygenase-1 and suppression of AKT. *International Journal of Molecular Medicine*, *30*(5), 1021–1028.

Chen, T.-Y., Chen, D.-Y., Wen, H.-W., Ou, J.-L., Chiou, S.-S., Chen, J.-M., Wong, M.-L. and Hsu, W.-L. (2013). Inhibition of enveloped viruses infectivity by curcumin. *PloS one*, *8*(5), e62482.

Dasgupta, A. (2019). Antiinflammatory herbal supplements. In *Translational Inflammation* (pp. 69–91). Elsevier.

El-Hawaz, R.F., Grace, M.H., Janbey, A., Lila, M.A. and Adelberg, J.W. (2018). *In vitro* mineral nutrition of Curcuma longa L. affects production of volatile compounds in rhizomes after transfer to the greenhouse. *BMC Plant Biology*, *18*(1), 1–13.

Ferreira, V.H., Nazli, A., Dizzell, S.E., Mueller, K. and Kaushic, C. (2015). The anti-inflammatory activity of curcumin protects the genital mucosal epithelial barrier from disruption and blocks replication of HIV-1 and HSV-2. *PloS one*, *10*(4), e0124903.

Gupta, A., Mahajan, S. and Sharma, R. (2015). Evaluation of antimicrobial activity of Curcuma longa rhizome extract against Staphylococcus aureus. *Biotechnology reports*, *6*, 51–55.

Hergenhahn, M., Soto, U., Weninger, A., Polack, A., Hsu, C.H., Cheng, A.L. and Rösl, F. (2002). The chemopreventive compound curcumin is an efficient inhibitor of Epstein-Barr virus BZLF1 transcription in Raji DR-LUC cells. *Molecular Carcinogenesis: Published in cooperation with the University of Texas MD Anderson Cancer Center*, *33*(3), 137–145.

Hussain, Y., Alam, W., Ullah, H., Dacrema, M., Daglia, M., Khan, H. and Arciola, C.R. (2022). Antimicrobial potential of curcumin: therapeutic potential and challenges to clinical applications. *Antibiotics*, *11*(3), 322.

Ibrahim, M.A., Abdelrahman, A.H., Hussien, T.A., Badr, E.A., Mohamed, T.A., El-Seedi, H.R., Pare, P. W., Efferth, T. and Hegazy, M.-E.F. (2020). In silico drug discovery of major metabolites from spices as SARS-CoV-2 main protease inhibitors. *Computers in biology and medicine*, *126*, 104046.

Jena, A.B., Kanungo, N., Nayak, V., Chainy, G. and Dandapat, J. (2021). Catechin and curcumin interact with S protein of SARS-CoV2 and ACE2 of human cell membrane: insights from computational studies. *Scientific Reports*, *11*(1), 1–14.

Jennings, M.R. and Parks, R.J. (2020). Antiviral effects of curcumin on adenovirus replication. *Microorganisms*, *8*(10), 1524.

Jeong, E.-H., Vaidya, B., Cho, S.-Y., Park, M.-A., Kaewintajuk, K., Kim, S.R., Oh, M.-J., Choi, J.-S., Kwon, J. and Kim, D. (2015). Identification of regulators of the early stage of viral hemorrhagic septicemia virus infection during curcumin treatment. *Fish & Shellfish Immunology*, *45*(1), 184–193.

Kim, H.J., Yoo, H.S., Kim, J.C., Park, C.S., Choi, M.S., Kim, M., Choi, H., Min, J.S., Kim, Y.S. and Yoon, S.W. (2009). Antiviral effect of Curcuma longa Linn extract against hepatitis B virus replication. *Journal of Ethnopharmacology*, *124*(2), 189–196.

Kim, J., Ha, H.-L., Moon, H.-B., Lee, Y.-W., Cho, C.-K., Yoo, H.-S. and Yu, D.-Y. (2011). Chemopreventive effect of Curcuma longa Linn on liver pathology in HBx transgenic mice. *Integrative Cancer Therapies*, *10*(2), 168–177.

Kumar, A., Dora, J. and Singh, A. (2011). A review on spice of life Curcuma longa (turmeric). *Int. J. Appl. Biol. Pharmaceut. Tech*, *2*, 371–379.

Kumar, M., Sodhi, K.K. and Singh, D.K. (2021). Addressing the potential role of curcumin in the prevention of COVID-19 by targeting the Nsp9 replicase protein through molecular docking. *Archives of Microbiology*, *203*, 1691–1696.

Kumar, N. and Sakhya, S.K. (2013). Ethnopharmacological properties of Curcuma longa: a review. *International Journal of Pharmaceutical Sciences and Research*, *4*(1), 103.

Kumar, S., Kashyap, P., Chowdhury, S., Kumar, S., Panwar, A. and Kumar, A. (2021). Identification of phytochemicals as potential therapeutic agents that binds to Nsp15 protein target of coronavirus (SARS-CoV-2) that are capable of inhibiting virus replication. *Phytomedicine*, *85*, 153317.

Kumari, N., Kulkarni, A.A., Lin, X., McLean, C., Ammosova, T., Ivanov, A., Hipolito, M., Nekhai, S. and Nwulia, E. (2015). Inhibition of HIV-1 by curcumin A, a novel curcumin analog. *Drug Design, Development and Therapy*, *9*, 5051.

Lee, S.-Y., Cho, S.-S., Li, Y., Bae, C.-S., Park, K.M. and Park, D.-H. (2020). Anti-inflammatory effect of Curcuma longa and Allium hookeri co-treatment via NF-κB and COX-2 pathways. *Scientific Reports*, *10*(1), 5718.

Liu, S.-T., Zheng, S.-W., Hou, A.-J., Zhang, J.-X., Wang, S., Wang, X.-J., Yu, H. and Yang, L. (2022). A review: The phytochemistry, pharmacology, and pharmacokinetics of Curcumae Longae Rhizoma (Turmeric). *World Journal of Traditional Chinese Medicine*, *8*(4), 463.

Lv, Y., An, Z., Chen, H., Wang, Z. and Liu, L. (2014). Mechanism of curcumin resistance to human cytomegalovirus in HELF cells. *BMC Complementary and Alternative Medicine*, *14*, 1–7.

Lv, Y., Gong, L., Wang, Z., Han, F., Liu, H., Lu, X. and Liu, L. (2015). Curcumin inhibits human cytomegalovirus by downregulating heat shock protein 90. *Molecular Medicine Reports*, *12*(3), 4789–4793.

Lv, Y., Lei, N., Wang, D., An, Z., Li, G., Han, F., Liu, H. and Liu, L. (2014). Protective effect of curcumin against cytomegalovirus infection in Balb/c mice. *Environmental Toxicology and Pharmacology*, *37*(3), 1140–1147.

Mathai, N.J., Sony, D., Mane, P.P., Shetty, C.B., Latheef, L., Kamath, K., Khaleed, M., Kochikuzhyil, B.M. and Baliga, M.S. (2018). Antiarthritic effects of turmeric and curcumin: A revisit. *Polyphenols: Prevention and Treatment of Human Disease*, 247–252.

Mathew, D. and Hsu, W.-L. (2018). Antiviral potential of curcumin. *Journal of Functional Foods*, *40*, 692–699.

Mohsin, S.A., Shaukat, S., Nawaz, M., Ur-Rehman, T., Irshad, N., Majid, M., ul Hassan, S.S., Bungau, S. and Fatima, H. (2022). Appraisal of selected ethnomedicinal plants as alternative therapies against onychomycosis: Evaluation of synergy and time-kill kinetics. *Frontiers in Pharmacology*, *13*.

Mounce, B.C., Cesaro, T., Carrau, L., Vallet, T. and Vignuzzi, M. (2017). Curcumin inhibits Zika and chikungunya virus infection by inhibiting cell binding. *Antiviral Research*, *142*, 148–157.

Mulu, A., Gajaa, M. and Woldekidan, H.B. (2021). The impact of curcumin derived polyphenols on the structure and flexibility COVID-19 main protease binding pocket: a molecular dynamics simulation study. *PeerJ*, *9*, e11590.

Nag, A., Paul, S., Banerjee, R. and Kundu, R. (2021). In silico study of some selective phytochemicals against a hypothetical SARS-CoV-2 spike RBD using molecular docking tools. *Computers in Biology and Medicine*, *137*, 104818.

Narayanan, A., Kehn-Hall, K., Senina, S., Lundberg, L., Van Duyne, R., Guendel, I., Das, R., Baer, A., Bethel, L. and Turell, M. (2012). Curcumin inhibits Rift Valley fever virus replication in human cells. *Journal of Biological Chemistry*, *287*(40), 33198–33214.

Nisar, T., Iqbal, M., Raza, A., Safdar, M., Iftikhar, F. and Waheed, M. (2015). Turmeric: A promising spice for phytochemical and antimicrobial activities. *Am Eur J Agric Environ Sci*, *15*(7), 1278–1288.

Obata, K., Kojima, T., Masaki, T., Okabayashi, T., Yokota, S., Hirakawa, S., Nomura, K., Takasawa, A., Murata, M. and Tanaka, S. (2013). Curcumin prevents replication of respiratory syncytial virus and the epithelial responses to it in human nasal epithelial cells. *PloS one*, *8*(9), e70225.

Ooko, E., Kadioglu, O., Greten, H.J. and Efferth, T. (2017). Pharmacogenomic characterization and isobologram analysis of the combination of ascorbic acid and curcumin—two main metabolites of Curcuma longa—in cancer cells. *Frontiers in Pharmacology*, *8*, 38.

Oso, B.J., Adeoye, A.O. and Olaoye, I.F. (2022). Pharmacoinformatics and hypothetical studies on allicin, curcumin and gingerol as potential candidates against COVID-19-associated proteases. *Journal of Biomolecular Structure and Dynamics*, *40*(1), 389–400.

Ou, J.L., Mizushina, Y., Wang, S.Y., Chuang, D.Y., Nadar, M. and Hsu, W.L. (2013). Structure–activity relationship analysis of curcumin analogues on anti-influenza virus activity. *The FEBS journal*, *280*(22), 5829–5840.

Pacho, M.N., Pugni, E.N., Díaz Sierra, J.B., Morell, M.L., Sepúlveda, C.S., Damonte, E.B., García, C.C. and D'Accorso, N.B. (2021). Antiviral activity against Zika virus of a new formulation of curcumin in poly lactic-co-glycolic acid nanoparticles. *Journal of Pharmacy and Pharmacology*, *73*(3), 357–365.

Pantharos, P., Sukcharoen, P., Phadungrakwittaya, R., Akarasereenont, P., Booranasubkajorn, S. and Lumlerdkij, N. (2022). Utilization of UPLC-PDA and GC–MS/MS coupled with metabolomics analysis to identify bioactive metabolites in medicinal turmeric at different ages for the quality assurance. *Phytomedicine*, *102*, 154157.

Pawar, K.S., Mastud, R.N., Pawar, S.K., Pawar, S.S., Bhoite, R.R., Bhoite, R.R., Kulkarni, M.V. and Deshpande, A.R. (2021). Oral curcumin with piperine as adjuvant therapy for the treatment of COVID-19: a randomized clinical trial. *Frontiers in Pharmacology*, *12*, 1056.

Permatananda, P.A.N.K., Aryastuti, A.A.S.A., Cahyawati, P.N., Udiyani, D.P.C., Wijaya, D., Pandit, I.G.S. and Wirajaya, A.A.N.M. (2021). Phytochemical and antioxidant capacity test on turmeric extract (Curcuma longa) traditionally processed in Bali. *Journal of Physics: Conference Series, 1869*(1), 012035

Prakash, J., Srivastava, S., Ray, R., Singh, N., Rajpali, R. and Singh, G.N. (2017). Current status of herbal drug standards in the Indian pharmacopoeia. *Phytotherapy Research*, *31*(12), 1817–1823.

Prasansuklab, A., Theerasri, A., Rangsinth, P., Sillapachaiyaporn, C., Chuchawankul, S. and Tencomnao, T. (2021). Anti-COVID-19 drug candidates: A review on potential biological activities of natural products in the management of new coronavirus infection. *Journal of Traditional and Complementary Medicine*, *11*(2), 144–157.

Praveen, A., Prasad, D., Mishra, S., Nagarajan, S. and Chaudhari, S.R. (2021). Facile NMR approach for profiling curcuminoids present in turmeric. *Food Chemistry*, *341*, 128646.

Prusty, B.K. and Das, B.C. (2005). Constitutive activation of transcription factor AP-1 in cervical cancer and suppression of human papillomavirus (HPV) transcription and AP-1 activity in HeLa cells by curcumin. *International Journal of Cancer*, *113*(6), 951–960.

Qin, Y., Lin, L., Chen, Y., Wu, S., Si, X., Wu, H., Zhai, X., Wang, Y., Tong, L. and Pan, B. (2014). Curcumin inhibits the replication of enterovirus 71 *in vitro*. *Acta Pharmaceutica Sinica B*, *4*(4), 284–294.

Ryan, J.L., Heckler, C.E., Ling, M., Katz, A., Williams, J.P., Pentland, A.P. and Morrow, G.R. (2013). Curcumin for radiation dermatitis: a randomized, double-blind, placebo-controlled clinical trial of thirty breast cancer patients. *Radiation Research*, *180*(1), 34–43.

Sabir, S., Zeb, A., Mahmood, M., Abbas, S., Ahmad, Z. and Iqbal, N. (2020). Phytochemical analysis and biological activities of ethanolic extract of Curcuma longa rhizome. *Brazilian Journal of Biology*, *81*, 737–740.

Salem, M.A., El-Shiekh, R.A., Fernie, A.R., Alseekh, S. and Zayed, A. (2022). Metabolomics-based profiling for quality assessment and revealing the impact of drying of Turmeric (Curcuma longa L.). *Scientific Reports*, *12*(1), 1–11.

Sankhe, A.P., Memane, N.S., Gawali, V.P., Memane, S.N., Ramakrishnan, G., Kundu, T., Bagul, M.N., Kumar, A., Bansal, V. and Tiwari, R. (2022). A randomized, controlled, blinded, parallel group, clinical trial to study the role of Ayurcov (AyurCoro3), one day regimen as an adjuvant therapy for COVID-19 disease management, at dedicated Covid Hospital (DCH) in India. *Complementary Therapies in Medicine*, *67*, 102824.

Setlur, A.S., Naik, S.Y. and Skariyachan, S. (2017). Herbal lead as ideal bioactive compounds against probable drug targets of Ebola virus in comparison with known chemical analogue: A computational drug discovery perspective. *Interdisciplinary Sciences: Computational Life Sciences*, *9*, 254–277.

Sharma, R.K., Cwiklinski, K., Aalinkeel, R., Reynolds, J.L., Sykes, D.E., Quaye, E., Oh, J., Mahajan, S.D. and Schwartz, S. A. (2017). Immunomodulatory activities of curcumin-stabilized silver nanoparticles: Efficacy as an antiretroviral therapeutic. *Immunological Investigations*, *46*(8), 833–846.

Sharma, S., Sharma, S., Chourasia, R., Pandey, A., Rai, A.K. and Sahoo, D. (2021). Alzheimer's disease: Ethanobotanical studies. In *Naturally Occurring Chemicals Against Alzheimer's Disease* (pp. 11–28). Elsevier.

Sillapachaiyaporn, C., Nilkhet, S., Ung, A. and Chuchawankul, S. (2019). Anti-HIV-1 protease activity of the crude extracts and isolated compounds from Auricularia polytricha. *BMC complementary and alternative medicine*, *19*(1), 1–10.

Singh, G., Kapoor, I., Singh, P., de Heluani, C.S., de Lampasona, M.P. and Catalan, C.A. (2010). Comparative study of chemical composition and antioxidant activity of fresh and dry rhizomes of turmeric (Curcuma longa Linn.). *Food and Chemical Toxicology*, *48*(4), 1026–1031.

Singh, R., Bhardwaj, V.K. and Purohit, R. (2021). Potential of turmeric-derived compounds against RNA-dependent RNA polymerase of SARS-CoV-2: An in-silico approach. *Computers in Biology and Medicine*, *139*, 104965.

Sivani, B.M., Azzeh, M., Patnaik, R., Pantea Stoian, A., Rizzo, M. and Banerjee, Y. (2022). Reconnoitering the therapeutic role of curcumin in disease prevention and treatment: Lessons learnt and future directions. *Metabolites*, *12*(7), 639.

Soleimani, V., Sahebkar, A. and Hosseinzadeh, H. (2018). Turmeric (Curcuma longa) and its major constituent (curcumin) as nontoxic and safe substances. *Phytotherapy Research*, *32*(6), 985–95.

Srivastava, B.B.L., Ripanda, A.S. and Mwanga, H.M. (2022). Ethnomedicinal, Phytochemistry and Antiviral Potential of Turmeric (Curcuma longa). *Compounds*, *2*(3), 200–221.

Taha, M.N., Krawinkel, M.B. and Morlock, G.E. (2015). High-performance thin-layer chromatography linked with (bio) assays and mass spectrometry–a suited method for discovery and quantification of bioactive components? Exemplarily shown for turmeric and milk thistle extracts. *Journal of Chromatography A*, *1394*, 137–147.

Umadevi, P., Manivannan, S., Fayad, A.M. and Shelvy, S. (2022). In silico analysis of phytochemicals as potential inhibitors of proteases involved in SARS-CoV-2 infection. *Journal of Biomolecular Structure and Dynamics*, *40*(11), 5053–5059.

Umashankar, V., Deshpande, S.H., Hegde, H.V., Singh, I. and Chattopadhyay, D. (2021). Phytochemical moieties from Indian traditional medicine for targeting dual hotspots on SARS-CoV-2 spike protein: an integrative in-silico approach. *Frontiers in Medicine*, *8*, 672629.

Vajragupta, O., Boonchoong, P., Morris, G.M. and Olson, A.J. (2005). Active site binding modes of curcumin in HIV-1 protease and integrase. *Bioorganic & Medicinal Chemistry Letters*, *15*(14), 3364–3368.

Van Dau, N., Ham, N.N., Khac, D.H., Lam, N.T., Son, P.T., Tan, N.T., Van, D.D., Dahlgren, S., Grabe, M. and Johansson, R. (1998). The effects of a traditional drug, turmeric (Curcuma longa), and placebo on the healing of duodenal ulcer. *Phytomedicine*, *5*(1), 29–34.

Verma, A.K., Kumar, V., Singh, S., Goswami, B.C., Camps, I., Sekar, A., Yoon, S. and Lee, K.W. (2021). Repurposing potential of Ayurvedic medicinal plants derived active principles against SARS-CoV-2 associated target proteins revealed by molecular docking, molecular dynamics and MM-PBSA studies. *Biomedicine & Pharmacotherapy*, *137*, 111356.

Wang, X.-P., Wang, Q.-X., Lin, H.-P. and Chang, N. (2017). Anti-tumor bioactivities of curcumin on mice loaded with gastric carcinoma. *Food & Function*, *8*(9), 3319–3326.

Wei, Z.-Q., Zhang, Y.-H., Ke, C.-Z., Chen, H.-X., Ren, P., He, Y.-L., Hu, P., Ma, D.-Q., Luo, J. and Meng, Z.-J. (2017). Curcumin inhibits hepatitis B virus infection by down-regulating cccDNA-bound histone acetylation. *World Journal of Gastroenterology*, *23*(34), 6252.

Wu, Y.-Q. and Tong, T. (2022). Curcumae Rhizoma: A botanical drug against infectious diseases. *Frontiers in Pharmacology*, *13*.

Yang, M., Lee, G., Si, J., Lee, S.-J., You, H.J. and Ko, G. (2016). Curcumin shows antiviral properties against norovirus. *Molecules*, *21*(10), 1401.

Yang, X.X., Li, C.M. and Huang, C.Z. (2016). Curcumin modified silver nanoparticles for highly efficient inhibition of respiratory syncytial virus infection. *Nanoscale*, *8*(5), 3040–3048.

Zhu, L., Ding, X., Zhang, D., Ch, Y., Wang, J., Ndegwa, E. and Zhu, G. (2015). Curcumin inhibits bovine herpesvirus type 1 entry into MDBK cells. *Acta Virologica*, *59*(3), 221–327.

Zupin, L., Fontana, F., Clemente, L., Borelli, V., Ricci, G., Ruscio, M. and Crovella, S. (2022). Optimization of Anti-SARS-CoV-2 Treatments Based on Curcumin, Used Alone or Employed as a Photosensitizer. *Viruses*, *14*(10), 2132.

3 Decoding the Antiviral Prowess of *Tinospora cordifolia* (Willd.) Hook. f. and Thomson

Parthasarathy Seethapathy[1*], Sengamuthu Kumaresan[1], Anu Pandita[2], Seyed Alireza Salami[3] and Deepu Pandita[4*]

1. Introduction

Plants used for their therapeutic and aromatic properties have been essential in easing human suffering. Ancient plant extracts have remarkable biological capabilities for their antibacterial, antioxidant, and anticancer effects. Several diseases have historically been treated with natural products—the mixtures of secondary metabolites that plants contain, give their goods and positive therapeutic benefits. Phenolic, alkaloids, flavonoids, and tannins are the most significant bioactive components (Kabera et al., 2014). The emergence of germs resistant to antibiotics, unfavorable side effects of some antimicrobial treatments, and minor adverse effects of natural medicines have necessitated the discovery of new bioactive compounds. Globally, infectious diseases are the leading causes of death. Among economically significant plants, medicinal and aromatic plants have been essential in easing human suffering. Throughout the beginning of time, both structured (Ayurveda, Siddha, and Unani) and unorganized (folk, tribal, aboriginal) systems have used plants as curative agents (Channaveeraiah and Entoori, 2022). *Tinospora cordifolia* (Willd.) Hook. f. & Thomson is an herbaceous vine belonging to the family Menispermaceae. T. cordifolia is endemic to the tropical regions of the Indian subcontinent, has a long history, and has been widely used from ancient civilizations to the present because of its promising therapeutic characteristics and substantial potential for enhancing health and well-being. It has been established that there are thirty-four species of *Tinospora* distributed across different geographical regions worldwide (Castillo et al., 2013).

Numerous studies have explored the potential medicinal properties associated with plants. This botanical specimen is commonly called "Giloy" or "Amrit" and is widely recognized as a highly adaptable rejuvenating vine in the Indian

[1] Amrita School of Agricultural Sciences, Amrita Vishwa Vidyapeetham, Coimbatore, Tamil Nadu, India
[2] Vatsalya Clinic, New Delhi, India
[3] Department of Horticultural Science, Faculty of Agriculture, University of Tehran, Karaj, Iran
[4] Government Department of School Education, Jammu, Jammu and Kashmir, India: ORCID ID- https://orcid.org/0000-0002-8361-1872.
* Corresponding author: spsarathyagri@gmail.com[1]; deepupandita@gmail.com[4]
DOI: 10.1201/9781003433583-3

Ayurvedic and Unani medical traditions. It is renowned for its ability to enhance and fortify the immune system. Ayurveda has historically employed this herb to address a range of disorders; however, despite undergoing clinical examination, the efficacy of these treatments could be more reliable (Jain et al., 2019). The ancient Ayurvedic textbooks, such as the Ashtanga Hridaya, Ashtanga Sangraha, Chakradatta, Charak Samhita and the Sushrut Samhita, contain descriptions of this giloy usage, which is advocated for managing diverse diseases and enhancing well-being. The comprehensive documentation regarding the species of the genus *Tinospora* in China can be found in the work titled "Bencao Gangmu Shiyi," authored by Zhao Xuemin during the reign of the Qing dynasty in the year 1765 (Chi et al., 2016). The present document constitutes an exhaustive examination of scholarly literature (Payyappallimana et al., 2022). This inquiry pertains to the subject of *T. cordifolia* and its potential advantages in the context of enhancing overall well-being. *T. cordifolia* has garnered significant attention from researchers worldwide for over fifty years, owing to its multifarious applications in conventional medicine. Research on this plant has encompassed a wide range of investigations, including pre-clinical experiments, clinical trials, botanical compound studies, and examinations of its pharmacological properties in various domains (Arunachalam et al., 2022).

T. cordifolia, a herb with a rich historical background, has been widely utilized from ancient civilizations to the present day due to its promising medicinal properties and significant potential in promoting health and well-being. The examination of various texts spanning different periods reveals consistent documentation of the utilization of giloy, as evidenced by descriptions of its actions and indications. This substantiates the enduring usage of this medicinal substance over an extended period. Later reports indicated that new actions of strength promotion were observed, which were found to be beneficial for eye health. Additionally, this herb was observed in the treatment of anemia, chronic diarrhea, cellulitis/erysipelas, cough, distress, dizziness, and worm infestations. The Charaka Samhita has included *Tinospora* as one of the four Medhya Rasayana drugs, known for their ability to enhance intellect. Other researchers consider Tinospora as an important therapeutic agent, separately and in combination with other drugs. Its efficacy has been acknowledged in treating conditions such as gout, dermatological ailments, jaundice, fever, ocular disorders, metabolic processes and musculoskeletal disorders, and excessive urination, among others (Hao et al., 2022).

For centuries, this remedy has been employed in the management of diverse medical conditions; *T. cordifolia* possesses a wide range of compounds with pharmaceutical importance, including alkaloids, diterpenoid lactones, glycosides, phenol compounds, steroid hormones, sesquiterpenoids and tinosporide (Modi et al., 2020). These diverse groups of substances contribute to the immense usefulness of *T. cordifolia*. These substances exhibit pharmacological characteristics that confer antiarthritic, antidiabetic, antiHIV, anti-infective, anti-inflammatory, antimicrobial, antioxidant, antiosteoporotic, antipyretic, antitoxic, hepatoprotective, and immuno-modulatory effects (Joshi et al., 2023). The benefits of immune modulation and rejuvenation are commonly associated with it. This chapter presents an extensive examination of scholarly literature regarding *T. cordifolia* and its potential advantages in enhancing overall well-being.

2. Taxonomy and Scientific Classification

The Menispermaceae (moonseed), a primarily tropical family consisting of dioecious and essentially climbing greenery with over 70 genera and 450 species of plants, has been the focus of several molecular investigations that have substantiated its monophyly. Additionally, these studies have revealed the para- and polyphyly of most groups presently acknowledged within this family. To clarify the evolutionary relationships within the Menispermaceae family, distinguished by its wide range of physical characteristics and limited understanding, the ndhF gene found in the chloroplast was utilized (Ortiz et al., 2007).

This family comprises primarily climbing plants that are dioecious and distributed worldwide. The taxonomic and geographic sampling of the Menispermaceae family, in this study, is more comprehensive than previous research, as it incorporates analyses of three genetic markers (matK, trnL-F, ndhF) and 41 morphological characteristics. *T. cordifolia*, also known as gurjo, heart-leaved moonseed, guduchi, or giloy, is an important herbaceous plant belonging to the Menispermaceae family. It is native to tropical areas of the Indian subcontinent (Ortiz et al., 2016). Over the past decade, molecular phylogenetics has significantly enhanced our comprehension of the genus. However, the precise delineation of its boundaries still needs to be solved. This study presents a comprehensive phylogenetic analysis of the Burasaieae taxonomic group, employing six genetic markers (atpB, ITS, matK, ndhF, rbcL, trnL-F). The primary emphasis of this analysis is placed on the genus *Tinospora* (Lian et al., 2019). The *Tinospora* genus encompasses 34 species, many of which have been utilized as traditional medicinal herbs by indigenous populations residing in the tropical and subtropical regions of Africa, Asia, and Australia. One of the most well-known species in the Tinospora genus is *T. cordifolia*, also called Guduchi or Giloy. Apart from *T. cordifolia*, there are other species within the *Tinospora* genus, such as *T. bakis, T. capillipes, T. cordifolia, T. crispa, T. sagittata, T. malabarika,* and *T. rumphii*, are also used in traditional medicine in some regions. However, their active principles, botanical descriptions, distribution, pharmacological uses and medicinal properties uses may vary.

The updated taxonomic classification of *Tinospora cordifolia* as per GBIF. org is presented here:

Kingdom	: Plantae
Phylum	: Tracheophyta
Class	: Magnoliopsida
Order	: Ranunculales
Family	: Menispermaceae

Genus : *Tinospora*
Species : *cordifolia*
Binomial nomenclature : *Tinospora cordifolia* (willd.) Hook. F. & Thomson
Basionym : *Menispermum cordifolium* Willd.

Synonyms

Chasmanthera cordifolia (DC.) Baill.
Cocculus convolvulaceus DC.
Chasmanthera cordifolia Baill.
Cocculus verrucosus Wall.
Cocculus cordifolius (Willd.) DC.
Menispermum glabrum Burm.fil.
Menispermum verrucosum Roxb.
Tinospora convolvulacea Miers
Tinospora fosbergii Kundu
Tinospora cordifolia var. *congesta* Mujaffar, Yasin & Solanki
Tinospora verrucosa (Fleming) W. Theob.

Other Local Names

Bengali : Gulancha
English : Gulancha/Indian tinospora
Gujarathi : Galo
Hindi : Giloya, Guduchi
Kannada : Amrita balli
Marathi : Shindilakodi
Sanskrit : Guduchi, Madhuparni, Amrita, Chinnaruha, Vatsadaani, Tantrika, Kundalini & Amritavalli
Tamil : Shindilkodi
Telugu : Tippatiga

3. Botanical Description of the Plant

T. cordifolia is a large, smooth, vine-like, deciduous shrub. A large, ligneous, succulent, climbing plant is under discussion. The grayish-brown bark is corky. Young shoots are hairless and flaky. Lenticels and longitudinal ridges decorate the stem. Ridged branches create long aerial roots that spread from host plants (Saeed et al., 2020). The stem structure exhibits a fibrous composition, and upon transverse sectioning, yellowish wedge-shaped wood bundles are observed. These bundles are radially arranged and contain prominent vessels interspersed by slender medullary rays (De Wet et al., 2016). The stem exhibits a configuration of lenticles resembling a rosette, while the bark displays a coloration ranging from creamy white to grey, characterized by a distinctively pronounced spiral pattern. Conversely, older branches have a light brown exterior with warty protrusions caused by oval lenticels. The diagonally smooth surface of older stems exhibits a radial structure characterized by prominent medullary rays that traverse porous tissues.

A bitter taste describes the flavor profile. The leaves possess a simple, alternate, membranous texture and exhibit a cordate to sinuate shape with a 4–17 × 4.5–14 cm size (**Fig. 1**). The blooming of flowers typically occurs during the summer season, whereas the ripening of fruits is commonly observed during the winter season (Kumar et al., 2019). Flowers exhibit unisexual characteristics, having small physical dimensions, pedicellate, and yellow coloration. The specimens are situated in the axillary position and exhibit racemes ranging from 2 to 9 cm on leaflet branches. Female flowers typically exhibit solitary growth, whereas male flowers aggregate in clusters. Fruits usually possess a single source and display a fleshy texture. The seeds exhibit a bent morphology, curved or half-moon shape, endospermic, cotyledons elongated, and leaflike with small radicles (Gautam et al., 2020).

Fig. 1 (a) Drawing of *Tinospora cordifolia* (b-c) Vine with cordate-shaped leaves, (d-e) Stem vine, (f) T.S. of the vine, (g) Inflorescence and (h) Fruits (*Source:* (a g-h) https://indiabiodiversity.org/species/show/231352; (b-f) Original photographs by author DP).

4. Geographical Distribution

The giloy is widespread and expansive geographically throughout the tropical subcontinent of India, ranging from the Himalayas to the southern point of Peninsular India. Additionally, it is prevalent in neighboring countries such as Bangladesh, China, Maldives, Nicobar Is., Andaman Islands, Vietnam, Myanmar, Indonesia, Pakistan, the Philippines, Thailand, and Sri Lanka. In addition, Malaysia, Indonesia, and others in Southeast Asia have reported seeing the plant (Fig. 2). *T. cordifolia* thrives in various soil conditions, from acidic to alkaline, and requires moderate soil precipitation. Seeds can be dispersed through barochory, or gravitational dispersal, zoochory, or dispersal by birds or animals, and anthropochory, or dispersal by humans. In tropical regions, the trees can achieve heights of up to 300 meters (Lade et al., 2018; https://powo.science.kew.org/).

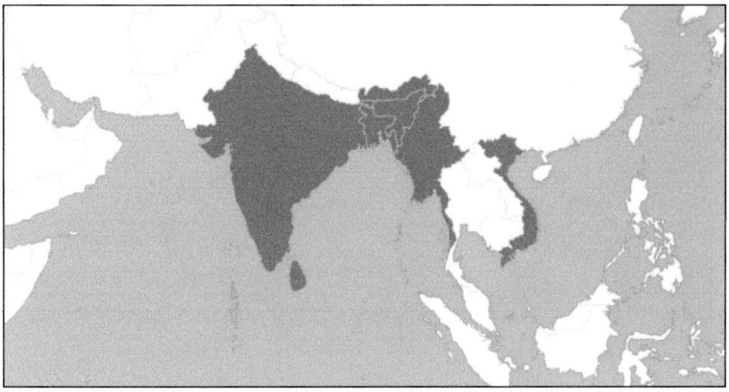

Fig. 2 The native range of *T. cordifolia* is Indian Subcontinent to Indo-China. (*Source:* https://powo.science.kew.org/)

5. Phytochemical Constituents and Metabolite Profiling

Nutrient content per 100 g of *T. cordifolia*, consists of protein (2.3 g), carbohydrates (3.3 g), fibre (11.3 g), calcium (85.2 mg), iron (5.8 mg), Vitamin A (303.7 mcg), and Vitamin C (56 mg) (Malla and Bista, 2021). *T. cordifolia* is known for its abundant presence of alkaloids, phenolics, sesquiterpenoids, norditerpenoids, furano diterpenoids, sterols, clerodane lignans, aliphatic substances, carbohydrates, essential oils, and fatty acids. The *Tinospora* plant is found to have a minimum content of 0.02% w/w of cordifolioside A, while the percentage of tinosporaside in Tinospora ranges from 0.03 to 0.04. Many bioactive compounds have been identified with therapeutic potential, including alkaloids such as berberine and bitter compounds like tinosporin, tinosporic acid, tinosporol, and lipids (Singh and Chaudhuri, 2017). Some constituents have been isolated from the plant mainly show that 1-2, substituted pyrrolidine, 18-norcleropdanedditerpine-O-glucoside, a diterpenoid furanolactone, amritosides, arabinogalactan polysaccharide, aryltetrahydrofuranolignan, β-sitosterol, berberine, bergenin, cordifolon, cordifolide, cordifolisides A to E, cordifol, cordioside, columbin, chasmanthin, crude giloininand, ecdysterone, giloin, gilenin, gilosterol, glucan polysaccharide, heptacosanol, hydroxyecdysone, isocolumbin, jatrorrhizine, makisterone A, magnoflorine, nonacosan-15-one, octacosonal, palmatine, palmarin, palmatosides C and F, phenolic lignan-3-(α,4-Dihydroxy-3-methoxybenzyl)-4-(4-hydroxy-3-methoxybenzyl)-tetrahydrofuran, picrotene, syringen, syringine, syringine apiosylglycoside, tinosponone, tinosporone, tinosporic acid, tinosporol, tinosporin, tinosporon, tinosporidine, tinosporide, tembetarine, tetrahydropalmaitine, and unosporin respectively (Saha and Ghosh, 2012).

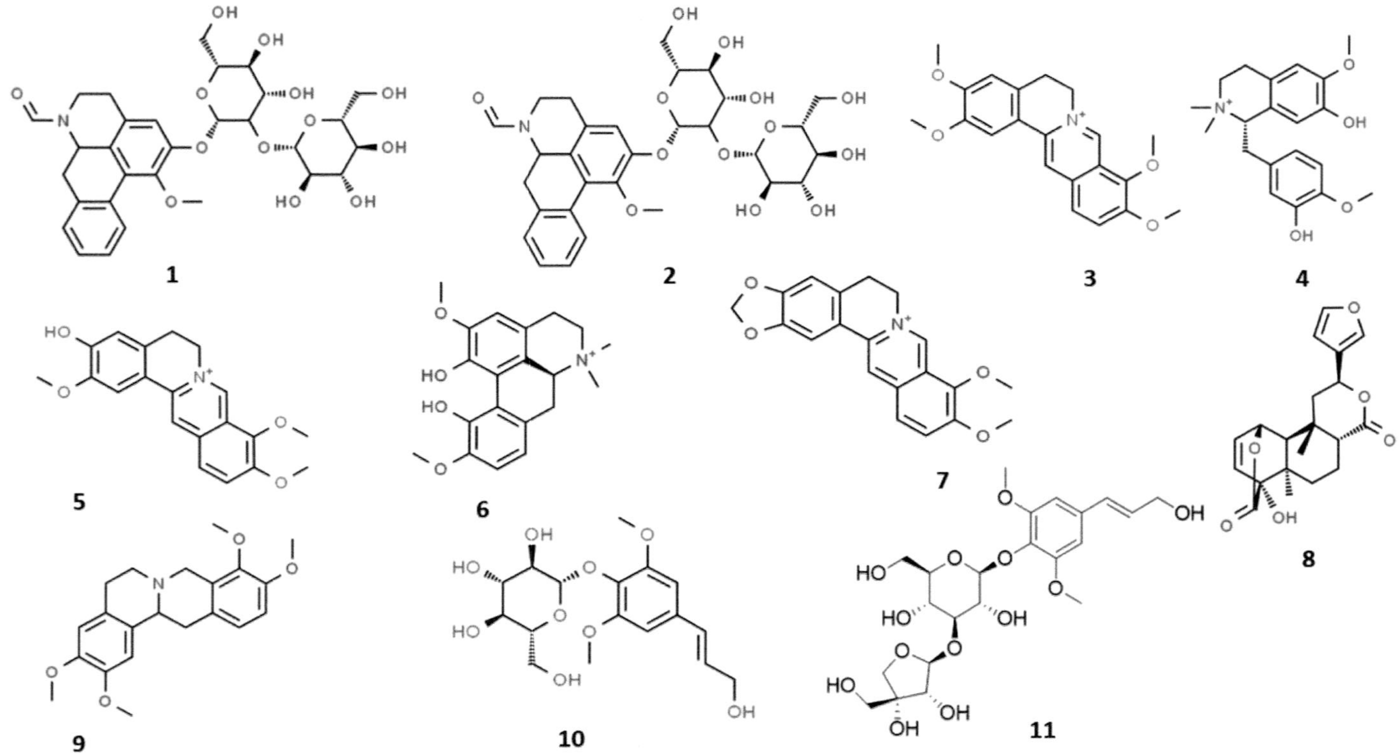

Fig. 3 Structures of some biologically active phytochemicals in *T. cordifolia* generated by ChemSpider: 1. Tinoscorside A 2. Tinoscorside B 3. Palmatine 4. Tembetarine 5. Jatrorrhizine 6. Magnoflorine 7. Berberine 8. Columbin 9. Tetrahydropalmatin 10. Syringin 11. Cordifolioside A.

Alkaloids reported from stem and roots of *T. cordifolia* include aporphine alkaloids, N-formylasimilobine 2-0-β-D-glucopyranosyl-(1›2)-β-D-glucopyr anoside (tinoscorside A,1), aporphine alkaloids, N-acetylasimilobine 2-0-β-D-glucopyranosyl-(1›2)-β-D-glucopyr anoside (tinoscorside B,2), berberine, choline, jatrorrhizine, magnoflorine, palmatine, tembetarine, tetrahydropalmatine, tinosporin and isocolumbin (Jagetia and Rao, 2006; Upadhyay et al., 2010; Patel and Mishra, 2011; Saha and Ghosh, 2012). Glycosides reported from stem of *T. cordifolia* are 18-norclerodane, cordioside, furanoid diterpene glucoside, palmatosides, tinocordiside, and tinocordifolioside (Saha and Ghosh, 2012). A total of 32 diterpenoids and their glycosides, characterized by clerodane and norclerodane skeletons, along with one monoterpenoid, five sesquiterpenoids, and cycloeuphordenol (single triterpenoid), were extracted from *T. cordifolia*. Flavonoids, saponins, and phytosterols (2,3,14,20,22,25-hexahydroxyl-5-cholest-7-en-6-one, giloinsterol, ecdysterone, makisterone A, and δ-sitosterol) are also present. Such bioactive constituents, alone or in combination, account for the antioxidant activity. The Giloy family is full of alkaloids and terpenes (Spandana et al., 2013). Sesquiterpenoid is extracted from the stem as tinocordiofolin (Duraipandiyan et al., 2012). The giloy leaves possess abundant proteins (11.2%) and are fairly rich sources of calcium and phosphorus (Kaur et al., 2016). The specifics of phytochemical types, distribution in plants, active principles in *T. cordifolia* and structures of some biologically active phytochemicals are given in **Table 1** and **Fig. 3**.

6. Medicinal/Therapeutic Uses

T. cordifolia has long been recognized as an imperative element of ancient Ayurvedic medicine, where it has been exploited for the treatment of various disease conditions, including asthma, bone fracture, cancer, chronic diarrhea, deadly bug bite, dysentery, eye diseases, fever, jaundice, pain, skin illness, and snake bite. Because of its anti-diabetic, anti-allergic, anti-arthritic, anti-microbial, anti-inflammatory, anti-osteoporotic, anti-periodic, and anti-spasmodic characteristics, *T. cordifolia* extracts are widely employed in a wide variety of herbal remedies for the treatment of a range of disorders (Parthipan et al., 2011; Sharma et al., 2019). Several medical systems in various nations use *T. cordifolia.* Studies have demonstrated that the plant possesses a diverse array of secondary metabolites, including alkaloids, cardiac glycosides, polyphenols, phytosterols, tannins, saponins, and triterpenoids. These compounds have been shown to possess significant medicinal properties, functioning as immunomodulators, antidiabetic agents, anti-allergic substances, anti-leprotic substances, and anti-inflammatory substances. Because of the presence of arabinogalactan, the aqueous extract of *T. cordifolia* stem has demonstrated the ability to elicit immunological activity. This plant is recognized for its hypolipidemic, hepatoprotective, antineoplastic, antispasmodic, antipyretic, and hypoglycemic effects. Moreover, it is useful for general weakness, digestive issues, children's fever and appetite loss, diarrhea,

Table 1 Distribution of Phytochemical constituents in *T. cordifolia.*

Type of Chemicals	*Active Principle*	*Distribution in Plant*	*Reference*
Alkaloids	Berberine	Stem	Upadhyay et al. (2010); Sharma et al. (2012)
	Tinosporin	Leaf	
	Tinosporic acid	Whole plant	
	Isocolumbin	Root	
	Palmitine	Stem, root	
Glycosides	Cordifoliside, tinocordiside, tinocordifoliside, cordioside, syringin, palmatosides, syringinapiosylglycoside, nonderodane glycoside	Stem	Kapil and Sharma (1997); Sinha et al. (2004); Bala et al. (2015)
Diterpenoid lactones	Furanolactone, tinosporon, clerodane derivatives, jateorine, tinosporisides	Whole plant	Choudhary et al. (2013); Wink (2015)
	Diterpenoid, columbin	Stem	
Steroids	Sitosterol	Aerial parts, stem	Singh and Chaudhuri (2017); Sharma et al. (2019)
	Octacosanol, heptacosanol, ecdysterone, giloinsterol	Stem	
Sesquiterpenoids	Tinocordifolin	Stem	Maurya and Handa (1998)
Other miscellaneous compounds	Jatrorrhizine	Root	Singh et al. (2003); Kumar et al. (2017)
	Tinosporidin, cordifelone, cordifol, giloin, giloinin	Whole plant	

gonorrhea, urinary infections, viral hepatitis, and anemia (Jan and Shrivastava, 2019). In a recent study, researchers discovered that an arabinogalactan polysaccharide derived from *T. cordifolia*, exhibited protective properties against iron-mediated lipid peroxidation that occurred in rat brain homogenate. This protection was observed through the inhibition of lipid hydroperoxide (LOOH). The polysaccharide demonstrated a high reactivity towards various reactive oxygen species, including DPPH, superoxide radicals, and hydroxyl radicals. These findings suggest that the arabinogalactan polysaccharide from *T. cordifolia* may have potential therapeutic applications in combating oxidative stress-related disorders (Subramanian et al., 2002).

According to Ayurvedic Pharmacopoeia of India, giloy is classified as a herbal remedy with anti-diabetic properties, attributed to the presence of diterpenoids, alkaloids, and glycosidic constituents. Multiple *in vitro* and *in vivo* investigations have demonstrated that the alkaloid compounds jatrorrhizine, palmatine, and magnoflorin exhibit a synergistic effect in inducing hyperglycemia. This effect is achieved through a mechanism involving the release of insulin, mimicking the action of insulin, and inhibiting gluconeogenesis. The anti-diabetic effects of palmatine were observed in L6 myocyte cells via an insulin-dependent mechanism, as well as through the upregulation of expression of the glucose transporter-4 (Glut-4) and peroxisome proliferator-activated receptor alpha (PPARα). It is worth noting that both berberine and palmatine exhibited comparable potential in terms of their antidiabetic properties (Patel and Mishra, 2012; Sangeetha et al.,2013; Sharma et al., 2015). Tinosporin, a diterpenoid compound, has demonstrated efficacy against various viral diseases, including retroviruses as Human Immunodeficiency Virus type 1 (HIV-1), HIV-2 and all subgroups, Human T-lymphotropic viruses (HTLV) due to its immunomodulatory properties and selective prevention of viruses, explicitly targeting T helper cell types (Chetan and Nakum, 2010). Palmatine, a bioactive compound found in *T. cordifolia*, has been shown to have the ability to decrease the activity of B-cell lymphoma-2 and induce apoptosis associated with the natural production of caspase-activated DNA in both the nucleus and cytoplasm. The compound palmatine possesses a pharmacophore group of octacosanol, an aliphatic alcohol. This particular pharmacophore has been observed to demonstrate angioinhibitory and antimetastatic implications (Dhama et al., 2016).

T. cordifolia affects the immune system and can be linked to the interactions of many different chemicals. These substances include low molecular weight alkaloids, clerodane diterpenoids, sesquiterpenoids, and phenyl propanoids. Magnoflorine, N-methyl-2-pyrrolidone, N-formylannonain, syringin, cordioside, cordial, tinocordiside, 11-hydroxymuskatone, cordifolioside A and cordifolioside B are some of the chemicals that fall into this category. Additionally, the immunomodulatory actions of *T. cordifolia* are contributed to an arabinogalactan with a high molecular weight known as G1-4A, 1,4--D-glucan (RR1) and enzyme thiol amylase (Singh et al., 2003; Mukherjee et al., 2010; Sharma et al., 2012; Pandey et al., 2014). In *in-vivo* studies, consumption of extract of *T. cordifolia* produced anti-tumor biomarkers such as reactive nitrogen intermediates, Tumour Necrosis Factor (TNF), and Interleukin-1 (IL-1) resulting in substantial cytotoxicity. Additionally, it showed apoptosis of the bone marrow cells (BMC) at higher dosages and leucocytosis in lesser amounts (Alexander et al., 2010). The use of *T. cordifolia* effectively lowered neutropenia caused by both single and multiple applications of cyclophosphamide (Singh et al., 2004). The antioxidant ascorbic acid was compared to the efficacy of *T. cordifolia* extracts to exhibit significant activity. The typical substance, on the other hand, exhibited robust reducing power at extremely low dosages. In general, a wide diversity of tested plant extracts demonstrated decreasing potency with increasing doses (Mishra et al., 2013).

According to Duraipandian et al. (2012), a chemical compound found in the stems of *T. cordifolia* is effective against many fungi and bacteria. *T. cordifolia* hydro-alcoholic extract lowered breast inflammation in a bovine model by increasing granulocyte activity. As *Staphylococcus aureus* bacterial infection is the cause of mastitis, the antibacterial activity was evidenced by its ability to stop this inflammation. The stem and foliage of *T. cordifolia* possess the highest inhibitory action against clinical urinary pathogens namely *Klebsiella pneumoniae* and *Pseudomonas aeruginosa* and inhibit urinary tract infections (Pradhan et al., 2013).

According to Singh et al. (2014), *Pseudomonas aeruginosa* was found in a patient with burn injuries, and silver nanoparticles derived from *T. cordifolia* stem showed effective antibacterial action. Many bacterial strains, including *S. typhi*, *K. peumoniae*, *E. coli,* and *P. aeruginosa*, exhibited potential antibacterial activity when tested against *T. cordifolia* extracts by either inhibition of their growth or diminishing their life. The antibacterial properties of *T. cordifolia* extract against *P. aeruginosa* were observed by Royani et al. (2023) across various methanol concentration ratios. The methanol extract of *T. cordifolia* exhibited a minimum inhibitory concentration (MIC) of 4096 g/mL. The antibacterial effectiveness against *P. aeruginosa* increases by 50%, 75%, and 100%. The structure compounds of *T. cordifolia* extract consist of derivative alkaloid compounds that contain a phenol aromatic group. The active compounds found in the methanol extract of *T. cordifolia* include Calopiptin, d-Lirioferine (Lirioferine), Moupinamide, Piperanine, and Yuanhunine (Singh et al., 2014).

6.1 Uses Described in Pharmacopoeias and Traditional Systems of Medicine

Ayurvedic texts including the Ashtanga Hridaya, Sushrut Samhita, Charak Samhita, and other accords such as Bhava Prakash and Dhanvantri Nighantu, refers *T. cordifolia* as Guduchi or Amrita and has the common names Amara, Amritvalli, Chinmarrhuha, and Chinnodebha (Tiwari et al., 2018). It is customarily asserted in the Tikta-Saka Varga section of Sushurta Samhita for the treatment of numerous ailments, including Svasa (asthma), Aruci (anorexia), Maha Jvara (fever), and Kustha (leprosy) (Marke et al., 2022). It has been portrayed in Dhanvantri Nighantu for treating bleeding piles, alleviating itching, treating erysipelas, and extending life. It has also been demonstrated to be Deepanam (ignites the digestive fire), Laghu (light), Chakshushyam (beneficial for the eyes), Dhatukrit (builds the seven bodily tissues), Bayasthaapankarakam (preserves youth and lifespan) and Medhayam (rejuvenating for mind). There are substantial signs for the treatment of various ailments including Vat Rakta (gout), Jvara (fever), and Kamala in the context of Ashtanga Hridaya and Charak Samhita (jaundice). It was included in Bengal Pharmacopoeia of 1868 and was regarded by European doctors practicing in India as a significant source of medications such as tonics, diuretics, and antiperiodics. As per Ayurvedic literature, *T. cordifolia* has the potential as a diuretic, astringent, bitter tonic, cure-all, and aphrodisiac for skin diseases, diabetes, chronic diarrhoea, dysentery, and jaundice (Upadhyay et al., 2010).

6.2 Traditional/Folklore Uses Described in Folk Medicine

The most popular plant in traditional medicine, *T. cordifolia*, has been utilized extensively by common people and tribal people as a medicinal herb for alleviating many ailments. *T. cordifolia* is highly prized for its wide range of therapeutic benefits, including its ability to detoxify the body and enhance the immune system. *T. cordifolia* medicinal characteristics are currently used in modern medicine to treat gout, arthritis, immunological support, gout, liver diseases, and recently to counteract the side effects of chemotherapy. Therefore, it is evident that *T. cordifolia* is the most significant therapeutic herb valued by the ancient Vedic periods, with a high potential for healing a wide range of diseases. It has already been noted that *T. cordifolia* has several therapeutic characteristics. It has been said to have osteoprotective properties used in Indian folk medicine to cure bone diseases (Mittal, 2014; Verma et al., 2021).

6.3 Applications supported by Clinical Data

According to Abiramasundari et al. (2018), the SERM (Selective Estrogen Receptor Modulators) properties of β-ecdysone from *T. cordifolia* have potential applications in the treatment of osteoporosis. The findings of this experiment demonstrate the impact of the extract of *T. cordifolia* on the viability and growth of two specific cancer cell lines, namely MCF-7 (breast adenocarcinoma) and HeLa (cervical adenocarcinoma). This was assessed through the utilization of the MTT (3-(4, 5-dimethylthiazolyl-2)-2, 5-diphenyltetrazolium bromide) assay and crystal violet test. There were no notable variations in the growth of MCF-7 and HeLa cells when exposed to varying concentrations (12.5, 25, 50, 75, and 100 mg/ml) of the plant extract, in comparison to the control group, in both experimental assessments. This observation suggests that *T. cordifolia* does not exhibit any stimulatory or inhibitory effects on the breast and cervical cells, indicating that it does not possess endocrine-disrupting properties. This suggests that the bioactive compounds of *T. cordifolia*, responsible for the osteoprotective effect, may function as a Selective estrogen receptor modulator (SERM). This compound exhibits agonistic effects on bone, promoting bone growth while having non-stimulatory effects on reproductive organs. A recent clinical study examined the safety profile of a product derived from *T. cordifolia* in Swiss albino mice. The study focused on evaluating the potential risks and adverse effects of this product in healthy mice. The assessment involved hematological and biochemical investigations to determine the effects of the product. The results of the study indicated that *T. cordifolia*-based product had the notable property of enhancing intestinal motility in mice. This finding suggests that the product could be beneficial in treating gastric problems without causing any adverse effects on the overall physiology of the mice (Patel et al., 2018).

In a recent study, Kuchewar et al. (2014) examined the effects of Ashwagandha and Tinospora on the antioxidant potential of healthy individuals. This investigation was designed as a randomized, double-blind trial, and participants were instructed to take a daily dosage of 500mg of the drug. The drug was administered in capsule form and taken orally with water for six months. In the assessment of the treatment, various parameters were evaluated both before and after the treatment. These parameters included haemoglobin percentage, Erythrocyte Sedimentation Rate

(ESR), Malondialdehyde (MDA), Super-Oxide Dismutase (SOD) level, and others. It was determined that certain herbs could act as antioxidants. This finding suggests that these herbs may have potential health benefits due to their antioxidant properties. In a recent study, the authors have concluded that both drugs are deemed safe for young adults.

6.4 Pharmacological Activities mainly Antiviral

Another study examined the *in vitro* antiviral effectiveness of an ethanolic extract derived from the entire *T. cordifolia* plant towards the hepatitis A virus (HAV). The preparation of the ethanolic extract from the chosen plant involved using a Soxhlet apparatus and a rotary evaporator. The toxicological impact of a chosen plant extract was evaluated for its antiviral activity against the Hepatitis A Virus using the MTT cell proliferation assay with Huh-7 cell lines. The usual dose employed in this study is Camptothecin (CPT) at a concentration of 25μG (Maddi et al., 2018). The emergence of the novel coronavirus, SARS-CoV-2, posed a momentous danger to human society. In light of its therapeutic importance, researchers have explored the chemical components found in *T. cordifolia* extract, belonging to diverse classes, for instance, alkaloids, lignans, steroids, and terpenoids (Sharma et al., 2021), as possibilities for COVID-19 treatment. Berberine, β-sitosterol, tetrahydropalmatine, octacosanol, and coline exhibit antiviral properties against severe acute respiratory syndrome coronavirus 2 (SARS-CoV-2) (Thakkar et al., 2021; Sapra et al., 2021). The primary chemical components of *T. cordifolia*, such as berberine and β-sitosterol, exhibit potential as antiviral agents against SARS-CoV-2. These constituents demonstrate notable inhibitory effects on SARS CoV-2 protein, surpassing the inhibitory capabilities of other known inhibitors. The constituents found in the stem substance of *T. cordifolia* demonstrate effective inhibition of 3CL protease activity, thereby exhibiting favorable effects in the regulation of virus growth and replication (Chowdhury et al., 2021). The computing molecular modeling technique was employed to assess the inhibitory potential of the suggested compounds against a viral spike protein and human receptor ACE2. Additionally, the ADME/T properties of these compounds were also evaluated. The findings from molecular docking and ADME/T investigations have identified six components that exhibit potential as potential drugs capable of inhibiting the interaction between the spike protein of SARS-CoV-2 and the ACE2 receptor protein in humans (Jena et al., 2021).

Devi (2021), conducted a study on *T. cordifolia* to assess its potential efficacy in reducing the resistance of HIV-positive individuals to retroviral treatment. The potential application of *T. cordifolia* in disease management is evident through its anti-HIV activity, which is characterized by the augmentation of CD4 T-cell count and reduction of eosinophil count in individuals living with HIV. The extract derived from *T. cordifolia* demonstrated a significant enhancement in both phagocytic and intracellular bactericidal properties. Furthermore, *T. cordifolia* exhibited the ability to activate peritoneal macrophages. In addition, *T. cordifolia* has been found to augment the process of phagocytosis and the ability to kill pathogens within cells. As reported, the plant extract in question has been observed to have a noteworthy stimulatory impact on B-lymphocytes, polymorphonuclear leukocytes, and macrophages (Arunachalam et al., 2022). Tinospora cordifolia exhibits inhibitory activity against HIV protease and has been observed to disrupt the interaction between gp120 and CD4, thereby inhibiting HIV-reverse transcriptase (Kalikar et al., 2008). These results propose that *T. cordifolia* possesses anti-viral potential against HIV (Sachan et al., 2019). The efficacy of an aqueous extract of *T. cordifolia* was assessed concerning Infectious Bursal Disease (IBD) in young chicks, which is caused by the Infectious Bursal Disease Virus (IBDV). The results indicated a significant reduction in mortality rate and an enhancement of the chicks' immune response (Krupanidhi et al., 2021). Through the utilization of virtual screening and molecular docking analysis, several phytochemical compounds present in *T. cordifolia*, specifically cardiofolioside B, tinosponone, xanosporic acid, tembetarine, and berberine, have been identified as potential lead molecules in the battle against SARS-CoV-2. The in-silico investigation demonstrated that tinosponone exhibits powerful, selective, and non-toxic inhibitory effects on the 3CL protease of SARS-CoV-2 (Goyal et al., 2022).

6.5 Future of Research and Drug Delivery

A comprehensive understanding of the bioactive compounds present in *T. cordifolia* is needed for uncovering the basis of its therapeutic properties. The use of modern chemical and pharmacological techniques, including bioassay-guided fractionation, chemical characterization, mechanisms of action, and structure-activity connection, could assist in identifying further active chemicals produced by this plant. In addition, the interactions that can occur between different chemicals, such as cohesion, additivity, and antagonism need to be investigated.

7. Toxicity and Safety

It has been found that the use of Ayurvedic interventions containing *T. cordifolia* in COVID-19 patients has not resulted in any reported cases of toxicity. This is particularly significant as these interventions were used with standard care treatment. It is believed to possess immunomodulatory properties, particularly useful in combating viral infections such as COVID-19. The study focused on assessing the safety of COVID-19 patients. It involved administering these interventions alongside standard care treatment and closely monitoring potential adverse effects. The results were encouraging, as no cases of toxicity were reported during the study period. This Ayurvedic medicine has demonstrated promising results in reducing the time to symptomatic recovery among individuals with unaffected and mildly indicative COVID-19 (Balkrishna et al., 2021; Kataria et al., 2022).

Conclusion

The pharmacologic data and numerous isolated phytochemicals from *T. cordifolia* unquestionably back its usage in herbal remedies and formulations. In addition to clinical and toxicological investigations, the possible mode of action was investigated for the formulations, crude extracts, and isolated bioactive compounds. *T. cordifolia* extracts contain a variety of phytochemical classes that have significant antibacterial, antioxidant, and anti-viral properties, and anticancer effects. *T. cordifolia* has been extensively researched for the biological action of diverse extracts, fractions, and compounds, with measurement of some bioactive compounds, as evidenced by documentation from various academic disciplines. In addition, substances isolated from *T. cordifolia* with anti-diabetic, anti-inflammatory, and immunomodulatory properties can be utilized as remedial mediators to cure these dreaded disorders, singly or in combination, in standardized forms. *T. cordifolia* abundance in subtropical Asian nations makes this plant an economically affordable and active herbal dietary supplement. Biological research and clinical studies on *T. cordifolia* indicate that this plant is safe and has effective therapeutic properties for use as a health supplement of commercial significance, besides a repository for forthcoming research on drugs in life-threatening disorders for which current medications have limited therapeutic value. More clinical trials based on current information should be carried out to support the therapeutic usage and acceptance of *T. cordifolia*.

References

Abiramasundari, G., Gowda, C.M. and Sreepriya, M. (2018). Selective Estrogen Receptor Modulator and prostimulatory effects of phytoestrogen β-ecdysone in Tinospora cordifolia on osteoblast cells. *Journal of Ayurveda and Integrative Medicine*, 9(3), pp.161–168.

Alexander, C.P., Kirubakaran, C.J.W. and Michael, R.D. (2010). Water soluble fraction of Tinospora cordifolia leaves enhanced the non-specific immune mechanisms and disease resistance in Oreochromis mossambicus. *Fish & shellfish immunology*, 29(5), pp.765–772.

Arunachalam, K., Yang, X. and San, T.T. (2022). Tinospora cordifolia (Willd.) Miers: Protection mechanisms and strategies against oxidative stress-related diseases. *Journal of Ethnopharmacology*, 283, pp.1–14.

Bala, M., Pratap, K., Verma, P.K., Singh, B. and Padwad, Y. (2015). Validation of ethnomedicinal potential of Tinospora cordifolia for anticancer and immunomodulatory activities and quantifcation of bioactive molecules by HPTLC. *Journal of Ethnopharmacology*, 175, 131–137.

Balkrishna, A., Bhatt, A.B., Singh, P., Haldar, S. and Varshney, A. (2021). Comparative retrospective open-label study of ayurvedic medicines and their combination with allopathic drugs on asymptomatic and mildly-symptomatic COVID-19 patients. *Journal of Herbal Medicine*, 29, pp.1–14.

Castillo, A.L., Osi, M.O., Ramos, J.D.A., De Francia, J.L., Dujunco, M.U. and Quilala, P.F. (2013). Efficacy and safety of Tinospora cordifolia lotion in Sarcoptes scabiei var hominis-infected pediatric patients: A single blind, randomized controlled trial. *Journal of pharmacology and pharmacotherapeutics*, 4(1), pp.39–46.

Channaveeraiah, S.K. and Entoori, K. (2022). Recent perspectives on impact of Ethnobotany in India: A review. *Journal of Medicinal Plants*, 10(4), pp.17–20.

Chetan, B. and Nakum, A. (2010). Use of natural compounds, chitin and tinosporin for the treatment of the targeted viruses (retroviruses) (HIV-1, HIV-2) all subgroups, HTLV and other viral disease. Indian patent Appl. IN 2010MU01350 A, 20100730.

Chi, S., She, G. and Han, D., (2016). Genus Tinospora: ethnopharmacology, phytochemistry, and pharmacology. *Evidence-Based Complementary and Alternative Medicine*, 20: 1–32.

Choudhary, N., Siddiqui, M. B., Azmat, S. and Khatoon, S. (2013). Tinospora cordifolia: ethnobotany, phytopharmacology and phytochemistry aspects. International *Journal of Pharmaceutical Sciences and Research*, 4(3), 891.

Chowdhury, P. (2021). *In silico* investigation of phytoconstituents from Indian medicinal herb 'Tinospora cordifolia (giloy)'against SARS-CoV-2 (COVID-19) by molecular dynamics approach. *Journal of Biomolecular Structure and Dynamics*, 39(17), pp. 6792–6809.

De Wet, H., Struwig, M. and Van Wyk, B.E. (2016). Taxonomic notes on the genera Tiliacora and Tinospora (Menispermaceae) in southern Africa. *South African Journal of Botany*, 103, pp. 283–294.

Devi, P. (2021). A review on Tinospora cordifolia: as an Immunomodulating agent. Himalayan Journal of Health Sciences, pp. 6–14.

Dhama, K., Sachan, S., Khandia, R., Munjal, A., MN Iqbal, H., K Latheef, S., Karthik, K., A Samad, H., Tiwari, R. and Dadar, M. (2016). Medicinal and beneficial health applications of Tinospora cordifolia (Guduchi): a miraculous herb countering various diseases/disorders and its immunomodulatory effects. *Recent patents on endocrine, metabolic & immune drug discovery,* 10(2), pp.96–111.

Duraipandiyan, V., Ignacimuthu, S., Balakrishna, K. and AL-Harbi, N.A. (2012). Antimicrobial activity of Tinospora cordifolia: an ethnomedicinal plant. *Asian Journal of Pharma*, 7(2), pp. 59–65.

Gautam, A., Kaur, H., Kaur, A., Prashar, P.K., Sood, A., Singh, S.K., Gulati, M., Pandey, N.K. and Kumar, B. (2020). Tinospora cordifolia: A Successful Story from Botanical background to Pharmaceutical Product. *Research Journal of Pharmacy and Technology*, 13(11), pp. 5620–5630.

Goyal, R., Bala, R., Sindhu, R.K., Zehravi, M., Madaan, R., Ramproshad, S., Mondal, B., Dey, A., Rahman, M.H. and Cavalu, S., (2022). Bioactive based nanocarriers for the treatment of viral infections and SARS-CoV-2. *Nanomaterials*, 12(9), p.1530.

Hao, D.C., Zhang, Y., He, C.N. and Xiao, P.G. (2022). Distribution of therapeutic efficacy of Ranunculales plants used by ethnic minorities on the phylogenetic tree of Chinese species. Evidence-based Complementary and Alternative Medicine, 2022.

Jagetia, G.C. and Rao, S.K. (2006). Evaluation of the antineoplastic activity of guduchi (Tinospora cordifolia) in Ehrlich ascites carcinoma bearing mice. *Biological and Pharmaceutical Bulletin*, 29(3), pp. 460–466.

Jain, S., Vaidya, A., Shah, K., Chauhan, D.N. and Chauhan, N.S. (2019). Anti-sickling Herbs. Plant and Human Health, Volume 3: *Pharmacology and Therapeutic Uses*, pp. 255–283.

Jan, M. and Shrivastava, M. (2019). Extraction of leaf pigments, carbohydrate and protein content from fresh leaves of Tinospora cordifolia (Willd.). *SurajPunj Journal for Multidisciplinary Research*, 9(1), pp.10–17.

Jena, S., Munusami, P., Mm, B. and Chanda, K. (2021). Computationally approached inhibition potential of Tinospora cordifolia towards COVID-19 targets. *Virus disease*, 32, pp. 65–77.

Joshi, Y., Rani, H., Kaur, G., Kumar, M., Sindhu, R.K., Kushwah, A.S. and Mittal, R. (2023). Herbal and Ayurvedic Plants as Remedial Approach for Viral Diseases with Focus on COVID-19: A Narrative Review. *Current Traditional Medicine*, 9(3), pp.66–86.

Kabera, J.N., Semana, E., Mussa, A.R. and He, X. (2014). Plant secondary metabolites: biosynthesis, classification, function and pharmacological properties. *J. Pharm. Pharmacol*, 2(7), pp.377–392.

Kalikar, M.V., Thawani, V.R., Varadpande, U.K., Sontakke, S.D., Singh, R.P. and Khiyani, R.K. (2008). Immunomodulatory effect of Tinospora cordifolia extract in human immuno-deficiency virus positive patients. *Indian Journal of Pharmacology*, 40(3), p.107.

Kapil, A. and Sharma, S., (1997). Immunopotentiating compounds from Tinospora cordifolia. *Journal of Ethnopharmacology*, 58(2), 89–95.

Kataria, S., Sharma, P., Ram, J.P., Deswal, V., Singh, M., Rana, R., Singhal, R., Tripathi, A., Kumar, K. and Trehan, N. (2022). A pilot clinical study of an add-on Ayurvedic formulation containing Tinospora cordifolia and Piper longum in mild to moderate COVID-19. *Journal of Ayurveda and Integrative Medicine*, 13(2), p.100454.

Kaur, G., Prabhakar, P.K., Lal, U.R. and Suttee, A. (2016). Phytochemical and biological analysis of Tinospora cordifolia. *International Journal of Toxicological and Pharmacological Research*, 8(4), pp.297–305.

Krupanidhi, S., Abraham Peele, K., Venkateswarulu, T.C., Ayyagari, V.S., Nazneen Bobby, M., John Babu, D., Venkata Narayana, A. and Aishwarya, G. (2021). Screening of phytochemical compounds of Tinospora cordifolia for their inhibitory activity on SARS-CoV-2: an *in silico* study. *Journal of Biomolecular Structure and Dynamics*, 39(15), pp.5799–5803.

Kuchewar, V.V., Borkar, M.A. and Nisargandha, M.A. (2014). Evaluation of antioxidant potential of Rasayana drugs in healthy human volunteers. *Ayu*, 35(1), p.46.

Kumar, D.V., Geethanjali, B., Avinash, K.O., Kumar, J.R. and Basalingappa, K.M. 2017. Tinospora cordifolia: the antimicrobial property of the leaves of amruthaballi. *Journal of Bacteriology and Mycology*, 5(5), 363–371.

Kumar, K.N., Kishore, D., Kumar, V.J., Balakrishnan, P. and Manoj, N.T., (2019). Tinospora cordifolia: a wonderful miracle herb of 21st century of India. *Medicinal Plants-International Journal of Phytomedicines and Related Industries*, 11(2), pp.117–122.

Lade, S., Sikarwar, P.S., Ansari, M.A., Khatoon, S., Kumar, N., Yadav, H.K. and Ranade, S.A. (2018). Diversity in a widely distributed dioecious medicinal plant, Tinospora cordifolia (Willd.) Miers. ex. Hook F. and Thomas. *Current Science*, pp.1520–1526.

Lian, L., Ortiz, R.D.C., Jabbour, F., Chen, Z.D. and Wang, W. (2019). Re-delimitation of Tinospora (Menispermaceae): Implications for character evolution and historical biogeography. *Taxon*, 68(5), pp.905–917.

Maddi, R., Kandula, V.L., Vallepu, B., Navuluri, H., Kolluri, H. and Vunnam, D.T. (2018). Preliminary phytochemical analysis and invitro anti–viral activity of ethanolic extract of whole plant of Tinospora cordifolia (Thunb.) miers against hepatitis-A virus. *International Journal of Scientific Research in Biological Sciences*, 5, p.3.

Malla, S. and Bista, L. (2021). Tinospora cordifolia: A Multipurpose Miracle Plant Having Medicinal Importance: A Review. *Matrix Science Pharma*, 5(3), pp.54–59.

Marke, V., Shete, P. and Dhundale, V. (2022). Tinospora cordifolia: A Valuable Plant in Ayurveda. *Indian Journal of Ecology*, 49(1), pp.38–41.

Maurya, R. and Handa, S.S. (1998). Tinocordifolin, a sesquiterpene from Tinospora cordifolia. *Phytochemistry*, 49(5), 1343–1345.

Mishra, A., Kumar, S. and Pandey, A.K. (2013). Scientific validation of the medicinal efficacy of Tinospora cordifolia. *The Scientific World Journal*, 2013.

Mittal, J., Sharma, M.M. and Batra, A. (2014). Tinospora cordifolia: a multipurpose medicinal plant—A. *Journal of Medicinal Plants*, 2(2).

Modi, B., Kumari Shah, K., Shrestha, J., Shrestha, P., Basnet, A., Tiwari, I. and Prasad Aryal, S. (2020). Morphology, biological activity, chemical composition, and medicinal value of Tinospora Cordifolia (willd.) Miers. *Advanced Journal of Chemistry-Section B*, 2020, pp.36–54.

Mukherjee, R., De, U.K. and Ram, G.C. (2010). Evaluation of mammary gland immunity and therapeutic potential of Tinospora cordifolia against bovine subclinical mastitis. *Tropical animal health and production*, 42, pp.645–651.

Ortiz, R.D.C., Kellogg, E.A. and Van Der Werff, H. (2007). Molecular phylogeny of the moonseed family (Menispermaceae): implications for morphological diversification. *American Journal of Botany*, 94(8), pp.1425–1438.

Ortiz, R.D.C., Wang, W., Jacques, F.M. and Chen, Z. (2016). Phylogeny and a revised tribal classification of Menispermaceae (moonseed family) based on molecular and morphological data. *Taxon*, 65(6), pp.1288–1312.

Pandey, V.K., Amin, P.J. and Shankar, B.S. (2014). G1-4A, a polysaccharide from Tinospora cordifolia induces peroxynitrite dependent killer dendritic cell (KDC) activity against tumor cells. *International Immunopharmacology*, 23(2), pp.480–488.

Parthipan, M., Aravindhan, V. and Rajendran, A. (2011). Medico-botanical study of Yercaud hills in the eastern Ghats of Tamil Nadu, India. *Ancient Science of Life*, 30(4), 104.

Patel, A.G., Patel, M.R. and Nariya, M.B. (2018). Assessment of Trisama, an ayurvedic formulation on intestinal transit time in swiss albino mice. *Ayu*, 39(1), p.46.

Patel, M.B. and Mishra, S. (2011). Hypoglycemic activity of alkaloidal fraction of Tinospora cordifolia. *Phytomedicine*, 18(12), pp.1045–1052.

Patel, M.B. and Mishra, S.M. (2012). Magnoflorine from Tinospora cordifolia stem inhibits α-glucosidase and is antiglycemic in rats. *Journal of Functional Foods*, 4(1), pp.79–86.

Payyappallimana, U., Ravikumar, K. and Venkatasubramanian, P. (2022). Can Guduchi (Tinospora cordifolia), a well-known ayurvedic hepato-protectant cause liver damage? *Journal of Ayurveda and Integrative Medicine*, p.100658.

Pradhan, D., Kar, D.M. and Ghosh, G. (2020). Bioassay-guided Fractionation of Anti-microbial principle from Ethanolic Extract of Tinospora cordifolia and its Characterization. *Research Journal of Pharmacy and Technology*, 13(3), pp.1415–1422.

Royani, A., Hanafi, M., Lotulung, P.D.N., Prastya, M.E., Verma, C., Manaf, A. and Alfantazi, A. (2023). Isolation and identification of bioactive compounds from Tinospora cordifolia stem extracts as antibacterial materials in seawater environments. *Arabian Journal of Chemistry, 16*(9), p.105014.

Sachan, S., Dhama, K., Latheef, S.K., Abdul Samad, H., Mariappan, A.K., Munuswamy, P., Singh, R., Singh, K.P., Malik, Y.S. and Singh, R.K. (2019). Immunomodulatory potential of Tinospora cordifolia and CpG ODN (TLR21 agonist) against the very virulent, infectious bursal disease virus in SPF chicks. *Vaccines*, 7(3), p.106.

Saeed, M., Naveed, M., Leskovec, J., Kakar, I., Ullah, K., Ahmad, F., Sharif, M., Javaid, A., Rauf, M., Abd El-Hack, M.E. and Abdel-Latif, M.A. (2020). Using Guduchi (Tinospora cordifolia) as an eco-friendly feed supplement in human and poultry nutrition. *Poultry Science*, 99(2), pp.801–811.

Saha, S. and Ghosh, S. (2012). Tinospora cordifolia: One plant, many roles. *Ancient Science of Life*, 31(4), p.151.

Sangeetha, M.K., Priya, C.M. and Vasanthi, H.R. (2013). Anti-diabetic property of Tinospora cordifolia and its active compound is mediated through the expression of Glut-4 in L6 myotubes. *Phytomedicine*, 20(3-4), pp.246–248.

Sapra, L., Bhardwaj, A., Azam, Z., Madhry, D., Verma, B., Rathore, S. and Srivastava, R.K. (2021). Phytotherapy for treatment of cytokine storm in COVID-19. *Frontiers in Bioscience-Landmark*, 26(5), pp.51–75.

Sharma, H., Rao, P.S. and Singh, A.K. (2021). Fifty years of research on Tinospora cordifolia: From botanical plant to functional ingredient in foods. *Trends in Food Science & Technology*, 118, pp.189–206.

Sharma, P., Dwivedee, B.P., Bisht, D., Dash, A.K. and Kumar, D. (2019). The chemical constituents and diverse pharmacological importance of Tinospora cordifolia. *Heliyon*, 5(9), e02437.

Sharma, R., Amin, H. and Prajapati, P.K. (2015). Antidiabetic claims of Tinospora cordifolia (Willd.) Miers: critical appraisal and role in therapy. *Asian Pacific Journal of Tropical Biomedicine*, 5(1), pp. 68–78.

Sharma, U., Bala, M., Kumar, N., Singh, B., Munshi, R.K. and Bhalerao, S. (2012). Immunomodulatory active compounds from Tinospora cordifolia. *Journal of Ethnopharmacology*, 141(3), 918–926.

Singh, D. and Chaudhuri, P.K. (2017). Chemistry and pharmacology of Tinospora cordifolia. *Natural Product Communications*, 12(2), p.1934578X1701200240.

Singh, K., Panghal, M., Kadyan, S., Chaudhary, U. and Yadav, J.P. (2014). Antibacterial activity of synthesized silver nanoparticles from Tinospora cordifolia against multi drug resistant strains of Pseudomonas aeruginosa isolated from burn patients. *Journal of Nanomedicine & Nanotechnology*, 5(2), p.1.

Singh, N., Singh, S.M. and Shrivastava, P., (2004). Immunomodulatory and antitumor actions of medicinal plant Tinospora cordifolia are mediated through activation of tumor-associated macrophages. *Immunopharmacology and Immunotoxicology*, 26(1), pp.145–162.

Singh, S.S., Pandey, S.C., Srivastava, S., Gupta, V.S., Patro, B. and Ghosh A.C. (2003). Chemistry and medicinal properties of Tinospora Cordifolia (Guduchi). *Indian Journal of Pharmacology*, 35(2), 83–91.

Sinha, K., Mishra, N.P., Singh, J. and Khanuja, S.P.S. (2004). Tinospora cordifolia (Guduchi), a reservoir plant for therapeutic applications: a review. *Indian Journal of Traditional Knowledge*, 3(3), 257–270.

Spandana, U., Ali, S.L., Nirmala, T., Santhi, M. and Babu, S.S. (2013). A review on Tinospora cordifolia. *International Journal of Current Pharmaceutical Review and Research*, 4(2), pp.61–68.

Subramanian, M., Chintalwar, G.J. and Chattopadhyay, S. 2002. Antioxidant properties of a Tinospora cordifolia polysaccharide against iron-mediated lipid damage and γ-ray induced protein damage. *Redox Report*, 7(3), pp.137–143.

Thakkar, S.S., Shelat, F. and Thakor, P. (2021). Magical bullets from an indigenous Indian medicinal plant Tinospora cordifolia: An *in silico* approach for the antidote of SARS-CoV-2. *Egyptian Journal of Petroleum*, 30(1), pp.53–66.

Tiwari, P., Nayak, P., Prusty, S.K. and Sahu, P.K. (2018). Phytochemistry and pharmacology of Tinospora cordifolia: A review. *Systematic Reviews in Pharmacy*, 9(1), pp.70–78.

Upadhyay, A.K., Kumar, K., Kumar, A. and Mishra, H.S. (2010). Tinospora cordifolia (Willd.) Hook. f. and Thoms.(Guduchi)–validation of the Ayurvedic pharmacology through experimental and clinical studies. *International Journal of Ayurveda Research*, 1(2), p.112.

Verma, D.K., Kumar, P. and El-Shazly, M. (2021). Unmasking the many faces of Giloy (Tinospora cordifolia L.): a fresh look on its phytochemical and medicinal properties. *Current Pharmaceutical Design*, 27(22), pp.2571–2581.

Wink, M. (2015). Modes of action of herbal medicines and plant secondary metabolites. *Medicines*, 2(3), 251–286.

4 *Acorus calamus* L.: Bioactive Compounds act Pharmacologically as Antiviral Agents

Anu Pandita[1#], Lepakshi Md. Bhakshu[2#], K. Venkata Ratnam[3], Pulala Raghuveer Yadav[4] and Deepu Pandita[5*]

1. Taxonomy and Classification

Acorus calamus L. belongs to the monocotyledonous family Acoraceae. The genus or generic name *Acorus* is a Latin derivative of the Greek word áchórou (eye pupil). The species or specific name *calamus* is derived from the Greek word kálamos (which means reed). The common name of the plant, Sweet Flag, is given due to the sweet scent of this plant and its similarity to the Iris species, referred to commonly as flags in English (https://en.wikipedia.org/wiki/Acorus_calamus). Acorus was first placed in the Araceae family, but later, based on morphological and anatomical features, it shifted to its own family Acoraceae (Hotta, 1970; Grayum, 1987).

A. calamus well known as 'Sweet-flag' and 'Vacha', is a popular traditional Indian and Chinese medicinal plant with rhizomes having been used in massive disorders. *A. calamus* has found use in Asia for at least preceding 2000 years (Packer and Ringius, 1984). The herbal extracts were therapeutically employed as an anti-depressant, anti-stroke, and for learning improvement (Sharma et al., 2014; Lam et al., 2016; Amritha et al., 2018; Parcha et al., 2022; Ashrafi et al., 2022). The rhizome has therapeutic characteristics that can be used to treat dyspepsia-related emetic stomach and pests like moths and lice. Dried roots have been used for a long time to flavor bitter liqueurs and appetizers, as well as in medicinal formulation as an antispasmodic and an antiulcer (Sharma et al., 2014; Amritha et al., 2018; Parcha et al., 2022; Ashrafi et al., 2022).

[1] Vatsalya Clinic, Krishna Nagar, New Delhi, India.
[2] Department of Botany, Dr. YSR Government Degree College, Vedurukuppam, Chittoor District 517 569, Andhra Pradesh, India. https://orcid.org/0000-0002-3159-2593
[3] Department of Botany, Rayalaseema University, Kurnool, Andhra Pradesh, India.
[4] Department of Biotechnology, Indian Institute of Technology Hyderabad, Kandi, Telangana State, India.
[5] Government Department of School Education, Jammu, Jammu and Kashmir, India: ORCID ID-https://orcid.org/0000-0002-8361-1872
[#] *Both these authors have contributed equally*
* Corresponding author: deepupandita@gmail.com
DOI: 10.1201/9781003433583-4

Systematic Position

Kingdom	:	Plantae
Division	:	Magnoliophyta
Class	:	Liliopsida
Order	:	Acorales
Family	:	Acoraceae
Genus	:	*Acorus*
Species	:	*calamus*
Binomial name	:	*Acorus calamus* L., 1753
Synonyms	:	*Calamus aromaticus* Garsault

Vernacular names: *A. calamus* is widely known as sweet rush, sweet flag, sweet myrtle, sweet cane, sweet root, gladdon, myrtle grass, cinnamon sedge, myrtle sedge, Bacch (in Unani), vacha (Ayurvedic), Bhutanashini/Jatila/Ugragandh (in Sanskrit) and Plant of Venus (in Italy) (Packer and Ringius, 1984; Parcha et al., 2022).

2. Botanical Description of the Plant

Acorus calamus L., perhaps indigenous to the Indian subcontinent, grows in marshy, sandy loam and light alluvial soil in both the temperate and sub temperate regions with temperatures in the range of 10.8°C to 38.8°C (Fig. 1). *A. calamus* plant is semi-aquatic, herbaceous and perennial lacking a proper stem and produces a bunch of sword-shaped leaves. The cluster of leaves arises from a widely-branched, creeping rhizome of almost 3 cm diameter. The underground rhizomatous stem is branched, extensive, brown in colour, curved, twisted, cylindrical, with short nodes. The roots white in colour, are long of 2 m in length spreading horizontally. Leaves are radiant green, long, erect, narrow, sharp-pointed, sword-like, and aromatic in structure with thicker middle potion and curvy margins and reach up to 150cm in length (Hotta, 1970; Quraishi et al., 2017).

The inflorescence is a leaf-like spathe and spike-like spadix. The flowers are very dense, cylindrical in shape, yellow and greenish-brown in colour. The fruits are trivial, berry-like with limited seed numbers. The flowering and fruiting of this plant take place in July–August (Hotta, 1970; Amrita et al., 2018).

3. Geographical Distribution

A. calamus is semiaquatic in nature and occurs in moist habitats of swamps, banks of ponds, rivers, and streams and nurtures in both high (1800 meters) and low (900 meters) altitudes throughout Asia, Europe and North America (Sharma et al., 2014) and in 42 countries (Quraishi et al., 2017). The Global Biodiversity Information Facility (https://www.gbif.org.), records the distribution of Acorus in several zones of the globe as well as India (Fig. 2). Acorus habitats include wetlands of temperate and subtropical regions and are dispersed throughout tropical Asia and the north temperate hemisphere (Agarwal et al., 1956). In India, *A. calamus* is habituated in

Fig. 1 (a) Drawing of *Acorus calamus* L. (Source: Adapted from http://plantillustrations.org/illustration.php?id_illustration=13110/ *Kohler,F.E., Medizinal Pflanzen, vol. 1: t. 17 (1887) [W. Müller]*) (b) Flowering plant (c) Inflorescence (d) Rhizome and roots (e) Seeds (f) Dried Root (f) ((b-f) *Source:* Adapted from https://tropical.theferns.info/image.php?id=Acorus+calamus).

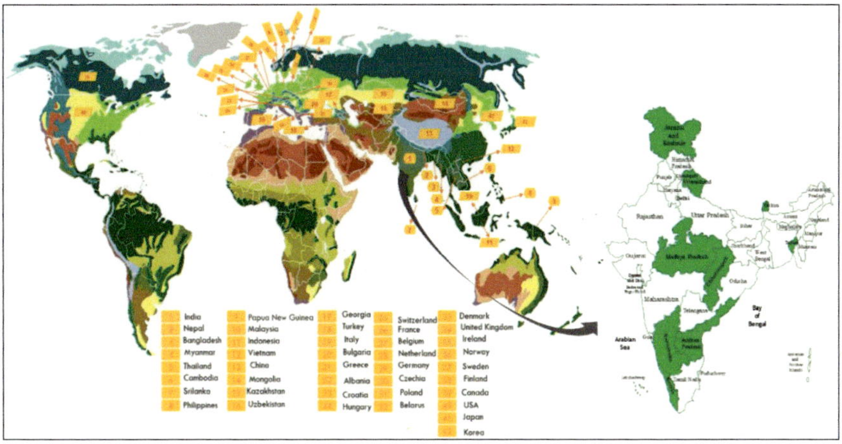

Fig. 2 Distribution of *A. calamus* in India and worldwide (Source: Sharma et al., 2020).

the marshes, and wild or cultivated Himalayan lands at an altitude of 2200 meters (Raina et al., 2003).

4. Phytochemical Constituents and Metabolite Profiling

Scientists have been investigating the phytochemicals of *Acorus calamus* since 1957. Around 145 compounds belonging to different groups were reported in rhizomes and leaves. The pharmacologically effective compounds include phenylpropanoids, triterpenoid saponins, sterols, monoterpenes alkaloids, triterpene glycosides, sesquiterpenoids, phenylpropanoids (asarone and eugenol) and sesquiterpenoids (Parcha et al., 2022). Various researchers of different countries reported arenes or aromatic compounds, sesquiterpenes, phyto-steroids, and aetheroleum or essential oils in Acorus (Bhattacharya and Kumar 2011). The α- asarone and β-asarones are the major bioactive chemical compounds in *Acorus* spp. (Bjomstad et al., 2009; Lam et al., 2016). Rhizomes and leaves of *A. calamus* in different global geographical zones revealed different chemical constituents. A few examples of these are summarized in Table 1.

Table 1 Location-wise Distribution of Major Bioactive Compounds in *A. calamus.*

Geographical Region	*Major compounds in A. calamus rhizomes*	*Major compounds in A. calamus leaves*	*References*
Himalayan region of India	β-asarone and α-asarone, caryophyllene, iso-asarone, methyl isoeugenol, safrole	b-asarone and linalool	(Raina et al., 2003)
Italy	Acorenone, β-asarone, dehydroxy-isocalamendiol, isoacorone, (Z)-sesquilavandulol, isoacorone (isoacoron), crypto-acorone		(Marongiu et al., 2005)
Lithuania		Phenolic constituents of (Z)-asarone, α-humulene, (E)-caryophyllene, (Z)-methyl isoeugenol, germacrene, linalool, camphor and isoborneol	(Radusiene et al., 2007)
Konya Province, Turkey	Camphor, pre-isocalamendiol, 1,4-(trans)-1,7(trans)-acorenone, isoshyobunone, 2,6-di-epi-shyobunone and b-gurjunene		(Ozcan et al., 2002)
Northern Kazakhstan and southern West Siberia	Acorafuran		(Tkachev et al., 2006)
Canada	Pre-isocalamenediol, acorenone, shyobunone and cryptoacorone		(Garneau et al., 2008)

Contd.

Table 1 *Contd.*

Geographical Region	*Major compounds in A. calamus rhizomes*	*Major compounds in A. calamus leaves*	*References*
Poland	(-)-Acorenone, shyobunone isomer sesquiterpenoid, (Z)-asarone phenylpropanoids		(Wilczewska et al., 2008)
Indian specimen	Asarone (82%), calamenol (5%), calamine (4%), calameone (1%), methyl eugenol (1%), eugenol (0.3%), acorin and acoretin		(Koul et al., 1989)

Tong et al. (2010) reported tatarinowin-A (cadinane-type sesquiterpenoid), tatarinoids A and B (phenylpropanoids), and tatarinoid C (trinorlignan) and other 15 recognized bioactive compounds in *A. calamus* whereas, Bains et al. (2005) reported lectins. The root aryl aldehyde and leaf asarone aromatic constituents give aroma to volatile oils (Venskutonis and Dagilyte, 2011). Calamenone (Tricyclic sesquiterpene), calamendiol, and isocalamendiol (two known sesquiterpenes) from Acorus roots and terpinene-γ, terpinolene, a- and b-pinenes, phellandrene-β, terpinen-α, myrcene, para cymene, thujane, and limonene monoterpenes were isolated from the plant volatile oils (Wu et al., 1994). Total phenolics and flavonoids in different leaf extracts revealed a major content of phenolics and flavonoids in the n-butanol extracts. Acoradin lignan was isolated from the Acorus rhizome (Singh, 2012). From *A. calamus* rhizome ethanol extract, a novel steroid glycoside known as α-sitosterol plus six other steroids were reported (Dong et al., 2008) and two new sesquiterpenes, cadinane with a propan-2-ylidene were reported in rhizomes (Dong et al., 2010). Sharma et al. (2020) isolated 145 compounds/constituents from different plant parts such as rhizomes, leaves, and roots, and elucidated their nature as phenylpropanoids, sterols, triterpenoid saponins, triterpene glycosides, monoterpenes, sesquiterpenoids, and alkaloids and also their chemical structures were elucidated (Table 2; Fig. 3). Among these asarone and eugenol phenylpropanoids and sesquiterpenoids are principal effective compounds (Sharma et al., 2020).

5. Differences in Phytoconstituents of *A. Calamus* at Varietal Levels

The beta-asarone in *Acorus* essential oil fluctuates with the polyploidy levels of various cytotypes (individuals with the same ploidy levels), sub-varieties, and/or biological species of *Acorus calamus*. Beta-asarone (90–96%) phenylpropanoid is abundant in Acorus tetraploid varieties. In the tetraploid cytotype of *Acorus calamus* var. *angustata* ENGER, approximately 80% of beta-asarone phenylpropanoid is

Table 2 Names, plant part sources, and classification of important 145 bioactive compounds/metabolites isolated from various parts of of A. calamus. (*Source:* Sharma et al., 2020)

Classification of Compounds	*Compound No.*	*Chemical Constituents*	*Plant Parts*
Phenylpropanoids	**1**	Alpha (α) (or trans) -Asarone	Rhizomes
	2	Beta (β) (or cis) -Asarone	
	3	Gamma (γ)-Asarone	
	4	Fragrance allergen (Eugenyl acetate)	Rhizomes
	5	Eugenol phenolic molecule	
	6	Isoeugenol (phenylpropanoid)	
	7	Methyl eugenol	Rhizomes
	8	Methyl isoeugenol	Rhizomes
	9	Calamol	Rhizomes
	10	Azulene (Azunamic or Cyclopentacycloheptene)	
	11	Methyl eugenol (Eugenol methyl ether)	
	12	Dipentene (D-limonene)	
	13	Asaronaldehyde (Asarylaldehyde)	
	14	Terpinolene (10-carbon monoterpene)	
	15	1,8-cineole (eucalyptol)	
	16	(*E*)-isoeugenol acetate (acetyl trans-isoeugenol)	
	17	(*E*)-methyl isoeugenol (MIE)	
	18	Cis-methyl isoeugenol	Rhizomes
	19	Euasarone ($C_{12}H_{16}O_3$)	
	20	Cinnamaldehyde (C_9H_8O)	
	21	Cyclohexanone	Rhizomes
	22	Acorin	Rhizomes
	23	Isoasarone	
	24	Safrole (4-allyl-1,2-methylenedioxy-benzene)	
	25	Z-3-(2,4,5-trimethoxyphenyl)-2-propenal ($C_{12}H_{14}O_4$)	Rhizomes
	26	2,3-dihydro-4,5,7-trimethoxy-1-ethyl-2-methyl-3 (2,4,5-trimethoxyphenyl) indene ($C_{24}H_{32}O_6$)	
	27	(Z)-asarone ($C_{12}H_{16}O_3$)	Leaves
	28	(E)-caryophyllene ($C_{15}H_{24}$)	
	29	Estragole (p-allylanisole, methyl chavicol)	Rhizomes

Contd.

Table 2 *Contd.*

Classification of Compounds	*Compound No.*	*Chemical Constituents*	*Plant Parts*
	30	Carvacrol (cymophenol)	
	31	2-cyclohexane-1-one (C6H8O)	
	32	Naphthalene ($C_{10}H_{8)}$	
	33	γ-Cadinene (gamma-Cadinene)	
	34	Aristolene (Aristol-9-ene)	
	35	1(5),3-aromadenedradiene	
	36	5-n-butyltetraline ($C_{14}H_{20}$)	
	37	4,5-dehydro-isolongifolene ($C_{15}H_{22}$)	
	38	Calarene (1(10)-Aristolene) ($C_{15}H_{24}$)	
	39	Isohomogenol (Trans-Methyl Isoeugenol)	
	40	Zingiberene ($C_{15}H_{24)}$	
	41	alpha-Calacorene (α-Calacorene)	
	42	5,8-dimethyl isoquinoline ($C_{11}H_{11}N$)	
	43	Cyclohexane methanol (Cyclohexylmethanol)	
	44	Longifolene (Junipene)	
	45	Isoelemicin ($C_{12}H_{16}O_3$)	
Sesquiterpenoids	**46**	Calamene (C_{15} H_{22})	
	47	Calamenenol	
	48	Calameone	
	49	Preisocalamendiol ($C_{15}H_{24}O$)	
	50	1,4-(trans)1,7(trans)-acorenone	
	51	1,4-(cis)-1,7-(trans)-acorenone	
	52	2,6 diepishyobunone ($C_{15}H_{24}O$)	
	53	α-Gurjunene ((-)-alpha-Gurjunene)	
	54	β-Gurjunene (beta-Gurjunene)	
	55	α-Cedrene ((-)-alpha-cedrene)	
	56	β-Elemene ((-)-beta-Elemene)	
	57	β-Cedrene ((+)-beta-Cedrene)	
	58	β-Caryophyllene (Caryophyllene)	
	59	Valencene ((+)-Valencene)	
	60	Viridiflorene ((+)-Ledene)	
	61	α-Selinene (Eudesma-3,11-diene)	
	62	δ-Cadinene ((+)-delta-Cadinene)	
	63	α-Curcumene (alpha-Curcumene)	

Contd.

Table 2 *Contd.*

Classification of Compounds	*Compound No.*	*Chemical Constituents*	*Plant Parts*
	64	Shyobunone ((+)-Shyobunone)	
	65	Isoshyobunone	
	66	Caryophyllene oxide (beta-Caryophyllene oxide)	
	67	Humulene oxide II (Humulene 6,7-epoxide)	
	68	Elemol ($C_{15}H_{26}O$)	
	69	Cedrol ((+)-Cedrol)	
	70	Spathulenol (espatulenol)	
	71	Acorenone ((-)-Acorenone)	
	72	α-Cadinol (. alpha. -Cadinol)	
	73	Humulene epoxide II ((-)-Humulene epoxide II)	
	74	α-Bisabolol (alpha-Bisabolol)	
	75	Asaronaldehyde (Asaraldehyde)	Rhizomes
	76	Calamusenone ($C_{15}H_{22}O$)	Rhizomes
	77	Isocalamendiol (Isocalamenediol)	
	78	Dehydroxyiso-calamendiol	
	79	Epishyobunone	
	80	Acorone (Isoacorone)	Rhizomes
	81	Neo-acorane A ($C_{15}H_{22}O_4$)	Rhizomes
	82	Acoric acid ($C_{15}H_{24}O_4$)	
	83	Calamusin D	
	84	1β,5α-Guaiane-4β,10α-diol-6-one	
	85	(+)-Dioxosarcoguaiacol (Dioxosarcoguaiacol)	Rhizomes
	86	7-tetracycloundecanol,4,4,11,11-tetramethyl	Rhizomes
	87	4α,7-Methano-4α-naphth[1,8a-b] oxirene	
	88	Espatulenol (Spathulenol)	Rhizomes
	89	Vulgarol B ($C_{15}H_{24}O$)	
	90	Tatanan A ($C_{36}H_{48}O_9$)	Rhizomes
	91	Acoramone (1-(2,4,5-trimethoxyphenyl) propan-2-one)	
	92	2-hydroxyacorenone (2-Acetylphenol)	
	93	4-(2-formyl-5-methoxymethyl pyrrol-1-yl) butyric acid methyl ester	

Contd.

Table 2 *Contd.*

Classification of Compounds	*Compound No.*	*Chemical Constituents*	*Plant Parts*
	94	2-acetoxyacorenone	
	95	Acoramol ((Z)-3-(2,4,5-trimethoxyphenyl) prop-2-en-1-ol)	
	96	N-transferuloyl tyramine (Moupinamide)	
	97	Tatarinoid A (Tatariid A)	
	98	Tatarinoid B (2-Propanone)	
	99	Acortatarin A (pollenopyrroside B)	
Monoterpenes	**100**	α-Pinene	Rhizomes, Roots
	101	β-Pinene	
	102	Camphene	
	103	o-Cymol	
	104	p-Cymene	
	105	gamma -terpinene	
	106	alpha-Terpinolene	
	107	trans-Anethole	
	108	Thyme camphor	
	109	Isoaromadendrene epoxide	
	110	2-Camphanone	Rhizomes, Leaves, Roots
	111	Sabinene	Roots
	112	trans-2-Hexenal	
	113	Cajeputene	
	114	Cis-linaloloxide (carbamate)	
	115	Cis-sabinene hydrate (4-Thujanol)	
	116	Trans-linalool oxide (trans-Linalool 3,6-oxide)	
	117	Linalool (3,7-Dimethylocta-1,6-dien-3-ol)	
	118	4-Carvomenthenol	
	119	α-Acoradiene	
	120	beta-Acoradiene	
	121	α-Terpineol	
	122	Isocamphol	Leaves
Xanthone glycosides	**123**	4,5,8-trimethoxy-xanthone-2-*O*-β-D-glucopyranosyl (1-2)-*O*-β-D-galactopyranoside	Rhizomes

Contd.

Table 2 *Contd.*

Classification of Compounds	*Compound No.*	*Chemical Constituents*	*Plant Parts*
Triterpenoid saponins	**124**	1β,2α,3β, 19α-Tetrahydroxyurs-12-en-28-oic acid-28-O- {(β-D-glucopyranosyl (1-2)}-β-D galactopyranoside	Root, Rhizomes
	125	3-β, 22-α-24,29-Tetrahydroxyolean-12-en-3-O-(β-Darabinosyl (1,3)}-β-D-arabinopyranoside	
Alkaloids	**126**	Trimethoxyamphetamine,2,3,5 (TMA-4, (R)-)	
	127	Pyrimidin-2-one,4-[N-methylureido]-1-[4methyl amino carbonloxy methy]	
Triterpene glycoside	**128**	22-[(6-deoxy-α-L-rhamnopyranosyl) oxy]-3,23-dihydroxy-, methyl ester, (3β,4β,20α,22β)	
Steroids/Sterols	**129**	β-daucosterol (Sitogluside)	
Amino acids	**130**	Arginine	Roots
	131	Lysine	
	132	Phenylalanine	
	133	Threonine	
	134	Tryptophan (h-Trp-oh)	
	135	α-alanine (L-alpha-alanine)	
	136	Asparagine (Aspartamic acid)	
	137	Aspartic acid (H-Asp-OH)	
	138	Norvaline ((S)-2-Aminopentanoic acid)	
	139	Proline ((2S)-pyrrolidine-2-carboxylic acid)	
	140	Tyrosine (p-Tyrosine)	
	141	Glutamic acid (L-glutamate)	
Fatty acids	**142**	Palmitic acid (palmitate)	Rhizomes
	143	Myristic acid (Tetradecanoic acid)	
	144	Palmitoleic acid ((Z)-Hexadec-9-enoic acid)	
	145	Stearic acid (Octadecanoic acid)	

present. In the triploid cytotype of *Acorus calamus* var. *calamus* L., 5% beta-asarone phenylpropanoid occurs in oil, whereas diploid cytotype of *Acorus calamus* var. *americanus* WULFF completely lacks beta-Asarone phenylpropanoid however, have higher monoterpenoid geranyl acetate content (Wagner et al., 2011; Rajput et al., 2014; Joshi, 2016).

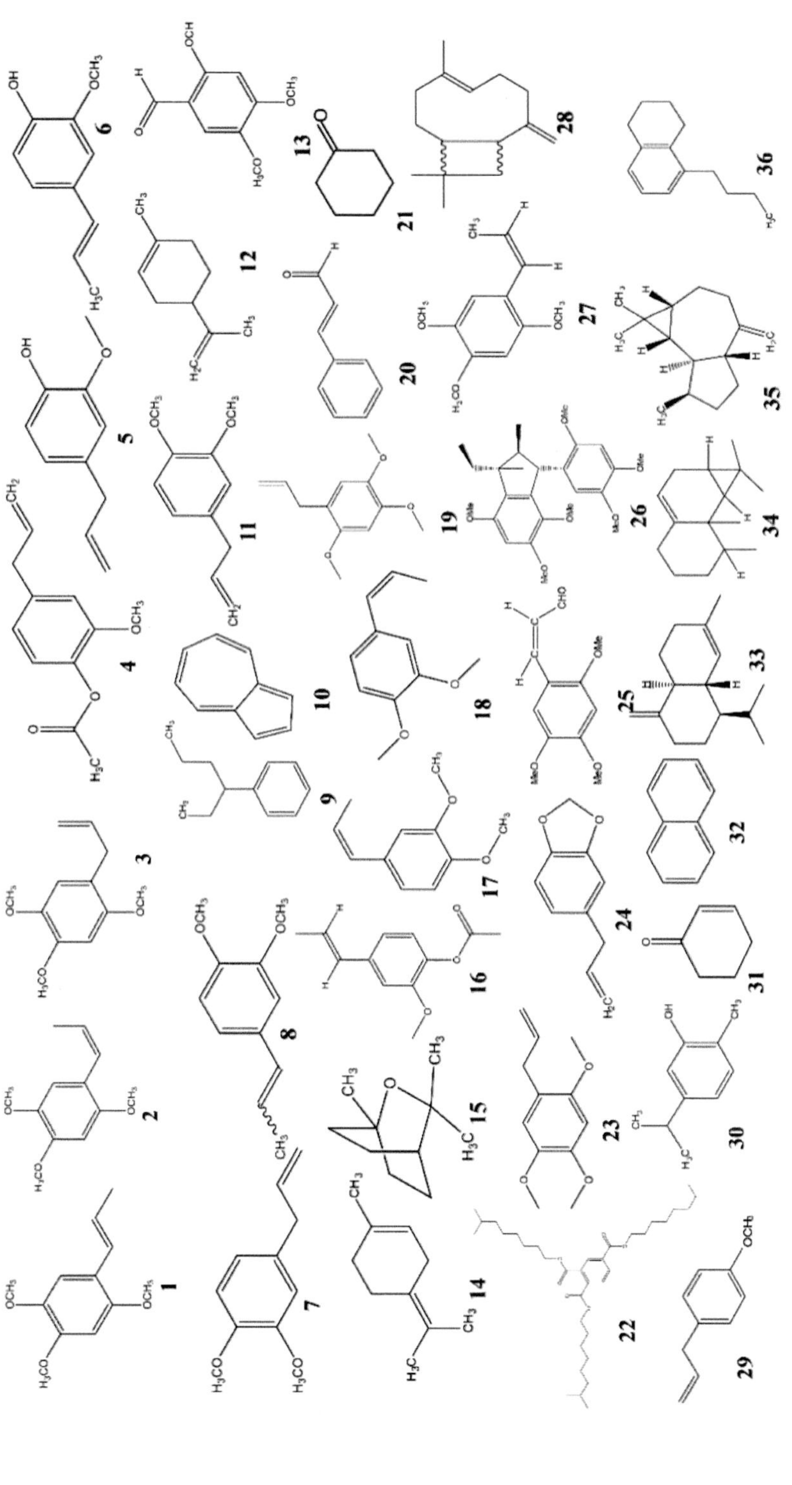

Fig. 3 *Contd.*

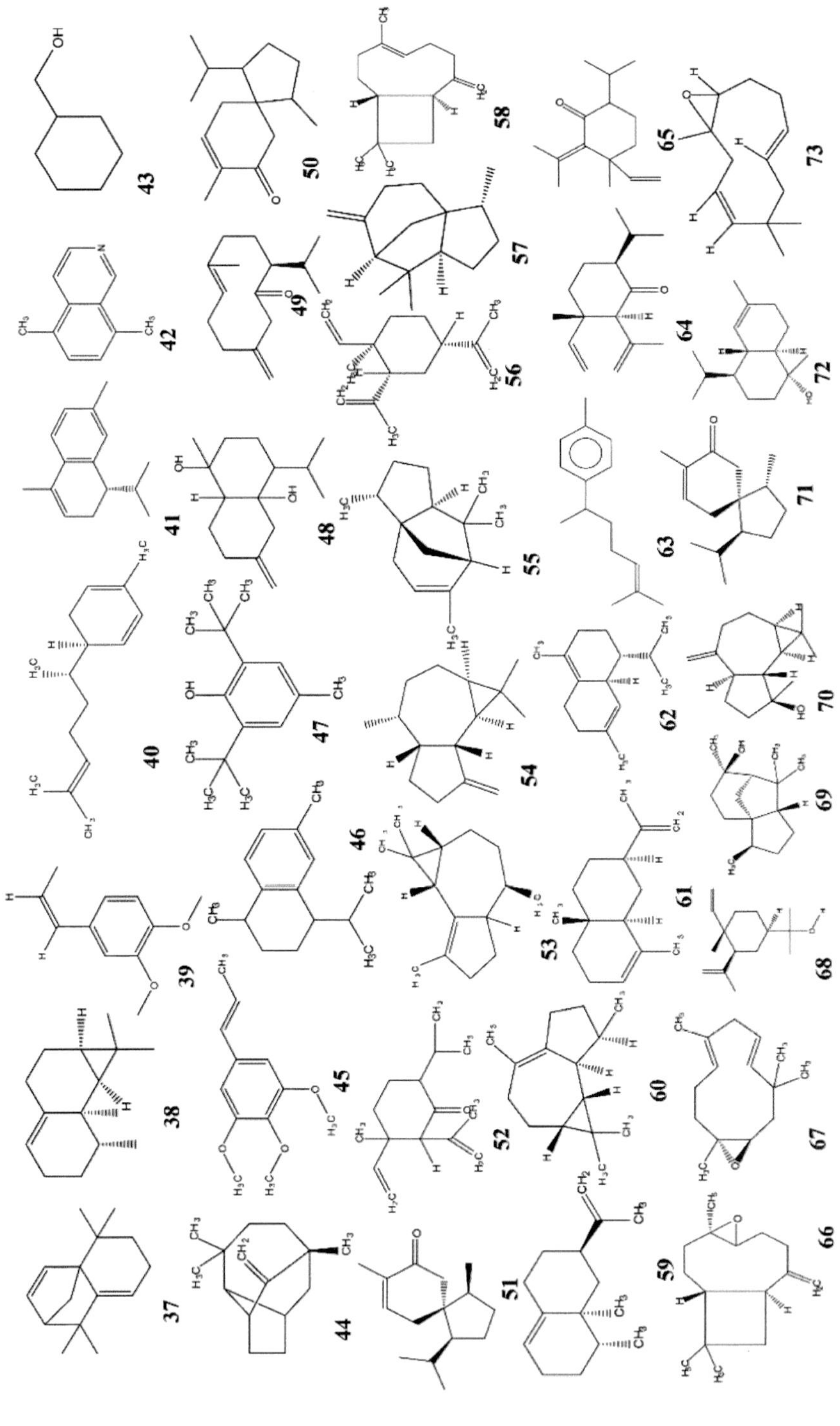
37
38
39
40
41
42
43
44
45
46
47
48
49
50
51
52
53
54
55
56
57
58
59
60
61
62
63
64
65
66
67
68
69
70
71
72
73

Fig. 3 *Contd.*

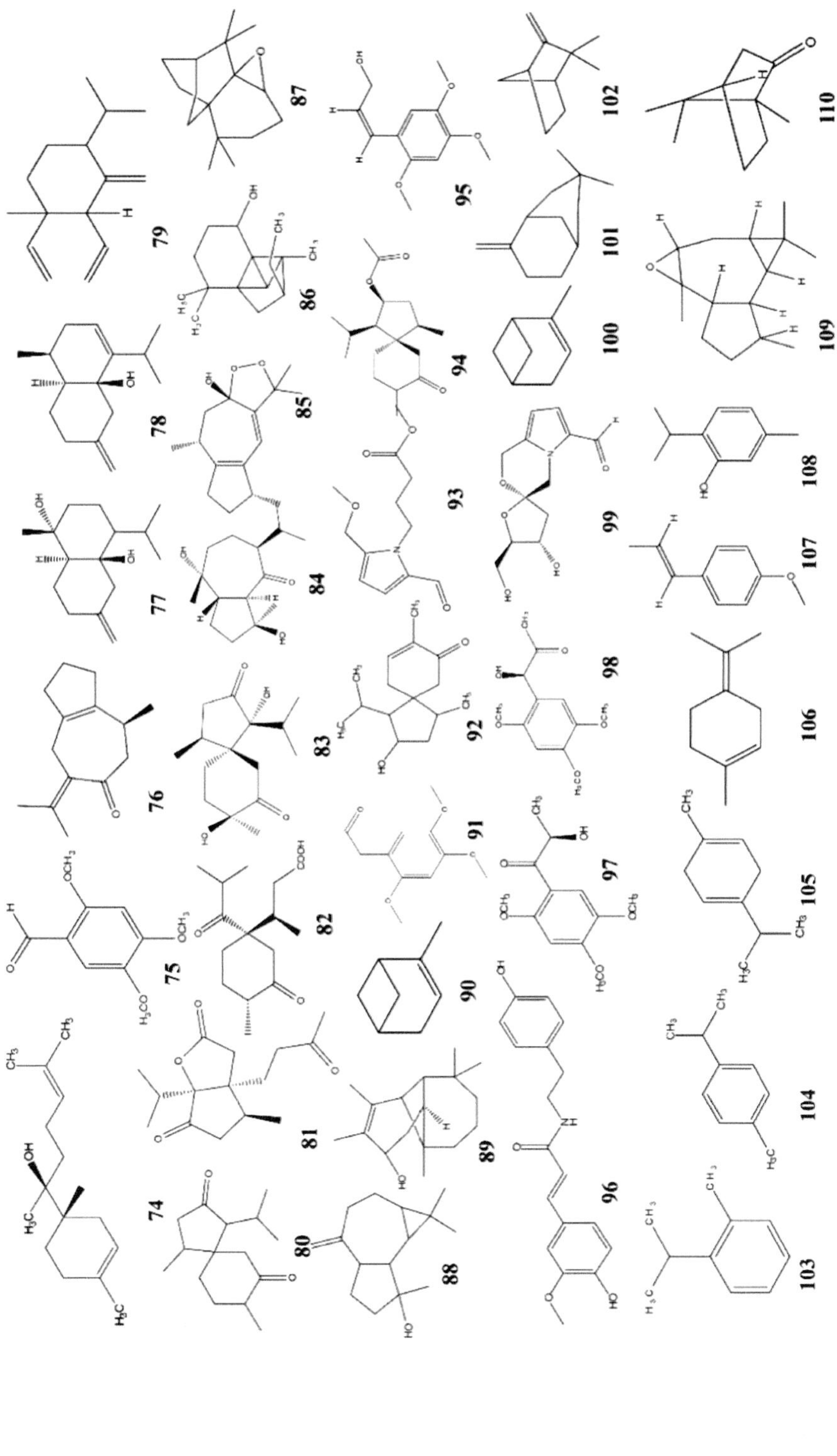

Fig. 3 *Contd.*

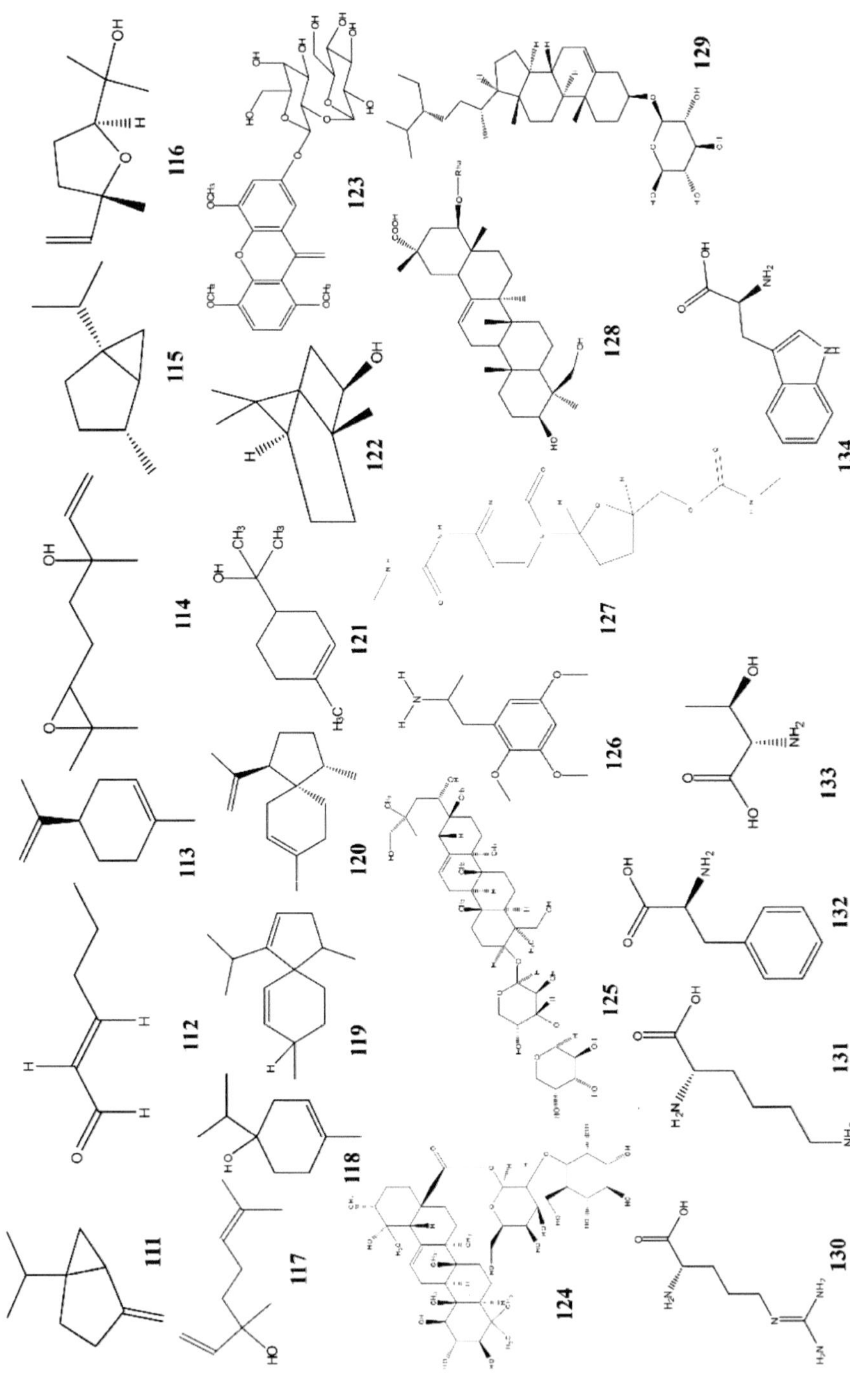

111
112
113
114
115
116
117
118
119
120
121
122
123
124
125
126
127
128
129
130
131
132
133
134

Fig. 3 *Contd.*

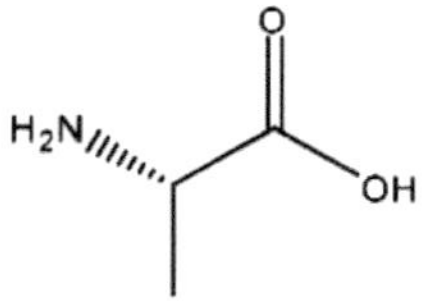

135

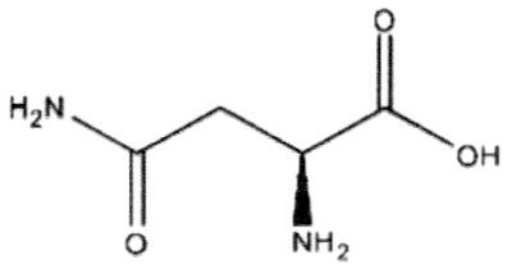

136

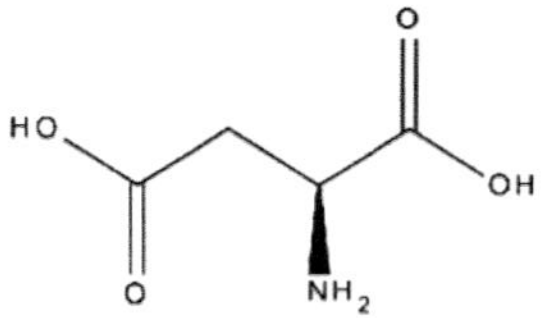

137

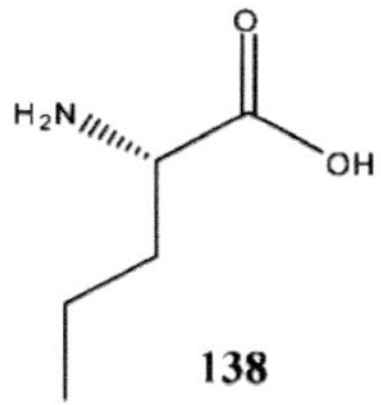

138

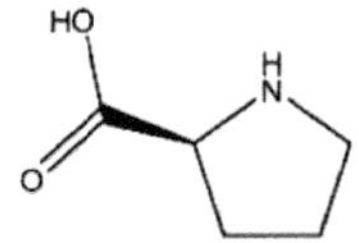

139

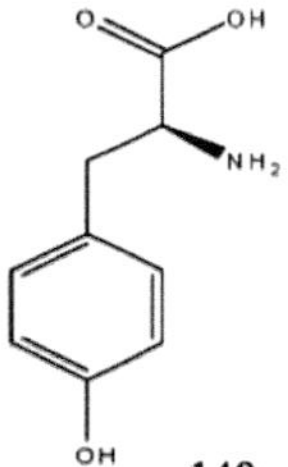

140

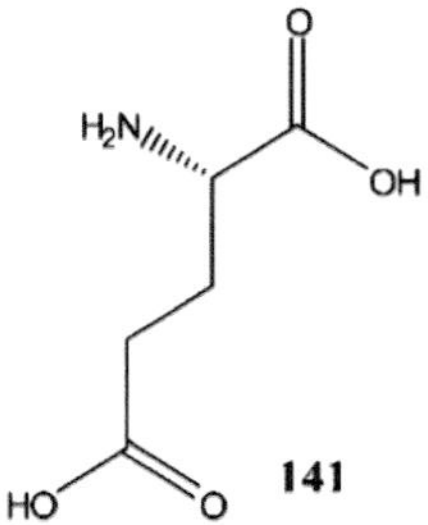

141

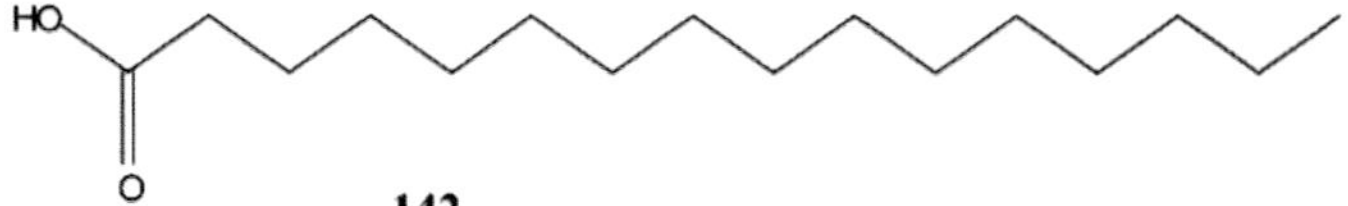

142

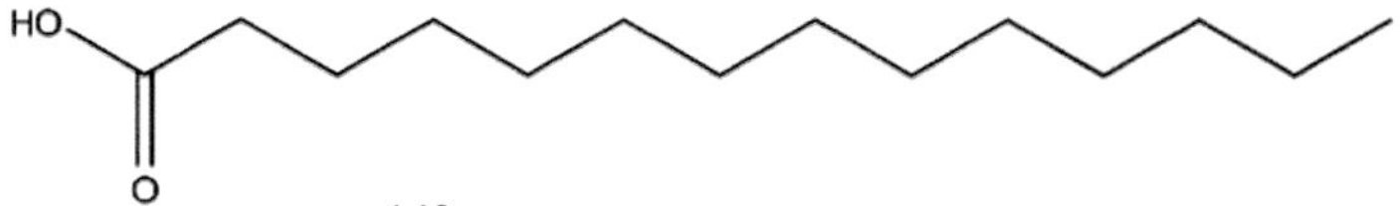

143

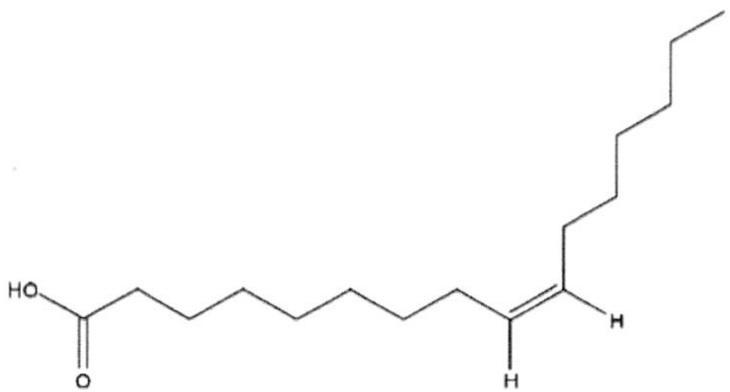

144

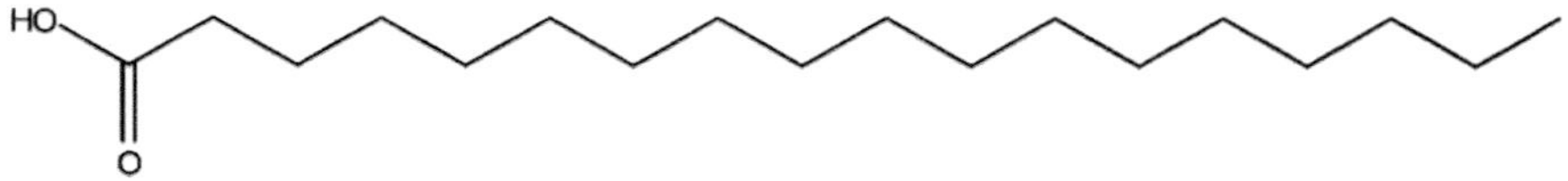

145

Fig. 3 Chemical structures of secondary metabolites (1-145) reported from various parts of *A. calamus* claimed for diverse biological activities

6. Therapeutic Potential of *A. Calamus*

6.1 As per Pharmacopoeias and Traditional Systems of Medicine

A. calamus rhizomes are used expansively by Chinese, Indian, American, and other countries (Phongpaichit et al., 2005). In China, people used Acorus against swellings, constipation, ear-related, gastrointestinal, respiratory, and memory disorders, haemorrhage, cancer, hallucinations, and as a neuroprotective, antipyretic, analgesic, aphrodisiac, contraceptive, in detoxification, increasing learning performance, anti-aging effect in senescence and to gain fair skin (Agarwal et al., 1956; Wallnofer and Rottauscher, 1965; Perry and Metzger, 1980; Boissya and Majumder, 1980; Duke and Ayensu, 1985; Dragendorff, 1898; Shih-Chen, 1973; Li, 1977; Zhang et al., 1994; Nishiyama et al., 1994 a, b). Roots and rhizomes of *A. calamus* have been used in the Indian systems of traditional medicine for centuries. In the Ayurvedic system of medicine, *Acorus* is used for insomnia sleep disorders, melancholia, remittent fevers, neurosis, epilepsy, hysteria, and memory loss (Kirtikar and Basu 1956; Singh and Arora 1956). In India, *A. calamus* rhizomes find usage in curing ailments like lung conditions (asthma and bronchitis), cough, cold, eczema, epilepsy, stuttering, ulcer, scab, sores, dermatitis, dysentery, diarrhoea, mental illnesses, gastrointestinal, kidney and liver disorders, wounds, fever, body pain, stomach ache, otitis externa, cancer, respiratory, emmenagogue, anthelmintic, paralysis, arthritis, herpangina, skin, respiratory, neurological, gastrointestinal disorder, increase menstrual flow and used as stimulant, analgesic, and contraceptive (Dastur 1951; Usher 1984; Shrestha and Dhillion 2003; Sharma et al., 2004; Dwivedi et al., 2006; Kingston et al., 2007; Pradhan and Badola, 2008; Ragupathy et al, 2008; Kingston et al., 2009; Tomar 2009; Poonam and Singh 2009; Barbhuiya et al., 2009; Balakumbahan et al., 2010; Nisha and Rajeshkumar 2010; Rajith and Ramachandran, 2010; Boktapa and Sharma, 2010; Sher et al., 2011; Khatun et al., 2011; Yabesh et al., 2014). Acorus is a brain and nervous system rejuvenator and is also used in cure of skin diseases, bronchitis, general debility, cough, anxiety, depression, fever, inflammation, mental fatigue, tumours, sinusitis, haemorrhoids, memory loss, flatulence, numbness, tension headache, joint pains and as poisoning antidotes (Panchal et al., 1989; Vaidyaratnam, 1994; Nadkarni, 1998; Sharma et al., 2004, Sharma and Arora, 2006). *A. calamus* is also defined in the Indian Siddha medicinal system for paediatric practices. Rhizome paste finds usage in the rural south India for improving child speech and memory (Meena et al., 2010). The rhizome is an expectorant, emmenagogue, bitter tonic, antispasmodic, emetic, and laxative, increases catabolic rate, and is used in curing kidney and liver troubles, chronic diarrhoea, dysentery, epilepsy, mental ailments, fever, abdominal tumours, and rheumatism (Arseculeratne et al., 1985; Sharma et al., 2004, Sharma and Arora, 2006). Mature leaves of Acorus behave as insect repellents, insecticides, antifungal and antibacterial agents. Dried *A. calamus* rhizomes comprise yellow aromatic volatile oil with flavones, asarone, choline, zinc, methanol, ethanol, camphor, and eugenol (Chauhan, 1999).

In the United States of America and Indonesia, Acorus finds use in traditional folk medicine as an abortifacient, stimulant, and tonic, in respiratory disorders and gastrointestinal disorders (Wren, 1956; Dobelis, 1986; Gilani et al., 2006; Si et al., 2010).

In Pakistan, *A. calamus* is used as medicine for fever, stomach complaints, colic and diarrhoea, and snakebites, and as an insect repellent (Satyavati et al., 1976; Tariq et al., 2010). In England, *A. calamus* finds use in gastrointestinal, neurological, dysentery, and chronic catarrh and as an antibacterial and analgesic (Grieve, 1971), and in Malaysia it is used against rheumatism, diarrhoea, dyspepsia, and hair loss (Kirtikar and Basu, 1975).

6.2 Biological Effects of Extracts of A. calamus

A. calamus is a medicinal and aromatic herb with varied pharmacological assets, for instance, analgesic, anti-inflammatory, immunomodulatory and antipyretic (Derle and Gujar 2001; Karthiga et al., 2016), anti-ulcer, anti-secretagogue, and cytoprotective (Rafatullh et al., 1994), antidiarrheal (Shoba and Thomas, 2001), neurotoxicity (Shukla et al., 2002), hypolipidemic (Parab and Mengi, 2002), anti-cellular and immunosuppressive (Mehrotra et al., 2003), anti-cholinesterase (Oh et al., 2004), antimicrobial (Kim et al., 2011; Devi and Ganjewala, 2009), antimutagenic (Aqil et al., 2008), larvicidal (larvicide) (Suryadevara and Khanam, 2002), antidepressant (Shashikala et al., 2015) and anticonvulsant effects (Kaushik et al., 2017), and prove effective in neurological and metabolic disorders (Tiwari et al., 2014; Sharma et al., 2020). The rhizome ethanolic extracts inhibit the proliferation of phytohemagglutinin mitogen and purified protein derivative antigen stimulated human peripheral blood mononuclear cells (PBMCs). The rhizome extracts repressed the growth of several cell lines (mouse and human) and production of nitric oxide, interleukin-2 (IL-2), and Tumor Necrosis Factor-α (TNF-α). *A. calamus* extracts regulated expression of Cluster of Differentiation 25 (CD25) demonstrating antiproliferative and immunosuppressive potential (Mehrothra et al., 2003).

6.3 Pharmacological Potential as an Antiviral agent

The rhizome alcohol extract showed effective antiviral action against Herpes Simplex Virus ((HSV) 1 and 2 (Badam 1995). Vero cells on pretreatment with extract inhibited replication of the HSV-1 and HSV-2 showing that the host cells remain. The phenylpropanoid β-asarone inhibits the replication of both viral types. The hexane extracts were demonstrated for the anti-HIV reverse transcriptase effect (Silprasit et al., 2011). Both, crude alcohol extract and *β*-asarone phenylpropanoid were toxic to the host cells. The ethanol extract of the root of *A. calamus* was demonstrated for its antiviral properties on the DENV2 (Dengue) virus using the mosquito larva C6/36 cell line as replication and transfection host while monkey kidney fibroblast cells (BHK-21) line has served as a target of infection. The inhibitory capacity of root extract was tested by standardized protocols such as cytopathic effect and plaque assay (Yao et al., 2018). Yao et al. (2018), studied the effect of root extracts and isolated twelve phytoconstituents on the proliferation and inhibition properties using *in vitro* cell system.

6.3.1 Anti-DENV2 outcome of root extracts of A. calamus

The root ethanol extracts showed an anti-DENV2 effect in a concentration dependent manner. Among the twelve phyto-constituents, Tatanan A showed a significant decrease in DENV2-induced LDH, better than the standard drug such as ribavirin (Yao et al., 2018). Tatanan A has reported the inhibition of viral RNA synthesis indicating the protection against the DENV2 (Yao et al., 2018). Saptawati et al. (2017) demonstrated that twelve compounds of Vacha affect DENV2 at lower concentrations. Rosamalena et al. (2019) investigated the effect of Tatanan A on the DENV2 virus (*in vitro*) and explained that Tatanin A, affected NS5 (non-structural proteins domain 5) of methyl transferase and other key viral proteins. Further, Tatanan A also affected viral infection, as evidenced in immunofluorescent studies. They also demonstrated that the root extracts inhibited viral replication in a dose-dependent manner. Bioactive compounds of Vaccha were tested using molecular docking studies of viral key proteins. The interaction and binding nature of phyto-compounds (Fig. 4) revealed their potential inhibition of viral replication (Rosamalena et al., 2019). The interaction of components was indicated with specific colors as shown in Fig. (4) and the binding interaction between NS5 protein with asarone, myristicin calamusin D, acoric acid, homoegonol, geranial, geranyl acetate, artesunic acid, and geraniol revealed their effectiveness (Rosmalena et al., 2019).

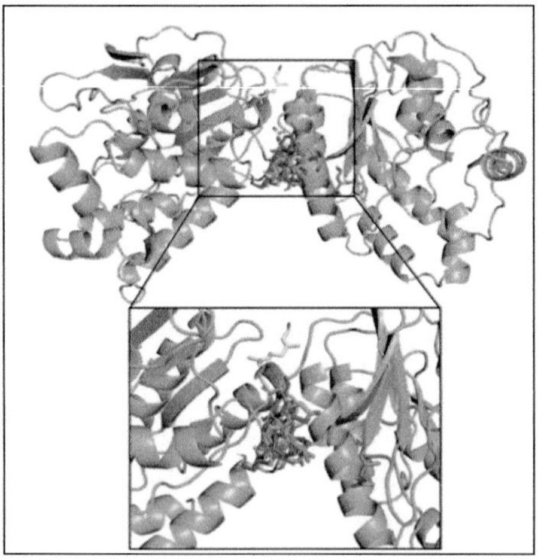

Fig. 4 Molecular Docking - Binding interaction between dengue virus NS5 protein with - calamusin D (magenta), artesunic acid (cyan), homoegonol (blue), geraniol (tints-wheat), acoric acid (light green), asarone (dark green), geranial (yellow), geranyl acetate (white), myristicin (orange). (*Source:* Rosmalena et al., 2019)

The *in-silico* studies of Tribudi et al. (2022), demonstrated binding with helicase (a key enzyme of replication) and inhibition of NS5 activity of Flavivirus. In this, connection the bioactive compounds of *A. calamus* subjected to *in silico* drug interaction/binding studies revealed that the five major compounds such as acorone, acoric acid, acoronone, and calmendiol exhibited properties supporting the earlier findings (Tribudi et al., 2022). Additionally, these drugs' interactions with the aforementioned constituents demonstrated that the NS5 helicase-protease had eliminated the water molecules, resulting in a strong binding that demonstrated its inhibitory capabilities (Tribudi et al., 2022). Based on the binding properties (Fig. 5) of phytocompounds with helicase, the docking studies revealed that five major compounds are highly interactive and interfered with the binding sites able to block the viral replication of Dengue viral strains, and these studies suggested further processing to develop the medicaments against dengue viral infections (Tribudi et al., 2022).

6.3.2 How A. calamus active compounds behave against SARS-CoV-2 Viral Proteases?

The Severe Acute Respiratory Syndrome Coronavirus (SARS-CoV), Middle East respiratory syndrome coronavirus (MERS-CoV), and novel coronavirus (SARS-CoV-2) are the source of severe respiratory human diseases (Jiao et al., 2021; Huang et al., 2024). COVID-19 (Coronavirus disease 2019) triggered by the novel coronavirus SARS-CoV-2 has emerged as the greatest damaging communicable disease of the contemporary era due to constant viral mutations and new variants (Huang et al., 2024). From 24 isolated compounds of *A. calamus* L. (Fig. 6), 8 active components were identified exhibiting inhibitory effects on SARS-CoV-2 proteases i.e., papain-like protease (PLpro) and main protease (Mpro) which are essential for coronavirus replication (Báez-Santos et al., 2014; Huang et al., 2024). The active compound 1R,5R,7S-guaiane-4R,10R-diol-6-one (1) proved best viral protease inhibitory activity against SARS-CoV-2 PLpro whereas neo-tatarine (23), menecubebane B sesquiterpenoids (13) and neo-acorane A sesquiterpene (18) exhibited significant inhibitory effects on both main protease and papain-like protease of SARS-CoV-2 and may serve as lead compounds for drug designing against coronaviruses (Huang et al., 2024).

Molecular docking (Fig. 7) confirmed stable conformations of active compounds of *A. calamus* with two key target enzymes i.e., PLpro and Mpro of SARS-CoV-2 (Báez-Santos et al., 2014; Huang et al., 2024). The 3D and 2-D plots of the crystal structure of papain-like protease of SARS CoV-2 (6W9C3 chain structure) molecular docking with *A. calamus* compounds revealed that these compounds can thwart SARS-CoV-2 after binding with PLpro and Mpro targets with stable conformations, and therefore hold potential for future anti-CoV drug designing (Huang et al., 2024).

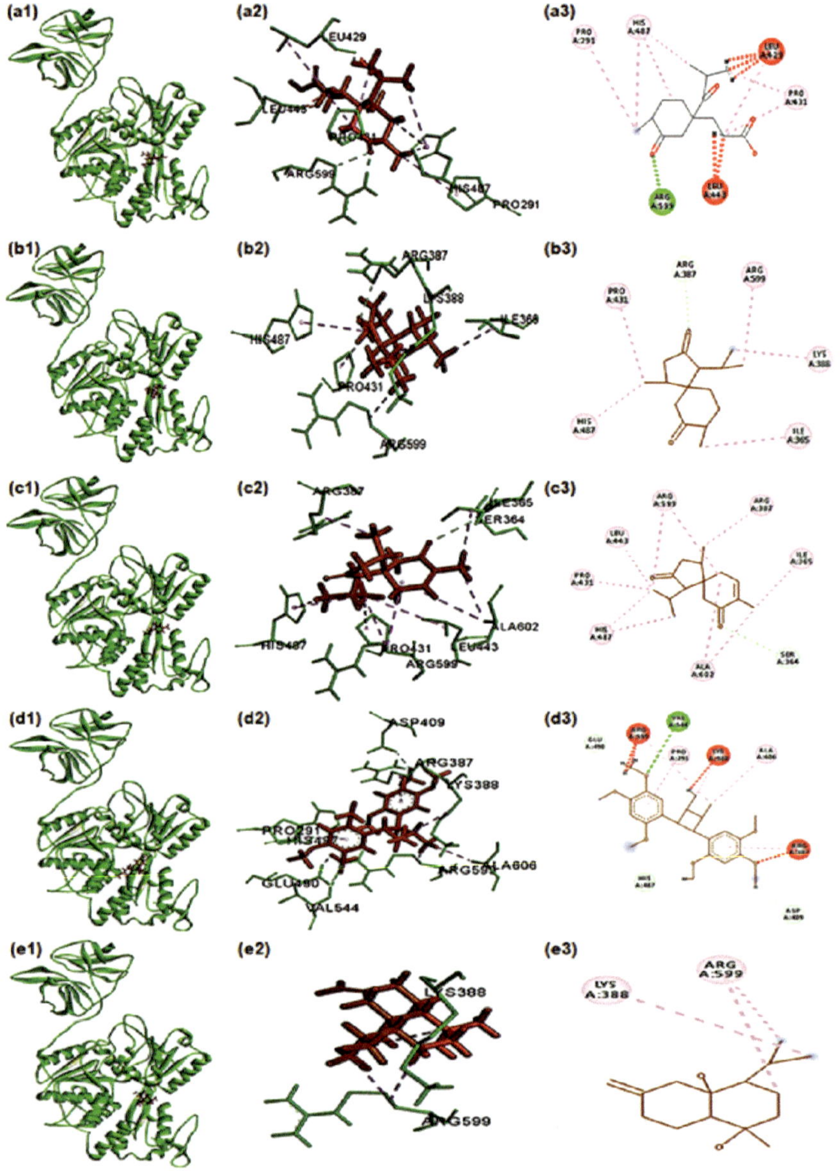

Fig. 5 Molecular Docking studies results from the interaction between (a) acoric acid, (b) acorone, (c) acoradin, (d) acoronene, and (e) calamendiol complexes with NS3 protease-helicase. Number 1 shows visualization of ligand-protein molecular complex. The 3D structure in 2 and 3 shows 2D structure (*Source:* Tribudi et al., 2022).

7. Toxicity and Safety

Goeggelmann and Schimmer (1983) suggested that the commercial samples of *A. calamus* should be devoid of β-asarone or at least concentration for human

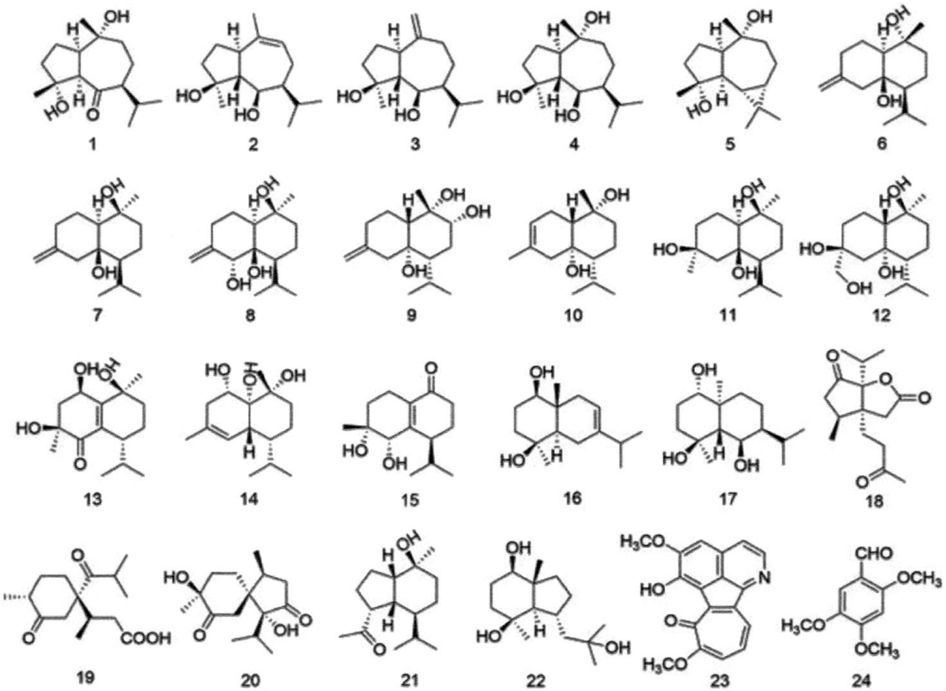

1. 1R,5R,7S-guaiane-4R,10R-diol-6-one
2. 4β,6β-dihydroxy-1α,5β(H)-guai-9-ene
3. 4β,6β-dihydroxy-1α,5β(H)-guai-10(14)-ene
4. Teuclatriol
5. (-)-alloaromadendrane-3β,9β-diol
6. Isocalamendiol
7. Calamendiol
8. Calamusin H
9. Calamentriol A
10. 1β,7β(H)-3-cadinadiene-6α,10α-diol
11. Calamentriol B
12. Calamentriol C
13. Menecubebane B
14. Calamentriol D
15. Oxyphyllenodiols A
16. Oplodiol
17. Ananosmin
18. Neo-acorane A
19. Acoric Acid
20. Calamusin D
21. Oplopanone A
22. Bullatantriol
23. Neo-tatarine
24. 2,4,5-trimethoxybenzaldehyde

Fig. 6 Structural formulae of 24 compounds isolated from *A. calamus* rhizomes for usage against SARS-CoV-2 (*Source:* Huang et al., 2024).

usage during phytotherapies. The extracts of Vaccha were reported to be mildly carcinogenic since β-asarone affected chromosomes of human lymphocytes in the *in vitro* studies. Further, β-asarone possesses a genotoxic effect and was suggested to be used at below-toxic levels (Abel, 1987). The Food and Drug Administration has banned it in the United States and Canada due to hazardous concerns (Motley, 1994). Singh et al. (2001) reported induction of cancer in tested animal systems due to beta-asarone. Even, alpha-asarone also reported a mutagenic effect on *Salmonella typhimurium* as similar to a mycotoxin (Jin et al., 1982; Mukherjee and Wahile, 2006). Further, it also reported ill effects such as gastroenteritis, disturbance in digestion, persistent constipation as well as diarrhoea, and the appearance of blood in faeces (Bhat et al., 2012). It is noticed that b-asarone occurs in triploid and tetraploid plant varieties. However, the American varieties lacking b-asarone in diploid plants are recommended for use in pharma industries (Bhat et al., 2012).

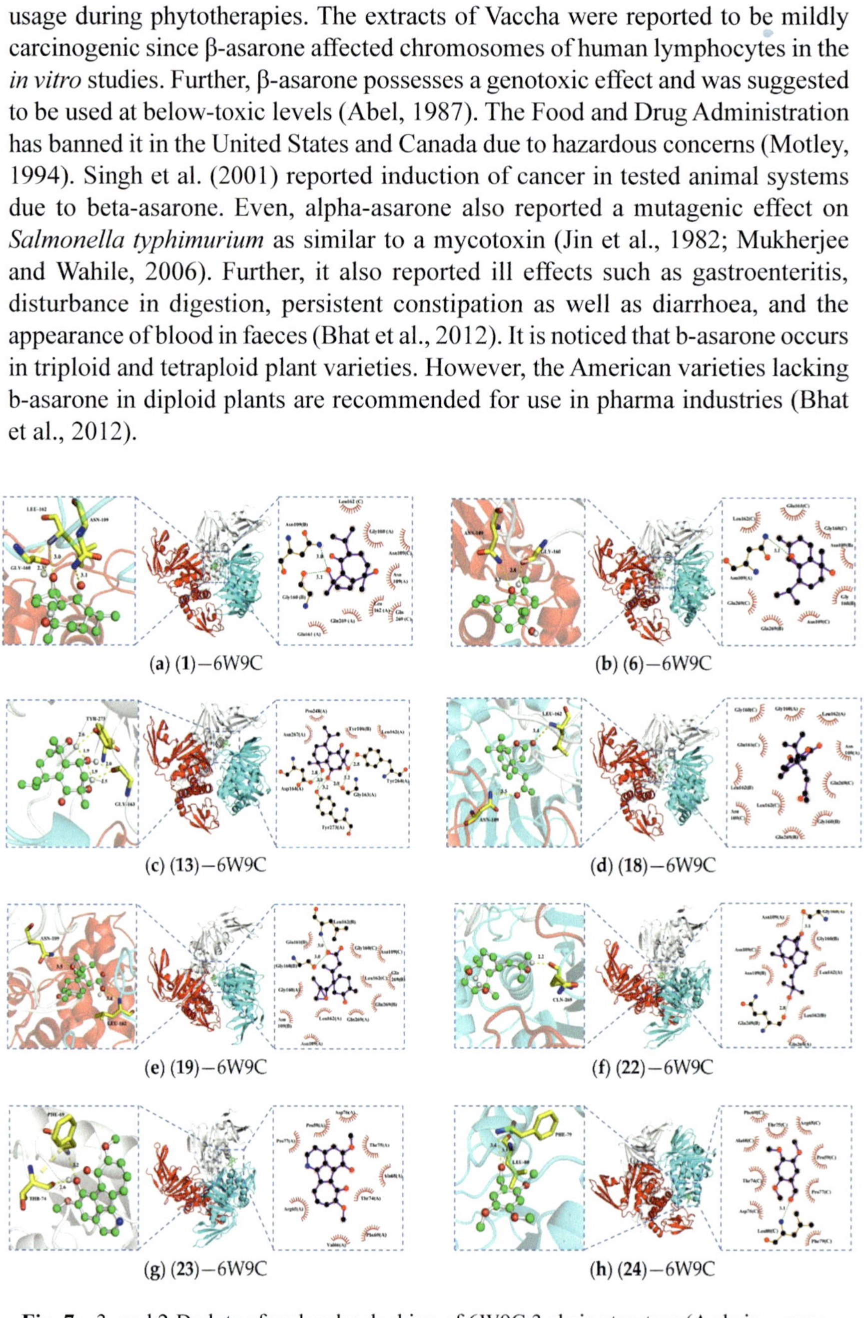

(a) (1)—6W9C (b) (6)—6W9C (c) (13)—6W9C (d) (18)—6W9C (e) (19)—6W9C (f) (22)—6W9C (g) (23)—6W9C (h) (24)—6W9C

Fig. 7 3- and 2-D plots of molecular docking of 6W9C 3 chain structure (A chain = gray, B chain = cyan, C chain = red) with active compounds (*Source:* Huang et al., 2024).

8. Conclusion and Future Prospectives

A. calamus finds use in traditional besides folk medicine against various diseases from times immemorial. It has also been proven for its curing abilities. However, it is also reported to the genotoxic effects. Vaccha is an aromatic and multi-faceted medicinal plant with antiviral qualities. The extracts and active constituents of the root or rhizome of *A. calamus* prove beneficial in the prevention of viral replication in DENV2, Flavivirus, and coronaviruses in both the *in-vitro* and *in-vivo* environments signifying their role in designing anti-viral drugs. *A. calamus* extracts and Tatanan A active principle has demonstrated antiviral effects on the Dengue virus and it has exhibited significant inhibition effect on the proliferation of viral particles in *in-vitro* and *in-silico* systems. However, the effect has to show its efficacy *in vivo* and requires clinical trials. The chemotypic estimation and construction of non-toxic genetical traits of *A. calamus* are also being developed to extemporize the safe and natural phytoconstituents for effective antiviral treatment. The molecular docking approved stable conformations of the active compounds of *A. calamus* with two crucial target enzymes *i.e.*, PLpro and Mpro of SARS-CoV-2, and revealed that these compounds can prevent SARS-CoV-2 infection after binding to PLpro and Mpro and consequently hold potential for forthcoming anti-CoV drug designing.

References

Abel, G. (1987). Chromosome damage in human lymphocytes induced by *β*-asarone. *Planta Med.* 53: 251–253.

Agarwal, S.L., Dandiya, P.C., Singh, K.P. and Arora, B.R. (1956). A note on the preliminary studies of certain pharmacological actions of *Acorus calamus*. *J Am Pharm Assoc* 45: 655–656. doi: 10.1002/jps.3030450921.

Amritha, D.K., Damayenti, Y.D., et al. (2018). Acorus calamus: a bio-reserve of medicinal values. *Journal of Basic and Clinical Physiology and Pharmacology*, 29(2):107–122. https://doi.org/10.1515/jbcpp-2016-0132.

Aqil, F., Zahin, M. and Ahmad, I. (2008). Antimutagenic activity of methnolic extracts of four ayurvedic medicinal plants. *Indian J. Exp. Biol.,* 46:668–672.

Arseculeratne, S.N., Gunatilaka, A.A.L. and Panabokke, R.G. (1985). Studies on medicinal plants of Sri Lanka. *J. Ethnopharmacology,* 13:323–335.

Ashrafi, S., Rahman, M., Ahmed, P. et al. (2022). Prospective Asian plants with corroborated antiviral potentials: Position standing in recent years. Beni-Suef Univ. J. Basic Appl Sci. 11: 47. https://doi.org/10.1186/s43088-022-00218-y.

Badam, L. (1995). *In vitro* studies on the effects of *Acorus calamus*. extract and *β*-asarone on Herpes viruses. *Deerghayu. Intl.* 11: 16– 18.

Báez-Santos Y.M., Barraza S.J., Wilson M.W., Agius M.P., Mielech, A.M., et al., (2014). X-ray structural and biological evaluation of a series of potent and highly selective inhibitors of human coronavirus papain-like proteases. *J. Med. Chem.* 57:2393–2412. doi: 10.1021/jm401712t.

Bains, J.S., Dhuna, V., Singh. J., Kamboj, S.S., Nijjar, K.K. and Agrewala, J.N. (2005). Novel lectins from rhizomes of Acorus with mitogenic activity and inhibitory potential towards murine cancer cell lines. *Int. J. Immunopharm.,* 5:1470– 1478.

Balakumbahan, R., Rajamani, K. and Kumanan, K. (2010). *Acorus calamus*: an overview. *J. Med. Plants Res.,* 4: 2740– 2745.

Barbhuiya A.R., Sharma G.D., Arunachalam A. and Deb, S. (2009). Diversity and conservation of medicinal plants in Barak valley, Northeast India. *Indian J. Tradit. Knowl.* 8:169–175.

Bhat, S.D., Ashok, B.K. and Acharya, R. (2012). Exploring the concept of vacha (*Acorus calamus* Linn.) Shodhana in Ayurveda. *Int. J. Res. Ayurveda Pharm.*, 3:341– 344.

Bhattacharya, A.K. and Kumar, V.S. (2011). Chemistry and biological activity of Bach (*Acorus calamus*). *J. Med. Aromat. Plant Sci.*, 33:270– 278.

Bjomstad, K., Helander, A., Hulten, P. and Beck, O. (2009). Bioanalytical investigation of asarone in connection with *Acorus calamus* oil intoxications. *J. Anal. Toxicol.*, 33: 604– 609.

Boissya C.L. and Majumder, R. (1980). Some folklore claims from the Brahmaputra Valley (Assam) *Ethnomedicine.* 6:139–145.

Boktapa N.R. and Sharma A.K. (2010). Wild medicinal plants used by local communities of Manali, Himachal Pradesh, India. *Ethnobot. Leafl.* 3:259–267.

Chauhan, N.S. (1999). *Medicinal and aromatic plants of Himachal Pradesh.* New Delhi: Indus Publishing Company, 72–76.

Dastur J.F. (1951). *Medicinal Plants of India and Pakistan.* D.B. Taraporevala Sons and Co. Ltd; Bombay, India: p. 12.

Derle, D.V. and Gujar, K.N. (2001). Anti-inflammatory, analgesic and antipyretic activity of *Acorus calamus* and *Curcuma amada. Indian Drugs,* 38:444–447.

Devi, A. and Ganjewala, D. (2009). Antimicrobial activity of *Acorus calamus* L. and leaf extract. *Acta Biol. Szeged.,* 53:45-49.

Dobelis, I.N. (1986). *Magic and Medicine of Plants.* The Reader's Digest Association, Inc.; Pleasantville, New York, NY, USA.

Dong, W., Li, M., Yang D. and Lu, R. (2010). Two new sesquiterpenes from *Acorus calamus*. *Planta Med.,* 76:1742–1745.

Dong, W.W., Jiao, W., Deng, M.C., Yang, C.B., Yue, J.M. and Lu, R.H. (2008). A new steroid glycoside derivative from *Acorus calamus*. *J. Chin. Chem. Soc.,* 55:1277–1279.

Dragendorff G. *Die Heilpflanzen der Verschie Denen Volker und Zeiten.* F. Enke; Stuttgart, Germany: 1898.

Duke, J.A. and Ayensu E.S. (1985). *Medicinal Plants of China.* Reference Publications, Inc; Algonac, MI, USA.

Dwivedi, S.N., Dwivedi S. and Patel P.C. (2006). Medicinal plants used by the tribal and rural people of Satna district, Madhya Pradesh for the treatment of gastrointestinal diseases and disorders. *Nat. Prod. Rad.* 5:60–63.

Garneau, F.X., Collin, G., Gagnon, H., Belanger, A., Lavoie, S., Savard, N. and Pichette, A. (2008). Aromas from Quebec. I. Composition of the essential oil of the rhizomes of *Acorus calamus* L. *J. Essent. Oil Res.,* 20: 250–254.

Gilani, A.U., Shah, A.J., Ahmad, M. and Shaheen, F. (2006). Antispasmodic effect of *Acorus calamus* is mediated through calcium channel blockade. *Phytotherapy Research,* 20: 1080–1084.

Global Biodiversity Information Facility. Available online: https://www.gbif.org/ (accessed on 10 February 2020).

Goeggelmann, W. and Schimmer, O. (1983). Mutagenicity testing of β-asarone and commercial calamus drugs with *Salmonella typhimurium*. *Mutat. Res.,* 121: 191–194.

Grayum, M.H. (1987). A summary of evidence and arguments supporting the removal of *Acorus* from the Araceae. *Taxon* 36: 723–729.

Grieve, M.A (1971). *Modern Herbal.* Volume II Dover Publications, Inc; New York, NY, USA.

Hotta, M. (1970). A system of the family Araceae in Japan and adjacent areas. Mem. Fac. Sci. Kyoto Imp. Univ. *Ser. Biol.* 4: 72–96.

Huang Y., Li Z, Ma Y., Wu Q., Kong J., Zhao L., Li S., and Li, J. (2024). Screening for Active Compounds of Acorus calamus against SARS-CoV-2 Viral Protease and Mechanism Prediction. Pharmaceuticals (Basel). 17(3):325. doi: 10.3390/ph17030325.

Jiao D., Dong X., Yu Y. and Wei, C. (2021). Gene Presence/Absence Variation analysis of coronavirus family displays its pan-genomic diversity. *Int. J. Biol. Sci.* 17:3717. doi: 10.7150/ijbs.58220.

Jin, Z., Qin, W. and Chen, X. (1982). Mutagenicity of α.–asarone from *Acorus calamus*. and its metabolic pathway *in vitro*. *Zhejiang Yike Daxue Xuebao,* 11: 1–3.

Joshi, R.K. (2016). *Acorus calamus* Linn.: phytoconstituents and bactericidal Property, *World J Microbiol Biotechnol.,* 32, 164.

Karthiga, T., Venkatalakshmi, P., Vadivel, V. and Brindha, P. (2016). *In vitro* anti-obesity, antioxidant and anti-inflammatory studies on the selected medicinal plants. *Int. J. Toxicol. Pharmacol. Res,* 8(5), 332–340.

Kaushik, R., Jain, J., Yadav, R., Singh, L., Gupta, D. and Gupta, A. (2017). Isolation of beta-asarone from Acorus calamus Linn. and evaluation of its anticonvulsant activity using MES and PTZ models in mice. *Pharmacology, Toxicology and Biomedical Reports*, 3(2).

Khatun M.A., Harun-Or-Rashid M. and Rahmatullah, M. (2011). Scientific validation of eight medicinal plants used in traditional medicinal systems of Malaysia: A review. *Am. Eurasian J. Sustain. Agric.* 5:67–75.

Kim, W.J., Hwang, K.H., Park, D.G., Kim, T.J., Kim, D.W., Choi, D.K., Moon, W.K. and Lee, K.H. (2011). Major constituents and antimicrobial activity of Korean herb *Acorus calamus*. *Nat. Prod. Res.,* 13:1278–1281.

Kingston C., Jeeva S., Jeeva G.M., Kiruba S., Mishra B.P. and Kannan, D. (2009). Indigenous knowledge of using medicinal plants in treating skin diseases in Kanyakumari district, Southern India. *Indian J. Tradit. Knowl* 8:196–200.

Kingston C., Nisha B.S., Kiruba S. and Jeeva, S. (2007). Ethnomedicinal plants used by indigenous community in a traditional healthcare system. *Ethnobot. Leafl.* 11:32–37.

Kirtikar K.R. and Basu, B.D. (1975). *Indian Medicinal Plants.* Volume IV M/S. Bishen Singh Mahendra Pal Singh; Dehradun, India.

Kirtikar, K.R. and Basu, B.D. (1956). In: Basu LM, editor. Indian medicinal plants, 1956. (vol. III):2045–8 [Allahabad, India]. Agarwal, S.L., Dandiya, P.C., Singh, K.P., Arora, R.B. A note on the preliminary studies of certain pharmacological actions of *Acorus calamus*. *J. Am. Pharm. Assoc.,* 45:655–656.

Koul, O., Smirle, M.J. and Isman, M.B. (1989). Asarones from *Acorus calamus* L. oil their effect on feeding behavior and dietary utilization in *Peridroma saucia*. *J. Chem. Ecol.,* 16:1–6.

Lam, K.Y., Chen, J., Lam, C.T., Wu, Q., Yao, P., Dong, T.T., Lin, H. and Tsim, K.W. (2016). Asarone from *Acori Tatarinowii* rhizoma potentiates the nerve growth factor-induced neuronal differentiation in cultured PC12 cells: a signaling mediated by protein kinase A. *PLOS ONE,* 11, e0163337.

Li H.L. (1977). Hallucinogenic plants in Chinese herbals. *Harv. Univ. Bot. Mus. Leafl.* 25:161–177.

Marongiu, B., Piras, A., Porcedda, S. and Scorciapino, A. (2005). Chemical composition of the essential oil and supercritical CO_2 extract of *Commiphora myrrha* (Nees) Engl. and of *Acorus calamus* L. *J. Agric. Food Chem.,* 53:7939–7943.

Meena, A.K., Rao, M.M., Singh, A. and Kumar, S. (2010). Physiochemical and preliminary Phytochemical studies on the rhizome of *Acorus calamus. International Journal of Pharmacy and Pharmaceutical Science,* 2: 130–131.

Mehrotra, S. Mishra, K.P., Maurya, R. and Srimal, R.C. (2003). Anticellular and immunosuppressive properties of ethanolic extract of *Acorus calamus* rhizome. *Int. Immunopharmacol.,* 3(1): 53–61.

Motley, T.J. (1994). The ethnobotany of sweet flag, *Acorus calamus* (Araceae). *Econ Bot.,* 48:397–412.

Mukherjee, P.K. and Wahile, A. (2006). Perspectives of safety for natural health products. In: Sharma RK, Arora A, eds., *Herbal Drugs–A Twenty first Century Perspective*, New Delhi, India, Jaypee Brothers Medicinal Publishers Ltd., pp. 50–59.

Nadkarni, K.M. (1998). *Indian Materia Medica*, 1st ed. Popular Prakashan, India.

Nisha, M.C. and Rajeshkumar, S. (2010). Survey of crude drugs from Coimbatore city. *Indian J. Nat. Prod. Resour.* 1:376–383.

Nishiyama, N., Zhou Y. and Saito, H. (1994b). Beneficial effects of DX-9386, a traditional Chinese prescription, on memory disorder produced by lesioning the amygdala in mice. *Biol. Pharm. Bull.,* 17:1679–1781.

Nishiyama, N., Zhou, Y. and Saito, H. (1994a). Ameliorative effects of chronic treatment using DX-9386, a traditional Chinese prescription on learning performance and lipid peroxide content in senescence accelerated mouse. *Biol. Pharm. Bull.*17: 1481–1484.

Oh, M.H., Houghton, P.J., Whang, W.K. and Cho, C.H. (2004). Screening of Korean herbal medicine used to improve cognitive function for anti-cholinesterase activity. *Phytomedicine,* 11:544–548.

Ozcan, M., Akgul, A. and Chalchat, J.C. (2002). Volatile constituents of the essential oil of *Acorus calamus* L. grown in Konya Province (Turkey). *J. Essent. Oil Res.,* 14:366–368.

Packer, J.G. and Ringius, G.S. (1984). The distribution and status of *Acorus* (Araceae) in Canada. *Canadian Journal of Botany* 62(11): 2248–2252.

Panchal, G.M., Venkatakrishna-Bhatt, H., Doctor R.B. and Vajpayee, S. (1989). Pharmacology of *Acorus calamus* L. *Indian J. Exp. Biol.,* 27(6): 561–567.

Parab, R.S. and Mengi, S.A. (2002). Hypolipidemic activity of *Acorus calamus* L. in rats. *Fitoterapia,* 73: 451–455.

Parcha, V., Chhetri, S., Kumar, D. and Rana, R. (2022). Overview: Insights on the Phytochemical, Pharmacological and Biological Aspects of *Acorus calamus* and *Artemisia roxburghiana*: Wild Aromatic Plant Species of Himalayan Region of Uttarakhand, *GIS SCIENCE JOURNAL*, 9(1), 1678–1704. https://doi.org/10.53879/id.58.09.12262

Perry L.M. and Metzger J. (1980). *Medicinal Plants of East and Southeast Asia.* MIT Press; Cambridge, UK.

Phongpaichit, S., Pujenjob, N., Rukachaisirikul, V. and Ongsakul, M. (2005). Antimicrobial activities of the crude methanol extract of *Acorus calamus* Linn. *J Sci Technol.,* 27: 517–523.

Poonam K. and Singh, G.S. (2009). Ethnobotanical study of medicinal plants used by the Taungya community in Terai Arc Landscape, India. *J. Ethnopharmacol.* 123:167–176. doi: 10.1016/j.jep.2009.02.037.

Pradhan B.K. and Badola, H.K. (2008). Ethnomedicinal plant use by Lepcha tribe of Dzongu valley, bordering Khangchendzonga Biosphere Reserve, in north Sikkim India. *J. Ethnobiol. Ethnomed.* 4:1–18. doi: 10.1186/1746-4269-4-22.

Quraishi, A., Mehar, S., Sahu, D. and Jadhav, S.K. (2017). *In vitro* mid-term conservation of *Acorus calamus L.* via cold storage of encapsulated microrhizome. *Braz. Arch. Biol. Technol.* 60:1–9. doi: 10.1590/1678-4324-2017160378.

Radusiene, J., Judzentiene, A., Peciulyte, D. and Janulis, V. (2007). Essential oil composition and antimicrobial assay of *Acorus calamus* leaves from different wild populations. *Plant Genet. Resour.,* 5:37–44.

Rafatullh, S., Tariq, M., Mossa, J.S., Al-yahya, M.A., Al-said, M.S. and Ageel, A.M. (1994) Anti-secretagogue, antiulcer and cytoprotective properties of *Acorus calamus* in rats. *Fitoterapia,* 65:19–23.

Ragupathy S., Steven N.G., Maruthakkutti M., Velusamy B. and Ul-Huda M.M. (2008). Consensus of the 'Malasars' traditional aboriginal knowledge of medicinal plants in the Velliangiri holy hills, India. *J. Ethnobiol. Ethnomed.* 4:8–16. doi: 10.1186/1746-4269-4-8.

Raina, V.K., Srivastava, S.K. and Syamasunder, K.V. (2003). Essential oil composition of *Acorus calamus* L. from the lower region of the Himalayas. *Flav.Fragr. J.,* 18:18–20.

Rajith N.P. and Ramachandran, V.S. (2010). Ethnomedicines of Kurichyas, Kannur district, Western Ghats, Kerala. *Indian J. Nat. Prod. Resour.* 1:249–253.

Rajput, S.B., Tonge, M.B. and Karuppayil, S.M. (2014). An overview on traditional uses and pharmacological profile of *Acorus calamus* Linn. (Sweet flag) and other *Acorus* species. *Phytomedicine*, 21(3):268–76. doi: 10.1016/j.phymed.2013.09.020.

Rosmalena, R., Elya, B., Dewi, B.E., Fithriyah, F., Desti, H., Angelina, M., et al. (2019). The Antiviral Effect of Indonesian Medicinal Plant Extracts Against Dengue Virus *In Vitro* and *In Silico*. *Pathogens*, *8*: 85. https://doi.org/10.3390/pathogens8020085.

Saptawati, L., Febrinasari, R.P., Yudhani, R.D., Yono, H., Faza, A.G., et al. (2017). *In vitro* study of eight Indonesian plants extracts as anti-dengue virus. *Health Sci. J. Indones.,* 8: 12–18.

Satyavati G.V., Raina M.K. and Sharmal, M. (1976). *Medicinal Plants of India.* Volume I. *Indian Council of Medical Research*; New Delhi, India: pp. 14–16.

Sharma, P.K., Chauhan and N.S., Lal, B. (2004). Observations on the traditional phytotherapy among the inhabitants of Parvati valley in western Himalaya, India. *J Ethnopharmacol.,* 92:167–176.

Sharma, V., Singh, I. and Chaudhary, P. (2014). Acorus calamus (The Healing Plant): a review on its medicinal potential, micropropagation and conservation. Nat Prod Res. 28(18):1454–1466. doi: 10.1080/14786419.2014.915827.

Sharma, V., Sharma, R., Gautam, D.S., Kuca, K., Nepovimova, E. and Martins, N. (2020). Role of Vacha (*Acorus calamus* Linn.) in Neurological and Metabolic Disorders: Evidence from Ethnopharmacology, Phytochemistry, Pharmacology and Clinical Study. *J. Clin. Med. 9*, 1176. https://doi.org/10.3390/jcm9041176

Sharma, R.K., and Arora, A. (2006). eds., *Herbal Drugs–A Twenty first Century Perspective*, New Delhi, India, Jaypee Brothers Medicinal Publishers Ltd., pp. 50–59.

Shashikala, G.H., Prashanth, D., Jyothi, C.H., Imran Maniyar, I.M. and Manjunath, H. (2015). Evaluation of antidepressant activity of aqueous extract of roots of *Acorus calamus* in albino mice.

Sher Z., Khan Z. and Hussain F. (2011). Ethnobotanical studies of some plants of Chagharzai valley, district Buner, Pakistan. *Pak. J. Bot.* 43:1445–1452.

Shih-Chen L. (1973). *Chinese Medicinal Herbs.* Georgetown Press; San Francisco, CA, USA.

Shoba, F.G. and Thomas, M. (2001). Study of antidiarrhoeal activity of four medicinal plants in castor oil induced diarrhoea. *J. Ethnopharmacol.,* 76:73–76.

Shrestha P.M. and Dhillion, S.S. (2003). Medicinal plant diversity and use in the highlands of Dolakha district Nepal. *J. Ethnopharmacol.* 86:81–96. doi: 10.1016/S0378-8741(03)00051-5.

Shukla, P.K., Khanna, V.K., Ali, M.M., Maurya, R.R., Handa, S.S. and Srimal, R.C. (2002). Protective effect of *Acorus calamus* against acrylamide induced neurotoxicity. *Phytother. Res.,* 16: 256–260.

Si, M.M., Lou, J.S., Zhou, C.X. and Shen, J.N. (2010). Insulin releasing and α-glucosidase inhibitory activity of ethyl acetate fraction of *Acorus calamus in vitro* and *in vivo*. *Journal of Ethnopharmacology,* 128, 154–159.

Silprasit, K., Seetaha, S. and P. Pongsanaraku et al. (2011). Anti-HIV-1 reverse transcriptase activities of hexane extracts from some Asian medicinal plants. *J. Med. Plant Res.* 5: 4899–4906.

Singh, C., Jamwal, U. and Singh, P. (2001). *Acorus calamus* (sweet flag), an overview of oil decomposition, biological activity and usage. *Int. J. Med. Arom. Plants.* 23:687–708.

Suryadevara, P. and Khanam, S. (2002). Screening of plant extracts for larvicidal activity against *Culex quinquefasciatus*. *J. Nat. Remedies.,* 2:186–188.

Tariq, M.R., Naqvi, S.N., Choudhary, M.I. and Abbas, A. (2010). Importance and implementation of essential oil of Pakistan *Acorus calamus* Linn. as a biopesticide. *Pak. J. Bot.,* 42:2043–2050.

Tiwari, N., Mishra, A., Bhatt, G. and Chaudhary, A. (2014). Isolation of Principle Active Compound of *Acorus Calamus. vivo* assessment of pharmacological activity in the treatment of neurobiological disorder (stress). *J. Med. Clin. Res*, 2(9): 2201–2212.

Tkachev, A.V., Gurev, A.M. and Yusubov, M.S. (2006). Acorafuran, a new sesquiterpenoid from *Acorus calamus* essential oil. *Chem. Nat. Compd.,* 42:696–698.

Tomar A. (2009). Folk medicinal uses of plant roots from Meerut district, Uttar Pradesh. *Indian J. Tradit. Knowl.* 8:298–301.

Tong, X.G., Zhou, L.L., Wang, Y.H., Xia, C., Wang, Y., Liang, M, Hou, F.F. and Cheng, Y.X. (2010). Cortatarins A and B, two novel antioxidative spiroalkaloids with a naturally unusual morpholine motif from *Acorus*. *J. Org. Lett.,* 12:1844–1847.

Tribudi, Y.A., Agustin, A.T., Setyaningtyas, D.E. and Gusmalawati, D. (2022). Bioactive compound profile and biological modeling reveals the potential role of purified methanolic extract of sweet flag (*Acorus calamus* L.) in inhibiting the Dengue Virus (DENV) NS3 Protease-Helicase. *infection*, 7: 8. DOI:10.22146/ijc.68317.

Usher, G. (1984). *Spilanthes Acmella, a Dictionary of Plants Used by Man.* CBS Publishers and Distributers; New Delhi, India: p. 38.

Vaidyaratnam, P.S.V. (1994). Indian Medicinal Plants–A Compendium of 500 Species. Orient Longman Ltd., Madras, 4: 59–64.

Venskutonis P.R. and Dagilyte, A. (2011). Composition of essential oil of sweet flag (*Acorus calamus* L.) leaves at different growing phases. *J. Essent. Oil Res.,* 15:313–318.

Wagner, H., Bauer, R., Melchart, D., Xiao, P.G. and Staudinger, A. (2011). Chromatographic Fingerprint Analysis of Herbal Medicines, 2nd ed. Springer Wien, New work.

Wallnofer H. and Rottauscher A. (1965). *Chinese Folk Medicine and Acupuncture.* Bell Publishing Co, Inc; New York, NY, USA.

Wilczewska, A.Z., Ulman, M., Chilmoriczyk, Z., Maj, J., Koprozicz, T., Tomczyk, M. and Tomczykowa, M. (2008). Comparison of volatile constituents of *Acorus calamus* and *Asarum europaeum* obtained by different techniques. *J. Essent. Oil Res.,* 20:390–395.

Wren, R.C. (1956). *Potter's New Cyclopaedia of Botanical Drugs and Preparations.* Sir Isaac Pitman and Sons, Ltd; London, UK.

Wu, L.J., Sun, L.L., Li, M., Yang, H., Jiang, Z.R., Lu, Y., Tian, Z., Zhenq, Q., Miyase, T. and A. Ueno. (1994). Studies on the constituents of the roots of *Acorus calamus* L. *J. Pharm. Soc. Jpn.,* 114:182–185.

Yabesh, J.M., Prabhu, S. and Vijayakumar, S. (2014). An ethnobotanical study of medicinal plants used by traditional healers in silent valley of Kerala, India. *J. Ethnopharmacol.* 154:774–789. doi: 10.1016/j.jep.2014.05.004.

Yao, X., Ling, Y., Guo, S., Wu, W., He, S., Zhang, Q., Zou, M., et al. (2018). Tatanan A from the *Acorus calamus* L. root inhibited dengue virus proliferation and infections. *Phytomedicine.* 42:258-267. doi: 10.1016/j.phymed.2018.03.018.

Zhang, Y., Takashina, K., Saito, H. and N. Nishiyama. (1994). Anti-aging effect of DX-9386 in senescence-accelerated mouse. *Biol. Pharm. Bull.* 17: 866–868.

5 Ethnopharmacological, Nutraceutical, and Therapeutic Potential of *Citrullus colocynthis* (L.) Schrad.

Bushra Hafeez Kiani[1*]

1. Introduction

Cucurbitacea is known to be one of the most genotypically diversified families in the plant kingdom (Zaini et al., 2011). Most Cucurbitacea species can survive extreme cold, dryness, and overwatering. *Momordica charantia, Cucumis melo, Citrullus colocynthis, Cucurbita moschata, Cucumis sativus,* and *Lagenaria siceraria*, are a few commonly known members of this group melons, as well as pumpkins (Robinson and Decker-Walters, 1997). The cultivation of cucurbit vegetables and fruits has improved over the years thanks to widely known health benefits. Asian countries including China and India have produced the most cucurbits in the past twenty years. Russia, the United States of America, Egypt, and Iran also grow these plants (Zaini et al., 2011).

C. colocynthis (L) Schrader is an important plant of the Cucurbitacea family (Table 1), due to its pharmaceutical and nutritional importance and is mostly found in arid areas (Sawaya et al., 1983; Asyaz et al., 2010). It is used conventionally to treat asthma, high blood sugar, and hyperbilirubinemia (Baquar and Tasnif, 1984; Kirtikar et al., 1984; Qureshi et al., 2010). Numerous investigations have been conducted on the pharmacological, toxicological, and phytochemical properties of C. colocynthis (Salama, 2012; Ali et al., 2013).

Currently, a revised overview of the properties of this plant is not publicly accessible. This chapter describes the plant's multifaceted value as a nutritional fruit, traditional treatment, and healthy dietary component. The toxicology, phytochemical, ethnopharmacology, and pharmaceutical activity of Citrullus colocynthis, are discussed in this chapter using new research findings.

2. Synonyms

Citrullus colocynthis has also been known as colocynth, egusi, Abu Jahl's melon, bitter apple, wild gourd, vine of Sodom, or bitter cucumber. The English term for *Citrullus colocynthis* is bitter apple or colocynth. Hanjal, Indrayan, Kattu, Anedri,

[1] Department of Biological Sciences (Female Campus), International Islamic University, Islamabad, Pakistan.

* Corresponding author: bushrahafeezkiani@gmail.com

DOI: 10.1201/9781003433583-5

Table 1 Classification of Citrullus colocynthis.

Kingdom	*Plantae*
Order	Cucurbitales
Family	*Cucurbitaceae*
Genus	*Citrullus*
Species	*colocynthis*
Binomial Name	*Citrullus colocynthis* (L.) Schrad.

Kattu vellari, Rakhal, and Pcitummatti are the common names of this fruit in the South Asian region (Kumar et al., 2009; Amamou et al., 2011).

3. Botanical Description

The family Cucurbitaceae includes *C. colocynthis* (L.) Schrader. Southern Islands, Sudan, Saharan and Arabian deserts, and South Asia are regions where this species is extensively distributed (Al-Ghaithi et al., 2004; Krishnaraju et al., 2005; Perveen et al., 2007; Gurudeeban et al., 2010; Marzouk et al., 2010b). *C. colocynthis* had been brought to Cyprus and Spain by Arabs in the medieval period (Bellakhdar et al., 1991; Wasfi et al., 1995; Abdel-Hassan et al., 2000).

C. colocynthis are perennials that produce small blossoms. The rough leaves are alternatively placed on the leaf stalks, measuring 1.5 to 2 centimeters in width, 5 to 10 centimeters long, and 3-7 deeper lobes. Yellow blossoms are produced on angled, hairy, rough stems (Savithramma et al., 2007; Amamou et al., 2011) which may also be spotted in the leaf axils. It is bisexual, singular, and attached to a peduncle. Every shrub bears 15 to 30 green circular fruit, with undulating yellowish lines, that turn completely yellow once dried. The fruit has a diameter of 7 to 10 centimeters.

The spherical fruits of *C. colocynthis* have a tough rind and smooth texture with 200 to 300 seeds in each fruit; however, it has a notable bitterness (Fig. 1). Once mature, the brown seeds appear smaller (6 millimeters long), smooth, oval, and flattened (Schafferman et al., 1998). Approximately 75 percent of the total fruit weight was in seeds. Certain morphological characteristics of this fruit, including their weight, volume, exocarp thickness, pulp fraction, fruit density, and fruit-to-

Fig. 1 *Citrullus colocynthis* plant parts (A) Plant (B) Fruit (C) Leaves.

seed ratio have been described (Aviara et al., 2007). The mean weight of the fruit is 506 grams, with the seeds making up 71.8 grams of this weight and the pulp making up about half (Aviara et al., 2007).

4. Geographical Distribution

C. colocynthis is a vine that thrives in desert soil. *C. colocynthis* was first detected in the Mediterranean Basin and Asia, but it is also found along the coastlines of the Caspian Sea, Northern Africa, the Sahara Desert, and India.

C. colocynthis is indigenous to the following countries: Pakistan, India, Afghanistan, Bangladesh, Sri Lanka, Libya, Morocco, Syria, Egypt, Kuwait, Algeria, Somalia, Djibouti, Lebanon, Oman, Sicilia, Saudi Arabia, Palestine, Socotra, Nigeria, Myanmar, Iran, Burkina Faso, Gulf States, Tunisia, Eritrea, Sinai, Yemen, Greece, Iraq, Senegal, Western Sahara, Canary Islands, Benin, Turkmenistan, Cyprus, Sudan, Kenya, Cape Verde, Turkey, Ethiopia, Spain, Mali, Ivory Coast, and Mauritania.

Among the countries where *C. colocynthis* has been adapted are the United States of America, Bulgaria, Argentina, Corsica, Hungary, Aegean Islands, Italy, United Kingdom, Primorye, Romania, Uzbekistan, Australia, Paraguay, and Transcaucasus.

5. Phytochemical Constituents and Metabolite Profiling

C. colocynthis contains several biologically active chemicals that are well-documented. These have been divided into six categories: triglycerides, bioflavonoids, carbohydrates, glycosides, alkaloids, and aetheroleum (Jayaraman *et. al*, 2009; Najafi *et al*., 2010; Salama, 2012). Publications on separating and identifying particular biochemical components are scarce, nevertheless. According to reports, the primary compounds of *C. colocynthis* include cucurbitacin.

5.1 Cucurbitacins

The cucurbitacins are a class of botanical compounds that are mostly triterpenes, oxygen-rich, and tetracyclic produced from cucurbitane skeletons [19-(10-9β)-abeo-10α-lanost-5-en]. Given that the methyl molecule on carbon 10 has shifted onto carbon 9, these compounds could not be regarded as steroids. The Cucurbitaceae family is where cucurbitacins are present in the majority. Based on their structural characteristics, cucurbitacins are divided into twelve groups; however, C. colocynthis does not contain any members of these groups. Cucurbitacins seem to play a significant part in pharmacological development, especially in the creation of antitumor drugs, as a function of their cytotoxic nature (Chen et al., 2005).

Cucurbitacin E is a type of cucurbitacin found abundantly in the mesocarp of *C. colocynthis* fruits (Chen et al., 2005; Ali et al., 2013). Colocynthosides B and A had been extracted from methanol extracts of the Egyptian fruit. Other Cucurbitacins extracted from BuOH fractions were cucurbitacin K 2-O-β-D glucopyranoside, hexanocucurbitacin I 2-O-β-D glucopyranoside, khekadaengoside E, and Cucurbitacin I glucoside (Dahmén and Leander, 1976; Hatam et al., 1989; Kanchanapoom et al., 2002; Yoshikawa et al., 2007; Kumar et al., 2009).

One more research (Torkey et al., 2009) identified 2-O-β-D-glucopyranosyl cucurbitacin E among the main cucurbitacins present in the pericarp of Citrullus colocynthis. Another study mentioned the presence of cucurbitacin E, D, C, B, and A in this fruit (Adams et al. (2001). From the *C. colocynthis* pericarp's EtOAc-soluble fraction, cucurbitacin E 2-O-β-D-glucopyranoside, 1-(3,4-Dihydroxyphenyl)-1-propanone, and beta-Hydroxypropiovanillone were the main extracted as major components. From the butanol soluble fractions, Colocynthoside B and A were extracted from the 1-Butanol soluble fraction, along with other cucurbitacins such as isoorientin 30-methyl ether, cucurbitacin J 2-O-β-D-glucopyranoside, 4-hydroxybenzyl β-D-glucopyranoside, benzyl β -D-glucopyranoside, cucurbitacin K 2-O-β-D-glucopyranoside, cucurbitacin I glucoside, cucurbitacin I 2-O-beta-D-glucopyranoside, hexanocucurbitacin I 2-O-β-D-glucopyranoside, aponaretin, khekadaengoside isovitexin 4'-O-glucoside, helicin, and gastrodin (Yoshikawa et al., 2007).

5.2 Glycosides, Phenolic Acids and Flavonoids

Cucurbitacin glucosides such as 2-O-β-D-glucopyranosyl cucurbitacin L and 2-O-β-D-glucopyranosyl cucurbitacin I, bioflavonoid glycosides such as isovitexin 4'-O-glucoside, isoorientin 3-O-methyl ether, and saponaretin had been extracted from *C. colocynthis* pericarp methanolic extracts using n-BuOH fractions (Delazar et al., 2006). A group of naturally occurring substances known as polyphenols operates as antioxidants. Secondary metabolites known as bioflavonoids have antioxidative and radical-terminating properties (Shahidi and Wanasundara, 1992). The *C. colocynthis* pericarp underwent an initial screening for phytochemicals revealing phenolic and bioflavonoid content. Following quantification, it was discovered that there were 3.07 mg/g of complete phenolic content (assessed as gallic acid) and 0.51 mg/g of total bioflavonoids (assessed as catechin equivalent) in the dried sample (Hussain et al., 2013). The ethanolic extractions of *C.* colocynthis pericarp contained 8 phenolcarboxylic acids, such as 4-hydroxybenzoic acid, 5-O-caffeoylquinic acid, 3,4-dihydroxy-cinnamic acid, 3,4,5-trihydroxybenzoic acid, 4-hydroxy-3-methoxy benzoic acid, 3,5-dimethoxy-4-hydroxycinnamic acid, 4-hydroxycinnamic acid, and 4-hydroxy-3-methoxy cinnamic acid, as well as bioflavonoids, such as catechuic acid, myricetol, sophoretin, and rhamnolutein (Hussain et al., 2013). Sophoretin was also found in *C. colocynthis* pericarp, according to reports (Meena and Patni, 2008).

5.3 Fatty Acids and Tochopherol

The primary saturated fatty acids found in the seed oils of *C. colocynthis* are octadecanoic acid (6.1–10.5%) and hexadecanoic acid (8.1–17.3%), (Sawaya et al., 1983; Sadou et al., 2007; Sebbagh et al., 2009) while the primary monounsaturated fatty acids are oleic acid and linoleic acid (50.6–61.1%). The seed oil's composition indicates that it is oleic-linoleic acid oil, similar to numerous plant oils (Sawaya et al., 1983; Sadou et al., 2007; Sebbagh et al., 2009). Given this, it is probable that the oil from *C. colocynthis*, similar to certain cucurbit seed oils, might be used for cooking. This oil's content is comparable to oil extracted from Carthamus tinctorius, having 80 to 85 percent unsaturated fatty acids (Schafferman et al. 1998). Reports studying the physical and chemical properties of the seed oil show that it contains 0.18 mg/kg β-carotene, 435 mg/kg γ-tocopherol, and 45.1 mg/kg α-tocopherol which are naturally occurring free radical-scavengers (Kalhoro et al., 2002).

5.4 Alkaloids

Numerous investigations have found alkaloid compounds throughout the *C. colocynthis* pericarp, however only a handful have isolated or identified specific these alkaloid compounds (Sayed et al., 1973; Najafi et al., 2010; Mukherjee and Patil, 2012; Ali et al., 2013). Vitamin J along with two unidentifiable alkaloid compounds was extracted by Sayed et al. in 1973 using the pericarp of *C. colocynthis.*

5.5 Volatile Compounds

Seventeen volatile organic compounds (VOCs) were extracted from the mesocarp of the fruit, with concentrations varying between 0.51 percent to 48.0 percent peak areas (Gurudeeban et al., 2011). The discovered alcohols were cyclopentane methanol 5-methoxy, 2-methyl,2-pentanol; 2-furan-methanol, tetrahydro-5-methyl- trans and cis and isomers; 4-(1-methyl) ethoxy-1-butanol. Ketone compounds identified were 3,4-dimethyl,2-hexa-none; 3-methyl,2-heptanone; 2-methyl,4-heptanone.1-propoxypentane and 2,3-epoxymethylpropionate were the epoxy compounds found (Gurudeeban et al., 2011). Hexadecane, tetradecane, pentadecane, and tridecane had been found on the exocarp in minute amounts.

3-methyl- 2-heptanone; 2-methyl, 4-heptanone; and 3, 4-dimethyl, 2-hexanone; were the ketones that were described. 1-propoxypentane, palmitic acid, and 2,3-epoxy methyl propionate were two epoxy compounds. The fruit's surface may have had small amounts of pentadecane, tetradecane, hexadecane, and tridecane hydrocarbons. The remnant chemicals were trimethyl silyl methanol and di-is octylester. The fatty acid route should have provided the majority of such molecules (Gurudeeban et al., 2011).

5.6 Nutritional Properties and Potential as Functional Foods

The most important factors in determining food quality are its dietary values. In addition to being used medicinally, *C. colocynthis* fruit has been widely used as a foodstuff for both people and animals (Sadou et al., 2007). After more than a decade of research into the nutritional and physiological properties of plant seeds, the United States Department of Agriculture (USDA) scientists came to the conclusion that they might eventually be employed commercially in the food sector (National Research Council, 2006).

5.7 Nutrients

Despite the extensive significance of *C. colocynthis* as an eatable, a global readership has limited access to nutritional information. Table 2 shows the approximate composition (protein, moisture, fat, and ash) of the seeds and fruits from various nations. Various agroecological conditions and agricultural techniques that vary by nation may be the cause of certain variances in the values. The kernels of the seeds have roughly 30 percent protein, 3 percent fiber, 10 percent carbohydrate, 50 percent oil, and 4 percent ash (National Research Council, 2006). The matured fruits have an increased moisture content, which accounts for over 90 percent of their consumable weights (Aviara et al., 2007). According to a different study, the moisture content of bitter apple seeds was 4.91 grams per 100 grams, whereas the amounts of ash and protein were 13.19 and 2.00 grams per 100 grams (Sadou et al., 2007). These seeds are now being looked into as potential oil sources for use in cooking and medicine. The output of whole seed oil of *C. colocynthis* was 26.60 grams per 100 grams, whereas the content of seed kernel oil was 56.5 grams per 100 grams (Sawaya et al., 1986; Sadou et al., 2007). Its oil yield is somewhat greater than *Glycine max* and *Gossypium herbaceum* and comparable to other crops like *Carthamus tinctorius* and *Helianthus annuus* (Swern, 1979).

Another Pakistani research indicated an oil production of nineteen percent (Sawaya et al., 1986), while seeds from India had a greater oil yield of thirty-six percent (Singh and Yadava, 1978). According to reports, the oil extracted from *C. colocynthis* seeds may be suitable for ingestion by both humans and animals (Sawaya et al.1983). The oil extraction had been discovered to have a mildly flavored and odorless yellowish color with green tints (Sawaya et al., 1983).

In addition to being used as a fuel source, *C. colocynthis* seeds have an acceptable amino acid profile and comprise about thirteen percent protein by mass of decorticated seeds (Sadou et al., 2007). The proteinaceous contents are favorably equivalent to those of the most popular legume seeds (National Research Council, 2006). The protein composition is unique due to the exceptionally high content of critical amino acids. To manufacture patties that replace meat, *C. colocynthis* seed meal is utilized after the oil has been removed in part. The amount of un-defatted meals used in specific nutritional preparations depends on the individual's dietary preferences. After being dry roasted, the seeds of *C. colocynthis* could be eaten.

5.8 Amino Acids

Methionine, tryptophan, and arginine are just a few of the many amino acids found in abundance in *C. colocynthis* (National Research Council, 2006). Its protein quality is rated as having biological indices "lower than *Glycine max*es but roughly equivalent to or more than many oilseeds". Lysine and threonine are the two limiting amino acids in nutrition (National Research Council, 2006). *C. colocynthis* protein's amino acid makeup was described and contrasted with other oilseed proteins by Sawaya et al. in 1986. The two primary amino acids, glutamic acid, and arginine were found to have 19.8 grams and 15.9 grams of amino acids per 100 g of proteins, respectively. Additional significant amino acids included aspartic acid, glycine, phenylalanine, leucine, serine, and alanine (45 grams of amino acids per 100 grams of protein). There haven't been many nutritional trials done yet. Yet, it was discovered that adding *C. colocynthis* flour to the diets of West Africans significantly improved growth. Also, the literature claims that its seeds could improve conventional and contemporary food compositions (National Research Council, 2006).

5.9 Minerals

The body needs minerals to perform its usual functions, making them vital nutrients. The fruits and seeds of *C. colocynthis* include many micronutrients (vitamins and minerals) that may greatly improve diet. The possibility of using *C. colocynthis* seed as a calcium and niacin source in lower West African regions with minimal milk consumption seems encouraging. The two main minerals found in the seeds are potassium (K^{+}), which has a concentration of 569 milligrams per 100 grams of seeds, and calcium (Ca^{++}), which has a concentration of 465 milligrams per 100 grams of seeds. The seed is a good supplier of PO_4^{3n-} and Mg^{++} as well. The lowest levels of Fe and Zn concentration are present. Both mineral elements (Potassium and Calcium) are well recognized to contribute to the body's alkalinization process and to maintaining the electrolytic balance of body fluids (Sadou et al., 2007; Zaini et al., 2011). According to a different study, the minerals $PO4^{3-}$, K^{+}, Mg^{++}, Mn^{++}, S^{2-}, Ca^{++}, Fe^{++}, and Zn^{++} make up the majority of *Citrullus colocynthis*' mineral composition (National Research Council, 2006).

6. Medicinal/Therapeutic Uses

Numerous pharmaceutical studies have been motivated by *Citrullus colocynthis* in conventional medical uses. The bioactivities of several extracts and phytoconstituents were studied, particularly focusing on their effects on diabetes and cancer. Owing to the abundant cucurbitacin concentrations in *C. colocynthis*, the use of this plant in cancer therapeutics can be further explored.

6.1 Antiviral Activity and its Mechanism

The antiviral properties of *Citrullus colocynthis* have not been extensively studied, and there is limited information available on its potential antiviral activity. Some preliminary studies suggest that extracts from the plant may have activity against some viral pathogens, including Hepatitis B and C viruses. The cytopathic effect (CPE) reduction assay was used to investigate the anti-viral effect of SNPs/*C. colocynthis* AEs, as well as the equivalent AEs of various plant parts against HAV,

HSV-I, and HSV-II. In a nutshell, Vero cell lines were grown in microtiter plates to form cell layers in confluence as previously explained. The cell line had been induced with efficient MNCCs of the SNPs and *Citrullus colocynthis* solutions or the matching AEs of various parts of the plant while being challenged with virus suspensions of 10-4 TCID50 per milliliters. After that, plates for tissue culture were cultured for 3 to 6 days at 37°C. In the absence of the tested substances, the infection control and untreated Vero cells control were included. The plates were examined under an inverted microscope daily until the virus inside the control wells displayed full CPE. Formalin was used to fix the monolayers, and then a 0.1 percent crystal violet solution was used to stain them (Dargan, 1998; Vijayan et al., 2004). Three separate studies, each carried out in triplicate, were used to determine the CPE using light microscopy. The CPE's percentage inhibition was measured, and the results were classified as follows: inconsequential (less than 25 percent), mild (25 percent to less than 50 percent), moderate (50 percent to less than 75 percent), and strong (greater than 75%).

Several potential viral species, including HIV-1 (Elechiguerra et al., 2005), hepatitis B (Lu et al., 2008), H1N1 influenza A virus (Xing et al., 2011 and Mori et al., 2013), and herpes simplex (Baram et al., 2009 to Baram-Pinto et al., 2009) are effectively inhibited by metallic nanoparticles, specifically gold and silver nanoparticles (Galdiero et al., 2011) (Baram et al., 2009). The anti-viral action of plant parts (root, foliage, seed, and mesocarp) in this investigation was considerably boosted by the SNPs produced in AEs of *Citrullus colocynthis*, which had no impact on any of the tested virus species. The key features of nanoparticle systems are particle size and size distribution. They significantly alter their toxicity and capacity to target (Lara et al., 2011). The research found that the antiviral action of silver nanoparticles was dependent on size. By reducing the CPE of HSV-II at 75 percent, HAV by 70 percent, and HSV-I by 73 percent, we discovered that the SNPs/*C. colocynthis* roots AEs (with an average size of SNP of 7.398 nanometers) had a strong antiviral effect. The CPE was reduced with HSV-II by 68 percent, HAV by 63 percent, and HSV-I by 67 percent through the SNPs/*C. colocynthis* in leaf AEs (with an average size of SNP being 13.376 nanometers). The lowest anti-viral action was noted for SNPs/*C. colocynthis* of fruit AEs in all virus types (average SNP size: 19.267 nanometers). This size-dependent antiviral action of SNPs, which has also been seen in other pieces of research (Elechiguerra et al., 2005; Lu et al., 2008; Mori et al., 2013), shows that the virus and SNPs are only able to attach in a certain location.

In addition to their direct interaction with viral surface glycoproteins (DNA or RNA), metal nanoparticles can also penetrate the intracellular matrix and show antiviral activity by coming into touch with the virus's genomic sequence. Another suggested mechanism of action is the interaction of metallic nanoparticles with the components of virus-infected cells necessary for viral replication. (Lara et al., 2011; Mori et al., 2013).

To confirm these results and assess their potential as a secure and efficient antiviral therapy, additional research is required.

6.2 Antidiabetic activity

One of the metabolic disorders with the quickest rate of growth is *diabetes mellitus*. The hunt for more efficient and safe anti-diabetic medications continues because the current therapy is symptomatic and necessitates lifelong use of pharmacological treatments, which have several negative impacts in addition to being expensive. *C. colocynthis* is a plant that has been widely utilized as an anti-diabetic in several nations (Al-Ghaithi et al., 2004). It is noteworthy that both animal and human studies have been conducted on this plant. By taking the aqueous extract orally, one may be able to lessen some harmful outcomes of streptozotocin and sugar levels (Abdel-Hassan et al., 2000; Al-Ghaithi et al., 2004). Moreover, studies have shown that fruit extracts have an insulinotropic effect (Nmila et al., 2000; Bnouham et al., 2006). Various seed extractions administered to islets and separated rat pancreas in the context of 8.3 millimolar blood sugar dramatically increased insulin production *in vitro*. In normoglycaemic rabbits, oral treatment of the aqueous extract (300 milligrams per kilogram) significantly decreased glycemia. When administered orally to normoglycemic rabbits at a dose of 50 milligrams per kilogram, compounds (saponins, glycosides, tertiary and quaternary alkaloids) extracted from *C. colocynthis* fruit were examined. None of the alkaloids had a hypoglycemic impact. The glycosidic component, however, considerably reduced glycemia. In alloxan-induced diabetic rabbits, the saponin component decreased glycemia at significantly lower dosages (10–20 mg/kg), suggesting that such glycosidic and saponin compounds may be in charge of the rind of *C. colocynthis*' ability to lower blood sugar levels (Bnouham et al., 2006).

Iranian herbalists frequently employ the fruit of this plant as an anti-diabetic medication (Huseini et al., 2009). The effectiveness of the fruit was examined in 50 type II diabetic patients throughout a 2-month clinical experiment. Three times a day, either 100 milligrams *C. colocynthis* fruit capsules or placebos were given to two batches of 25 people with everyone receiving normal diabetes treatment. Fasting blood sugar, high-density lipoprotein (HDL), creatinine levels, low-density lipoprotein (LDL), alanine transaminase, glycated hemoglobin (HbA1c), alkaline phosphatase, total cholesterol, triglyceride, aspartate transaminase, and urea were assessed before and after two months of treatment in the patients. The outcomes demonstrated a substantial reduction in HbA1c and sugar levels during fasting in patients receiving *C. colocynthis*. No difference was reported in the serology of both groups. No significant gastrointestinal side effects for either category were reported (Huseini et al., 2009).

C. colocynthis has been additionally utilized for its anti-diabetic properties in the United Arab Emirates. The impact of the seeds' aqueous extracts on the biological and chemical factors was evaluated using rats, and both standard and diabetes induced by STZ were assessed (Al-Ghaithi et al., 2004). A single intraperitoneal STZ injection at 60 milligrams per kilogram body weight dosage caused diabetes mellitus. In rats with diabetes, orally dosing plant extracts (300 milligrams per kilogram, two times daily for two weeks) significantly reduced the concentrations of AST and LDH in the plasma, however, the elevated blood levels of ALP and GGT were not lowered (Al-Ghaithi et al., 2004). Streptozotocin (STZ)-induced diabetes in rats was used to assess the impact of diets supplemented with *C. colocynthis*, sunflower, or olive oils on the pancreatic M-cell mass (Tannin-Spitz et al., 2007a). All animals experienced acute hyperglycemia following STZ administration. In contrast to other rat groups, the hyperglycemia in the rat group given a diet supplemented with *C. colocynthis* oil was noticeably less severe 2 months later. The rat models given a meal rich in this oil showed reduced insulin resistance than the other rats when their insulin sensitivity was evaluated utilizing the homeostasis model assessment approach. The M-cell mass in the pancreas of diabetic rats induced with STZ who were given a diet containing colocynthis oil as well as the control group given the exact dietary plan was comparable two months after STZ injection. In contrast, compared to the *C. colocynthis* group, the M-cell mass in the pancreas remains smaller in the diabetic rat models treated with STZ which had been provided diets rich in sunflower and olive oils. Supplementing with *C. colocynthis* oil may be advantageous since it partially maintains or restores pancreatic M-cell mass inside the diabetic rats treated with STZ (Sebbagh et al., 2009).

When the effects of *C. colocynthis* hydro-ethanol extractions were analyzed on hyperlipidemic diabetic rat models induced with alloxan, findings revealed a notable decrease in the overall amounts of free fatty acids, phospholipids, triglycerides, and cholesterol in the liver, as well as serum, once weighed against diabetic untreated rat models (Dallak, 2011). All these studies revealed that *C. colocynthis* was a safe anti-diabetic and validated its traditional use in treating diabetes.

6.3 Antilipidemic Activity

The models of hyperlipidemic human and hyperlipidemia individuals were used to study *C. colocynthis* ability to reduce lipid levels. In hyperlipidemic rabbits, the plant's ethanol extract (1.2 grams per kilogram/day) returned the serum cholesterol level near normal (Daradka et al., 2007). Daily consumption of 300mg of plant seed powder decreases cholesterol and triglyceride levels (Rahbar and Nabipour, 2010).

6.4 Antimicrobial Activity

Numerous pieces of research show that extracts of *C. colocynthis* have biocidal antifungal properties with variance across different populations (Marzouk et al., 2009; Maezouk et al., 2010a, 2010b). Extracts of *C. colocynthis* have an antibacterial effect on diderm as well as monoderm bacteria (Marzouk et al., 2010a, 2010b). Pakistani research showed that alcoholic extracts of *C. colocynthis* inhibited the growth of *Bacillus subtilis* (Memon et al., 2003) whereas a poor inhibitory effect was reported for *Escherichia coli* and *Pseudomonas aeruginosa* strains. This antibacterial activity against these bacterial strains may likely be attributed to the bioflavonoids, tannin, and glycosidic components in *C. colocynthis* (Memon et al., 2003).

Another research examined the inhibitory effect of *C. colocynthis* extracts created using different solvents against eight bacterial strains, *Salmonella typhimurium, Bacillus cents, Escherichia coli, Staphylococcus epidermidis, Pseudomonas aeruginosa, Proteus vulgaris, Mycobacterium smegmatis,* and *Staphylococcus aureus* (Sharma et al., 2010). Most of the prepared plant extractions had an average MIC ranging between 20 and 100 μg per mL toward the eight bacteria strains. A group of researchers in Iranian studied the antibacterial properties of both water and alcoholic extracts of *C. colocynthis* fruit against the wild-type strain of *Staphylococcus aureus* (ATCC 25923) as well as the strain isolates from patients receiving novobiocin treatments (Najafi et al., 2010). These extracts had an antibacterial activity which can be compared with the standard novobiocin.

6.5 Anti-inflammatory Activity

Non-steroidal anti-inflammatory drugs or NSAIDs have been known to cause gastric ulcers (Graham et al., 1988). NSAIDS which show inhibition towards selective COX-2 are known to be less harmful but may cause fatality due to heart-related cytotoxicity. This has led to the retraction of this particular type of NSAID for commercial use. Hence, a need for safe NSAIDs exists. *C. colocynthis* is one of the most popular conventional therapeutic plants because of its anti-inflammation (Wasfi et al., 1995; Marzouk et al., 2010a). The liquid fruit extracts of 4 mg per kg were screened for anti-inflammatory properties using paw edema albino rat models induced with carrageenan (Marzouk et al., 2010a). This verifies *C. colocynthis*' therapeutic effects on rheumatoid arthritis, anti-inflammation, and a painkiller. Researchers have also studied the anti-inflammatory effect of various extracts of *C. colocynthis* in paw edema albino rat models induced with carrageenan (Aly and Naddaf, 2006). Similar research also analyzed the crude extracts of unripened Tunisian fruits by the same paw edema albino rat models induced with carrageenan to study anti-inflammatory properties (Marzouk et al., 2011).

Many metabolites and phytochemicals prevent ulcers from developing, however, few researchers have examined plants' potential to prevent ulcers (Zaidi et al., 2012). Based on this assumption, *C. colocynthis* may be anti-ulcer and anti-inflammation.

6.6 Antioxidant Activity

Polyphenols are linked to the medicinal properties of various conventionally used plants (Shahidi and Wanasundara, 1992; Hussain et al., 2008; Hussain et al., 2013). The plant extract of *C. colocynthis* is rich in antioxidants including plant sterol and polyphenol (Sebbagh et al., 2009). *C. colocynthis* extracts showed antioxidative properties in various cell line research (Kumar et al., 2008). The radical-terminating properties rise with the increase in extract content with the highest antioxidative property at 2.5 grams per milliliter (Kumar et al., 2008). Antioxidative properties of isoorientin 3-O-methyl ether, isovitexin, and isosaponarin, extracted from

C. colocynthis had been reported using RC50 and DPPH test which are the concentrations at which 50 percent of the DPPH radicals have been scavenged. These amounts were reported to be approximately 5.62 ~ 10–4 the inhibition levels of the squamous larynx cancer cells were 32.43 percent for low dose, 43.24 percent for moderate dose, and 70.27 percent for high dose groups. The cellular growth and programmed cell death of Hep-2 lines by cucurbitacin B involve the down-regulation of cyclin B1 and Bcl-2 proteins through the suppression of the STAT3 mechanism (Liu et al., 2008b).

The commonest brain tumor in humans is glioblastoma multiforme (GBM), which nearly always results in death (Yin et al., 2008). Cucurbitacin B evaluated upon adult glioblastoma cultures in aqueous media decreased proliferation by 50 percent at 0.1 M. (ED50). Soft-gel experiments revealed that cucurbitacin B blocked virtually all glioblastoma multiforme clonogenic cells at 10 to 8 M, suggesting that the medication would be a potential choice for therapeutic trials (Yin et al., 2008). Much interest has been developed in cucurbitacin E due to its cytotoxicity and antitumor properties. While lesser effective as compared to cucurbitacin I isolated in *C. colocynthis, cucurbitacin L* was still lethal to the KB and HeLa cell cultures (Lavie et al., 1964; Cassady and Suffness, 1990; Chen et al., 2005).

6.7 Anticancer Activity

There are several publications on the anti-proliferative action of *C. colocynthis* extracts as well as its isolates mostly in research (Chen et al., 2005; Tannin-Spitz et al., 2007a; Liu et al., 2008a, b; Yin et al., 2008). Certain female breast carcinoma cell lines were unable to proliferate when exposed to the cucurbitacin glucosides (20 M) found in this plant's foliage (Tannin-Spitz et al., 2007b). Once tested on the Hep-2 human larynx cancer cell lines, cucurbitacin B (1-100 M) suppressed cell growth depending on time and dosage (Liu et al., 2008b). Cucurbitacin B administration led to an increase of cells in the mitotic and growth phase as well as cell death depending on time and dosage according to fluorescence-activated cell sorting. With the use of Hoechst 33258 staining, distinct signs of cellular death, such as nuclear fragmentation, chromatin condensation, and apoptotic bodies, could be seen. A Western blot examination showed that the production of the protein molecules p-STAT3 (23,065 ppm), Bcl-2 (11,033 ppm), and cyclin B1 (17,328 ppm), had been markedly reduced according to *in-vivo* assays.

Several *C. colocynthis* extract acts as a larvicide against mosquitos in separate research (Rahuman et al., 2008). (9Z)-Octadec-9-enoic acid and linoleic acid act as larvicides in that research, and they were effective towards the larvae of the fourth insect—*Anopheles stephensi Liston, Aedes aegypti, Culex quinquefasciatus* through fractionation of extracts involving petroleum ether and bioassays. The research's conclusions clearly showed that *C. colocynthis* fractions and extracts, such as linoleic acid and (9Z)-octadec-9-enoic acid, exhibit a considerable larval mortality rate (Rahuman et al., 2008).

6.8 Other Pharmacological Properties

Citrullus colocynthis' impact on keratinous filament development has been investigated in albino rats because of its traditional Ayurvedic usage as a hair stimulant (Roy et al., 2007). In comparison, to control treatment rats, the findings showed that treating with petroleum ether extracts decreased the time it took for hair to start growing to half. Also, compared to the usual medication, minoxidil (67 percent), the therapy was effective in getting a higher percentage of hair follicles (470 percent) into the anagenic stage.

The anti-allergenic component of cucurbitacin E, obtained from methanol extracts of *C. colocynthis* fruit, inhibited the development of passive cutaneous anaphylaxis in mouse models. The cucurbitacin E at 100 milligrams per kg dose along with its aglycone at a 125 mg per kg dosage showed anti-allergenic activities (Yoshikawa et al., 2007). *C. colocynthis* fruit extracts created using dichloromethane; n-hexane, dichloromethane, and ethyl alcohol were found to have pesticidal action against the cowpea aphid (Torkey et al., 2009). The lethal dose (LD50) is 7.13 and 10 to 2 mg per milliliter. In a similar research, quercetin, the positive control, had RC50 values of 2.78 10 to 5 mg per milliliter (Kumar et al., 2008). The extremely potent 2,2-diphenylpicrylhydrazyl antioxidant properties of the aqueous extracts of Tunisian C. colocynthis seeds have an IC50 of 0.021 mg per milliliter (Marzouk et al. 2010b). *C. colocynthis* cucurbitacin glycoside acts as an ABTS scavenger (IC50, 145 M), likely due to its immediate impact on a variety of free radical species (Tannin-Spitz et al., 2007a). These outcomes provided substantial evidence in favor of *C. colocynthis* as a source of naturally derived antioxidants.

7. Toxicity and Safety

Despite *C. colocynthis* being well-established as a conventional pharmaceutical, studies thoroughly assessing its tolerability and toxicity are scarce. Diwan et al. (2000) determined the median fatal dosage (LD50) of saponin (derived from the entire *C. colocynthis* plant) to be 200 mg per kilogram. This indicates that the examined phytoconstituents do not show toxicity when compared to the fatal dosages of many biologically active drugs used in the pharmaceutical industry (Sharma, 1998).

8. Conclusion and Future Prospects

Conventional, therapeutic, and pharmacological uses of *Citrullus colocynthis*, as well as the nutritional, phytochemical, and pharmacological importance of the fruit and seeds have been outlined in this chapter. Contemporary pharmaceutical and phytoconstituents investigations have helped to substantiate several of the conventional applications of this plant, and yet several remain to be substantiated. These phytochemicals and extracts have been reported to show activity against tumors, inflammation, diabetes, microbes, and pests. It is noteworthy that *C.*

colocynthis has great nutritional value because it contains a lot of proteins, edible seed oils, and various critical components including calcium, potassium, and magnesium, all of which are known to have important medical benefits.

The phytochemical analysis and pharmacological use of *C. colocynthis* have been studied; however, some aspects of this plant can be further researched. Quantifiable data is required regarding these specialized metabolic compounds and alkaloids. Cucurbitacins are abundant in the *C. colocynthis* fruits, most of which share a similar structural framework. Consequently, it could be intriguing to look into the connections between these structures and their activities. Contemporary pharmaceutical research has supported a range of conventional applications for *C. colocynthis* albeit a portion of this research has validated this in cultured cells. Therefore, scientific testing and field trials are necessary to properly examine the efficiency and reliability of these plant extracts and phytochemicals.

Despite the lack of research about gastrointestinal illnesses, *C. colocynthis* is thought to be effective in treating a variety of stomach diseases including dyspepsia, stomach flu, dyschezia, and colicky pain. The use of *C. colocynthis* as a gastrointestinal treatment has potential; as such disorders typically involve calcium channel blockers and/or gastrointestinal stimulating components. Additionally, this can mitigate negative effects, preventing the gastrointestinal stimulating components from surpassing a reasonable level. Thus, there is a chance to investigate this plant's medicinal qualities further.

C. colocynthis is proven to fight inflammation, however, in comparison to NSAIDs, it might not cause ulcers in the gastrointestinal tract. This has yet to be scientifically verified however it is deduced based on the historical use of this plant for the treatment of stomach flu. Therefore, if proven to have ulcer-treating properties, it might be a better choice for treating inflammatory illnesses. There is potential to scientifically explore the efficacy of *C. colocynthis* for treating respiratory problems, including coughing, bronchial asthma, and chest cold, especially given that broncho-dilating medications in the market have significant adverse effects such as over-stimulation of the heart. On the other hand, it was found that plant-based therapies have special bronchodilation pathways that can both improve and lessen negative effects (Gilani and Atta-ur-Rahman, 2005). Muscarinic antagonists combined with calcium channel blockers for example, in *Carum roxburghianum* (Khan et al., 2012), *Fumaria parviflora* (Rehman et al., 2012), and *Hyoscyamus niger* (Gillani et al., 2008). Similarly, calcium channel blockers with phosphodiesterase inhibitors in *Lepidium sativum* (Rehman et al., 2012a), *Acorus calamus* (Shah and Gilani, 2010), *Nepeta cataria* (Gilani et al., 2009), and *Andropogon muricatus* (Shah and Gilani, 2012) are probable for preventing over-stimulation of the heart which is commonly a side-effect observed when using oral muscarinic antagonists and phosphodiesterase inhibitory constituents. High blood sugar and high blood pressure are becoming widespread and *C. colocynthis* has been used as a traditional medicine to treat these diseases. Despite the extensive research about the plant's usefulness in lowering blood sugar, only a few initial studies demonstrated its anti-cholesterol activity (Daradka et al., 2007; Rahbar and Nabipour, 2010). Increased blood pressure, sugar, and cholesterol frequently occur within the same patient, and present-day treatments require the lifetime usage of various medicines. These medicines are not only expensive but also have several adverse effects. Therefore, nonclinical as well as clinical research on *C. colocynthis* is needed for treating cardiac and metabolic illnesses.

In light of the aforementioned findings and highlighted research gaps, it is possible that various pairings could help reduce harmful effects or aid in boosting the performance of various medicinal drugs used for treating diseases related to the gastrointestinal tract, air passageway, metabolism, cardiovascular conditions, and autoimmune conditions. The seeds and fruits of *Citrullus colocynthis* have a promising future as phyto-pharmaceutics and nutraceuticals.

References

Abdel-Hassan, I.A., Abdel-Barry, J.A. and Tariq Mohammeda, S. (2000). The hypoglycaemic and antihyperglycaemic effect of Citrullus colocynthis fruit aqueous extract in normal and alloxan diabetic rabbits. *Journal of Ethnopharmacology* 71, 325–330.

Adam, S.E.I., Al-Yahya, M.A. and Al-Farhan, A.H. (2001). Response of Najdi sheep to oral administration of Citrullus colocynthis fruits, Nerium oleander leaves or their mixture. *Small Ruminants Research* 40, 239–244.

Al-Ghaithi, F., El-Ridi, M.R., Adeghate, E. and Amiri, M.H. (2004). Biochemical effects of Citrullus colocynthis in normal and diabetic rats. *Molecular and Cellular Biochemistry* 261, 143–149.

Ali, A.A., Alian, M.A. and Elmahi, H.A. (2013). Phytochemical analysis of some chemical metabolites of Colocynth plant (Citrullus colocynthis L.) and its activities as antimicrobial and antiplasmidial. *Journal of Basic and Applied Scitific Research* 3, 228–236.

Aly, A.M. and Naddaf, A. 2006. Anti-inflammatory activities of Colocynth topical gel. *Journal of Medical Sciences* 6, 216–221.

Amamou, F., Bouafia, M., Chabane-Sari, D., Meziane, R.K. and Nani, A. (2011). Citrullus colocynthis: a desert plant native in Algeria, effects of fixed oil on blood homeostasis in Wistar rat. *Journal of Natural Product and Plant Resources* 1, 1–7.

ARS, (2009). Systematic Botany and Mycology Laboratory. Islamin Republic of Iran, Nishabur <http://eol.org/data_objects/1381267>.

Asyaz, S., Khan, F.U., Hussain, I., Khan, M.A. and Khan, I.U. (2010). Evaluation of chemical analysis profile of Citrullus colocynthis growing in Southern area of Khyber Pukhtunkhwa, Pakistan. *World Applied Sciences Journal* 10, 402–405.

Aviara, N.A., Shittu, S.K. and Haque, M.A. (2007). Physical properties of guna fruits relevant in bulk handling and mechanical processing. *International Agrophysics* 21, 7–16.

Baquar, S.R. and Tasnif, M. (1984). Medicinal Plants of Southern West Pakistan, D-42. Periodical Expert Book Agency, Delhi.

Baram et al., 2009. Inhibition of herpes simplex virus type 1 infection by silver nanoparticles capped with mercaptoethane sulfonate. *Bioconjugate Chem* 20, 1497–1502.

Bellakhdar, J., Claisse, R., Fleurentin, J. and Younos, C. (1991). Repertory of standard herbal drugs in the Moroccan pharmacopoeia. *Journal of Ethnopharmacology* 35, 123–143.

Bnouham, M., Ziyyat, A., Mekhfi, H., Tahri, A. and Legssyer, A. (2006). Medicinal plants with potential anti-diabetic activity: a review of ten years of herbal medicine research (1990–2000). International *Journal of Diabetes and Metabolism* 14, 1–25.

Cassady, J.M. and Suffness, M. (1990). Terpenoid antitumor agents. *Medicinal Chemistry* 16, 245–269.

Chen, J.C., Chiu, M.H., Nie, R.L., Cordell, G.A. and Qiu, S.X. (2005). Cucurbitacins and cucurbitane glycosides: structures and biological activities. *Natural Product Reports* 22 (3), 386–399.

Chopra, R.N., (1958). Indigenous Drugs of India. U.N Dhur and Sons, Ltd., Calcutta, India.

Dahmén, J. and Leander, K., (1976). The structure of parishin, a glucoside from Vanda parishii. *Phytochemistry* 15, 1986–1987.

Dallak, M. (2011). *In vivo*, hypolipidemic and antioxidant effects of Citrullus colocynthis pulp extract in alloxan-induced diabetic rats. *African Journal of Biotechnology* 10, 9898–9903.

Daradka, H., Almasad, M.M., Qazan, W.S., El-Banna, N.M. and Samara, O.H. (2007). Hypolipidamic effects of Citrullus colocynthis L. in rabbits. *Pakistan Journal of Biological Sciences* 10, 2768–2771.

Dargan, D.J. (1998). Investigation of the anti-HSV activity of candidate antiviral agents, *Herpes simplex Virus Protocols*. Springer, pp. 387– 405.

Delazar, A., Gibbons, S., Kosari, A.R., Nazemiyeh, H., Modarresi, M., Nahar, L. and Sarker, S.D. (2006). Flavone C-glycosides and cucurbitacin glycosides from Citrullus colocynthis. Daru 14, 109–114.

Diwan, F.H., Abdel-Hassan, I.A. and Mohammed, S.T. (2000). Effect of saponin on mortality and histopathological changes in mice. *Eastern Mediterranean Health Journal* 6, 345–351.

Elechiguerra, J.L., Burt, J.L., Morones, J.R., Camacho-Bragado, A., Gao, X., Lara, H.H. and Yacaman, M.J. (2005). Interaction of silver nanoparticles with HIV-1. *J Nanobiotechnol* 3, 1–10.

Galdiero, S., Falanga, A., Vitiello, M., Cantisani, M., Marra, V. and Galdiero, M., (2011). Silver nanoparticles as potential antiviral agents. *Molecules* 16, 8894–8918.

Gilani, A.H., Rahman, A.U. (2005). Trends in ethnopharmocology. *J Ethnopharmacol. 222*;100(1-2): 43-49. doi: 10.1016/j.jep.2005.06.001. PMID: 16127805.

Gilani, A.H., Shah, A.J., Zubair, A., Khalid, S., Kiani, J., Ahmed, A., Rasheed, M. and Ahmed, V.U. (2009). Chemical composition and mechanisms underlying the spasmolytic and bronchodilatory properties of the essential oil of Nepeta cataria L. *Journal of Ethnopharmacology* 121, 405–411.

Gillani, A.H., Khan, A., Raoof, M., Jabeen, Q., Ghayur, M.N. and Siddiqui, B.S. Vohra, W. (2008). Gastrointistinal selective airways and urinary bladder relaxant effects of Hyoscymus niger are mediated through dual blockade of muscarinic receptor and Caþþ influx. Fundamental and Clinical Pharmacology 22, 87–99.

Graham, D.Y., Agrawal, N.M. and Roth, S.H. (1988). Prevention of NSAID induced gastric ulcer with misoprostol: multicentre, double-blind placebo-controlled trial. *Lancet* 2, 1277–1280.

Gurudeeban, S., Ramanathan, T. and Satyavani, K., (2011). Characterization of volatile compounds from bitter apple (Citrullus colocynthis) using GC–MS. *International Journal of Chemical and Analytical Science* 2, 108–110.

Gurudeeban, S., Satyavani, K. and Ramanathan, T. (2010). Bitter apple (Citrullus colocynthis): an overview of chemical composition and biomedical potentials. *Asian Journal of Plant Sciences* 9, 394–401.

Hatam, N.A.R., Whiting, D.A. and Yousif, N.J. (1989). Cucurbitacin glycosides from Citrullus colocynthis. Phytochemistry 28, 1268–1271.

Huseini, H.F., Darvishzadeh, F., Heshmat, R., Jafariazar, Z., Raza, M. and Larijani, B. (2009). The clinical investigation of Citrullus colocynthis (L.) schrad fruit in treatment of Type II diabetic patients: a randomized, double blind, placebo-controlled clinical trial. Phytotherapy Research 23, 1186–1189.

Hussain, A.I., Anwar, F., Hussain Sherazi, S.T. and Przybylski, R. (2008). Chemical composition, antioxidant and antimicrobial activities of basil (Ocimum basili-cum) essential oils depends on seasonal variations. Food Chemistry 108, 986–995.

Hussain, A.I., Rathore, H.A., Sattar, M.Z.A., Chatha, S.A.S., Ahmad, F., Ahmad, A. and Johns, E.J. (2013). Phenolic profile and antioxidant activity of various extracts from Citrullus colocynthis (L.) from the Pakistani flora. *Industrial Crops and Products* 45, 416–422.

Jayaraman, R., Shivakumar, A., Anitha, T., Joshi, V.D. and Palei, N.N. (2009). Anti-diabetic effect of petroleum ether extract of Citrullus colocynthis fruits against streptozotocin-induced hyperglycemic rats. *Romanian Journal of Biology-Plant Biology* 54, 127–134.

Kalhoro, M.A., Afza, N., Saleem, M. and Malik, A. (2002). Pharmacochemical studies of the oil, aerial parts, pulp and peel of Citrullus colocynthis. *Journal of Chemical Society of Pakistan* 24, 274–276.

Kanchanapoom, T., Kasai, R. and Yamasaki, K. (2002). Cucurbitane, hexanorcucurbitane and octanorcucurbitane glycosides from fruits of Trichosanthes tricuspidata. *Phytochemistry* 59, 215–228.

Khan, M., Khan, A.U. and Gilani, A.H. (2012). Gut and airways relaxant effects of Carum roxburghianum. *Journal of Ethnopharmacology* 141, 938–946.

Kirtikar, K.R., Basu, B.D., Blatter, E., Caius, J.F. and Mhaskar, K.S. (1984). *Indian Medicinal Plants.* Lalit Mohan Basu, Allahabad, India.

Krishnaraju, A.V., Rao, T.V.N., Sundararaju, D., Vanisree, M., Tsay, H.S. and Subbaraju, G.V. (2005). Assessment of bioactivity of indian medicinal plants using brine shrimp (Artemia salina) lethality assay. *International Journal of Applied Science and Engineering* 3, 125–134.

Kumar, D., Kashyap, S.K. and Maherchandani, S. (2009). Antibacterial activity of some plant extracts. Veterinary Practitioner 10 (2), 148–151.

Kumar, S., Kumar, D., Manjusha, Saroha K., Singh, N. and Vashishta, B. (2008). Antioxidant and free radical scavenging potential of Citrullus colocynthis (L.) Schrad. methanolic fruit extract. *Acta Pharmaceutica* 58, 215–220.

Lavie, D., Willner, D. and Merenlender, Z. (1964). Constituents of Citrullus colocynthis (L.) *Schrad. Phytochemistry* 3 (1), 51–56.

Liu, T., Zhang, M., Zhang, H., Sun, C., Yang, X., Deng, Y. and Ji, W. (2008a). Combined antitumor activity of curcurbitacin B and docetaxel in laryngeal cancer. *European Journal of Pharmacology* 587, 78–84.

Liu, T., Zhang, M., Zhang, H., Sun, C., Yang, X., Deng, Y. and Ji, W. (2008b). Effects of cucurbitacin B on cell proliferation and apoptosis in Hep-2 cells. Lin Chung Er Bi Yan Hou Tou Jing Wai Ke Za Zhi 22, 403–407.

Lu, L., Sun, R.W., Chen, R., Hui, C.K., Ho, C.M., Luk, J.M., Lau, G.K. and Che, C.M. (2008). Silver nanoparticles inhibit hepatitis B virus replication. *Antivir Ther.* 13:253–262.

Marzouk, B., Marzouk, Z., Daccor, R., Edziri, H., Haloui, E., Fenina, N. and Aouni, M. (2009). Antibacterial and anticandidal screening of Tunisian Citrullus colocynthis Schrad. from medenine. *Journal of Ethnopharmacology* 125, 344–349.

Marzouk, B., Marzouk, Z., Haloui, E., Fenina, N., Bouraoui, A. and Aouni, M. (2010a). Screening of analgesic and anti-inflammatory activities of Citrullus colocynthis from southern Tunisia. *Journal of Ethnopharmacology* 128, 15–19.

Marzouk, Z., Marzouk, B., Mahjoub, M.A., Haloui, E., Mighri, Z., Aouni, M. and Fenina, N. (2010b). Screening of the antioxidant and the free radical scavenging potential of Tunisian Citrullus colocynthis Schrad. from mednine. *Journal of Food, AgricultureAgriculture and Environment* 8, 261–265.

Marzouk, B., Marzouk, Z., Haloui, E., Turki, M., Bouraoui, A., Aouni and M., Fenina, N. (2011). Anti-inflammatory evaluation of immature fruit and seed aqueous extracts from several populations of Tunisian Citrullus colocynthis Schrad. *African Journal of Biotechnology* 10, 4217–4225.

Meena, M.C. and Patni, V. (2008). Isolation and identification of flavonoid "Quercetin" from Citrullus colocynthis (Linn.) Schrad. *Asian Journal of Experimental Sciences* 22, 137–142.

Memon, U., Brohi, A.H., Ahmed, S.W., Azhar, I. and Bano, H. (2003). Antibacterial screening of Citrullus colocynthis. *Pakistan Journal of Pharmaceutical Sciences* 16, 1–6.

Mori, Y., Ono, T., Miyahira, Y., Nguyen, V.Q., Matsui, T. and Ishihara, M., (2013) Antiviral activity of silver nanoparticle/chitosan composites against H1N1 influenza A virus. *Nanoscale Research Letters* 8, 93.

Mukherjee, A. and Patil, S.D. (2012). Effects of alkaloid rich extract of Citrullus colocynthis fruit on Artemia salina and human cancerous (MCF-7 and HEPG-2) cells. *Journal of PharmaSciTEch* 1, 15–19.

Najafi, S., Sanadgol, N., Nejad, B.S., Beiragi, M.A. and Sanadgol, E. (2010). Phytochemical screening and antibacterial activity of Citrullus colocynthis (Linn.) schrad against Staphylococcus aureus. *Journal of Medicinal Plant Research* 4, 2321–2325.

National Research Council, (2006). Lost Crops of Africa. National Academies Press, Washington, DC, (ISBN 978-0-309–10333-6).

Nmila, R., Gross, R., Rchid, H., Roye, M., Manteghetti, M., Petit, P., Tijane, M., Ribes, G. and Sauvaire, Y. (2000). Insulinotropic effect of Citrullus colocynthis fruit extracts. *Planta Medica* 66, 418–423.

Perveen, Upadhyay B., Roy and S., Kumar, A. (2007). Traditional uses of medicinal plants among the rural communities of Churu district in the Thar desert, India. *Journal of Ethnopharmacology* 113, 387–399.

Qureshi, R., Bhatti, G.R. and Memon, R.A. (2010). Ethnomedicinal uses of herbs from northern part of Nara desert, Pakistan. *Pakistan Journal of Botany* 42, 839–851.

Rahbar, A.R. and Nabipour, I., (2010). The hypolipidemic effect of Citrullus colocynthis on patients with hyperlipidemia. Pakistan Journal of Biological Sciences 13, 1202–1207.

Rahuman, A.A., Venkatesan and P., Gopsalakrishnan, G. (2008). Mosquito larvicidal activity of oleic and linoleic acids isolated from Citrullus colocynthis (L.) Schrad. *Parasitology Research* 103, 1383–1390.

Rehman, N.U., Bashir, S., Al-Rehaily, A.J. and Gilani, A.H. (2012). Mechanisms underlying the antidiarrheal, antispasmodic and bronchodilator activities of Fumaria parviflora and involvement of tissue and species specificity. *Journal of Ethno-pharmacology* 144 (1), 128–137.

Rehman, N.U., Khan, A., AlKharfy, K.M. and Gilani, A.H. (2012a). Pharmacological basis for the medicinal use of Lepidium sativum in airways disorders. Evidence Based Complementary and Alternative Medicine 2012 (2012), 596524, http://dx.doi. org/10.1155/2012/596524.

Robinson, R.W. and Decker-Walters, D.S. (1997). Crop production science in horticulture. Cucurbit, 6, CAB International. Wallingford, Oxon, U.K., New York, N.Y.

Roy, R.K., Thakur, M. and Dixit, V.K. (2007). Effect of Citrullus colocynthis on hair growth in albino rats. *Pharmaceutical Biology* 45, 739–744.

Sadou, H., Sabo, H., Alma, M.M., Saadou, M. and Leger, C.L. (2007). Chemical content of the seeds and physico-chemical characteristic of the seed oils from Citrullus colocynthis, Coccinia grandis, Cucumis metuliferus and Cucumis prophetarum of Niger. *Bulletin of the Chemical Society of Ethiopia* 21, 323–330.

Salama, H.M.H. 2012. Alkaloids and flavonoids from the air dried aerial parts of Citrullus colocynthis. *Journal of Medicinal Plant Research* 6, 5150–5155.

Savithramma, N., Sulochana, C., Rao and K.N., (2007). Ethnobotanical survey of plants used to treat asthma in Andhra Pradesh, India. *Journal of Ethnopharmacology* 113, 54–61.

Sawaya, W.N., Daghir, N.J. and Khalil, J.K. (1986). Citrullus colocynthis seeds as a potential source of protein for food and feed. *Journal of Agricultural and Food Chemistry* 34, 285–288.

Sawaya, W.N., Daghir, N.J. and Khan, P., (1983). Chemical characterization and edibility of the oil extracted from Citrullus colocynthis seeds. *Journal of Food Science* 48, 104–106.

Sayed, M.D., Balbaa, S.I., Afifi and M.S.A., (1973). Nitrogenous bases of the different organs of Citrullus colocynthis. *Planta Medica* 24, 260–265.

Schafferman, D., Beharav, A., Shabelsky, E. and Yaniv, Z. (1998). Evaluation of Citrullus colocynthis, a desert plant native in Israel, as a potential source of edible oil. *Journal of Arid Environments* 40, 431–439.

Sebbagh, N., Cruciani-Guglielmacci, C., Ouali, F., Berthault, M.F., Rouch, C. and Sari, D.C., Magnan, C. (2009). Comparative effects of Citrullus colocynthis, sunflower and olive oil-enriched diet in streptozotocin-induced diabetes in rats. *Diabetes and Metabolism* 35, 178–184.

Shah, A.J. and Gilani, A.H. (2010). Bronchodilatory effect of Acorus calamus (Linn.) is mediated through multiple pathways. *Journal of Ethnopharmacology* 131, 471–477.

Shah, A.J. and Gilani, A.H. (2012). The calcium channel blocking and phosphodiesterase inhibitory activities of the extract of Andropogon muricatus explain its medicinal use in airways disorders. *Phytotherapy Research* 26 (8), 1256–1258.

Shahidi, F. and Wanasundara, P.K. (1992). Phenolic antioxidants. *Critical Reviews in Food Science and Nutrition* 32, 67–103.

Sharma, A., Singh, S. and Nag, T.N. (2010). Antibacterial activity of *Citrullus colocynthis* and Tribulus terrestris against some pathogenic bacteria. *Asian Journal of Microbiology and Biotechnology and Environmental Sciences* 12, 633–637.

Sharma, S.K., (1998). Medicinal Plants Used in Ayurveda. National Academy of Ayurveda, Ministry of Health and Family Welfare Govt. of India, New Delhi, India.

Singh, A.K. and Yadava, K.S., (1978). Note on the fat content in *Citrullus colocynthis. Indian Journal of Agricultural Sciences* 48, 766 (766).

Swern, D. (1979). Bailey's Industrial Oil and fat Products, 4th ed. Wiley Interscience, New York.

Tannin-Spitz, T., Bergman, M. and Fgrossman, S. (2007a). Cucurbitacin glycosides: antioxidant and free-radical scavenging activities. *Biochemical and Biophysical Research Communications* 364, 181–186.

Tannin-Spitz, T., Grossman, S., Dovrat, S., Gottlieb, H.E. and Bergman, M. (2007b). Growth inhibitory activity of cucurbitacin glucosides isolated from Citrullus colocynthis on human breast cancer cells. *Biochemical Pharmacology* 73, 56–67.

Torkey, H.M., Abou-Yousef, H.M., Azeiz, A. and Farid, H.E.A. (2009). Insecticidal effect of Cucurbitacin E Glycoside isolated from Citrullus colocynthis against Aphis craccivora. *Australian Journal of Basic and Applied Sciences* 3, 4060–4066.

Vijayan, P., Raghu, C., Ashok. G., Dhanaraj, S. and Suresh. B. (2004). Antiviral activity of medicinal plants of Nilgiris. *Indian Journal of Medical Research* 120, 24–29

Wasfi, I.A., Bashir, A.K., Abdalla, A.A., Banna, N.R. and Tanira, M.O.M. (1995). Antiin- flammatory activity of some medicinal plants of the United Arab Emirates. *International Journal of Pharmaceutics* 33, 124–128.

Yin, D., Wakimoto, N., Xing, H., Lu, D., Huynh, T., Wang, X., Black, K.L. and Koeffler, H.P. (2008). Cucurbitacin B markedly inhibits growth and rapidly affects the cytoske-leton in glioblastoma multiforme. *International Journal of Cancer* 123, 1364–1375.

Yoshikawa, M., Morikawa, T., Kobayashi, H., Nakamura, A., Matsuhira, K., Nakamura, S. and Matsuda, H. (2007). Bioactive saponins and glycosides, XXVII. Structures of new cucurbitane-type triterpene glycosides and antiallergic constituents from Citrullus colocynthis. *Chemical and Pharmaceutical Bulletin* 55, 428–434.

Zaidi, S.F., Muhammad, J.S., Shahryar, S., Usmanghani, K., Gilani, A.H., Jafri, W. and Sugiyama, T. (2012). Anti-inflammatory and cryoprotective effects of selected Pakistani medicinal plants in Helicobacter pylori-infected gastric epithelial cells. *Journal of Ethnopharmacology* 141 (1), 403–410.

Zaini, N.A.M., Anwar, F., Hamid, A.A. and Saari, N. (2011). Kundur [Benincasa hispida (Thunb.) Cogn.]: a potential source for valuable nutrients and functional foods. *Food Research International* 44, 2368–2376.

6 Antiviral Effects of *Zingiber officinale* Roscoe and its Phytochemical Constituents

Chanin Sillapachaiyaporn[1,2], Mani Iyer Prasanth[1,2], Tewin Tencomnao[1,2] and Siriporn Chuchawankul[3,4*]

1. Introduction

Zingiber officinale Roscoe (*Z. officinale*), commonly known as ginger, belongs to the Family Zingiberaceae. It is a tropical plant, growing well in warm and humid climates. *Z. officinale* is mostly cultivated in China, Nepal, the US, India, Bangladesh, Taiwan, Jamaica, Nigeria, and other parts of the world (Imtiyaz et al., 2013). The botanical characteristics of *Z. officinale* are a one-meter height, narrow green leaves, yellow flowers, and pale-yellow rhizomes (Pandey et al., 2022). The rhizomes of this plant are globally used for spice, culinary, and herbal medicine (Mbaveng and Kuete, 2017). At present, several pharmaceutical properties of *Z. officinale* have been addressed, including antioxidant (Stoilova et al., 2007), anti-inflammation (Penna et al., 2003), anti-arthritis (Fouda and Berika, 2009), anticancer (Ajayi and Adeshina, 2022), antidiabetic (Al-Amin et al., 2006), anti-bacteria (Malu et al., 2009) and antivirus (Perera et al., 2021).

2. Taxonomy and Scientific Classification

Kingdom : Plantae
Phylum : Tracheophyta
Class : Liliopsida
Order : Zingiberales
Family : Zingiberaceae
Genus : *Zingiber*
Species : *Zingiber officinale* Roscoe

[1] Department of Clinical Chemistry, Faculty of Allied Health Sciences, Chulalongkorn University, Bangkok, Thailand.
[2] Center of Excellence on Natural Products for Neuroprotection and Anti-ageing, Chulalongkorn University, Bangkok, Thailand.
[3] Department of Transfusion Medicine and Clinical Microbiology, Faculty of Allied Health Sciences, Chulalongkorn University, Bangkok, Thailand.
[4] Immunomodulation of Natural Products Research Unit, Faculty of Allied Health Sciences, Chulalongkorn University, Bangkok, Thailand.
* Corresponding author: siriporn.ch@chula.ac.th

DOI: 10.1201/9781003433583-6

3. Phytochemical Constituents and Metabolite Profiling

Z. officinale contains phenolic compounds, flavonoids, terpenoids, and steroids. The list of reported phytochemical constituents of *Z. officinale* is demonstrated in **Table 1**. The secondary metabolites of *Z. officinale* can be divided into two groups, including volatile oil and non-volatile compounds. The *Z. officinale* volatile oil is mainly composed of sesquiterpenes and monoterpenes such as zingiberene, ar-curcumene, β-bisabolene, β-sesquiphellandrene, α-farnesene, α-copaene, bisacurone epoxide, and caryophyllene (Abdullahi et al., 2020; Ashraf et al., 2017; Yousfi et al., 2021). For non-volatile compounds, *Z. officinale* comprises gingerols, gingerdiols, gingerdiones, paradols, and shogaols, which are phenolic compounds. Comparative studies demonstrated that different drying methods affected phenolic and flavonoid contents in *Z. officinale*. The vacuum oven drying process yielded the highest amounts of phenolic and flavonoid compounds, followed by freeze drying, shade drying, and fresh rhizomes (Ghasemzadeh et al., 2016). Moreover, different extraction solvents yield different phenolic and flavonoid constituents. Aqueous or ethanolic extract (high polarity index) provided a greater amount of phenolic content than flavonoids, while petroleum ether extract (low polarity index) showed a higher yield of flavonoids than phenolic content (Osabor et al., 2015; Fatima et al., 2021).

4. Medicinal/Therapeutic Uses

4.1 Uses Described in Pharmacopoeias and in Traditional Systems of Medicine

Z. officinale is one of the most utilized herbal plants in traditional systems, for example, traditional Indian Ayurvedic and Chinese medicines (Mbaveng and Kuete, 2017). In traditional Ayurvedic remedies, they drink a mixture of juices from fresh ginger, garlic, and honey for cough and asthma suppressants (Kumar et al., 2011). In traditional Chinese medicine, the prescription, Yakammaoto, is composed of nine herbal plants, including *Z. officinale*, and has been utilized to manage flu-like symptoms such as fever and cough. Interestingly, this traditional Chinese medicine effectively provided antiviral activity against human coxsackievirus B4 (CVB4) (Yen et al., 2014).

4.2 Traditional/Folklore Uses Described in Folk Medicine

In folk medicine, *Z. officinale* has been traditionally used to relieve several abnormal symptoms such as abdominal pain, cough, vomiting, and inflammation (Yasmin et al., 2020). Decoction of *Z. officinale* rhizomes is orally medicated to treat cancers in several cultures such as India (Silambarasan and Ayyanar, 2015), Singapore (Siew et al., 2014), Palestine (Jaradat et al., 2016), Morocco (Kabbaj et al., 2012), and Gana (Agyare et al., 2018). Moreover, in Singapore, they consume *Z. officinale* rhizomes for enhanced digestion, immunity, and health (Siew et al., 2014).

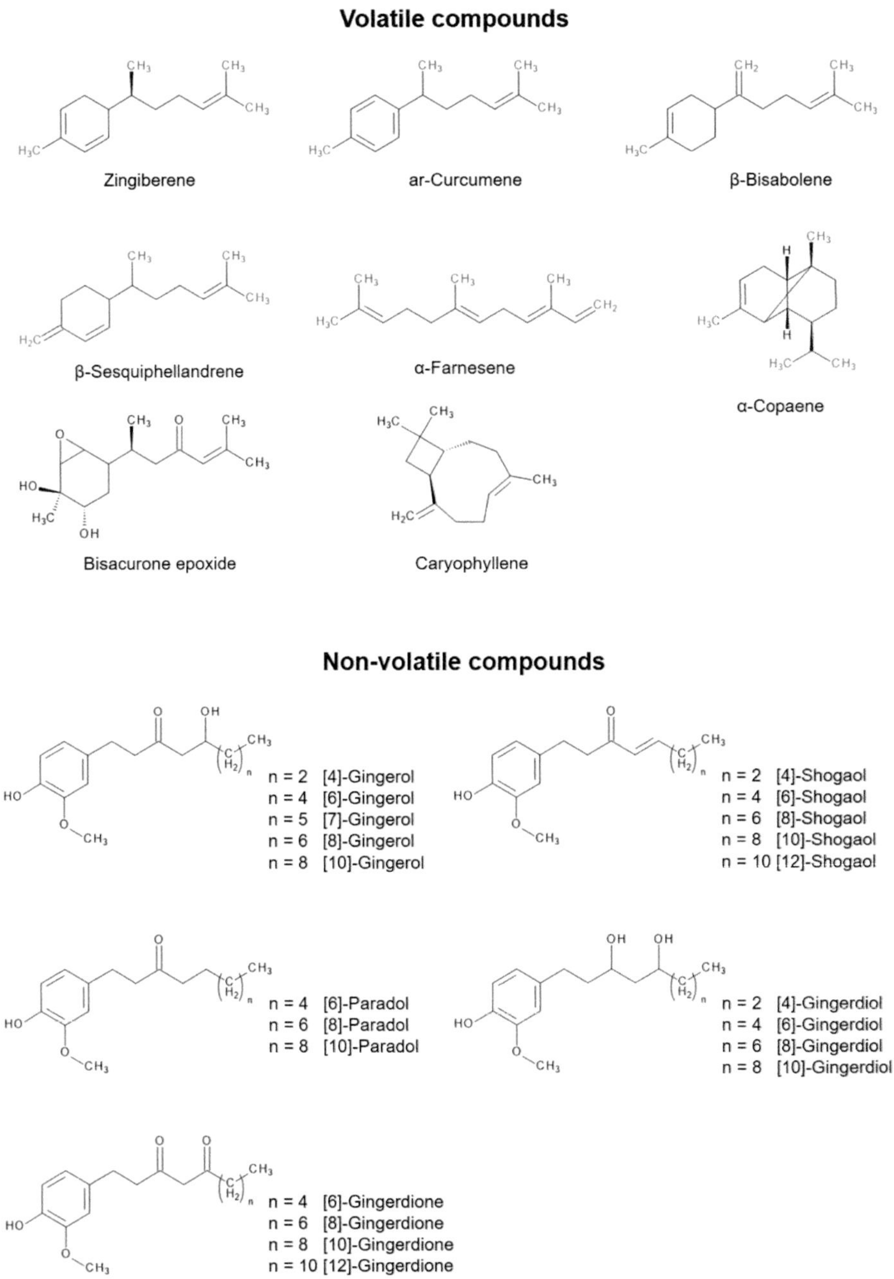

Fig. 1 Chemical structures of phytochemicals of *Zingiber officinale.*

Table 1 Reported phytochemical constituents of *Zingiber officinale.*

Compound	*Molecular Formula*	*Molecular Mass*	*Classification*	*References*
(3S,5S)-3,5-Diacetoxy-1,7-*bis*(4-hydroxy-3-mrthoxy)-heptane	$C_{25}H_{32}O_8$	460.5	Diarylheptanoids	(Ali et al., 2008)
[10]-Gingerdiol	$C_{21}H_{36}O_4$	352.5	Phenols	(Ashraf et al., 2017)
[10]-Gingerdione	$C_{21}H_{32}O_4$	348.5	Phenols	(Ashraf et al., 2017)
[10]-Gingerol	$C_{21}H_{34}O_4$	350.5	Phenols	(Ali et al., 2008; Ashraf et al., 2017)
[10]-Paradol	$C_{21}H_{34}O_3$	334.5	Phenols	(Ashraf et al., 2017)
[10]-Shogaol	$C_{21}H_{32}O_3$	332.5	Phenols	(Ali et al., 2008)
[12]-Gingerdione	$C_{23}H_{36}O_4$	376.5	Phenols	(Ashraf et al., 2017)
[12]-Shogaol	$C_{23}H_{36}O_3$	360.5	Phenols	(Ali et al., 2008)
[4]-Gingerdiol	$C_{15}H_{24}O_4$	268.35	Phenols	(Ashraf et al., 2017)
[4]-Gingerol	$C_{15}H_{22}O_4$	266.33	Phenols	(Ali et al., 2008)
[4]-Shogaol	$C_{15}H_{20}O_3$	248.32	Phenols	(Ali et al., 2008)
[6]-Gingerdiol	$C_{17}H_{28}O_4$	296.4	Phenols	(Ashraf et al., 2017)
[6]-Gingerdione	$C_{17}H_{24}O_4$	292.4	Phenols	(Ali et al., 2008; Ashraf et al., 2017)
[6]-Gingerol	$C_{17}H_{26}O_4$	294.4	Phenols	(Ali et al., 2008; Ashraf et al., 2017)
[6]-Isoshogaol	$C_{17}H_{24}O_3$	276.4	Phenols	(Ali et al., 2008)
[6]-Paradol	$C_{17}H_{26}O_3$	278.4	Phenols	(Ali et al., 2008; Ashraf et al., 2017)
[6]-Shogaol	$C_{17}H_{24}O_3$	276.4	Phenols	(Ali et al., 2008; Ashraf et al., 2017)
[7]-Gingerol	$C_{18}H_{28}O_4$	308.4	Phenols	(Ali et al., 2008)
[8]-Gingerdiol	$C_{23}H_{36}O_6$	408.5	Phenols	(Ashraf et al., 2017)
[8]-Gingerdione	$C_{19}H_{28}O_4$	320.4	Phenols	(Ashraf et al., 2017)
[8]-Gingerol	$C_{19}H_{30}O_4$	322.4	Phenols	(Ali et al., 2008; Ashraf et al., 2017)

Contd.

Table 1 *(Contd.)*

Compound	*Molecular Formula*	*Molecular Mass*	*Classification*	*References*
Vanillic acid	$C_8H_8O_4$	168.15	Phenols	(Mohsin et al., 2022)
Zingerone	$C_{11}H_{14}O_3$	194.23	Phenols	(Dhanik et al., 2017)
Zingiberene	$C_{15}H_{24}$	204.35	Sesquiterpenes	(Ashraf et al., 2017; Yousfi et al., 2021)
α-Copaene	$C_{15}H_{24}$	204.35	Sesquiterpenes	(Ashraf et al., 2017)
α-Farnesene	$C_{15}H_{24}$	204.35	Sesquiterpenes	(Ashraf et al., 2017; Yousfi et al., 2021)
α-Phellandrene	$C_{10}H_{16}$	136.23	Monoterpenes	(Ashraf et al., 2017)
α-Pinene	$C_{10}H_{16}$	136.23	Monoterpenes	(Abdullahi et al., 2020; Ashraf et al., 2017)
α-Zingiberene	$C_{15}H_{24}$	204.35	Sesquiterpenes	(Abdullahi et al., 2020)
β-Bisabolene	$C_{15}H_{24}$	204.35	Sesquiterpenes	(Abdullahi et al., 2020; Ashraf et al., 2017; Yousfi et al., 2021)
β-Phellandrene	$C_{10}H_{16}$	136.23	Monoterpenes	(Abdullahi et al., 2020; Ashraf et al., 2017)
β-Pinene	$C_{10}H_{16}$	136.23	Monoterpenes	(Abdullahi et al., 2020)
β-Sesquiphellandrene	$C_{15}H_{24}$	204.35	Sesquiterpenes	(Abdullahi et al., 2020; Ashraf et al., 2017; Yousfi et al., 2021)

4.3 Applications Supported by Clinical Data

Among the hundreds of studies reported based on the clinical studies conducted using ginger, the treatment objectives were the most prominent ones (47.2%), followed by dietary supplements (40.4%) and different drugs (27%). The individual trials exhibited promising outcomes , based on the effects on health-related quality of life, including blood pressure variation, energy expenditure, and reduction of xerostomia (Matin et al., 2024). Network pharmacology-based analysis highlighted that ginger compounds could target and bind to human genes involved in the viral infection, including SARS-CoV-2 (Samy et al., 2024).

4.4 Pharmacological Activities Mainly Antiviral

Z. officinale has been widely studied for its pharmacological activities such as antioxidant (Stoilova et al., 2007), anti-inflammation (Penna et al., 2003), anti-arthritis (Fouda and Berika, 2009), anticancer (Ajayi and Adeshina, 2022), antidiabetic (Al-Amin et al., 2006), anti-bacteria (Malu et al., 2009) and antivirus (Perera et al., 2021). In Egypt, the hepatitis C virus (HCV)-infected patients who orally received ethanolic *Z. officinale* extract (500 mg/twice a day for a month) showed a significant reduction of viral load, compared to the healthy control group (Abdel-Moneim et al., 2013). Moreover, *in vitro* cell-based experiments demonstrated that *Z. officinale* extract could suppress the SARS-CoV-2 expression nucleoprotein in Vero E6 cells with IC_{50} at 29.19 μg/mL; and the extract inhibited viral-induced plaque formation with the IC_{50} at 1.45 μg/mL when the extract was treated post-infectious phase (Kanjanasirirat et al., 2020). The hot water extract of *Z. officinale* fresh rhizome effectively reduced human respiratory syncytial virus (HRSV)-induced plaque formation on Hep-2 (IC_{50} = 81.4 μg/mL) and A549 (28.2 μg/mL) cells. In contrast, the *Z. officinale* extracts from dried rhizomes did not show significant inhibition against HRSV (San Chang et al., 2013). Furthermore, essential oils isolated from *Z. officinale* showed antiviral activity against non-enveloped viruses: enterovirus and poliovirus type I with IC_{50} ranging from 9-13 μg/mL (José-Rita et al., 2022). Traditional Chinese medicine, namely Yakammaoto, which is composed of nine medicinal plants, including *E. sinica* (stem), *P. ternate* (tube), *Z. officinale* (rhizome), *T. farfara* (flower), *A. tataricus* (root and rhizome), *Z. jujuba* (fruit), *B. chinensis* (rhizome), *A. sieboldii* (whole plant), and *S. chinensis* (fruit), could inhibit CVB4 infection (Yen et al., 2014). Besides, [6]-Gingerol, a constituent of *Z. officinale* showed antiviral activities against chikungunya virus (CHIKV) (Hayati et al., 2021) and SARS-CoV-2 (Nag et al., 2021; Yen et al., 2014). Based on the molecular docking analysis, phytochemical compounds derived from *Z. officinale*: 24-methylcholesta-7-en-3β-on, geraniin, and zingiberene provided possible inhibitory activity on SARS-CoV-2 infection by interacting with the main protease (Mpro), spike protein (S protein), and host angiotensin-converting enzyme 2 (ACE2) (Nag et al., 2021; Umashankar et al., 2021; Zubair et al., 2021). The antiviral activities of *Z. officinale* and its chemical constituent are listed in **Table 2**.

4.5 Mode of Action of Antiviral Compounds

Z. officinale potentially suppresses viral replication via inhibiting processes in the viral replication cycle. Commonly, viruses invade targeted cells through the binding to the host receptor, called viral attachment. Then, the viruses need to copy the viral genome and produce viral proteins within the host cells. After that, the viral

Table 2 Antiviral activities and mode of actions of active compounds from *Zingiber officinale*.

Virus	*Active Compounds*	*Dosage*	*Mode of Action*	*Experimental Model*	*References*
Chikungunya virus (CHIKV)	[6]-Gingerol	IC_{50} = 38 μM (post-treatment) IC_{50} = 31 μM (full treatment)	N/A	CHIKV infected HepG2 hepatocytes	(Hayati et al., 2021)
Enterovirus	Essential oils of *Z. officinale*	IC_{50} = 9.86 and 9 μg/mL (48 and 72 h of incubation)	N/A	Enterovirus infected human rhabdomyosarcoma cells	(José-Rita et al., 2022)
Hepatitis C virus (HCV)	*Z. officinale* extract	500 mg/ twice a day	N/A	HCV infected patients	(Abdel-Moneim et al., 2013)
Human coxsackievirus B4 (CVB4)	Traditional plant combination (Yakammaoto) containing *Z. officinale* ([6]-gingerol = 0.33%)	IC_{50} = 119.9 μg/mL (CVB4 infected A549 cells) IC_{50} = 90.1 μg/mL (CVB4 infected HK-2 cells) IC_{50} = > 300 μg/mL (CVB4 infected Hep-2 cells)	Inhibit viral attachment, internalization, and replication	CVB4 infected HK-2 (kidney), A549 (lung), and Hep2 (upper and lower respiratory tract) cell lines	(Yen et al., 2014)
Human respiratory syncytial virus (HRSV)	Hot water extract of fresh *Z. officinale* rhizome	IC_{50} = 81.4 μg/mL (in Hep-2 cells) IC_{50} = 28.2 μg/mL (in A549 cells)	Inhibit viral aattachment	HRSV infected Hep-2 and A549 cells	(San Chang et al., 2013)
Poliovirus type I	Essential oils of *Z. officinale*	IC_{50} = 12.70 and 9.13 μg/mL (48 and 72 h of incubation)	N/A	Poliovirus type I infected L20B cells	(José-Rita et al., 2022)
Severe acute respiratory syndrome coronavirus 2 (SARS-CoV-2)	[6]-Gingerol	N/A	Inhibit S protein	Molecular docking	(Nag et al., 2021)
		IC_{50} = > 100 μM (nucleoprotein expression) IC_{50} = 1.38 μg/mL (plaque formation)	N/A	SARS-CoV-2 infected Vero E6 cells	(Kanjanasirirat et al., 2020)

Table 2 *Contd.*

Virus	*Active Compounds*	*Dosage*	*Mode of Action*	*Experimental Model*	*References*
(SARS-CoV-2)	24-Methylcholesta-7-en-3β-on	200 μg/mL (75% inhibition)	Inhibit 3CL protease	Molecular docking and *in vitro* enzyme inhibition assay	(Zubair et al., 2021)
	Geraniin	N/A	Inhibit S protein	Molecular docking	(Umashankar et al., 2021)
	Z. officinale extract	IC_{50} = 29.19 μg/mL (nucleoprotein expression) IC_{50} = 1.45 μg/mL (plaque formation)	N/A	SARS-CoV-2 infected Vero E6 cells	(Kanjanasirirat et al., 2020)
	Zingiberene	N/A	Inhibit S protein	Molecular docking	(Nag et al., 2021)
		N/A	Interact with ACE2 of host cells	Molecular docking	(Nag et al., 2021)

particles are assembled and released as new viruses. Moreover, blocking the host receptors that facilitate viral attachment could inhibit viral entry, and subsequently suppress viral replication. Herein, the modes of action of *Z. officinale*'s antiviral compounds are divided into three major groups, including viral attachment and entry inhibition, viral protein synthesis and modification inhibition, and interaction with host cellular proteins.

Inhibition of Viral Attachment and Entry

The viral attachment and entry step marks the beginning of the viral life cycle. The *in silico* screening experiments showed that [6]-gingerol, geraniin, and zingiberene, phytochemicals isolated from *Z. officinale* can inhibit the S protein of SARS-CoV-2 (Nag et al., 2021; Umashankar et al., 2021). Besides, the mixture of Chinese traditional medicine (Yakammaoto) contained approximately 0.33% of [6]-gingerol could suppress CVB4-induced plaque formation in A549, HK-2, and HEp-2 cells. Moreover, the inhibitory mechanisms were evaluated, and the Yakammaoto could inhibit CVB4 attachment at IC_{50} of 273.2 μg/mL (Yen et al., 2014).

Inhibition of Viral Protein Synthesis and Modification

Viral protein synthesis and modification are important processes of viral replication. Some viruses, such as human immunodeficiency virus (HIV) and SARS-CoV-2, need a maturation step to cleave the immature proteins into functional proteins using protease enzymes (Rangsinth et al., 2021; Sillapachaiyaporn et al., 2019). In this regard, the protease activity inhibition leads to suppression of viral replication. *In silico* molecular docking demonstrated that 24-methylcholesta-7-en-3β-on, a phytosterol isolated from hexane extract of *Z. officinale* pseudostems, strongly blocks the enzyme for SARS-CoV-2 maturation. Moreover, *in vitro* assay exhibited that 24-methylcholesta-7-en-3β-on at a concentration of 200 μg/mL inhibited SARS-CoV-2 3CL protease activity with a percentage of inhibition at 75% (Zubair et al., 2021).

Interaction with Host Cellular Proteins

SARS-CoV-2 uses the viral S protein to detect and bind to host cells, successfully transmitting the viral DNA. Several studies suggest that the inhibition of human ACE2 could provide a protective effect against the SARS-CoV-2 infection (Prasansuklab et al., 2021). Based on the computational analysis, zingiberne, a sesquiterpene found in *Z. officinale* has been reported as a candidate compound to prevent infection by interacting with human ACE2 (Nag et al., 2021). However, the effect of zingiberene against ACE2 still needs to be investigated using *in vitro* or *in vivo* models .

4.6 Future of Research and Drug Delivery

An effective drug delivery system with maximum utilization of delivery dose and minimal loss along with increased permeability is still not available as the present models face poor bioavailability of ginger extracts and their phytochemicals. Recent studies have utilized transdermal administration using nanoemulsions, nanoemulgels, or hydrogels (Maleki et al., 2023). The skin permeability was also improved, exhibiting enhanced bioavailability (Attri et al., 2023).

Network pharmacology-related studies have reported that the phytocompounds in ginger could block the viral survival in the host via multi-biological processes including cell cycle, cell communication, apoptosis, and inflammatory responses. Molecular docking simulations identified (6)-gingerdiacetate to inhibit SARS-CoV-2 spike RBD and human ACE2 in the host cells (Samy et al., 2024). However, detailed studies are essential on each phytocompound to confirm the exact effect against the virus, followed by pre-clinical and clinical trials. With the advent of nanotechnology, drug delivery can be executed effectively , however, pharmacokinetics and pharmacodynamics- related studies have to be implemented to understand the molecular pathways and bioavailability in enhancing antiviral properties along with other diseases (Szymczak et al., 2024).

A network including phytocompounds of ginger, along with human and SARS-CoV-2 proteins, reported that the anti-SARS-CoV-2 activity of ginger was mediated by 20 phytocompounds, against 18 potential human targets, and 12 SARS-CoV-2 proteins and thereby exerts antiviral activity (Chen and Zhang, 2023).

5. Toxicity and Safety

Plengsuriyakarn et al. analyzed cytotoxicity, toxicity, and anticancer activity of ginger ethanolic extract by feeding it to cholangiocarcinoma-induced hamsters. Acute and subacute toxicity analysis suggested that the extract had no toxic effects in cholangiocarcinoma-induced hamsters (Plengsuriyakarn et al., 2012). Another study was done in rats recently, wherein 28-day repeated administration of ethanol extracts was done in different phases. Interestingly, the rats did not exhibit toxicity and mortality. However, the liver had distorted hepatocytes and pyknosis, and the kidney had renal tubular distortion. Taken together, ginger extract is less toxic, and may not induce mortality, however, long-term administration could have an impact on liver and kidney functioning (Zirahei et al., 2022).

Conclusion

Z. officinale, or ginger, is widely utilized in traditional Indian Ayurvedic and Chinese medicine and has been shown to improve the host's general health. Interestingly, ginger has exhibited antiviral activity against many potent viruses including SARS-CoV-2 by inhibiting viral attachment and entry, inhibiting viral protein synthesis and modification, and interacting with host cellular proteins.

Ginger extract normally does not induce any toxicity. However, persistent use of high doses may induce some damage to the internal organs. However, the need of the hour is an efficient drug delivery system, which could not only limit the intake of ginger but also can improve its potential activity. The application of nanoencapsulation of ginger has improved the intake into the cells. Further incorporation of the approaches into antiviral studies would help researchers develop a better and potentially preventive strategy against viral diseases, thereby improving healthcare.

References

Abdel-Moneim, A., Morsy, B.M., Mahmoud, A.M., Abo-Seif, M.A. and Zanaty, M.I. (2013). Beneficial therapeutic effects of Nigella sativa and/or Zingiber officinale in HCV patients in Egypt. *EXCLI journal*, *12*, 943.

Abdullahi, A., Khairulmazmi, A., Yasmeen, S., Ismail, I., Norhayu, A., Sulaiman, M., Ahmed, O. and Ismail, M. (2020). Phytochemical profiling and antimicrobial activity of ginger (Zingiber officinale) essential oils against important phytopathogens. *Arabian Journal of Chemistry*, *13*(11), 8012–8025.

Agyare, C., Spiegler, V., Asase, A., Scholz, M., Hempel, G. and Hensel, A. (2018). An ethnopharmacological survey of medicinal plants traditionally used for cancer treatment in the Ashanti region, Ghana. *Journal of ethnopharmacology*, *212*, 137–152.

Ajayi, B. and Adeshina, A. (2022). 6-gingerol Suppresses Angiogenesis and Promotes T-cell cytotoxicity in Mice Model of Colorectal Cancer. *European Journal of Cancer*, *174*, S12.

Al-Amin, Z.M., Thomson, M., Al-Qattan, K.K., Peltonen-Shalaby, R. and Ali, M. (2006). Anti-diabetic and hypolipidaemic properties of ginger (Zingiber officinale) in streptozotocin-induced diabetic rats. *British Journal of Nutrition*, *96*(4), 660–666.

Ali, B.H., Blunden, G., Tanira, M.O. and Nemmar, A. (2008). Some phytochemical, pharmacological and toxicological properties of ginger (Zingiber officinale Roscoe): a review of recent research. *Food and Chemical Toxicology*, *46*(2), 409–420.

Ashraf, K., Sultan, S. and Shah, S. (2017). Phychemistry, phytochemical, pharmacological and molecular study of Zingiber officinale Roscoe: a review. *Int J Pharm Pharm Sci*, *9*(11), 8–16.

Attri, D.S., Rathour, A., Ray, R.K. and Kumar, V. (2023). Formulation and Evaluation of Hydrogel for Topical Drug Delivery of Zingiber officinale Rosc. and Withania somnifera (L.) Dunal to Increase the Bioavailability of Oils for the Treatment of Arthritis. *Ann. Phytomed*, *12*, 1–10.

Chen, S. and Zhang, L. (2023). Unraveling the Potential Anti-Severe Acute Respiratory Syndrome Coronavirus 2 Mechanisms of Ginger by Computational Target Fishing. *Journal of Medicinal Food*, *26*(10), 721–731.

Dhanik, J., Arya, N. and Nand, V. (2017). A review on Zingiber officinale. *Journal of Pharmacognosy and phytochemistry*, *6*(3), 174–184.

Fatima, H., Shahid, M., Jamil, A. and Naveed, M. (2021). Therapeutic potential of selected medicinal plants against carrageenan induced inflammation in rats. *Dose-Response*, *19*(4), 15593258211058028.

Fouda, A.M.M. and Berika, M.Y. (2009). Evaluation of the Effect of Hydroalcoholic Extract of Zingiber officinale Rhizomes in Rat Collagen-induced Arthritis. *Basic & Clinical Pharmacology & Toxicology*, *104*(3), 262–271.

Ghasemzadeh, A., Jaafar, H.Z. and Rahmat, A. (2016). Variation of the phytochemical constituents and antioxidant activities of Zingiber officinale var. rubrum Theilade associated with different drying methods and polyphenol oxidase activity. *Molecules*, *21*(6), 780.

Hayati, R.F., Better, C.D., Denis, D., Komarudin, A.G., Bowolaksono, A., Yohan, B. and Sasmono, R.T. (2021). [6]-Gingerol inhibits chikungunya virus infection by suppressing viral replication. *BioMed Research International, 2021*.

Imtiyaz, S., Rahman, K., Sultana, A., Tariq, M. and Chaudhary, S.S. (2013). Zingiber officinale Rosc.: a traditional herb with medicinal properties. *CELLMED*, *3*(4), 26.21–26.27.

Jaradat, N.A., Shawahna, R., Eid, A.M., Al-Ramahi, R., Asma, M.K. and Zaid, A. dN. (2016). Herbal remedies used by breast cancer patients in the West Bank of Palestine. *Journal of Ethnopharmacology*, *178*, 1–8.

José-Rita, B.J., Bertin, G.K., Ibrahime, S.K., Yannick, K., Erick-Kévin, B.G., Riphin, K.L., Ceylan, R., Zengin, G. and Mireille, D. (2022). Study of the chemical and *in vitro* cytotoxic activities of essential oils (EOs) of two plants from the Ivorian flora (Lippia multiflora and Zingiber officinale) and their antiviral activities against non-enveloped viruses. *South African Journal of Botany*, *151*, 387–393.

Kabbaj, F., Meddah, B., Cherrah, Y. and Faouzi, E. (2012). Ethnopharmacological profile of traditional plants used in Morocco by cancer patients as herbal therapeutics. *Phytopharmacology*, *2*(2), 243–256.

Kanjanasirirat, P., Suksatu, A., Manopwisedjaroen, S., Munyoo, B., Tuchinda, P., Jearawuttanakul, K., Seemakhan, S., Charoensutthivarakul, S., Wongtrakoongate, P. and Rangkasenee, N. (2020). High-content screening of Thai medicinal plants reveals Boesenbergia rotunda extract and its component Panduratin A as anti-SARS-CoV-2 agents. *Scientific reports*, *10*(1), 19963.

Kumar, G., Karthik, L. and Rao, K. B. (2011). A review on pharmacological and phytochemical properties of Zingiber officinale Roscoe (Zingiberaceae). *Journal of Pharmacy Research*, *4*(9), 2963–2966.

Maleki, H., Azadi, H., Yousefpoor, Y., Doostan, M., Doostan, M. and Farzaei, M.H. (2023). Encapsulation of ginger extract in nanoemulsions: preparation, characterization and *in vivo* evaluation in rheumatoid arthritis. *Journal of Pharmaceutical Sciences*, *112*(6), 1687–1697.

Malu, S., Obochi, G., Tawo, E. and Nyong, B. (2009). Antibacterial activity and medicinal properties of ginger (Zingiber officinale). *Global Journal of Pure and Applied Sciences*, *15*(3–4).

Matin, M., Matin, F.B., Ksepka, N., Wysocki, K., Mickael, M.-E., Wieczorek, M., Horbańczuk, J.O., Jóźwik, A. and Atanasov, A. (2024). The clinical research on ginger (Zingiber officinale): Insights from ClinicalTrials. gov analysis. *Planta medica*.

Mbaveng, A. and Kuete, V. (2017). Zingiber officinale. In *Medicinal Spices and Vegetables from Africa* (pp. 627–639). Elsevier.

Mohsin, S.A., Shaukat, S., Nawaz, M., Ur-Rehman, T., Irshad, N., Majid, M., ul Hassan, S.S., Bungau, S. and Fatima, H. (2022). Appraisal of selected ethnomedicinal plants as alternative therapies against onychomycosis: Evaluation of synergy and time-kill kinetics. *Frontiers in Pharmacology*, *13*.

Nag, A., Paul, S., Banerjee, R. and Kundu, R. (2021). *In silico* study of some selective phytochemicals against a hypothetical SARS-CoV-2 spike RBD using molecular docking tools. *Computers in Biology and Medicine*, *137*, 104818.

Osabor, V., Bassey, F. and Umoh, U. (2015). Phytochemical screening and quantitative evaluation of nutritional values of Zingiber officinale (Ginger). *American Chemical Science Journal*, *8*(4), 1–6.

Pandey, P., Awasthi, R., Dhiman, N., Sharma, B. and Kulkarni, G. T. (2022). Ethnopharmacological reports on herbs used in the management of tuberculosis. In *Herbal Medicines* (pp. 501–523). Elsevier.

Penna, S.C., Medeiros, M.V., Aimbire, F., Faria-Neto, H., Sertié, J.A.A. and Lopes-Martins, R. (2003). Anti-inflammatory effect of the hydralcoholic extract of Zingiber officinale rhizomes on rat paw and skin edema. *Phytomedicine*, *10*(5), 381–385.

Perera, W., Liyanage, J. A., Dissanayake, K., Gunathilaka, H., Weerakoon, W., Wanigasekara, D., Fernando, W., Rajapaksha, R., Liyanage, R. and Perera, B.T. (2021). Antiviral Potential of

Selected Medicinal Herbs and their Isolated Natural Products. *BioMed Research International, 2021.*

Plengsuriyakarn, T., Viyanant, V., Eursitthichai, V., Tesana, S., Chaijaroenkul, W., Itharat, A. and Na-Bangchang, K. (2012). Cytotoxicity, toxicity, and anticancer activity of Zingiber officinale Roscoe against cholangiocarcinoma. *Asian Pacific Journal of Cancer Prevention*, *13*(9), 4597-4606.

Prasansuklab, A., Theerasri, A., Rangsinth, P., Sillapachaiyaporn, C., Chuchawankul, S. and Tencomnao, T. (2021). Anti-COVID-19 drug candidates: A review on potential biological activities of natural products in the management of new coronavirus infection. *Journal of Traditional and Complementary Medicine*, *11*(2), 144–157.

Rangsinth, P., Sillapachaiyaporn, C., Nilkhet, S., Tencomnao, T., Ung, A.T. and Chuchawankul, S. (2021). Mushroom-derived bioactive compounds potentially serve as the inhibitors of SARS-CoV-2 main protease: An *in silico* approach. *Journal of Traditional and Complementary Medicine*, *11*(2), 158–172.

Samy, A., Hassan, A., Hegazi, N.M., Farid, M. and Elshafei, M. (2024). Network pharmacology, molecular docking, and dynamics analyses to predict the antiviral activity of ginger constituents against coronavirus infection. *Scientific reports*, *14*(1), 12059.

San Chang, J., Wang, K.C., Yeh, C.F., Shieh, D.E. and Chiang, L.C. (2013). Fresh ginger (Zingiber officinale) has anti-viral activity against human respiratory syncytial virus in human respiratory tract cell lines. *Journal of Ethnopharmacology*, *145*(1), 146–151.

Siew, Y.-Y., Zareisedehizadeh, S., Seetoh, W.-G., Neo, S.-Y., Tan, C.-H. and Koh, H.-L. (2014). Ethnobotanical survey of usage of fresh medicinal plants in Singapore. *Journal of Ethnopharmacology*, *155*(3), 1450–1466.

Silambarasan, R. and Ayyanar, M. (2015). An ethnobotanical study of medicinal plants in Palamalai region of Eastern Ghats, India. *Journal of Ethnopharmacology*, *172*, 162–178.

Sillapachaiyaporn, C., Nilkhet, S., Ung, A.T. and Chuchawankul, S. (2019). Anti-HIV-1 protease activity of the crude extracts and isolated compounds from Auricularia polytricha. *BMC Complementary and Alternative Medicine*, *19*(1), 1–10.

Stoilova, I., Krastanov, A., Stoyanova, A., Denev, P. and Gargova, S. (2007). Antioxidant activity of a ginger extract (Zingiber officinale). *Food Chemistry*, *102*(3), 764–770.

Szymczak, J., Grygiel-Górniak, B. and Cielecka-Piontek, J. (2024). Zingiber Officinale Roscoe: The Antiarthritic Potential of a Popular Spice—Preclinical and Clinical Evidence. *Nutrients*, *16*(5), 741.

Umashankar, V., Deshpande, S.H., Hegde, H.V., Singh, I. and Chattopadhyay, D. (2021). Phytochemical moieties from Indian traditional medicine for targeting dual hotspots on SARS-CoV-2 spike protein: an integrative *in-silico* approach. *Frontiers in Medicine*, *8*, 672629.

Yasmin, A., Chia, S., Looi, Q., Omar, A., Noordin, M. and Ideris, A. (2020). Herbal extracts as antiviral agents. In *Feed additives* (pp. 115–132). Elsevier.

Yen, M.H., Lee, J.J., Yeh, C.F., Wang, K.C., Chiang, Y.W., Chiang, L.C. and San Chang, J. (2014). Yakammaoto inhibited human coxsackievirus B4 (CVB4)-induced airway and renal tubular injuries by preventing viral attachment, internalization, and replication. *Journal of Ethnopharmacology*, *151*(3), 1056–1063.

Yousfi, F., Abrigach, F., Petrovic, J., Sokovic, M. and Ramdani, M. (2021). Phytochemical screening and evaluation of the antioxidant and antibacterial potential of Zingiber officinale extracts. *South African Journal of Botany*, *142*, 433–440.

Zirahei, V., Dibal, N.I. and Attah, O.O. (2022). Acute Toxicity and Effect of Prolonged Oral administration of Zingiber officinale ethanol extract on Liver and Kidney Histology in Rats.

Zubair, M.S., Maulana, S., Widodo, A., Pitopang, R., Arba, M. and Hariono, M. (2021). GC-MS, LC-MS/MS, Docking and molecular dynamics approaches to identify potential SARS-CoV-2 3-chymotrypsin-like protease inhibitors from Zingiber officinale Roscoe. *Molecules*, *26*(17), 5230.

7 An Investigation into the Antiviral Properties of *Calendula officinalis* L.

Quazi Mohammad Imranul Haq[1*], Sayed Sartaj Sohrab[2] and Issa Sulaiman Al-Amri[3]

1. Taxonomy and Scientific Classification (Dhingra et al., 2022)

Kingdom : Plantae
Subkingdom : Tracheobionta
Division : Magnoliophyta
Class : Magnoliopsida
Subclass : Asteridae
Order : Asterales
Family : Asteraceae
Tribe : Calenduleae
Genus : Calendula
Species : Officinalis

2. Synonyms (McCoy et al., 2005)

There are seven (10) synonyms available globally. 1. Calendula officinalis L.; 2. Calendula officinalis; 3. Common marigold; 4. Pot marigold; 5. English marigold; 6. Bride of the Sun; 7. Bull flower; 8. Butterwort; 9. Ruddles; 10. Scotch marigold; 11. Calendula.

3. Botanical Description of the Plant

An aromatic herbaceous perennial with lax or upright stems, *C. officinalis* grows up to 80 cm (31 in) tall. Leaves are usually oblong-lanceolate, hairy on both sides, and five to seventeen centimeters (2 to 7 inches) in length with complete margins or occasionally wavy margins or weak teeth along the margins. The inflorescences

[1] Department of Biological Sciences and Chemistry, College of Arts and Sciences, University of Nizwa, Birkat Al-Mouz, Nizwa, Sultanate of Oman.
[2] Special Infectious Agents Unit, King Fahd Medical Research Center, King Abdulaziz University, Jeddah, Saudi Arabia.
[3] Department of Biological Sciences and Chemistry, College of Arts and Sciences, University of Nizwa, Birkat Al-Mouz, Nizwa, Sultanate of Oman.
* Corresponding author: qmihaq@gmail.com

DOI: 10.1201/9781003433583-7

of plants with wild inflorescences have numerous disc florets seated around a single ray floret ring at the center, with two rows of bracts covered in yellow hairs; the inflorescences are yellow with two rackets of hairy bracts. On the other hand, the disc florets are tubular and hermaphrodite as opposed to the female, tridentate, peripheral ray florets. Flowers can only appear at all times of the year under favorable conditions. The fruit's name indicates that it contains thorns and curving achenes. (Walters and Cullen, 1984; Smith et al., 2007), with an average weight of 10.1 mg (n=50).

Calendulas are considered a valuable resource by var Lepidoptera species including cabbage moths, gothic moths, large yellow underwings, and setaceous Hebrew characters that use calendulas as food plants, calendulas are considered to be a valuable resource. This plant produces prickly achene-like fruits that are arched in shape. An aroma described as light and fragrant can be found in *C. officinalis*. In terms of taste, it has a bitter taste (Singh et al., 2011).

Fig. 1 Graphic picture of Leaves of *C. officinalis* Linn. (Dhingra et al., 2022)

Fig. 2 Graphic picture of Stem of *C. officinalis* Linn. (Dhingra et al., 2022)

Fig. 3 Graphic picture of Flower of *C. officinalis* Linn. (Dhingra et al., 2022)

3.1 Cultivation

Several species of this are native to Central and Southern Europe, Western Asia, and the United States. Plants of *C. officinalis* are commonly implanted throughout the world. This is especially true in warm summer climates where it cannot even support itself (Khalid and da Silva, 2010). There is no doubt that calendulas can tolerate a wide variety of soil types, which is why they are so easy to plant in a garden. This is because they grow in a wide variety of climates. Flowers bloom from spring to autumn in mild climates. It is best to plant winter seeds in autumn where the temperature will not fall below freezing during the winter months (Moghaddasi and Kashani, 2012). Warmer subtropical months deteriorate plants. In sunny areas, seeds proliferate voluntarily, whereas, in half-sunny areas, they do best in bright areas with rich, well-drained compost. Marigold seeds grow into vibrant yellow, gold, and orange blooms in less than two months (Junghanns, 2000).

3.2 Calendula Care

There are only a few temperate zones where calendulas grow as evergreens, but most can be found in zones 9–11. Calendula may wilt on hot days unless it receives some shade during the daytime hours (Dhingra et al., 2022). Like most Asteraceae plants it grows best in organically rich, well-drained soil. (Wilen et al., 2004). Many soil pH levels can be compatible with this plant, but it is like mildly acidic soil (6.0 to 7.0). Ensure that the plants receive regular watering as soon as they are established. Only infrequent watering is needed for mature plants. These plants should not be overwatered. In hot environments, calendula generally dies at the end of the growing season due to its intolerance of higher daytime temperatures. These plants will be destroyed by a winter storm (Gomes et al., 2007). A winter can protect the plants overnight if cold weather is forecast for the next few days. Afterward, when the sun begins to warm the wind, it is best to expose the plants. Three to four inches of compost will also protect plants from low temperatures. Calendula does not require nutrition or hydration (Gushchina et al. 2018). If it is grown in nutrient-rich soil, it does not require any additional nutrients. For marginal soils, a stable, liquid-soluble fertilizer may be necessary; nevertheless, overfeeding might result in tall, slender plants. Mollafilabi and Hosseini (2012) recommend applying a dissolved and balanced fertilizer to plants in vessels every month.

Its seeds sprout quickly and fertilize. These plants are often self-seed, the flowers' seeds can be saved and regrown. Start calendula seeds in a seed booster mix during the final six to eight weeks of winter. Seeds can also be sown directly in the yard right before low-freezing temperatures (Ratajczak et al., 2016). It typically takes two months for plants to flourish after being seeded. You do not want to confuse self-seeding plants with undergrowth in the yard. While calendula is traditionally used as a food preparation plant it is also widely grown in pots, where it thrives. Usually, dwarf varieties, as well as some variants, develop well in vessels

(Elhindi et al., 2020). Use well-draining, organic compost soil, or mix potting soil and organic manure. Several holes are drilled in the bottom of the pot, as this plant dislikes water. An equitable fertilizer must be applied constantly to potted plants. Pollinating insects like bees and bumblebees and beneficial organisms are captivated by calendula, an outstanding veggie garden assistance plant (Betül and Memet, 2021). A vibrant bloom attracts butterflies. Arthropods and diseases are not a major problem with calendula. Good airflow can help prevent powdery wilt on plants. The plants may be consumed by drones and crawlers, particularly when they are energetic. Debris should not be left on the floor to reduce this damage. Whiteflies and larvae can sometimes be a nuisance. Insecticidal washing-up liquid or water can be sprinkled on them to keep them away (Anderson et al., 2016).

3.3 *Calendula Varieties* (Duda et al., 2010)

Numerous cultivars of *C. officinalis* exist.

- With pink edges and dark apricot centers.
- An orange and red color mixture with a red tip on the petal.
- Greenheart oranges are unusual-looking plants with orange petals encircling lime-green cores.
- Orange petals surround lime-green centers on Greenheart Orange plants; they're very unusual-looking.
- It grows well in vessels; and has a compact, short form with yellow and orange blossoms.
- An excellent container plant, Dwarf Gem features orange, yellow, and apricot double petals.
- Prince: Has orange and yellow flowers and is heat-tolerant.
- A vibrant yellow flower with a black center blooms on a Golden Princess plant.

3.4 *Edible Uses*

A solid lacerating saline flavor is initially introduced when consumed. Compared to *Taraxacum officinale* (Dandelion), they contain several vitamins and minerals. The petals of fresh flowers are finely chopped and added to salads, while dried petals are used as a flavoring agent in soups, pastries, and other dishes. There is an abundance of vitamins A and C (Pires et al., 2017). A yellow dye can be made from the petals. As an alternative to saffron, it enhances the flavor of rice, soups, and other dishes. Additionally, brown or crimson hair can be highlighted with golden highlights using this product. The petals of the flowers are less bitter than those of the petals when used to make tea. The seed contains up to 37% protein and 46% oil, but it is not recommended for human consumption (de Lima et al., 2019).

3.5 *Other Uses*

As a result of these plants, the soil eelworm population is reduced as they act as insect repellents. The flowers are used in manufacturing cosmetics. They are useful for enhancing the color of hair in products such as hair products and skin moisturizers. The compost activator Quick Return contains optional ingredients such as flowers. A dried and powdered herb combination increases the bacterial load in compost piles, reducing the amount of time taken for compost to be made (Reddy et al., 2013). The flowers are boiled to produce yellow dye and used to color clothing and other items. This plant is capable of producing a human-usable essential oil. "In addition to its difficulty of finding, perfumes that have a sharp aroma are less likely to contain it" with "Further, perfumes with a strong aroma are less likely to contain it". The flowering ends when wet conditions are expected, and they can thus be used as a harsh weather prediction tool (Sumit et al., 2012).

4. Geographical Distribution

As of now, *C. officinalis has not been detected in the wild*; however, it may be a Moroccan species or a descendant from one that is common in southern Europe spread into Persia, and is cultivated in Arabia and Persia. Since ancient times, it has been cultivated everywhere for its recurrent blossoming, which originates in November, making it perfect for ornamental drives, but specimens gone wild can be found in the Mediterranean area between 0 and 600 m. above sea level (Conti and Scassellati, 2005).

5. Phytochemical Constituents and Metabolite Profiling

Triterpenoids and carotenoids (antioxidants and source of yellow-orange coloring) are found in the pollen and petals of *C. officinalis*. Also, additional carotenoids are contained in the leaves and stems. Most are lutein (80%), zeaxanthin (5%), and beta-carotene. In addition to flavonoids, triterpene oligoglycoside, oleanane-type triterpene glycosides, saponins, and sesquiterpene glucoside, *C. officinalis* flowers have antitumor and antioxidant properties (Spasov et al., 2012; Ukiya et al., 2006). A 100g serving of calendula flowers contains 29.8 mg of lutein (Yoshikawa et al., 2001).

It is well-known that the plant contains many chemical compounds, including terpenoids, flavonoids, coumarins, quinines, volatile oils, carotenoids, and amino acids. Table 1 provides information about these phytochemical constituents' pharmacological activities.

Quinones: The chloroplasts, mitochondria, and leaves of *C. officinalis* contain phytoquinone, plastoquinone, and tocopherol (Manke et al., 2010).

Flavonoids: An ethanolic extract of *C. officinalis* flowers contains several flavonoids. Additionally, there are quercetin, isorhamnetin, and isoquercetin, etc. (Chakraborty, 2010).

Carbohydrates: An ethanolic extract of the plant's inflorescences were found to contain polysaccharides, PS-I, II, and III with (1-3)-D-galactam backbones and alpha-L-rhamnan-(1-3)-araban side chains at C-6 (Pandey and Tripathi, 2014).

Coumarins: Extracts of *C. officinalis* inflorescence were found to contain coumarins, including scopoletin, umbelliferone, and esculetin (Lovecka et al., 2017).

Amino acids: The blossoms of the plant contain 15 amino acids in their pure form, including alanine, arginine, and aspartic acid, etc. (Chaleshtori et al., 2016).

Terpenoids: A diversity of terpenoids have been found in the petroleum ether extract of *C. officinalis* flowers. Several of them are sterols, stigmasterols, and diesters of diols, etc. Flower residues were used to synthesize cornulacic acid acetate, a new triterpenic ester in the oleanane series (Verma et al., 2018).

Lipids: An extract of seeds, leaves, and flowers of *C. officinalis* was studied for its lipid content. Neutral lipids constituted 15.7% of the seeds, while phospholipids constituted 0.6 percent, and glycolipids constituted 0.9 percent. Nine-hydroxy-18:2-dimorphecolic acid constitutes approximately 5 percent of the fatty acids in Marigold seed fatty acids. One more oxygenated fatty acid found in *C. officinalis* seed oil is D-(+)-9-hydroxy-10,12-octadecanoic acid (Nand et al., 2012).

Volatile oil: In the complete flowering stage (0.97 percent), *C. officinalis* flowers contain the most volatile oil, while in the pre-flowering stage (0.13 percent), they carry the least. As the vegetative loop progressed, the proportion showed various trends. There are several monoterpenes and sesquiterpenes in the volatile oil. At postflowering stages, the oil contained higher levels of cadinene, cadinol, t-muurolol, limonene, and 1,8-cineol, with lower levels of p-cymene (Hussain et al., 2012).

Carotenoids: A methanolic extract of *C. officinalis* flowers revealed the presence of a variety of carotenoids in their leaves, petals, and pollens. Pollen and petals had 1.61 percent and 7.71 percent carotenoid content, respectively (mg/g dry weight). It was found that the carotenoid composition of the leaves and stems contained Neoxanthin, 9Zneoxanthin, and violaxanthin, etc. (Deuschle et al., 2015). Leaf and stem carotenoids contain 0.85 and 0.18 percent, respectively (mg/g dry weight). Beta-carotene and lutein were the most abundant carotenoids, while quercetin and isorhamnetin glycosides dominated the flavonoids. *C. officinalis* petals were studied for their carotenoid content. Extracts of petals from orange and yellow flowered calendula varieties were found to contain nineteen carotenoids (Verma et al., 2014).

The orange-flowered varieties were the only ones to contain ten carotenoids. It has been determined that the ten carotenoids are responsible for the coral-like color of calendula petals since their ultraviolet absorption maxima are at a longer wavelength than flavoxanthin, the main carotenoid of the petals. This suggests that the (5Z)-carotenoids undergo catalytic isomerization at C-5 in a different pathway than the primary carotenoid metabolism (Meenatchisundaram et al., 2009).

Table 1 Pharmacological activities of phytochemical constituents.

Phytochemical Component	*Pharmacological Action*	*Mentions*
Quinones	Anti-cancer	(Sausserde and Kampuss, 2014)
Flavonoids	Antioxidant Anti-inflammatory Hepatoprotective	(Zhang et al., 2021)
Carbohydrates	Anti-inflammatory Immunostimulant	(Agrawal, 2011)
Coumarins	Antioxidant Anti-inflammatory Anti-cancer	(Tiwari et al., 2012)
Amino acids	Anti-cancer	(Tiwari et al., 2012)
Terpenoids	Cardioprotective Anti-diabetic Anti-inflammatory Anti-tumor	(Nishitani et al., 2004)
Volatile oils	Antioxidant	(Ludwiczuk et al., 2017)
Carotenoids	Antioxidant Anti-cancer Spasmolytic Anti-helminthic Wound healing Genotoxic	(Ak et al., 2021)

6. Medicinal/Therapeutic Uses

Known for its adaptability in herbal medicine, pot marigold is used in many home remedies. Insect bites, torn ligaments, eye irritation, etc. are some of the most common skin circumstances cured with it. In addition to treating high fevers and chronic infections, it is also a purifying and exfoliating herb (Milani et al., 2017). Medicinal benefits are believed to be associated with only the deep-orange colored flower range. The plant as a whole is antiphlogistic, disinfectant, antispasmodic, aperient, and diaphoretic. Fresh or dried leaves can be used, and they grow better if they are grown in the morning after precipitation has dispersed on a sunny day (Kodiyan and Amber, 2015). Flower buds should be wide open when cultivated and dried rapidly in the dark. They can also be used fresh or dried and should be dried rapidly when wide open. When taken continuously, tea made from the petals can ease varicose veins and improve blood circulation. Corns and blemishes can be easily removed using crumbled stems (Bunghez and Ion, 2011). Leaf, flower, and bud extracts are used to make homeopathic remedies. Inflammation and swelling are reduced by ingesting it. In the German Commission E Monographs, a remedial manual to herbal medicine, *C. officinalis* is approved for inflammation of the mouth and pharynx (throat), lacerations, and burns (Nicolaus et al., 2017).

6.1 Uses described in Pharmacopoeias and in Traditional Systems of Medicine

This plant belongs to the Calendula family. The plant has been shown to treat inflammations of internal organs, gastric ulcers, menstrual irregularities, and seizures, and act as a diuretic and sweat stimulant. Besides treating oropharyngeal mucosal inflammations, it is also used to treat lacerations and burns. There are many medicinal uses for calendula, including its use as an exfoliator and a purifying herb (Basch et al., 2006). In addition to their antipyretic and anticancer properties, the dried blossoms conceal odors as well. Topical floral infusions are used as antifungals and antiseptics for injuries, scars, wrinkles, sprains, and eye infections. Inflammatory conditions of the skin and mucous membranes can be treated with calendula tea, such as eyewashes, gargles, undies rashes, and other inflammations of the skin. Mental stress and sleeplessness can be treated with *C. officinalis* mother tincture in homeopathy (Ukiya et al., 2006).

The blossoms of *C. officinalis* are antipyretic, antiinflammatory, antiepileptic, and antibacterial, according to Ayurvedic and Unani systems of medicine. Several conditions have been treated with *C. officinalis* in orthodox and homeopathic remedies, including poor vision, varicosities, hemorrhoids, etc. (Safdar et al., 2010). Hepatic obstructions, snake bites, and heart problems were treated with calendula flowers in the Middle Ages. Throughout the eighteenth century, this remedy was used for headaches, jaundice, and red eyes. Jaundice, measles, chicken pox, and other diseases were treated with this plant during the civil conflict (Rehecho et al., 2011).

6.2 An Overview of Folklore uses Supported by Clinical Data can be found in Folk Medicine and Applications

Wound Healing Activity

Based on a study conducted on rats, Calendula flower extracts were established to be significantly active in remedying thermal burns by increasing collagen hydroxyproline and hexosamine levels, with a significant reduction in tissue impairment marker enzymes, and acute phase proteins decreased significantly. The antioxidant activity of Calendula was demonstrated by a decrease in lipid peroxidation (Carvalho et al., 2018; Rittapalli et al., 2019; Ashwlayan and Verma, 2018). Due to its antioxidant and antimicrobial properties, Calendula gel (2%) heals wounds significantly (Chandran and Kuttan, 2008; Leach, 2008).

The granulation tissue formation and improved circulation may be caused by angiogenesis, epithelialization, nucleoprotein, glycoprotein, and collagen metabolism from calendula (Hamad, et al., 2017), as well as reducing discomfort during dressing changes, Calendula treatment has been more effective than other medicines. When skin lesions, burns, and venous ulcers are cleaned with 10% Calendula solution and 2% Calendula gel, healing time is reduced, and many healed wounds are achieved (Aro et al., 2015). By increasing collagen and noncollagen protein concentrations and organizing collagen proteins, the Achilles tendon can be healed by topical application of Calendula cream (Varljen et al., 1989).

Immuno-stimulant Activity

There is immunostimulant action in the polysaccharide part of *C. officinalis*. There was a 40% to 57% phagocytosis associated with Polysaccharide-I, 20 to 30% with Polysaccharide-II, and 54 to 100% associated with Polysaccharide-III (Wagner et al., 1985; Bashir et al., 2006).

Spasmogenic and Spasmolytic Activity

Spasmogenic activity has been observed in aqueous/ethanolic plant extracts. Calendula flowers aqueous/ethanolic extract caused spontaneous contractions to relax and K^+ to induce contractions. Using dichloromethane in fractionated extract inhibited spontaneous contractions. The spasmolytic activity was mediated by calcium channel blockade (CCB) and can be explained under the basic physiology (Rusu et al., 2005).

Hepatoprotective Activity

The glutamo-pyruvate and glutamo-oxalate transaminases are reduced due to the drop in glutamo-pyruvate transaminase. calendula flower hydro-alcoholic extract reduced hepatocyte lysis in CCl_4-intoxicated rats by 28.5% (Varljen et al., 1989; Maysa et al., 2015). In addition to their antioxidant activity, flavonoids and terpenoids also scavenge free radicals. The extract of calendula flowers improved liver function in CCl_4-intoxicated rats (Iatsyno et al., 1978).

Anti-Inflammatory and Anti-Oedematous Activity

The sedative and anti-phlogistic properties of Calenduloside B (trioside of oleinic acid) have been demonstrated in Calendula roots (Preethi et al., 2010). The inflorescence extract of *C. officinalis* exhibited anti-inflammatory properties against mice with acute paw oedema induced by dextran and carrageenan (Hamad et al., 2011).

Anti-Bacterial and Antifungal Activity

Calendula has antibacterial and antifungal properties (Gazim et al., 2008). Antibacterial activity of *C. officinalis* flower extract is observed against a variety of bacteria. It was found that Pseudomonas aeruginosa was the most susceptible to inhibition by the essential oil of flowers. These effects showed that the essential oil of flowers prevented the growth of gram-positive bacteria such as Staphylococcus aureus and Bacillus subtilis *in vitro*, and inhibited gram-negative bacteria such as *Pseudomonas aeruginosa*. There was greater activity in the petals than in the reproductive parts of the plant (Szopa et al., 2020). A glycoside of oleanolic acid and saponin of *C. officinalis* showed anti-parasitic activity against Heligmosomo idespolygyrus.21Nerolidol of *C. officinalis* inhibited coenzyme Q synthesis in all intraerythrocytic stages of the parasite (Dhingra et al., 2022).

Anti-Cancer Activity

The extract of Calendula flower exhibits cytotoxic effects, it contains two main compounds: calenduloside G'6-O-methylester and calenduloside F'6-O-butylester (*in vitro*). That showed anti-cancerous activity by both of them in s everal cancer cell lines. A modified liquid incubation technique of the Salmonella/microsomal assay showed mutagenic and antimutagenic activity for thirteen saponins isolated and identified from *C. officinalis*, *C. arvensis*, and Hedera helix. In this experiment, the Salmonella tester strain TA98±S9 mix was used. A mutagenic urine concentrate from a smoker and a known promutagen were screened for antimutagenic activity. In addition, chlorophyllin's antimutagenic activity was compared with that of chlorophyll. A dose of 400ug/kg of all saponins was found to be non-toxic and non-mutagenic.

Antidiabetic and Anti-Hyperlipidaemic Activity

To induce diabetes in rats, alloxan (150mg/kg body weight) was administered intraperitoneally as a single injection. When compared to normal rats, diabetic rats had significantly higher blood glucose and urine glucose levels. When administered orally at doses of 25 and 50 mg/kg body weight, *C. officinalis* hydroalcoholic extract significantly lowers blood sugar and urine sugar levels in diabetic rats. Alloxan diabetic rats were treated with 100mg/kg body weight of *C. officinalis* hydroalcoholic extract, and all parameters were returned to normal levels of blood glucose, urine sugar, and serum lipid. The extract increased the total hemoglobin levels. There was a similarity between the extract and insulin. As a result, the investigation demonstrated that *C. officinalis* hydroalcoholic extract is both anti-diabetic and anti-hyperlipidemic (Varljen et al., 1989).

Cardioprotective Activity

C. officinalis may prevent Coronary heart disease. It increases cardiac evolved pressure and aortic flow and decreases myocardial infarction size and cardiomyocyte apoptosis in the heart by cumulative left ventricular growth pressure and aortic flow. Due to the two main compounds in calendula extract, calenduloside F'6-O-butylester, and calenduloside G'6-O-methylester, exhibited cytotoxic effects (Braga et al., 2009).

Anthelminthic activity

The dehydrated leaves and flowers of *C. officinalis* have anthelmintic properties. *C. officinalis* dehydrated flowers and leaves were distilled to produce an aqueous extract. It was carried out on Indian adult earthworms due to its biological similarities with the human gut roundworm parasite. *C. officinalis* blossom and leaf extracts demonstrated unconsciousness and worm destruction at 56.5 and 111.2 minutes, respectively. According to previous findings (Pérez-Carreón et al., 2002), saponins produce anthelmintic properties in plants, confirming previous findings that saponins possess anthelmintic properties.

Antioxidant Activity

Female Swiss albino mice were administered Calendula alcoholic extract at doses of 100 and 250 mg/kg body weight, respectively, and their superoxide production was inhibited by 12.6% and 38.7%, respectively. A month of treatment with *C. officinalis* significantly increased catalase activity in mice. Glutathione levels in the blood and liver were significantly increased by the extract. After taking Calendula extract, glutathione reductase was elevated while glutathione peroxidase was decreased (Ramos et al., 1998).

Genotoxic and Anti-genotoxic Activity

Tests were conducted in rat liver culture to determine if *C. officinalis* flower extracts activated and reversed unscheduled DNA synthesis produced by diethylnitrosamine. Using diethylnitrosamine in liver culture, a 40% integration goal of 3-Hthymidine was achieved at an absorption of 1.25 mole, while using aqueous and aqueous-ethanol extracts at concentrations ranging from 50 ng/mL to 16 ng/mL, diethylnitrosamine completely reversed its effects (Mishra et al., 2011). Aqueous and aqueous-ethanol polar extracts activated UDS in rat liver cultured cells without diethylnitrosamine when it was unavailable at concentrations of 25 and 3.7–100 g/mL, respectively. Thus, these polar excerpts had an antigenotoxic action at low doses, whereas they had a genotoxic action at higher doses. *C. officinalis* propylene glycol extract also demonstrated antigenotoxic effects against young growing pigs, evaluated by lymphocyte DNA division and urinary 8-hydroxy-2′-deoxyguanosine after 24 hours (Verma et al., 2016).

A wide spectrum of pharmacological effects has been demonstrated for *C. officinalis*, some of which have great potential for further development. In Table 2, we present some of the cellular mechanisms related to some of the pharmacological properties.

Table 2 Pharmacologically related cellular mechanisms.

Pharmacological Activities	*Mechanism of action*	*References*
Spasmolytic and Spasmogenic	• Calcium channel blocker • Cholinergic property	(Hamzawy et al., 2013)
Antidiabetic and Antihyperlipidemic	• Lowers blood serum and lipid levels • Activates increased insulin secretion from the pancreas	(Cruceriu et al., 2020)

Contd.

Table 2 *(Contd.)*

Pharmacological Activities	*Mechanism of action*	*References*
Hepatoprotective	• Decreases hepatocytolysis • Reduces enzyme changes	(Herold et al., 2003)
Anti-cancer	Leutin increases tumor latency	(Parente et al., 2012)
Antioxidant	Prevents DNA damage by scavenging free radicals	(Talhouk et al., 2007)
Wound healing	Increases collagen synthesis	(Ali et al., 2017)
Anti-inflammatory	• Antioxidant action against macrophage • Modulates cytokines • Inhibits COX-2 gene	(Ghaima et al., 2013)
Antifungal	• Inhibits growth • Comparable to fluconazole	(Alamgir and Uddin, 2010)
Antimicrobial	• Inhibits growth • Comparable to ciprofloxacin	(Kalvatchev et al., 1997)
Immunomodulatory	This compound may inhibit growth factors or molecule expression on lymphocyte surfaces	(de Santana et al., 2020)
Antiviral	Reverse transcriptase activity of HIV-1 is reduced by extract	(Kowalczyk et al., 2022)

In different studies, *C. officinalis* has been shown to exhibit therapeutic effects, thereby contributing to the advancement of therapeutics. Using *C. officinalis* to treat wounds, inflammation, and sepsis has shown high therapeutic efficacy (Efstratio et al., 2012).

In different concentrations, the antimicrobial effects of crude ethanol extracts of flower petals and reproductive parts were assessed inconsistency of eight different kinds of bacteria. *Pseudomonas aeruginosa* (25 mm inhibition zone at 100 mg/ml concentration), and *Staphylococcus aureus* (14 mm inhibition zone at 50 mg/ml concentration) were inhibited by the petals part extracts; however, reproductive part extracts were not as effective as petals part extracts (Barnes et al., 2007; Vanden et al., 1986).

An antimicrobial test was conducted on *C. officinalis* petal extracts against bacteria and fungi. Compared to ethanol extract, the methanol extract of *C. officinalis* demonstrated superior antibacterial action inconsistency to the maximum of the verified bacteria. Several fungi strains were successfully treated with methanol and ethanol extracts (Kaviena, 2017; Bogdanova, 1970; May and Willuhn, 1978). Both humans and animals are vulnerable to viruses. Their genome and proteins are produced in large amounts by redirecting body metabolism. Currently, available antiviral drugs are ineffective in treating diseases caused by these viruses. *C. officinalis* contains many active compounds that can be developed into antiviral drugs. There are several natural antiviral agents found in plants (Kalvatchev et al., 1997).

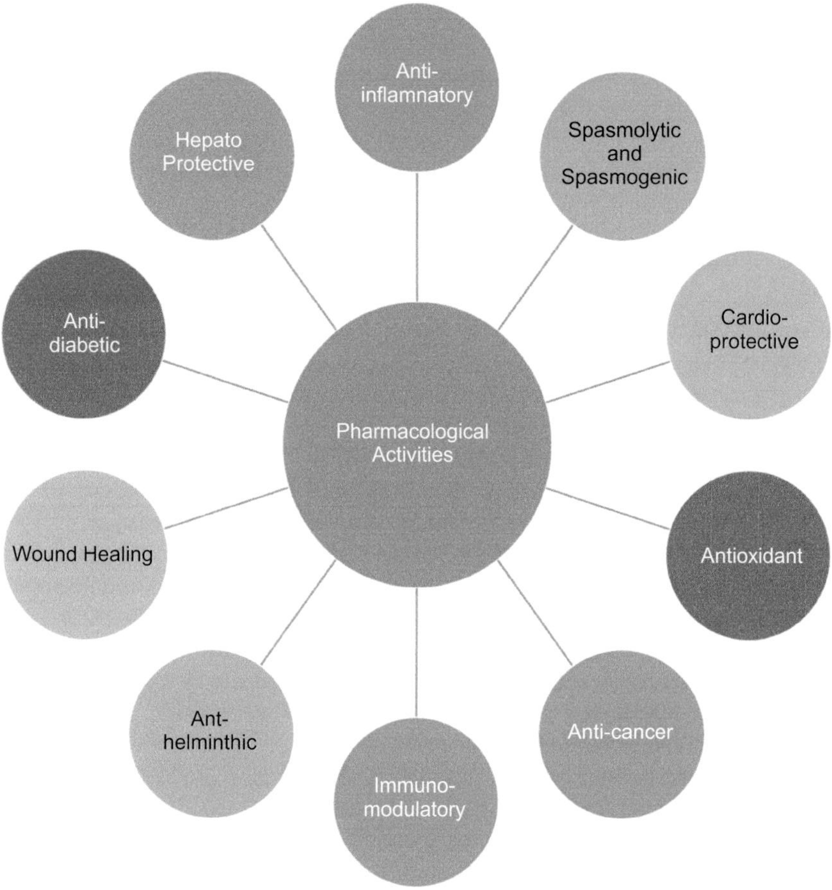

Fig. 4 *C. officinalis* Linn's pharmacological activities. (Dhingra et al., 2022)

6.3 Pharmacological Activities Mainly Antiviral and Mode of Action of Antiviral Compounds

Herpes simplex, influenza A2, and influenza APR-8 viruses were suppressed *in vitro* by a tincture of the flowers (Preethi et al., 2010). HIV-1 replication was inhibited *in vitro* by a chloroform extract of the flowers (IC500.4 mg/ml) (Jan et al., 2017). HIV-1 reverse transcriptase activity was also inhibited by chloroform extract in a dose-dependent manner (ED50 51.0 mg/ml). Mice treated with a 5% hot aqueous extract of the flowers, administered intraperitoneally, were inhibited in their replication of the encephalitis virus (Silva et al., 2007).

IC50 of the chloroform extract of the flowers was 0.4 mg/ml in lymphocytic MOLT4 cells infected with HIV-1 *in vitro*. The effects of chloroform extract on HIV-1 reverse transcriptase activity were also dose-dependent (ED50 51.0 mg/ml) (Al-Snafi, 2016; Szakiel et al., 2008). It was found that dichloromethane-methanol extract obtained from *C. officinalis* flowers exhibited significant anti-HIV activity when tested *in vitro* using MTT/tetrazolium assays. At a concentration of 1,000 g/mL, it inhibited HIV1-RT and suppressed HIV-mediated fusion (Ashwlayan and Verma, 2018). Various influenza APR-8, influenza A2, and herpes simplex viruses were suppressed by flower tincture *in vitro* (Mironescu et al., 2021; Jiang et al., 2020). Calendula flowers were shown to inhibit HIV1-RT and suppress HIV-mediated fusion *in vitro* by dichloromethane-methanol extracts (Mironescu et al., 2021).

C. officinalis leaves were extracted in methanol and ethanol at concentrations of 5, 10, and 15 mg/ml for *in vitro* testing against the Indian adult earthworm, *Pheretima posthuma*. Compared to albendazole (10 mg/ml), both extracts displayed significant anthelmintic activity, although higher concentrations displayed better anthelmintic activity *in vitro* (Jiang et al., 2020; Muley, et al., 2009). *Heligmosomoides polygyrus* larvae, the infective stage of this intestinal parasitic nematode, were inhibited by glycosides of oleanolic acid isolated from marigold (*C. officinalis*). Furthermore, both oleanolic acid and its glycosides reduced L3 survival during prolonged storage, but only oleanolic acid glucuronides affected nematode infectivity (Kashani et al., 2019). Human immunodeficiency virus type 1 (HIV-1) replication was inhibited by extracts of dried flowers from *C. officinalis*. In an *in vitro* MTT ketrazolium-based assay, only the organic extract exhibited potent anti-HIV activity against human lymphocytic Molt-4 cells. Moreover, Molt-4 cells that were co-cultivated with persistently infected U-937/HIV-1 cells for up to 24 h were completely protected from fusion and subsequent death by the organic extract (500 pg/mL). Furthermore, *C. officinalis* organic extract significantly decreased HIV-I reverse transcription activity in a dose- and time-dependent manner. A cell-free system treated with partially purified enzyme for 30 minutes inhibited RT by 85%. According to these findings, organic extracts of *C. oflicinalis* flowers have anti-HIV properties that may be therapeutic (Ashwlayan and Verma, 2018).

The organic extract from *C. officinalis* flowers may be useful for the treatment of viral infections. Further study is needed to isolate the active compound. IC50 value for chloroform extract is 0.4mg/ml, which inhibits HIV-1 replication in lymphocytic MOLT-4 cells *in vitro*. HIV-1 reverse transcriptase activity was also inhibited by chloroform extract in a dose-dependent manner (ED50 51.0mg/ml).70 (Ashwlayan and Verma, 2018). Herpes simplex virus, influenza APR-8, and influenza A2 were suppressed by calendula flower tincture.

6.4 Future of Research and Drug Delivery

C. officinalis has a variety of bioactive compounds, such as flavonoids, triterpenoids, carotenoids, and polysaccharides, that have been shown to have various pharmacological activities. These properties have prompted researchers to investigate its potential use in drug delivery systems, as well as its possible therapeutic applications.

One recent study published in the International Journal of Biological Macromolecules explored the potential use of *C. officinalis* polysaccharides as a carrier for drug delivery. The study found that the polysaccharides exhibited good biocompatibility, high stability, and sustained drug release, making them a promising candidate for drug delivery systems (Mironescu et al., 2021).

Another study published in the Journal of Ethnopharmacology investigated the anti-inflammatory and antioxidant properties of *C. officinalis* extracts, finding that they could potentially be used as a natural anti-inflammatory and antioxidant agent in various diseases (Jiang et al., 2020).

Furthermore, a review published in the Journal of Traditional and Complementary Medicine summarized the medicinal uses of *C. officinalis* in various cultures and highlighted the need for further research to understand its mechanisms of action and potential clinical applications (Muley et al., 2009).

In summary, *C. officinalis* has a long history of use in traditional medicine, and its potential therapeutic applications have led to a growing interest in its use in modern medicine. Further research is needed to explore its mechanisms of action and potential clinical applications, as well as its use as a carrier for drug delivery systems.

7. Toxicity and Safety

C. officinalis, or marigold, is generally considered to be safe for use in medicinal applications, and there is a limited number of reported adverse effects. However, as with any medicinal herb, it is important to be aware of potential safety concerns and to use it under appropriate conditions.

A review of studies published in the Journal of Ethnopharmacology in 2019 concluded that *C. officinalis* is generally safe for human consumption and topical use. The review found no evidence of significant toxicity but noted that it may cause allergic reactions in some individuals (Kashani et al., 2019). One study published in the Journal of Traditional and Complementary Medicine in 2014 examined the toxicity of *C. officinalis* extract in rats and found that it was safe at high doses (Muley et al., 2009). Another study published in the Journal of Environmental Science and Health in 2018 found that *C. officinalis* oil was not toxic to human skin cells and did not cause irritation or sensitization (Silva et al., 2018). It is worth noting that while *C. officinalis* is generally considered safe for use, it may interact with other medications and should be used with caution in individuals who are pregnant or breastfeeding.

It is possible to obtain *C. officinalis* extract from the flowers of *C. officinalis*, or *C. officinalis* as plant material derived from the flowers of *C. officinalis*. Butylene glycol and propylene glycol extractions were also reported. Various ingredients have been reported as components of these ingredients, including sugars, carotenoids, phenolic acids, sterols, saponins, flavonoids, resins, sterins, quinones, mucilages, vitamins, and polyprenylquinones. Almost 200 cosmetic formulations containing *C. officinalis* Extract have been reported (Andersen, 2001).

In rats and mice, acute toxicology studies have shown that the extract is not toxic. A minimal amount of skin irritation was observed in animal tests, and neither sensitization nor phototoxicity was observed. A formulation caused minimal irritation to the eyes, while others did not. A test showed that six saponins isolated from *C. officinalis* flowers were not mutagenic, while a tea derived from *C. officinalis* was non-genotoxic to *Drosophila melanogaster*. The extract caused little irritation or sensitization when tested in clinical trials. In the absence of any basis for extrapolating data from one botanical ingredient group to another, or to the same ingredient extracted differently, the data were considered inadequate for assessing the safety of these ingredients (Andersen, 2001).

According to the CIR (Cosmetic Ingredient Review) Expert Panel, there is insufficient evidence to support the safety of *C. officinalis* Extract and *C. Officinalis*. This extract is derived from the calendula plant, *C. officinalis*, and *C. officinalis* is a plant material derived from calendula flowers. *C. officinalis* Extract was reported used in 178 cosmetic formulations in 1998, and at concentrations of 10% and unknown concentrations in 1984. It was reported that one cosmetics company uses (Andersen, 2001).

In summary, *C. officinalis* is generally considered to be safe for use in medicinal applications, but it may cause allergic reactions in some individuals. It is important to use it under appropriate conditions and to consult with a healthcare professional before use.

Conclusion

In conclusion, *C. officinalis* (marigold) is a medicinal plant with a long history of use in traditional medicine for various purposes, including wound healing, anti-inflammatory, and antimicrobial properties. Its potential therapeutic applications and use as a carrier for drug delivery systems have led to a growing interest in its use in modern medicine.

Studies have found that *C. officinalis* is generally safe for human consumption and topical use, with a limited number of reported adverse effects. However, like any medicinal herb, it is important to use it under appropriate conditions and to be aware of potential safety concerns. While more research is needed to fully understand the mechanisms of action and potential clinical applications of *C. officinalis*, the existing evidence suggests that it may have various therapeutic properties and could be a promising natural remedy for certain conditions. It is important to consult with a healthcare professional before using *C. officinalis*, especially if you are pregnant, breastfeeding, or taking other medications.

References

Agrawal, A.D. (2011). Pharmacological activities of flavonoids: a review. *Int. J. Pharm. Sci. Nanotechnol*, *4*(2), pp.1394–1398.

Ak, G., Zengin, G., Ceylan, R., Fawzi Mahomoodally, M., Jugreet, S., Mollica, A. and Stefanucci, A. (2021). Chemical composition and biological activities of essential oils from Calendula officinalis L. flowers and leaves. *Flavour and Fragrance Journal*, *36*(5), pp.554–563.

Alamgir, M. and Uddin, S.J. (2010). Recent advances on the ethnomedicinal plants as immunomodulatory agents. *Ethnomedicine: a source of complementary therapeutics*, *37*(661), p.2.

Ali, E.M. and Abd El-Moaty, H. (2017). Antifungal activity of Achillea santolina L. and Calendula officinalis L. Essential oils and their constituents against fungal infection of liver as complication of cyclophosphamide therapy. *Journal of Essential Oil Bearing Plants*, *20*(4), pp.1030–1043.

Al-Snafi, A.E. (2016). Antiparasitic effects of medicinal plants (part 1)-A review. *IOSR Journal of Pharmacy*, *6*(10), pp.51–66.

Andersen, F.A. (2001). Final report on the safety assessment of Calendula officinalis extract and Calendula officinalis. *International Journal of Toxicology*, *20*, pp.13–20.

Anderson, V.M., Archbold, D.D., Geneve, R.L., Ingram, D.L. and Jacobsen, K.L. (2016). Fertility source and drought stress effects on plant growth and essential oil production of Calendula officinalis. *HortScience*, *51*(4), pp.342–348.

Aro A. A., Perez M. O., Vieira C. P., Esquisatto M. A. M., Rodrigues R.A.F, Gomes L. and Pimentel E.R. (2015). Effect of Calendula officinalis cream on achilles tendon healing.*Anat Rec*, 298, 428–435.

AshwlayanVD, K.A. and Verma, M. (2018). Therapeutic potential of Calendula officinalis. *Pharm Pharmacol Int J*, *6*(2), pp.149–155.

AshwlayanVD, K.A. and Verma, M. (2018). Therapeutic potential of Calendula officinalis. *Pharm Pharmacol Int J*, *6*(2), pp.149–155.

Barnes, J., Anderson, L.A. and Phillipson, J.D. (2007). *Herbal medicines*. vol 3rd Edition.

Basch, E., Bent, S., Foppa, I., Haskmi, S., Kroll, D., Mele, M., Szapary, P., Ulbricht, C., Vora, M. and Yong, S. 2006. Marigold (Calendula officinalis L.) an evidence-based systematic review by the natural standard research collaboration. *Journal of Herbal Pharmacotherapy*, *6*(3–4), pp.135–159.

Bashir, S., Janbaz, K.H., Jabeen, Q. and Gilani, A.H. (2006). Studies on spasmogenic and spasmolytic activities of Calendula officinalis flowers. *Phytotherapy Research: An International Journal Devoted to Pharmacological and Toxicological Evaluation of Natural Product Derivatives*, *20*(10), pp.906–910.

Betül, A.V.C.I. and Memet, İ.N.A.N. (2021). Comparing of Cultivated Annual and Perennial Calendula officinalis L. Species. *Kahramanmaraş Sütçü İmam Üniversitesi Tarım ve Doğa Dergisi*, *24*(3), pp.579–585.

Bogdanova NS et al. (1970). Study of antiviral properties of Calendula officinalis. Farmakol Toksikol (Moscow), 33:349.

Braga, P.C., Dal Sasso, M., Culici, M., Spallino, A., Falchi, M., Bertelli, A., Morelli, R. and Scalzo, R.L. (2009). Antioxidant activity of Calendula officinalis extract: inhibitory effects on chemiluminescence of human neutrophil bursts and electron paramagnetic resonance spectroscopy. *Pharmacology*, *83*(6), pp.348–355.

Bunghez, I.R. and Ion, R.M. (2011). Complex spectral characterization of active principles from marigold (Calendula officinalis). *Journal of Science and Arts*, *1*(14), pp.59–64.

Carvalho Jr, A.R., Diniz, R.M., Suarez, M.A., Figueiredo, C.S.E.S., Zagmignan, A., Grisotto, M.A., Fernandes, E.S. and Da Silva, L.C., (2018). Use of some asteraceae plants for the treatment of wounds: from ethnopharmacological studies to scientific evidences. *Frontiers in Pharmacology*, *9*, p.784.

Chakraborthy, G.S., (2010). Phytochemical screening of Calendula officinalis Linn leaf extract by TLC. *International Journal of Research in Ayurveda and Pharmacy (IJRAP)*, *1*(1), pp.131–134.

Chaleshtori, S.H., Kachoie, M.A. and Pirbalouti, A.G. (2016). Phytochemical analysis and antibacterial effects of Calendula officinalis essential oil. *Biosci. Biotechnol. Res. Commun*, *9*(3), pp.517–22.

Chandran, P.K. and Kuttan, R. (2008). Effect of Calendula officinalis flower extract on acute phase proteins, antioxidant defense mechanism and granuloma formation during thermal burns. *Journal of Clinical Biochemistry and Nutrition*, *43*(2), pp.58–64.

Mironescu, M., Lazea-Stoyanova, A., Barbinta-Patrascu, M.E., Virchea, L.I., Rexhepi, D., Mathe, E. and Georgescu, C. (2021). Green design of novel starch-based packaging materials sustaining human and environmental health. *Polymers*, *13*(8), p.1190.

Mishra, A., Mishra, A. and Chattopadhyay, P. (2011). A pilot study on in vitro evaluation of flowers of Calendula officinalis (L) as a natural anti-solar agent. *Journal of Natural Pharmaceuticals*, *2*(2), pp.77–77.

Moghaddasi Mohammad, S. and Kashani, H.H. (2012). Pot marigold (Calendula officinalis) medicinal usage and cultivation. *Scientific Research and Essays*, *7*(14), pp.1468–1472.

Mollafilabi, A. and Hosseini, H. (2012), March. Effect of different seed rates and nitrogen fertilizer on yield and yield components of calendula (Calendula officinalis). In *International symposium on Medicinal and Aromatic Plants-SIPAM 2012 997* (pp. 43–50).

Muley, B.P., Khadabadi, S.S. and Banarase, N.B. (2009). Phytochemical constituents and pharmacological activities of Calendula officinalis Linn (Asteraceae): a review. *Tropical journal of Pharmaceutical Research*, *8*(5).

Nand, P., Drabu, S. and Gupta, R.K. (2012). Phytochemical and antimicrobial screening of medicinal plants for the treatment of acne.

Nicolaus, C., Junghanns, S., Hartmann, A., Murillo, R., Ganzera, M. and Merfort, I. (2017). *In vitro* studies to evaluate the wound healing properties of Calendula officinalis extracts. *Journal of Ethnopharmacology*, *196*, pp.94–103.

Nishitani, S., Ijichi, C., Takehana, K., Fujitani, S. and Sonaka, I. (2004). Pharmacological activities of branched-chain amino acids: specificity of tissue and signal transduction. *Biochemical and Biophysical Research Communications*, *313*(2), pp.387–389.

Pandey, A. and Tripathi, S. (2014). Concept of standardization, extraction and pre phytochemical screening strategies for herbal drug. *Journal of Pharmacognosy and Phytochemistry*, *2*(5), pp.115–119.

Parente, L.M.L., Lino Júnior, R.D.S., Tresvenzol, L.M.F., Vinaud, M.C., de Paula, J.R. and Paulo, N.M. (2012). Wound healing and anti-inflammatory effect in animal models of Calendula officinalis L. growing in Brazil. *Evidence-based complementary and alternative medicine*, *2012*.

Pérez-Carreón, J.I., Cruz-Jiménez, G., Licea-Vega, J.A., Popoca, E.A., Fazenda, S.F. and Villa-Treviño, S. (2002). Genotoxic and anti-genotoxic properties of Calendula officinalis extracts in rat liver cell cultures treated with diethylnitrosamine. *Toxicology in vitro*, *16*(3), pp.253–258.

Pires, T.C., Dias, M.I., Barros, L. and Ferreira, I.C. (2017). Nutritional and chemical characterization of edible petals and corresponding infusions: Valorization as new food ingredients. *Food Chemistry*, *220*, pp.337–343.

Preethi, K.C., Kuttan, G. and Kuttan, R. (2010). Anti-inflammatory and Antitumor Activities of *Calendula officinalis*. *Pharmaceutical Biology*, 47(5), 490–495. https://doi.org/10.1080/13880200902988479.

Ramos, A., Edreira, A., Vizoso, A., Betancourt, J., López, M. and Décalo, M. (1998). Genotoxicity of an extract of Calendula officinalis L. *Journal of Ethnopharmacology*, *61*(1), pp.49–55.

Ratajczak, K., Sulewska, H., Szymańska, G., Wolna-Maruwka, A. and Faligowska, A. (2016). The effect of soil type and soil additives on the selected growth parameters and yield of flowerheads of L. *Herba Polonica*, *62*(2), pp.17–30.

Reddy, K.K., Grossman, L. and Rogers, G.S. (2013). Common complementary and alternative therapies with potential use in dermatologic surgery: risks and benefits. *Journal of the American Academy of Dermatology*, *68*(4), pp.e127–e135.

Rehecho, S., Uriarte-Pueyo, I., Calvo, J., Vivas, L.A. and Calvo, M.I. (2011). Ethnopharmacological survey of medicinal plants in Nor-Yauyos, a part of the Landscape Reserve Nor-Yauyos-Cochas, Peru. *Journal of Ethnopharmacology*, *133*(1), pp.75–85.

Rittapalli B.J., Lakshmi N., Boya C., Akula N.V.S., Gupta V.S.S.S. A., Sharma J.V.C. (2019). Various Pharmacological Actions of Calendula officinalis, Tagetes erecta, Carica papaya, Hypericum perforatum and Salvia officinalis. *Int. J. Pharm. Sci. Rev. Res.*, 59(1), Article No. 08, Pages: 42–51.

Rusu, M.A., Tamas, M., Puica, C., Roman, I. and Sabadas, M. (2005). The hepatoprotective action of ten herbal extracts in CCl4 intoxicated liver. *Phytotherapy Research: An International Journal Devoted to Pharmacological and Toxicological Evaluation of Natural Product Derivatives*, *19*(9), pp.744–749.

Safdar, W., Majeed, H., Naveed, I., Kayani, W.K., Ahmed, H., Hussain, S. and Kamal, A. (2010). Pharmacognostical study of the medicinal plant Calendula officinalis L.(family Compositae). *Int J Cell Mol Biol*, *1*(2), pp.108–116.

Sausserde, R. and Kampuss, K., (2014), May. Composition of carotenoids in calendula (Calendula officinalis L.) flowers. In *Proceedings of the 9th Baltic Conference on Food Science and Technology "Food for Consumer Well-Being", Jelgava, Latvia* (pp. 13–18).

Silva, A.F.A., Santos, A.C.A., Oliveira, M.L., Peixoto Sobrinho, T.J.S., Lima, G.R.M., and Matias, W.N. (2018). Evaluation of the safety of Calendula officinalis oil extract in humans. *Journal of Environmental Science and Health, Part A*, 53(2), 167–172.

Silva, E.J., Gonçalves, E.S., Aguiar, F., Evêncio, L.B., Lyra, M.M., Coelho, M.C.O., Fraga, M.D.C.C. and Wanderley, A.G. (2007). Toxicological studies on hydroalcohol extract of Calendula officinalis L. *Phytotherapy Research: An International Journal Devoted to Pharmacological and Toxicological Evaluation of Natural Product Derivatives*, *21*(4), pp.332–336.

Singh, M.K., Sahu, P., Nagori, K., Dewangan, D., Kumar, T., Alexander, A., Badwaik, H. and Tripathi, D.K. (2011). Organoleptic properties in-vitro and in-vivo pharmacological activities of Calendula officinalis Linn: An over review. *J Chem Pharm Res*, *3*(4), pp.655–663.

Smith, R.M., Baker, R.H., Malumphy, C.P., Hockland, S., Hammon, R.P., Ostojá-Starzewski, J.C., and Collins, D.W. (2007). Recent non-native invertebrate plant pest establishments in Great Britain: origins, pathways, and trends. *Agricultural and Forest Entomology*, *9*(4), 307–326.

Spasov, H., Ilieva, F., Gargova, E., Blagoeva, N., Stoyanov, N., Mitev, P. and Tchobanov, I. (2012). Study newly yeast strains for wine production from the area of Demir Kapija, Macedonia. *Scientific Works of Food Science, Engineering and Technologies, 18–19 Oct 2012. Plovdiv, Bulgaria*, *59*(59), 403–408.

Sumit, K., Vivek, S., Sujata, S. and Ashish, B. (2012). Herbal cosmetics: used for skin and hair. *Inven. J*, *2012*, pp.1–7.

Szakiel, A., Ruszkowski, D., Grudniak, A., Kurek, A., Wolska, K.I., Doligalska, M. and Janiszowska, W. (2008). Antibacterial and antiparasitic activity of oleanolic acid and its glycosides isolated from marigold (Calendula officinalis). *Planta medica*, *74*(14), pp.1709–1715.

Szopa, A., Klimek-Szczykutowicz, M., Jafernik, K., Koc, K. and Ekiert, H. (2020). Pot marigold (Calendula officinalis L.): a position in classical phytotherapy and newly documented activities. *Acta Scientiarum Polonorum. Hortorum Cultus*, *19*(3).

Talhouk, R.S., Karam, C., Fostok, S., El-Jouni, W. and Barbour, E.K. (2007). Anti-inflammatory bioactivities in plant extracts. *Journal of Medicinal Food*, *10*(1), pp.1–10.

Ukiya, M., Akihisa, T., Yasukawa, K., Tokuda, H., Suzuki, T. and Kimura, Y. (2006). Anti-inflammatory, anti-tumor-promoting, and cytotoxic activities of constituents of marigold (Calendula officinalis) flowers. *Journal of Natural Products*, *69*(12), pp.1692–1696.

Vanden Berghe, D.A., Vhetinck, A.J. and Van Hoof, L. (1986). Plant products as potential antiviral agents. *Bulletin de l'Institut Pasteur (France)*.

Varljen, J., Lipták, A. and Wagner, H. (1989). Structural analysis of a rhamnoarabinogalactan and arabinogalactans with immuno-stimulating activity from Calendula officinalis. *Phytochemistry*, *28*(9), pp.2379–2383.

Verma, P.K., Raina, R. and Sultana, M. (2014). Phytochemical constituents and antioxidant potential in floral extracts of Calendula officinalis Linn. *World J Pharmaceut Res*, *3*, pp.2067–83.

Verma, P.K., Raina, R., Agarwal, S. and Kaur, H. (2018). Phytochemical ingredients and Pharmacological potential of Calendula officinalis Linn. *Pharmaceutical and Biomedical Research*, *4*(2), pp.1–17.

Verma, P.K., Raina, R., Sultana, M. and Singh, M. (2016). Modulatory effect of Calendula officinalis on altered antioxidant status and renal parameters in diabetic rats. *Pharmaceutical and Biomedical Research*, *2*(4), pp.52–64.

Wagner, H., Proksch, A., Riess-Maurer, I., Vollmar, A., Odenthal, S. and Stuppner, H. (1985). Immunstimulierend wirkende polysaccharide (heteroglykane) aus höheren pflanzen. *Arzneimittel-Forschung*, *35*(7), pp.1069–1075.

Walters, S.M., and Cullen, J. (Eds.). (1984). *The European garden flora: a manual for the identification of plants cultivated in Europe, both out-of-doors and under glass* (Vol. 6). Cambridge University Press.

Wilen, R.W., Barl, B., Slinkard, A.E. and Bandara, M.S. (2004). Feasibility of cultivation calendula as a dual purpose industrial oilseed and medicinal crop. *Acta Horticulturae*, pp.199–206.

Yoshikawa, M., Murakami, T., Kishi, A., Kageura, T. and Matsuda, H. (2001). Medicinal flowers. III. Marigold. (1): hypoglycemic, gastric emptying inhibitory, and gastroprotective principles and new oleanane-type triterpene oligoglycosides, calendasaponins A, B, C, and D, from Egyptian Calendula officinalis. *Chemical and Pharmaceutical Bulletin*, *49*(7), pp.863–870.

Zhang, L., Zhang, G., Xu, S. and Song, Y. (2021). Recent advances of quinones as a privileged structure in drug discovery. *European Journal of Medicinal Chemistry*, *223*, p.113632.

8 Unravelling the Diverse Antiviral Therapeutic Potential of *Camellia sinensis* (L.) Kuntze

Sudip Prasad Jena[1], Priya Godara[1] and Dhaneswar Prusty[1*]

1. Taxonomy and Scientific Classification

Kingdom	:	Plantae
Subkingdom	:	Tracheobionta
Super-division	:	Spermatophyte
Division	:	Magnoliophyta
Class	:	Magnoliopsida
Sub-class	:	Dillenidea
Order	:	Theales
Family	:	Theaceae
Genus	:	*Camellia*
Species	:	*sinensis*
Binomial name	:	*Camellia sinensis L.*

2. Common Names

Tea, also known as *Camellia sinensis*, is a globally consumed drink with various names depending on the location.

In India, it is commonly referred to as "Chai," the Hindi term for tea, particularly the traditional spiced tea commonly consumed in India. In Japan, it is known as "Matcha," a unique variety of green tea ground into a powder and consumed during tea ceremonies (Dreher, 2018). Russia refers to it as "Caravan," the traditional name for the tea transported from China and Central Asia to Russia (Lee, 2014). Sri Lanka is recognized for producing high-grade tea, with many varieties classified as "Ceylon tea." This denomination is derived from the former name of the country, "Ceylon", which was in use until 1972. The favourable climate and soil conditions, coupled with the expertise of local tea cultivators, contribute to the exceptional quality of the tea produced in Sri Lanka. (Lennon, 2021). China simply refers to it as "Cha," encompassing all types of tea grown and consumed in the country (Benn, 2015). In Turkey, tea is commonly known as "çay." It is a popular beverage and is typically served in small glasses. The traditional method of brewing tea in Turkey

[1] Department of Biochemistry, School of Life Sciences, Central University of Rajasthan, NH-8, Bandarsindri, Kishangarh, Ajmer, Rajasthan, India.

* Corresponding author: dhaneswarprusty@curaj.ac.in

DOI: 10.1201/9781003433583-8

is by using a double teapot, called a "çaydanlık," which allows the tea to steep for a longer time before serving (Honda, 2012). Despite the various names, tea is universally appreciated as a refreshing, comforting, and healthy drink.

3. Botanical Characteristics

The tea plant, *Camellia sinensis*, is an evergreen or perennial shrub/small tree widely grown for tea production. The plant has a typical height of 3 to 4 ft. and is known for its profusely branched stem. The tea plant leaves are leathery, elliptical, and have serrated margins. The tea plant leaves are alternatively arranged and possess numerous oil glands containing essential aromatic oil (Pinto et al., 2021). These oil glands are responsible for the characteristic aroma and flavor of the tea. The axils of the leaves contain 2–3 whitish or pinkish flowers, which eventually develop into a 1–4 celled capsular fruit.

Tea plants are constantly trimmed to develop tender shoots called flushes, which are commercially important. These flushes are used to make a variety of teas, including green, black, oolong, and white tea (Zee et al., 2003). The tea plants are raised from seeds and propagated via grafting, budding, and layering. These cultivation methods produce plants resistant to disease and pests and ensure consistent quality and yield of tea leaves.

4. Geographical Distribution

The long-standing popularity of tea is because it has a wide variety of therapeutic uses made from the leaves of the *Camellia sinensis* plant. This plant has undergone a range of processing techniques used to create different varieties of tea (Fig. 1). One type, in particular, green tea, is made by steaming or pan-frying the leaves just after they have been harvested to stop them from oxidizing (Ahmed and Stepp, 2013). There are two main types of tea, one with narrow leaves and the other with broad leaves.

The narrow leaves variety, also known as Chinese tea, is believed to have originated in China and was first introduced to Turkey by traders in the 16th century (Choffnes, 2016). It quickly became popular in the Ottoman Empire and eventually spread to other parts of Europe. On the other hand, the broad leaves variety, also known as Japanese tea, was first introduced to Japan in the 18th century. Since then, Japan has become the world's largest producer of green tea and is known for its high-quality, premium teas such as Sencha, Gyokuro, and Matcha (Koide and Kimura, 2022). Japanese green tea is characterized by its bold, vegetal flavor and is often enjoyed with sushi and other traditional Japanese dishes.

England leads other countries in tea imports and is the world's largest tea consumer. The British have a strong tea-drinking culture, and tea is an important part of daily life in the country (Jingxia and Zhaodan, 2012). The popularity of green tea has been on the rise in recent years, and it has become a popular alternative to traditional black tea.

India is one of the world's largest tea producers, with a rich history of tea cultivation that dates back to the 19th century (Besky, 2014). The main tea-cultivating regions of India include the Assam Brahmaputra and Suma valleys and the Darjiling and Jalpaiguri districts of north Bengal. These account for a significant portion of India's tea production, typically around 70% or more (Kumar et al., 2004). The Suma Valley, located in the foothills of the Himalayas, is another major tea-growing region in India, known for its high-quality teas such as Darjeeling and Kangra (Kapadia, 2002).

The Darjeeling and Jalpaiguri districts of north Bengal, also known as the "Darjeeling Tea Belt", is a famous tea-growing region in India. The area is known for its high altitude and cool climate, which creates ideal conditions for producing Darjeeling tea. Darjeeling tea is known for its musky, floral aroma and unique taste. The remaining 30% of India's total tea production comes from the mountain slopes of the Nilgiri and Annamalai Kerela, Karnataka, Ranchi, Kangra, and Kumaon regions. These regions are characterized by their rocky terrain and cool climate, creating ideal tea cultivation conditions. The teas produced in these regions are known for their unique flavor profiles. Nilgiri tea is known for its bright, fruity taste, and Annamalai tea is known for its strong, rich flavor.

Overall, India is home to various teas, each with its unique flavor profile and characteristics. From the malty, robust Assam tea to the musky, floral Darjeeling tea, India's tea-growing regions offer something for every tea lover.

Tea is a delicate plant that requires specific climatic conditions to flourish. It thrives in mountainous regions with subtropical or tropical climates, where the temperature, rainfall, and altitude are just right. The ideal rainfall for tea growth is around 100–150cm annually. This ensures the tea plants receive adequate water for optimal growth and development. Temperature plays a crucial role in the growth of tea plants. Generally, 24–35°C is ideal for leaves to mature slowly and develop a rich flavour (Carr, 1972). Tea plants are not tolerant to dry conditions, so a consistent water supply is essential. A dry period can cause the plants to wilt and the leaves to lose flavor. Some shade is beneficial to lift growth and protect the leaves from the harsh sun (Han et al., 2018).

5. Phytochemical Constituents and Metabolic Profiling

Phytochemical analysis of *Camellia sinensis* has revealed a wide range of active compounds, including catechins, flavonoids, theanine, caffeine, and carotenoids. More than 4000 bioactive chemicals are thought to be present in tea. However, only about one-third of them are considered beneficial to health (Parmar et al., 2012). The most abundant catechin in green tea is epigallocatechin-3-gallate (EGCG) (Almatroodi et al., 2020). Other catechins found in green tea include epicatechin (EC), Epigallocatechin (EGC), and epicatechin gallate(ECG) (Tallei et al., 2021). The flavonoids found in tea include flavonols, flavones, and flavanones. Theanine, an amino acid, is also found in tea and is responsible for its unique flavor and taste (Juneja et al., 1999).

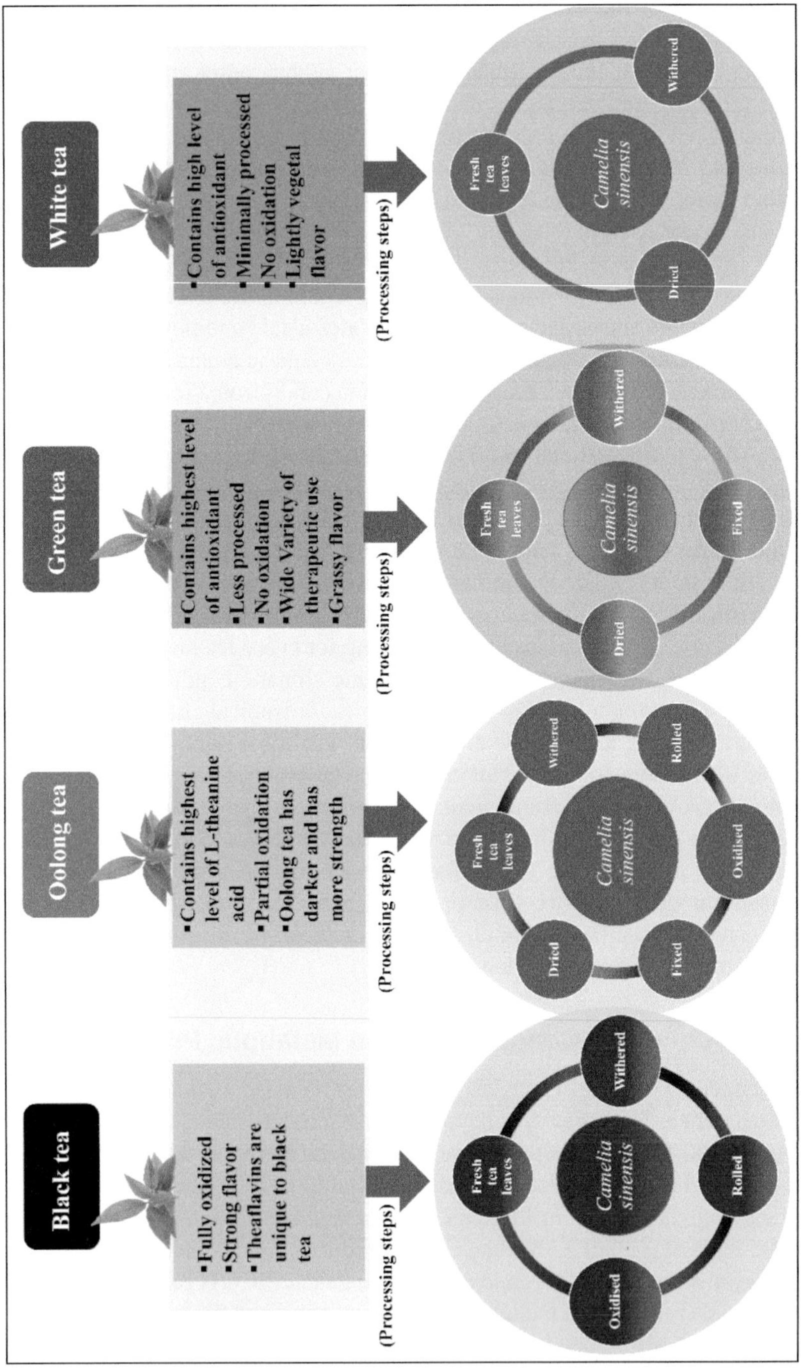

Fig. 1 The range of processing steps followed for obtaining different varieties of tea from *Camellia sinensis*.

Metabolic Profiling of green tea has been done by using various analytical techniques such as high-performance liquid chromatography (HPLC), gas chromatography-mass spectrometry (GC-MS), and nuclear magnetic resonance spectroscopy (NMR) (Le Gall et al., 2004). These techniques help us identify and quantify tea compounds and provide a wide range of understanding about their chemical composition.

Tea can be metabolically profiled by analyzing its impacts on several biological systems, such as antioxidant activity, anti-inflammatory activity, and cancer-fighting characteristics (Kumar et al., 2018). This can be done by performing *in vitro* assays, animal studies, and human clinical trials. These studies' results can help us understand the mechanism of action of green tea compounds based on which drugs may be designed using tea catechines to fight against different diseases.

Tree green tea leaf varieties have been analysed using 1H-NMR spectroscopy to identify the different metabolites present in each variety (Le Gall et al., 2004). The ^{1}H-NMR spectrum is divided into three parts: the first part represents the amino acids and other organic peaks, which are found in the region between δ 0.8 –4.0 ppm; the second part represents the anomeric protons of carbohydrates, which are found in the region of δ 4.0–5.5 ppm; and the third part represents the phenolic region, which is found in the area of δ 5.5–8.0 ppm (Wahyuni et al., 2017). Apart from the ^{1}H-NMR spectrum, additional 2D NMR techniques like J-resolved and COSY (Correlated Spectroscopy) spectra were utilized to recognize and analyze the metabolites found in the tea leaves (Lee et al., 2010).

The tea leaves contain identified metabolites such as amino acids, carbohydrates, organic acids, alkaloids, and flavonoids (Lee et al., 2015). The use of 2D NMR techniques such as J-resolved and COSY allowed for the identification of more specific metabolites, as these techniques provide more detailed information about the chemical structure of the compounds present in the tea leaves (Lee et al., 2015). Overall, the results of this study provide a complete understanding of the metabolic profile of the three varieties of tea leaves and can be used to differentiate between different cultivars or growing conditions.

6. Medicinal/Therapeutic Uses

6.1 Pharmacopoeias and in Traditional Systems of Medicine

Anti-oxidative activity

Green tea has been revered for its health benefits for thousands of years, and recent scientific research has confirmed its medicinal properties. One key factor that makes green tea so beneficial is its high content of antioxidants, specifically polyphenols, which have powerful anti-aging effects.

The anti-oxidative properties of green tea catechins have been studied extensively, and it has been discovered that they have a wide range of biological impacts. The major catechins found in green tea are (–)-epicatechin (EC), (–)-epigallocatechin (EGC), (–)-epigallocatechin gallate (ECG), and (–)-epigallocatechin gallate (EGCg). These compounds are highly reactive to the reactive oxygen species, created by oxidative stress and can cause damage to cells and tissues (Lee et al., 2014). These natural and safe catechins are more effective in neutralizing free radicals than other antioxidants, such as alpha-tocopherol (Hsu, 2005).

Anti-aging Activity

Researchers believe that certain plant flavonoids found in green tea extract may be beneficial in managing age-related muscle deterioration. Herzog et al. reported that two separate diets enriched with flavonoids, one containing green tea extracts and the other containing cocoa flavonols, were found to have improved the survival rate of mice. Both green tea extract and cocoa supplement diets positively impacted the mice's survival (Herzog et al., 2004). It was also found that GTE, but not coca, enhances the density of VAChT and VGluT2 afferent synapses on motor neurons. These synapses are known to be absent in the spinal cords of elderly individuals (Herzog et al., 2004). However, it was also found that GTE alone improved the fraction of neuroprotective microglial phenotypes and decreased age-associated microgliosis (Gras et al., 2021). These findings suggest that green tea extract may have potential benefits in managing age-related muscle deterioration.

Anti-atherosclerosis and Inflammation

Atherosclerosis can be triggered by a high saturated-fat diet, smoking, hypertension, hyperglycemia, obesity, and insulin resistance, which can trigger ECs to produce adhesion molecules, allowing leukocytes to adhere to the artery wall (Libby, 2006). NF-κB, can regulate two types of adhesion molecules, intercellular adhesion molecule-1 (ICAM-1) and vascular cell adhesion molecule-1 (VCAM-1), that bind leukocytes at the site of inflammation (Marui et al., 1993). Tumor necrosis factor-alpha (TNF-α), a potent pro-inflammatory cytokine, contributes to atherosclerosis and chronic inflammation (Brånén et al., 2004). GTC inhibits several inflammatory factors, including NF-kB, cytokines, and adhesion molecules, by producing an anti-inflammatory response. GTC, especially EGCG, helps lower the LDL and increase the HDL to prevent atherosclerosis (Maron et al., 2003).

6.2 Traditional /Folklore Uses Described in Folk Medicine

Anti-diabetic

Xiaoke tea is a traditional Chinese remedy for treating diabetes mellitus. 3–4 grams of Xiaoke tea brewed with hot, boiled water and taken 3–4 times daily to treat diabetic patients (Bailey et al., 1987) effectively. Studies have shown that Xiaoke tea has a gradual anti-hyperglycemic effect, which doesn't affect the insulin level, but it was unable to lower blood glucose levels in diabetic patients who were not using insulin (Hale et al., 1989). In comparison, black tea also has an anti-diabetic effect (Hale et al., 1989). Xiaoke tea has different ingredients and preparation methods, which a Chinese practitioner developed for greater efficacy in diabetic patients.

It's important to note that Xiaoke tea should not be considered a substitute for the medical treatment of diabetes, and it's always important to consult with a medical professional before using any traditional remedies to treat diabetes or any other health condition.

Oral Mucosal Lesion Treatment

Tea has been used for centuries as a therapeutic treatment for various bacteria that cause oral mucosal lesions. To prevent colds, one traditional method of using tea is to gargle with green tea, garlic, and lemon. (Salehi et al., 2019). The nutrients found in tea, specifically catechins, effectively inhibit and kill certain types of bacteria that live in the oral gut. Studies have shown that drinking green tea can also help to eliminate bacteria that cause bad breath (Narotzki et al., 2012). When the levels of these bacteria are reduced, there is also a reduction in plaque. In addition to its oral health benefits, green tea has been found to improve cognitive performance, increase mental alertness, and reduce the time needed to fall asleep. The fluoride in green tea has also been found to inhibit tooth decay (Awadalla et al., 2011).

It's important to note that drinking green tea alone will not replace good oral hygiene practices like brushing and flossing, and it's always recommended to consult with a dental professional before using green tea as a treatment for oral health.

Anti-dandruff Effect

Green tea has been a traditional remedy for anti-dandruff treatment for many years. The fungus *Malassezia Globosa* is the primary cause of dandruff, and it thrives on the sebum gland that is present on the scalp (Kumari et al., 2022). When this fungus consumes sebum, it metabolizes it into oleic acid, which in turn causes dandruff-like symptoms on the scalp (Nocentini et al., 2017). Using green tea as a treatment can help to keep the scalp moisturized, reduce the concentration of oleic acid, and prevent inflammation caused by the fungus (Vishnoi, et al., 2018). This is due to the polyphenols present in green tea, which have been found to have an antifungal effect (Venkateswara et al., 2011).

One way to use green tea for dandruff treatment is by applying it to the scalp and leaving it on for 15–20 minutes before washing it off with a mild shampoo. Drinking green tea is also beneficial for the hair and scalp.

Effect on the Skin

Green tea has long been used as a traditional remedy for treating skin conditions such as wounds, swelling, itching, and burns (Hoffmann et al., 2020). The flavonoids and tannins in green tea act as natural antiseptics, making it an effective treatment for stopping or slowing down bleeding and relieving itchy rashes and insect bites (Tabassum and Hamdani, 2014). Many hair and skin care products now incorporate tea tree oil, derived from the the tea leaf, for its anti-hair fall and sunburn treatment properties. A homemade green tea paste can be applied externally to the affected area for added benefits (Hoffmann et al., 2020).

6.3 Applications Supported by Clinical Data

Anti-carcinogenic Activity

Green tea has long been touted for its potential health benefits, and recent studies have confirmed that the ancient brew may hold the key to fighting cancer. In laboratory experiments, researchers have found that the chemical compounds found in green tea, specifically epigallocatechin gallate (EGCG), have powerful anti-carcinogenic properties that can fight off a variety of cancer cells (Zhang et al., 2007). Between the years 2004 and 2005, a case-control study was carried out in southeastern China (Zhang et al., 2007). The data has been collected from the breast cancer hospital from the age gap between 20 and 87-year-old female patients with histologically confirmed breast cancers. Data on the quantity, frequency, preparation, and kind of tea consumed, as well as diet and lifestyle, were gathered through in-person interviews with interviewers who asked correct and legitimate questions (Chacko et al., 2010). Green tea consumption was found to be associated with a reduced risk of breast cancer. The time spent drinking green tea, the quantity drunk, and the number of fresh batches made daily showed similar dose-response associations.

Despite the uncertainty, one thing is clear green tea is packed with powerful antioxidants and anti-inflammatory compounds that can help protect our cells from damage.

Effect of Green Tea Extract on Obese Women

A clinical trial was conducted with 78 overweight women to evaluate the effect of green tea extract on weight loss and cholesterol levels (Hsu et al., 2008). The study found that the group that received green tea extract experienced a small but significant reduction in body weight, a decrease in LDL cholesterol and triglycerides levels, and an increase in HDL cholesterol levels (Hsu et al., 2008). However, it's important to note that the study had a small sample size, and the results must be confirmed in larger clinical trials. Green tea extract should not be used as a sole treatment for weight loss, but it may benefit weight management.

EGCG decreases glucose production in the Hepatoma cell. The study conducted by Waltner-Law et al. (2002) provides strong in vitro evidence that EGCG (epigallocatechin gallate), a compound found in green tea, reduces glucose synthesis by H4IIE rat hepatoma cells. This study suggests that insulin may have an effect in this area due to its capacity to control the transcription of the two main gluconeogenic enzymes, glucose-6-phosphatase (G6Pase) and phosphoenolpyruvate carboxykinase (PEPCK).

The researchers discovered that EGCG mimics insulin, and boosts the tyrosine phosphorylation of the insulin receptor and the insulin receptor substrate. Additionally, it reduces the production of the gluconeogenic enzyme phosphoenolpyruvate carboxykinase (Anton et al., 2007). To study glucose synthesis in response to insulin or EGCG, H4IIE rat hepatoma cells were incubated in media containing pyruvate and lactate as substrates for gluconeogenesis. These cells were chosen for the study as they respond to hormones in a physiologically expected way (Wolfram et al., 2006). The cells were treated with the presence and absence of insulin concerning EGCG concentrations and found that EGCG mimics insulin at particular concentrations. However, higher concentrations of EGCG may or may not show the same effect of insulin activity (Mereles and Hunstein, 2011).

6.4 Pharmacological Activities Mainly Antiviral

6.4.1 Inhibitory Effect Against Dengue Virus Infection

Recent studies have shown that a compound, epigallocatechin gallate (EGCG), found in green tea may be used against the dengue virus (Al Aboody and Mickymaray, 2020). Flavonoids, secondary plant metabolites derived from tea, are particularly fascinating in their potential to enhance current drug discovery efforts. Specifically, EGCG has been shown to inhibit dengue virus infection with minimal toxicity (Raekiansyah et al., 2018). The compound EGCG, a tea catechin, is effective against the Dengue Virus Type 2 (DENV-2) virus, with inhibition percentages reaching approximately 90%. Studies have revealed that the mechanism of action of EGCG on Dengue Virus Type 2 (DENV-2) involves direct action on virus particles to reduce viral growth, as well as reducing infection by affecting virus internalization or by modulating the endosomal pH (Zhong et al., 2015), which can impact pH-dependent viral fusion (Huang et al., 2014; Raekiansyah et al., 2018).

6.4.2 Influenza Virus

The influenza A and B viruses infect the mucous membrane of the cells by using two proteins, hemagglutinin (HA) and neuraminidase (NA), found on the surface of the virus particle. The mechanism of infection involves the recognition of the HA protein of the virus by N-acetylneuraminic acid in the host cell membrane, followed by entry of the virus into the host cell through endosomal trafficking and uncoating of the virus (Furushima et al., 2018). The uncoated viral genome is then transported to the nucleus for replication. EGCG has been shown to inhibit the interaction of the virus with the cell membrane and affect the acidification of intracellular compartments such as endosomes and lysosomes, which in turn inhibits the growth of the influenza virus (Song et al., 2005). They also exert an inhibitory effect on neuraminidase and affect viral RNA synthesis, leading to the inhibition of infection (Furushima et al., 2018). EGCG interacts explicitly with the hemagglutinin envelope glycoprotein, potentially altering the viral envelope's physical properties (Nakayama et al., 1993).

In conclusion, green tea extract may potentially be a preventive agent against influenza A and B viruses by inhibiting the adsorption of the viruses to the host cells. Further studies are needed to investigate the mechanism of action and to confirm its efficacy as a treatment for influenza.

6.4.3 Human Immunodeficiency Virus (HIV)

Studies have shown that the green tea compound epigallocatechin gallate (EGCG) may have the potential as an anti-HIV medication. Research has found that EGCG can prevent the HIV envelope glycoprotein from attaching to human CD4 molecules on T cells (Chang et al., 1994). This is important because the CD4 molecule serves as a binding site for HIV vesicles and is crucial for the rapid spread of the infection by inhibiting helper T cells.

EGCG was found to be effective at the early stages of HIV infection. It was discovered that EGCG has a high affinity for CD4 molecules and can block the binding of the HIV glycoprotein envelope by attaching to them (Williamson et al., 2006). These findings provide new opportunities for the clinical use of EGCG as a potential anti-HIV medication. However, more research is needed to fully understand the mechanism of action and confirm the safety and efficacy of EGCG as a treatment for HIV.

6.4.4 SARS-COV-2 VIRUS

SARS-COV-2 enters human cells by binding to a cell membrane receptor called ACE2 using its spike protein (Liu et al., 2021). EGCG has been discovered to hinder the attachment of the spike protein to ACE2, which consequently impedes the virus from infiltrating host cells. This is done by EGCG binding to the spike protein receptor binding domain (RBD) on the viral membrane and competing with ACE2 for binding, thus inhibiting the formation of the RBD-ACE2 complex (Liu et al., 2021). The inhibition effect of EGCG on SARS-CoV-2 occurs through its action on ACE2 receptor and the main protease (Mpro). The SARS-CoV-2 virus attaches to the ACE2 protein found in intestinal cells, leading to respiratory distress and the production of a significant number of pro-inflammatory factors (Wang et al., 2021). This activation of High mobility group box 1 (HMGB1) results in its release into the cytoplasm, promoting the synthesis of inflammatory cytokines. The administration of EGCG has been shown to protect mice against lethal endotoxins, stimulate autophagy, and reduce cytoplasmic HMGB1 levels (Singh et al., 2021). At a concentration as low as 10 μg/ml, green tea extract (EGCG) effectively inhibited HMGB1 in macrophages with minimal cytotoxicity.

6.4.5 *Herpes Simplex Virus*

Herpes simplex virus (HSV) encodes for 11 or 12 glycoproteins found on the virus's surface (Campadelli-Fiume et al., 2007). Out of these, four glycoproteins, gD, Gb, and gH/gL, are essential for the virus to enter the host cells. These four glycoproteins interact with cell surface receptors to facilitate the fusion process between the viruses and host cell membrane (Subramanian and Geraghty, 2007). Epigallocatechin-3-gallate (EGCG), a compound found in green tea, has been shown to have antiviral properties against HSV (Campadelli-Fiume et al., 2007).

EGCG's mechanism of action against HSV involves its ability to inhibit the function of these viral glycoproteins, thereby disrupting the fusion process and inactivating the virion. Studies have shown that the inhibition of HSV infection by EGCG is concentration-dependent and only effective before virus adsorption (De Oliveira et al., 2013). However, it has been reported that EGCG can damage the viral envelope of HSV-1. Additionally, it has been observed that EGCG doesn't affect overall virus production (De Oliveira et al., 2013).

6.4.6 *Adenoviruses*

Adenoviruses are non-enveloped viruses consisting of a nucleocapsid and a double-stranded linear DNA genome. They cause a wide range of responses in human infections, ranging from asymptomatic to severe symptoms. Studies have revealed that the polyphenolic compound epigallocatechin-3-gallate (EGCG), found in green tea, possesses antiviral properties against adenovirus (Steinmann et al., 2013). The mechanism of action of EGCG against adenovirus is multi-faceted, including inhibition of viral replication by blocking viral DNA synthesis and protein expression (Steinmann et al., 2013). EGCG is the most effective when added to cells during the transition from the early to the late phase of viral infection, indicating that it may inhibit one or more late steps in the viral infection process (Weber et al., 2003). Additionally, EGCG has been demonstrated to render purified adenoviruses inactive and obstruct viral protease activity. Despite this, its potential as a therapy for adenovirus infections remains restricted (Steinmann et al., 2013).

6.4.7 *Enterovirus 71 (EV71)*

Enterovirus 71 (EV71) is a non-enveloped RNA virus belonging to the family *Picornaviridae*. The largest outbreak of EV71 to date occurred in Taiwan in 1998. According to recent studies, the lack of glucose-6-phosphate dehydrogenase (G6PD) has been found to increase susceptibility to EV71 infection (Ho et al., 2008). G6PD-deficient cells exhibit increased susceptibility to oxidative stress and facilitate EV71 replication more effectively. N-acetyl-cysteine treatment has been shown to alleviate EV71 propagation and provide cytoprotective effects on infected cells. It has been reported that epigallocatechin gallate (EGCG) has inhibitory roles in the replication of EV71 and the creation of infectious progeny (Ho et al., 2009). The antiviral effect of EGCG is associated with its antioxidant activity (Ho et al., 2009). Mechanistically, EGCG significantly reduces EV71-induced oxidative stress.

Furthermore, EGCG has been shown to inhibit enhanced EV71 replication in G6PD-deficient cells, suggesting that the antiviral effect of EGCG may be mediated through its antioxidant properties.

6.4.8 *Hepatitis B virus*

Tea catechins, particularly those found in green tea extract (GTE), have been found to be effective in reducing inflammation in various models of liver injury (Huang et al., 2014). Studies have shown that GTE has an inhibitory effect on the expression of hepatitis B surface antigen (HBs Ag) and hepatitis Be antigen (HBe Ag), which are both associated with viral persistence (Arbuthnot and Kew, 2001). Additionally, GTE has been found to inhibit the activity of intestinal alpha-glycosidase, which is important in processing glycoproteins and glycolipids in viruses. One specific catechin, epigallocatechin gallate (EGCG), has been shown to significantly reduce the replicative intermediate DNA, thereby reducing the production of closed circular DNA and displaying strong anti-HBV activity through the reduction of HBs Ag and HBe Ag secretion *in vitro* (Mahmood et al., 2016).

6.4.9 *Human Papilloma Virus*

Epigallocatechin-3-gallate (EGCG), a polyphenol found in green tea, has been shown to have an inhibitory effect on cervical carcinoma cell lines infected with human papillomavirus (HPV) and to regulate the expression of micro RNA (miRNA). Research has demonstrated that EGCG exhibits inhibitory effects on cervical carcinoma cell proliferation, which may be mediated by the regulation of micro RNA (miRNA) expression (Zhu et al., 2019). miRNAs are small non-coding RNA molecules that regulate protein-coding target genes in eukaryotes (Lee et al., 1993). Studies have suggested that miRNA may be used as biomarkers to assess disease progression or characteristics and may have applications in the treatment of various cancer types (Di Leva et al., 2014). Additionally, recent studies have also shown that tea consumption has been associated with the inhibition of tumor growth both *in vitro* and *in vivo* (Wang et al., 2018). However, further research is needed to establish the mechanism underlying the anti-cancer effects of EGCG and to determine its potential as a novel anti-cervical cancer drug. The different antiviral modes of action of Epigallocatechin-3-gallate (EGCG) is shown in Fig. 2.

6.5 *Overall Antiviral Mode of Action of Catechins*

The main antiviral compounds in *Camellia sinensis* are catechins, including epicatechin, epicatechin gallate, and epigallocatechin. These compounds have been found to inhibit viral replication, including influenza, herpes simplex virus, human immunodeficiency virus, and SARS-COV-2 virus (Yamaguchi et al., 2002; Isaacs et al., 2008; Kim et al., 2013; Jang et al., 2020). The catechins have been

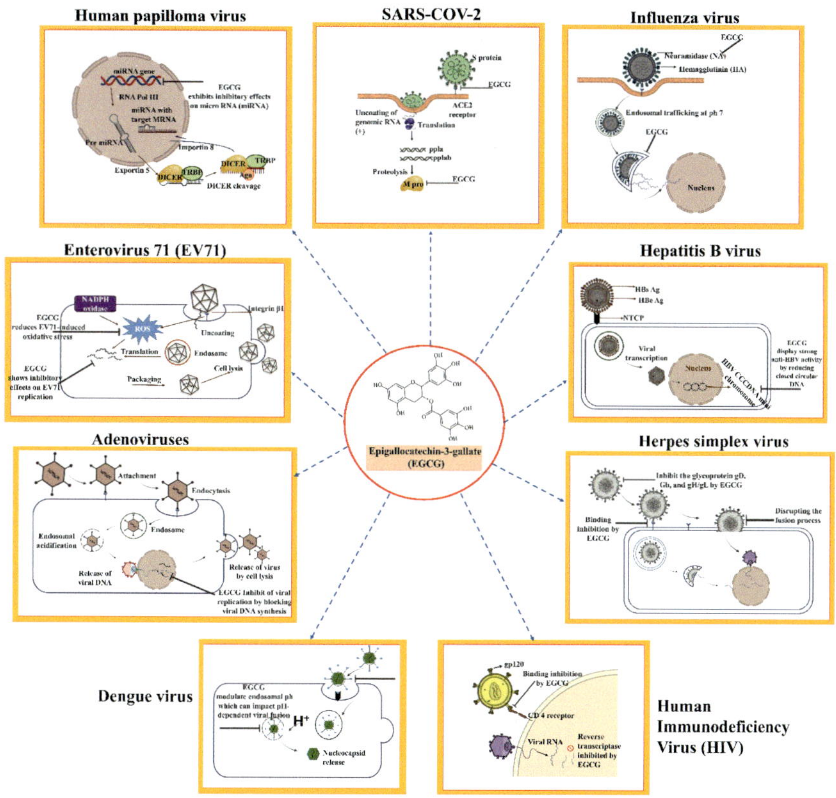

Fig. 2 The different antiviral modes of action of epigallocatechin-3-gallate (EGCG), a green tea catechin.

found to interact with the viral envelope, inhibiting the attachment of the virus to host cells, modifying the target receptor, and preventing the virus entry into the host cell (Xu and Zheng, 2017). They also have been shown to inhibit the activity of viral enzymes, such as reverse transcriptase and protease, which are essential for viral replication (Kitazato et al., 2007). Tea catechins have been found to stimulate the host immune system, boosting its ability to fight off viral infections. This is achieved by increasing the production of antiviral cytokine signaling molecules that play a critical role in activating the immune system and activating immune cells to target and eliminate infected cells (Chowdhury and Barooah, 2020).

The presence of glycoproteins on the surface of viruses is critical for their interaction with host cell receptors and subsequent entry into the host cell. This is a vital step in the replication cycle of the virus.

Epigallocatechin gallate (EGCG) has been shown to exhibit antiviral properties through its ability to inhibit the glycoproteins on the surface of the virus (Calland et al., 2015). This inhibition leads to the disruption of the fusion process between the virus and host cell, preventing the entry of the virus into the host cell.

Furthermore, EGCG has been shown to inhibit the endosomal trafficking process during phagocytosis, a process by which the immune system recognizes and engulfs foreign invaders, including viruses (Colpitts et al., 2015). The endosomal trafficking process is a crucial step in the phagocytic process and determines the fate of the engulfed virus. The inhibition of this process by EGCG can restrict the replication and spread of the virus within the host.

6.6 Future of Research and Drug Delivery

Tea has been known for its health benefits for years, and researchers are now uncovering even more potential uses for this beverage. The key to understanding the power of green tea lies in the chemical compounds found in its leaves, specifically a class of compounds called epigallocatechin gallate (EGCG) or polyphenols having wide therapeutic uses. These compounds are responsible for the bitter taste of green tea (Musial et al., 2020).

Future research in green tea may involve utilizing tea catechins as a basis for designing new drugs. One potential approach would be to modify target receptors, so they do not bind to host cell receptors. This could lead to the development of personalized medicine for treating various diseases. Scientists may find biomarkers that indicate how a person will react to green tea, which would help them figure out when and how much green tea is best for a particular person to provide more effective treatment.

It is expected that encapsulating the anti-cancer drug epigallocatechin-gallate (EGCG) in solid lipid nanoparticles and conjugating it with a peptide that targets a receptor commonly present in breast cancer cells, may increase the stability and bioavailability of EGCG (Radhakrishnan et al., 2019). Results from experiments on mice suggested that this technology would be a potential drug delivery strategy for breast cancer therapy since it was more successful than plain EGCG or non-conjugated nanoparticles at lowering cancer cell proliferation and tumor growth (Radhakrishnan et al., 2019).

Age-related macular degeneration (AMD) is a common condition that affects the retina, leading to vision loss and blindness. Abnormal growth of blood vessels in the retina, a process known as angiogenesis, is a key pathological feature of AMD (Gariano and Gardner, 2005). Vascular endothelial growth factor (VEGF) is a key regulator of angiogenesis, and its overexpression has been linked to the development of AMD. Green tea, specifically the compound Epigallocatechin-3-gallate (EGCG), has been studied for its potential benefits in preventing and treating AMD due to its antioxidant and anti-inflammatory properties. Research has shown that EGCG can reduce the production of VEGF, independent of oxygen concentration (Rosenfeld et al., 2006) This suggests that EGCG or a derivative of it could be used to design drugs that target the inhibition of VEGF production in specific cells or tissues. However, more research is needed to confirm green tea's protective effects and understand the exact mechanisms by which it may protect against AMD.

These forms of drug delivery are still under research, and more studies are needed to confirm their efficacy and safety for human use.

7. Toxicity and Safety

Green tea is generally considered safe for most people, but consuming it in high amounts can lead to some adverse side effects. People with iron deficiency should limit their intake of green tea or avoid it completely because it contains compounds called tannins, which may bind to iron and reduce its absorption in the body. This effect, known as iron chelation, can significantly impair the body's ability to absorb iron from food sources (Fan, 2016). Pregnant women and breastfeeding mothers should also avoid green tea due to its caffeine content, which can increase the risk of birth defects and miscarriage (Bedrood et al., 2018). Additionally, green tea may be avoided if we consume medication for anxiety disorder (Banji et al., 2011).

Another consideration is the effect of green tea on heart conditions. Patients with a heart condition should avoid taking green tea and aspirin together as they both can affect the functioning of the blood clotting factor and cause blood thinning (Chopade et al., 2008). Some people may experience difficulty sleeping, stomach upset, or even insomnia after consuming green tea. Taking other stimulants with green tea may lead to an increase in blood pressure and heart rate.

Green tea and its main constituents have a wide variety of drug interactions, and when taken with other medications, it may cause toxicity to the body. Studies have also found that high doses of green tea can lead to liver dysfunction in some individuals (Murakami, 2014).

It's worth noting that tannin is usually present in tea. So tannin intake on an empty stomach can increase acid, leading to stomach pain, burning sensation, or even constipation. To avoid this, when consuming green tea, it's best to eat something with protein, such as cereal biscuits, popcorn, peanuts , rice, or wheat puffs. Green tea is most effective when consumed with protein. Overall, green tea is generally considered safe when consumed in moderation by healthy adults. However, being aware of potential side effects and interactions with other medications is important.

Conclusion

As we continue to explore the world of tea, we also face the potential benefits and complications. The future of green tea research is optimistic but, at the same time, unpredictable and complex. The invention of biomarkers and molecular markers for tea could help reveal the mysteries of this ancient beverage, which could be used to better understand how it works with medication and can be exploited to upgrade our health. The path ahead may be winding, but the promise of green tea's health benefits makes it a journey worth taking. With each drink, we keep in mind the fragile balance between the positive and adverse consequences of this natural marvel and are motivated to keep looking for methods to take advantage of its full capability.

References

Ahmed, S. and Stepp, J.R. (2013). Green tea: The plants, processing, manufacturing and production. *Tea in health and disease prevention*, 19–31. DOI: 10.1016/B978-0-12-384937-3.00002-1.

Al Aboody, M.S. and Mickymaray, S. (2020). Anti-fungal efficacy and mechanisms of flavonoids. *Antibiotics, 9*(2), 45. DOI: 10.3390/antibiotics9020045.

Almatroodi, S.A., Almatroudi, A., Khan, A.A., Alhumaydhi, F.A., Alsahli, M.A. and Rahmani, A.H. (2020). Potential therapeutic targets of epigallocatechin gallate (EGCG), the most abundant catechin in green tea and its role in the therapy of various types of cancer. *Molecules, 25*(14), 3146. doi: 10.3390/molecules25143146.

Anton, S., Melville, L. and Rena, G. (2007). Epigallocatechin gallate (EGCG) mimics insulin action on the transcription factor FOXO1a and elicits cellular responses in the presence and absence of insulin. *Cellular signalling, 19*(2), 378–383. DOI: 10.1016/j.cellsig.2006.07.008.

Arbuthnot, P. and Kew, M. (2001). Hepatitis B virus and hepatocellular carcinoma. *Int J Exp Pathol, 82*(2), 77–100. doi:10.1111/j.1365–2613.2001.iep0082-0077-x.

Awadalla, H., Ragab, M., Bassuoni, M., Fayed, M. and Abbas, M. (2011). A pilot study of the role of green tea use on oral health. *International Journal of Dental Hygiene, 9*(2), 110–116. DOI: 10.1111/j.1601-5037.2009.00440.x.

Bailey, C.J., Brown, D., Smith, W. and Bone, A.J. (1987). Xiaoke, a traditional Chinese treatment for diabetes. Studies in streptozotocin diabetic mice and spontaneously diabetic BB/E rats. *Diabetes Res, 4*(1), 15–18.

Banji, D., Banji, O.J., Abbagoni, S., Hayath, M.S., Kambam, S. and Chiluka, V.L. (2011). Amelioration of behavioral aberrations and oxidative markers by green tea extract in valproate induced autism in animals. *Brain Research, 1410*, 141–151. DOI: 10.1016/j.brainres.2011.06.063.

Bedrood, Z., Rameshrad, M. and Hosseinzadeh, H. (2018). Toxicological effects of Camellia sinensis (green tea): A review. *Phytotherapy Research, 32*(7), 1163–1180. DOI: 10.1002/ptr.6063.

Benn, J.A. (2015). *Tea in China: A religious and cultural history*: University of Hawaii Press. DOI: 10.21313/hawaii/9780824839635.001.0001.

Besky, S. (2013). *The Darjeeling distinction: Labor and justice on fair-trade tea plantations in India*. Univ of California Press, Vol. 47. DOI: 10.1525/9780520957602.

Brånén, L., Hovgaard, L., Nitulescu, M., Bengtsson, E., Nilsson, J. and Jovinge, S. (2004). Inhibition of tumor necrosis factor-α reduces atherosclerosis in apolipoprotein E knockout mice. *Arteriosclerosis, thrombosis, and vascular biology, 24*(11), 2137-2142. DOI: 10.1161/01.ATV.0000143933.20616.1b.

Calland, N., Sahuc, M.-E., Belouzard, S., Pène, V., Bonnafous, P., Mesalam, A.A., . . . Wychowski, C. (2015). Polyphenols inhibit hepatitis C virus entry by a new mechanism of action. *Journal of Virology, 89*(19), 10053–10063. DOI: 10.1128/JVI.01473–15.

Campadelli-Fiume, G., Amasio, M., Avitabile, E., Cerretani, A., Forghieri, C., Gianni, T. and Menotti, L. (2007). The multipartite system that mediates entry of herpes simplex virus into the cell. *Rev Med Virol, 17*(5), 313–326. doi:10.1002/rmv.546.

Carr, M. (1972). The climatic requirements of the tea plant: A review. *Experimental Agriculture, 8*(1), 1-14. DOI:10.1017/S0014479700023449.

Chacko, S.M., Thambi, P.T., Kuttan, R. and Nishigaki, I. (2010). Beneficial effects of green tea: a literature review. *Chinese Medicine, 5*(1), 1–9. DOI: 10.1186/1749-8546-5-13.

Chang, C., Hsu, F. and Lin, J. (1994). Inhibitory effects of polyphenolic catechins from Chinese green tea on HIV reverse transcriptase activity. *Journal of Biomedical Science, 1*(3), 163–166. DOI: 10.1007/BF02253344.

Choffnes, D. (2017). Tea. In *Nature's Pharmacopeia*.Columbia University Press, 152–167. DOI: 10.14237/ebl.8.1.2017.952.

Chopade, V., Phatak, A., Upaganlawar, A. and Tankar, A. (2008). Green tea (Camellia sinensis): Chemistry, traditional, medicinal uses and its pharmacological activities-A review. *Pharmacognosy reviews, 2*(3), 157. https://api.semanticscholar.org/CorpusID:68384582.

Chowdhury, P. and Barooah, A.K. (2020). Tea bioactive modulate innate immunity: In perception to COVID-19 pandemic. *Frontiers in Immunology, 11*, 590716. DOI: 10.3389/fimmu.2020.590716.

Colpitts, C.C., Verrier, E. R. and Baumert, T. F. (2015). Targeting viral entry for treatment of hepatitis B and C virus infections. *ACS infectious diseases, 1*(9), 420–427. https://doi.org/10.1021/acsinfecdis.5b00039.

De Oliveira, A., Adams, S.D., Lee, L.H., Murray, S.R., Hsu, S.D., Hammond, J.R., . . . Chu, T.-C. (2013). Inhibition of herpes simplex virus type 1 with the modified green tea polyphenol palmitoyl-epigallocatechin gallate. *Food and Chemical Toxicology, 52*, 207–215. DOI: 10.1016/j.fct.2012.11.006.

Di Leva, G., Garofalo, M. and Croce, C.M. (2014). MicroRNAs in cancer. *Annu Rev Pathol, 9*, 287–314. DOI:10.1146/annurev-pathol-012513-104715.

Dreher, N. (2018). Food from nowhere: complicating cultural food colonialism to understand matcha as superfood. *Grad J Food Stud, 5*. DOI:10.21428/92775833.61bff69f.

Fan, F.S. (2016). Iron deficiency anemia due to excessive green tea drinking. *Clinical case reports, 4*(11), 1053. DOI: 10.1002/ccr3.707.

Furushima, D., Ide, K. and Yamada, H. (2018). Effect of Tea Catechins on Influenza Infection and the Common Cold with a Focus on Epidemiological/Clinical Studies. *Molecules, 23*(7). DOI:10.3390/molecules23071795.

Gariano, R.F. and Gardner, T.W. (2005). Retinal angiogenesis in development and disease. *Nature, 438*(7070), 960–966. DOI: 10.1038/nature04482.

Gras, S., Blasco, A., Mòdol-Caballero, G., Tarabal, O., Casanovas, A., Piedrafita, L., . . . Pereira, S. L. (2021). Beneficial effects of dietary supplementation with green tea catechins and cocoa flavanols on aging-related regressive changes in the mouse neuromuscular system. *Aging (Albany NY), 13*(14), 18051. DOI: 10.18632/aging.203336.

Hale, P., Horrocks, P., Wright, A., Fitzgerald, M., Nattrass, M. and Bailey, C. (1989). Xiaoke tea, a Chinese herbal treatment for diabetes mellitus. *Diabetic medicine, 6*(8), 675–676. DOI: 10.1111/j.1464-5491.1989.tb01255.x.

Han, W., Li, X., Yan, P., Zhang, L. and Ahammed, G.J. (2018). Tea cultivation under changing climatic conditions, 455–472. DOI: 10.19103/AS.2017.0036.19.

Herzog, E., Landry, M., Buhler, E., Bouali-Benazzouz, R., Legay, C., Henderson, C., . . . El Mestikawy, S. (2004). Expression of vesicular glutamate transporters, VGLUT1 and VGLUT2, in cholinergic spinal motoneurons. *European Journal of Neuroscience, 20*(7), 1752–1760. DOI: 10.1111/j.1460-9568.2004.03628.x.

Ho, H.Y., Cheng, M.L., Weng, S.F., Chang, L., Yeh, T.T., Shih, S.R. and Chiu, D.T.Y. (2008). Glucose-6-phosphate dehydrogenase deficiency enhances enterovirus 71 infection. *Journal of General Virology, 89*(9), 2080–2089. DOI: 10.1099/vir.0.2008/001404-0.

Ho, H.Y., Cheng, M.L., Weng, S.F., Leu, Y.L. and Chiu, D. (2009). Antiviral Effect of Epigallocatechin Gallate on Enterovirus 71. *Journal of Agricultural and Food Chemistry, 57*, 6140–6147. doi:10.1021/jf901128u.

Hoffmann, J., Gendrisch, F., Schempp, C.M. and Wölfle, U. (2020). New herbal biomedicines for the topical treatment of dermatological disorders. *Biomedicines, 8*(2), 27. DOI: 10.3390/biomedicines8020027.

Honda, Y. 2012. *History of turkish tea drinking habits.* The Proceedings of Turkish Tea Culture Lecture Meeting. https://www.ocha-festival.jp/archive/english/conference/ICOS2001/files/PROC/I-022.pdf.

Hsu, C.H., Tsai, T.H., Kao, Y.H., Hwang, K.C., Tseng, T.Y. and Chou, P. (2008). Effect of green tea extract on obese women: a randomized, double-blind, placebo-controlled clinical trial. *Clinical Nutrition, 27*(3), 363–370. DOI:https://doi.org/10.1016/j.clnu.2008.03.007.

Hsu, S. (2005). Green tea and the skin. *Journal of the American Academy of Dermatology, 52*(6), 1049–1059. DOI: 10.1016/j.jaad.2004.12.044.

Huang, H. C., Tao, M.H., Hung, T.M., Chen, J.C., Lin, Z.J. and Huang, C. (2014). (–)-Epigallocatechin-3-gallate inhibits entry of hepatitis B virus into hepatocytes. *Antiviral Res, 111*, 100–111. DOI:10.1016/j.antiviral.2014.09.009.

Isaacs, C.E., Wen, G.Y., Xu, W., Jia, J.H., Rohan, L., Corbo, C., . . . Hillier, S. (2008). Epigallocatechin gallate inactivates clinical isolates of herpes simplex virus. *Antimicrobial agents and chemotherapy, 52*(3), 962–970. DOI: 10.1128/AAC.00825-07.

Jang, M., Park, Y.-I., Cha, Y.-E., Park, R., Namkoong, S., Lee, J.I. and Park, J. (2020). Tea polyphenols EGCG and theaflavin inhibit the activity of SARS-CoV-2 3CL-protease *in vitro*. *Evidence-based Complementary and Alternative Medicine, 2020*. DOI: 10.1155/2020/5630838.

Jingxia, L. and Zhaodan, S. (2012). A comparative study of tea customs. *Cross-Cultural Communication, 8*(2), 128–133. DOI: http://dx.doi.org/10.3968/j.ccc.1923670020120802.2917.

Juneja, L.R., Chu, D.C., Okubo, T., Nagato, Y. and Yokogoshi, H. (1999). L-theanine—a unique amino acid of green tea and its relaxation effect in humans. *Trends in Food Science & Technology, 10*(6-7), 199–204. https://doi.org/10.1016/S0924-2244(99)00044-8.

Kapadia, H. (2002). *High Himalaya unknown valleys*. Indus Publishing.

Kim, M., Kim, S.-Y., Lee, H.W., Shin, J.S., Kim, P., Jung, Y.-S., . . . Lee, C.-K. (2013). Inhibition of influenza virus internalization by (−)-epigallocatechin-3-gallate. *Antiviral Research, 100*(2), 460–472. DOI: 10.1016/j.antiviral.2013.08.002.

Kitazato, K., Wang, Y. and Kobayashi, N. (2007). Viral infectious disease and natural products with antiviral activity. *Drug Discov Ther, 1*(1), 14–22. https://pubmed.ncbi.nlm.nih.gov/22504360/.

Koide, S. and Kimura, K.-i. (2022). 5 Japanese Green Tea. *Tea as a Food Ingredient: Properties, Processing, and Health Aspects*, 81. DOI: https://doi.org/10.1201/9781003152828.

Krishna Kumar, K., Rupa Kumar, K., Ashrit, R., Deshpande, N. and Hansen, J.W. (2004). Climate impacts on Indian agriculture. *International Journal of Climatology: A Journal of the Royal Meteorological Society, 24*(11), 1375–1393. https://doi.org/10.1002/joc.1081.

Kumar, A., Mosa, K.A., Ji, L., Kage, U., Dhokane, D., Karre, S., . . . Pathania, N. (2018). Metabolomics-assisted biotechnological interventions for developing plant-based functional foods and nutraceuticals. *Critical reviews in food science and nutrition, 58*(11), 1791–1807. DOI: 10.1080/10408398.2017.1285752.

Kumari, K., Yadav, N.P. and Luqman, S. (2022). Promising essential oils/plant extracts in the prevention and treatment of dandruff pathogenesis. *Current Topics in Medicinal Chemistry, 22*(13), 1104–1133. DOI: 10.2174/1568026622666220531120226.

Le Gall, G., Colquhoun, I.J. and Defernez, M. (2004). Metabolite profiling using 1H NMR spectroscopy for quality assessment of green tea, Camellia sinensis (L.). *Journal of Agricultural and Food Chemistry, 52*(4), 692–700. https://doi.org/10.1021/jf034828r.

Lee, C. (2014). From Kiachta to Vladivostok: Russian merchants and the tea trade. *Region: Regional Studies of Russia, Eastern Europe, and Central Asia, 3*(2), 195–218. DOI: 10.1353/reg.2014.0017.

Lee, J.E., Lee, B.J., Chung, J.O., Hwang, J.A., Lee, S.J., Lee, C.H. and Hong, Y.S. (2010). Geographical and climatic dependencies of green tea (Camellia sinensis) metabolites: a 1H NMR-based metabolomics study. *Journal of Agricultural and Food Chemistry, 58*(19), 10582–10589. DOI: 10.1021/jf102415m.

Lee, J.E., Lee, B.J., Chung, J.O., Kim, H.N., Kim, E.H., Jung, S., . . . Hong, Y.S. (2015). Metabolomic unveiling of a diverse range of green tea (Camellia sinensis) metabolites dependent on geography. *Food Chemistry, 174*, 452–459. DOI: 10.1016/j.foodchem.2014.11.086.

Lee, L.S., Kim, S.H., Kim, Y.B. and Kim, Y.C. (2014). Quantitative analysis of major constituents in green tea with different plucking periods and their antioxidant activity. *Molecules, 19*(7), 9173–9186. DOI: 10.3390/molecules19079173.

Lee, R.C., Feinbaum, R.L. and Ambros, V. (1993). The C. elegans heterochronic gene lin-4 encodes small RNAs with antisense complementarity to lin-14. *Cell, 75*(5), 843–854. doi:10.1016/0092-8674(93)90529-y.

Lennon, J. (2021). Tea Landscapes in Sri Lanka. *International Council on Monuments and Sites 11*, 303. https://openarchive.icomos.org/id/eprint/2530.

Libby, P. (2006). Inflammation and cardiovascular disease mechanisms. *The American Journal of Clinical Nutrition, 83*(2), 456S–460S. DOI: 10.1093/ajcn/83.2.456S.

Liu, J., Bodnar, B.H., Meng, F., Khan, A.I., Wang, X., Saribas, S., . . . Ho, W. (2021). Epigallocatechin gallate from green tea effectively blocks infection of SARS-CoV-2 and new variants by inhibiting spike binding to ACE2 receptor. *Cell & Bioscience, 11*(1), 168. DOI:10.1186/s13578-021-00680-8.

Mahmood, M., Martinez, J., Aslam, A., Rafique, A., Vinet, R., Laurido, C., . . . Ali, S. (2016). Antiviral effects of green tea (Camellia sinensis) against pathogenic viruses in human and animals (a mini-review). *African Journal of Traditional, Complementary and Alternative Medicines, 13*, 176. doi:10.4314/ajtcam.v13i2.21.

Maron, D.J., Lu, G.P., Cai, N.S., Wu, Z.G., Li, Y.H., Chen, H., . . . Zhao, J. (2003). Cholesterol-lowering effect of a theaflavin-enriched green tea extract: a randomized controlled trial. *Archives of Internal Medicine, 163*(12), 1448–1453. DOI: 10.1001/archinte.163.12.1448.

Marui, N., Offermann, M., Swerlick, R., Kunsch, C., Rosen, C., Ahmad, M., . . . Medford, R. (1993). Vascular cell adhesion molecule-1 (VCAM-1) gene transcription and expression are regulated through an antioxidant-sensitive mechanism in human vascular endothelial cells. *The Journal of Clinical Investigation, 92*(4), 1866–1874. DOI: 10.1172/JCI116778.

Mereles, D. and Hunstein, W. (2011). Epigallocatechin-3-gallate (EGCG) for clinical trials: more pitfalls than promises? *International Journal of Molecular Sciences, 12*(9), 5592–5603. DOI: 10.3390/ijms12095592.

Murakami, A. (2014). Dose-dependent functionality and toxicity of green tea polyphenols in experimental rodents. *Archives of Biochemistry and Biophysics, 557*, 3–10. DOI: 10.1016/j.abb.2014.04.018.

Musial, C., Kuban-Jankowska, A. and Gorska-Ponikowska, M. (2020). Beneficial properties of green tea catechins. *International Journal of Molecular Sciences, 21*(5), 1744. DOI: 10.3390/ijms21051744.

Nakayama, M., Suzuki, K., Toda, M., Okubo, S., Hara, Y. and Shimamura, T. (1993). Inhibition of the infectivity of influenza virus by tea polyphenols. *Antiviral Res, 21*(4), 289–299. doi:10.1016/0166-3542(93)90008-7.

Parmar., N., Rawat., M. and Kumar, V. (2012). Camellia sinensis (green tea): a review. *Global Journal of Pharmacology, 6*(2), 52–59. https://api.semanticscholar.org/CorpusID:43669711.

Narotzki, B., Reznick, A. Z., Aizenbud, D. and Levy, Y. (2012). Green tea: a promising natural product in oral health. *Archives of Oral Biology, 57*(5), 429–435. DOI: 10.1016/j.archoralbio.2011.11.017.

Nocentini, A., Vullo, D., Del Prete, S., Osman, S.M., Alasmary, F.A., AlOthman, Z., . . . Supuran, C.T. (2017). Inhibition of the β-carbonic anhydrase from the dandruff-producing fungus Malassezia globosa with monothiocarbamates. *Journal of Enzyme Inhibition and Medicinal Chemistry, 32*(1), 1064–1070. DOI: 10.1080/14756366.2017.1355307.

Pinto, T., Aires, A., Cosme, F., Bacelar, E., Morais, M.C., Oliveira, I., . . . Gonçalves, B. (2021). Bioactive (Poly) phenols, volatile compounds from vegetables, medicinal and aromatic plants. *Foods, 10*(1), 106. DOI: 10.3390/foods10010106.

Radhakrishnan, R., Pooja, D., Kulhari, H., Gudem, S., Ravuri, H. G., Bhargava, S. and Ramakrishna, S. (2019). Bombesin conjugated solid lipid nanoparticles for improved delivery of epigallocatechin gallate for breast cancer treatment. *Chemistry and Physics of Lipids, 224*, 104770. DOI: 10.1016/j.chemphyslip.2019.04.005.

Raekiansyah, M., Buerano, C.C., Luz, M.A.D. and Morita, K. (2018). Inhibitory effect of the green tea molecule EGCG against dengue virus infection. *Archives of virology, 163*(6), 1649–1655. DOI: 10.1007/s00705-018-3769-y.

Rosenfeld, P.J., Brown, D.M., Heier, J.S., Boyer, D.S., Kaiser, P.K., Chung, C.Y. and Kim, R.Y. (2006). Ranibizumab for neovascular age-related macular degeneration. *New England Journal of Medicine, 355*(14), 1419–1431. DOI: 10.1056/NEJMoa054481.

Salehi, B., Jornet, P.L., López, E.P.-F., Calina, D., Sharifi-Rad, M., Ramírez-Alarcón, K., . . . Setzer, W.N. (2019). Plant-derived bioactives in oral mucosal lesions: a key emphasis to curcumin, lycopene, chamomile, aloe vera, green tea and coffee properties. *Biomolecules, 9*(3), 106. DOI: 10.3390/biom9030106.

Singh, S., S.K ., M.F., Sonawane, A., Kar, P. and Sadhukhan, S. (2021). Plant-derived natural polyphenols as potential antiviral drugs against SARS-CoV-2 via RNA-dependent RNA polymerase (RdRp) inhibition: an in-silico analysis. *Journal of Biomolecular Structure and Dynamics, 39*(16), 6249–6264. DOI: 10.1080/07391102.2020.1796810.

Song, J.-M., Lee, K.-H. and Seong, B.-L. (2005). Antiviral effect of catechins in green tea on influenza virus. *Antiviral research, 68*(2), 66–74. DOI: 10.1080/07391102.2020.1796810.

Steinmann, J., Buer, J., Pietschmann, T. and Steinmann, E. (2013). Anti-infective properties of epigallocatechin-3-gallate (EGCG), a component of green tea. *Br J Pharmacol, 168*(5), 1059–1073. doi:10.1111/bph.12009.

Subramanian, R.P. and Geraghty, R.J. (2007). Herpes simplex virus type 1 mediates fusion through a hemifusion intermediate by sequential activity of glycoproteins D, H, L and B. *Proc Natl Acad Sci U S A, 104*(8), 2903–2908. doi:10.1073/pnas.0608374104.

Tabassum, N. and Hamdani, M. (2014). Plants used to treat skin diseases. *Pharmacognosy reviews, 8*(15), 52. DOI: 10.4103/0973-7847.125531.

Tallei, T.E., Niode, N.J., Idroes, R., Zidan, B., Mitra, S., Celik, I., . . . Capasso, R. (2021). A comprehensive review of the potential use of green tea polyphenols in the management of COVID-19. *Evidence-based Complementary and Alternative Medicine, 2021*. DOI: 10.1155/2021/7170736.

Venkateswara, B., Sirisha, K. and Chava, V.K. (2011). Green tea extract for periodontal health. *Journal of Indian Society of Periodontology, 15*(1), 18. DOI: 10.4103/0972-124X.82258.

Vishnoi, H., Bodla, R.B., Kant, R. and Bodla, R. (2018). Green Tea (Camellia sinensis) and its antioxidant property: A review. *Int J Pharm Sci Res, 9*(5), 1723–1736. DOI: 10.13040/IJPSR.0975-8232.9(5).

Wahyuni, D., Kristanti, M., Putri, R. and Rinanto, Y. (2017). NMR Metabolic profiling of green tea (Camellia sinensis L.) leaves grown at Kemuning, Indonesia. *Journal of Physics: Conference Series, 795*, 012013. DOI:10.1088/1742-6596/795/1/012013.

Waltner-Law, M.E., Wang, X.L., Law, B.K., Hall, R.K., Nawano, M. and Granner, D.K. (2002). Epigallocatechin gallate, a constituent of green tea, represses hepatic glucose production. *Journal of Biological Chemistry, 277*(38), 34933–34940. DOI: 10.1074/jbc.M204672200.

Wang, Y.Q., Li, Q.S., Zheng, X.Q., Lu, J.L. and Liang, Y.R. (2021). Antiviral effects of green tea EGCG and its potential application against COVID-19. *Molecules, 26*(13), 3962. DOI: 10.3390/molecules26133962.

Wang, Y.Q., Lu, J.L., Liang, Y.R. and Li, Q.S. (2018). Suppressive Effects of EGCG on Cervical Cancer. *Molecules, 23*(9). DOI:10.3390/molecules23092334.

Weber, J.M., Ruzindana-Umunyana, A., Imbeault, L. and Sircar, S. (2003). Inhibition of adenovirus infection and adenain by green tea catechins. *Antiviral Res, 58*(2), 167–173. doi:10.1016/s0166-3542(02)00212-7. DOI: 10.1016/s0166-3542(02)00212-7.

Williamson, M.P., McCormick, T.G., Nance, C.L. and Shearer, W.T. (2006). Epigallocatechin gallate, the main polyphenol in green tea, binds to the T-cell receptor, CD4: Potential for HIV-1 therapy. *Journal of Allergy and Clinical Immunology, 118*(6), 1369–1374. DOI: 10.1016/j.jaci.2006.08.016.

Wolfram, S., Raederstorff, D., Preller, M., Wang, Y., Teixeira, S.R., Riegger, C. and Weber, P. (2006). Epigallocatechin gallate supplementation alleviates diabetes in rodents. *The Journal of Nutrition, 136*(10), 2512–2518. DOI: 10.1093/jn/136.10.2512.

Xu, J., Xu, Z., and Zheng, W. (2017). A review of the antiviral role of green tea catechins. *Molecules, 22*(8), 1337. DOI: 10.3390/molecules22081337.

Yamaguchi, K., Honda, M., Ikigai, H., Hara, Y. and Shimamura, T. (2002). Inhibitory effects of (−)-epigallocatechin gallate on the life cycle of human immunodeficiency virus type 1 (HIV-1). *Antiviral Research, 53*(1), 19–34. DOI: 10.1016/s0166-3542(01)00189-9.

Zee, F., Sato, D., Keith, L., Follett, P. and Hamasaki, R.T. (2003). Small-scale tea growing and processing in Hawaii. http://hdl.handle.net/10125/12180.

Zhang, M., Holman, C.D.A.J., Huang, J.-p. and Xie, X. (2007). Green tea and the prevention of breast cancer: a case–control study in Southeast China. *Carcinogenesis, 28*(5), 1074–1078. DOI: 10.1093/carcin/bgl252.

Zhong, L., Hu, J., Shu, W., Gao, B. and Xiong, S. (2015). Epigallocatechin-3-gallate opposes HBV-induced incomplete autophagy by enhancing lysosomal acidification, which is unfavorable for HBV replication. *Cell Death Dis, 6*(5), e1770. doi:10.1038/cddis.2015.136.

Zhu, Y., Huang, Y., Liu, M., Yan, Q., Zhao, W., Yang, P., . . . Ma, L. (2019). Epigallocatechin gallate inhibits cell growth and regulates miRNA expression in cervical carcinoma cell lines infected with different high-risk human papillomavirus subtypes. *Experimental and Therapeutic Medicine, 17*(3), 1742–1748. DOI: 10.3892/etm.2018.7131.

9 Bioactive Compounds of *Ficus religiosa* L. and its Antiviral Activity

Mubeen Fatima[1], Safdar Hussain[1], Noman Ahmad[2], Hamaad Raza Ahmad[2], Ghousia Liaquat[3], Hafiza Salma[3], Muhammad Nadeem[3], Ghulam Hassan Abbasi[3], Muhammad Ali[3], Tahira Jameel[3], Muhammad Imran[3] and Muhammad Ashar Ayub[3*]

1. Taxonomy and Scientific Classification

Ficus religiosa Linn, (Table 1) often known as the peepal, is a member of the Moraceae family. Common names of this plant are Ahant (Assamese), Pipal tree (English), Pippala (Sanskrit), Asvattha, Ashud, Ashvattha (Bengali), Piplo, Piparo, Pipala, Pipalo (Gujrati), Pipal (Hindi), Basari, Ashvattha (Kannada), Bad (Kashmiri), Pimpal, Pippal, Pipal (Marathi), Aswatha (Oriya), (see Table 2 for a detailed list) Ravichettu (Telugu), Pippal (Punjabi), Arara (Tamil) (Singh and Jaiswal, 2014). One thousand species belong to the genus Ficus, which is part of the family Moraceae and is mostly found in warm and humid climates (Hesami and Rahmati-Joneidabad, 2018). Many of these species are not only useful for their therapeutic properties but also as attractive decorations. In addition to it being a native India, the evergreen, big, fuel-wood, medicinal, attractive, and scaly-leaved *Ficus religiosa L.* tree may be found in Burma/Myanmar, Pakistan, Nepal, Ceylon, China, Thailand, Bangladesh, Vietnam, Iraq, and Iran (Hesami et al., 2018).

India has long recognized it for its mythological, religious, and medicinal significance (Berg et al., 2005). A tall deciduous tree *F. religiosa* is a native to Indian species with no aerial roots (Table 4). It can grow up to 98 feet tall and has a trunk up to 10 feet in diameter. It is indigenous to Southeast Asia and India. A few roots are united by the stem, which is a pale yellow color. The 5–7 alternating, long, petiolate, serrate, or heart-shaped occasionally rounded at the base. As the plant matures the leaves change from pink to copper and then to green. The bark can be either brown or white. The enclosed fig inflorescences bear spherical fig-shaped, fruits known as figs. During the summer fruits are raw and green in color, but once matured, they become black during the rainy season (Singh and Jaiswal, 2014).

[1] Centre for Applied Molecular Biology (CAMB), University of the Punjab, Lahore, Pakistan.
[2] Institute of Soil and Environmental Sciences, University of Agriculture Faisalabad, Pakistan.
[3] Institute of Agro-Industry and Environment, The Islamia University of Bahawalpur, Pakistan.
* Corresponding author: ashar.ayub@iub.edu.pk

DOI: 10.1201/9781003433583-9

Table 1 Taxonomic Classification of *Ficus religiosa.*

Taxonomic Levels		*Reference(s)*
Kingdom	Plantae	(Chandrasekar, 2010; Singh and Jaiswal, 2014)
Phylum	Tracheophyte	
Subphylum	Euphyllophytina	
Infra-phylum	Radiotopses	
Class	Magnoliopsida	
Subclass	Dilleniidae	
Superorder	Urticanae	
Order	Uritcales	
Family	Moraceae	
Genus	Ficus	
Species	*Ficus religiosa*	

1.1 Different Names and Synonyms used for Ficus religiosa

Vernacular names for *Ficus religiosa* include Arayal (Malayalam), bobaum, heiliger feigenbaum, pepulbaum (German), sacred fig (Bengali), Peepal (Hindi), Ravi (Telgu), Asvatthah (Sanskrit), fico del diavolo (Italian), figueira-dospagodes (Portuguese) and Arasu (Table 2). Asvatthah, Peepal, Arayal, and Ravi are among the most prevalent names. Many Ayurvedic remedies, including Nyagrodhadi churna, Nalpamaradi tailam, Chandanasavam, and Saribadyasavam, rely on the barks of *F. religiosa* (Simha and Laxminarayana, 2007). Because it contains substances useful for the treatment of a variety of ailments—including diabetes, skin diseases, central nervous system disorders, gastrointestinal difficulties, respiratory disorders, etc. Therefore, F. religiosa is gaining a lot of interest in the medical field. (Sirisha, *et al.*, 2010). Ficus abu lesan (also known as teen mukadas, teen Asnam, and shajarat bebal) in Arabic. There are many other names for the fig: pu ti shu (in Chinese); botree (in English); arbre de Dieu (in French); bobaum (in German); peepultree (in English); tempelfikus (in Swedish) (Al-Snafi, 2017). Table 3 presents the synonyms of Ficus religiosa cited in different scholarly publications. It has been given over 150 different names in different parts of the globe (Kapile et al., 2022).

Table 2 Colloquial names of *F. religiosa.*

S.No.	*Country*	*Local Name*	*Reference(s)*
1	Brazil	figueira-dos-pagodes; figueira-religiosa	
2	Chinese	pu ti shu; putishu	
3	Cuba	Alamo	
4	Dominican Republic	Alamo; higuillo	
5	English	bo tree; Bodhi; botree fig; peepul tree; pipal tree; sacred fig; sacred tree	

Contd.

Table 2 *Contd.*

S.No.	*Country*	*Local Name*	*Reference(s)*
6	French	arbre bo; arbre de Die; figuier des pagodes	(Datil and Acevedo-Rodríguez, 2014)
7	Germany	Bobaum; heiliger Feigenbaum; indischer Pepulbaum;Pepulbaum	
8	Haiti	Laurel	
9	Israel	ficus kadosh	
10	Italy	fico del diavolo	
11	Myanmar	bawdi-nyaung	
12	Puerto Rico	Lagat	(Tsai and Yokoyama, 2019)
13	Sweden	mai-nyawng	
14	Assam	Ahant	(Singh and Jaiswal, 2014)
15	Bengali	Aswat,Aswatha	
16	English	peepal tree, Pipul	
17	Gujrati	pipul, Piplo, jari	
18	Hindi	Pipal, pipar, pipali	
19	Kannada	aswaththa, Arali, raji, basari	
20	Kashmiri	Bad	
21	Malayalam	Arasu, thullal	
22	Marathi	Pimpal	
23	Nepal	Pipal	
24	Oriya	Jari, aswatha	
25	Punjabi	Pipal	
26	Sanskrit	pippala, Bodhivriksha, ashwatha	
27	Telgu	Ravichettu	
28	Tamil	Kanavam	
29	Urdu	Peepal	
30	Arabic	Teen mukdas, teen asnam	(Kariyawasam, 1995; Al-Snafi, 2017)
31	Spanish	Higuera de agva	
32	Sri Lanka	Bodhi tree	

Table 3 Synonyms cited in different publications for *F. religiosa.*

S.No.	*Names*	*Source*
1	Ficus superstitiosa	The Plant List (2010)
2	Ficus peepul Griff.	The Plant List (2010)
3	Ficus rhynchophylla Steud.	The Plant List (2010)
4	Urostigma religiosum (L.) Gasp.	The Plant List (2010)

Contd.

Table 3 *Contd.*

5	Ficus caudata Stokes	The Plant List (2010)
6	Ficus religiosa var. rhynchophylla Miq.	The Plant List (2010)
7	Ficus religiosa var. cordata Miq.	(Chantarasuwan, 2013)
8	Urostigma affine Miq.	The Plant List (2010)
9	Ficus caudate	(Chantarasuwan, 2013)
10	Ficus affinior	(Wijaya *et al.*, 2021)
11	Ficus peepul	(Krishnamurthy, 2016)
12	Urostigma affine	(Chantarasuwan, 2013)

2. Botanical Description of *Ficus religiosa*

An evergreen or perennial tree is the *Ficus religiosa* which may grow up to 20 meters tall and has a diameter of between 1.5 and 2 meters. It has an uneven form, broad branches, and no aerial roots growing from its branches. The trunk has a regular form and often has buttresses at the base. The bark is a smooth gray color with brownish flecks scattered throughout, and it peels off in irregular, spherical pieces. The leaves are in alternating patterns , which are arranged spirally and broadly ovate, glossy, and coriaceous and are in dark green. The leaves measure 10–18 by 7.5–10 cm and have distinctive tail-like ends that are pink when the plant is young (Table 4). The leaves are stipulate and base-cordate. Petioles are typically between 7.5 and 10 centimeters in length (Singh and Jaiswal, 2014). There are galls on the leaves of the *Ficus religiosa* plant. Flowers are axillary and sessile and only produce male or female offspring. Figs that are in pairs, spherical, have flat tops or are green in color, measure up to 1.5 centimeters wide, are axillary, sessile, smooth, mature to a purple hue with red spots, and have broad, three-lobed basal bracts. This tree is given the special epithet "religiosa" because of the religious value that is associated with it. It is said that Prince Siddhartha sat and meditated beneath this tree, and while doing so, he found enlightenment. After this, he became known as the Buddha. Because of this, Buddhists consider the tree to be holy, and it is often seen growing near temples (Tiwari *et al.*, 2009).

2.1 Tree of Ficus religiosa

One of the most common types of trees found in India is the peepal tree. It expands at a rapid rate. The roots are linked to the trunk in such a way that they seem to be pillars for the support of the trunk. The bark of this tree has a pale gray color and peels off in places. This is the native fig tree that grows to be the biggest of its kind in the United States. It may reach quite huge dimensions. In its earlier phases, it is often epiphytic, which means that it grows on other trees and eventually causes them to die off by suffocating them with its rope-like roots. Another possibility is that the tree will grow via fractures in the wall, which will slowly but inevitably widen as a result of the developing roots. It is estimated that the Peepal tree in Sri

Lanka has been around for 2147 years. It is one of the trees that have the longest lifespan in the world. The peepal tree can withstand dryness as well as cold temperatures (Pokharel and Pokharel, 2021).

2.2 Leaves of *Ficus religiosa*

The leaves and immature twigs are glossy, smooth, and slightly leathery. They have a widely oval form that is abruptly compressed in the form of a long tail at the apex. The base of the leaf is either spherical or heart-shaped. The central nerve of the leaf is rather robust and should be given some consideration. In addition, there are five to nine lateral pairs that join together at their extremities to create an undulating line close to the leaf's border. In general, the leaves have a pendulous habit, which means they hang downward. After a rainstorm, the long, pointed shape of leaf tips contributes to the drainage of water off of the leaves and in the drying of the tree. They fall off throughout March and April, as well as in certain regions during the autumn months. When the young leaves emerge, they are typically pink in color, but as they mature, they deepen to a coppery brown and eventually become green (Sethuramani *et al.,* 2021).

2.3 Flowers of *Ficus religiosa*

The blooms are not visible because the figs are covering them up. Each fig has a few male flowers located at the aperture at the apex; each flower is made up of a single stamen that is supported by three-minute 'petals' that are colorless. Figs appear in pairs at the juncture of the leaf stem and the branch; they are green and smooth at first, but become purple when fully mature, and inside their flesh are a few male flowers that, when reaching maturity, also turn purple. The female flower has five 'petals' that envelop the pistil.

2.4 Fruits of *Ficus religiosa*

The fruits are often referred to as figs. Figs typically reach their peak maturity during May and June, but depending on where you live, you may be able to get this fruit at any time of the year. These fruits, as well as Banyan figs, attract the fig wasp, which may also be found on Banyan figs. Peepal fruits are favored by birds and bats, and when these animals eat them, the seeds are unable to be digested and are expelled into the environment, where they may establish new plants anywhere from a gutter to the side of a building. These seeds need circumstances of adequate wetness to germinate, and they may do so in the unlikeliest of settings. The little plant derives all of its nutrition from the air and the water, yet it clings to the wall or the gutter for support.

2.5 Biophysical Altitude Limits

The needed temperature for the *Ficus religiosa* tree ranges from 16 to 35 degrees Celsius; receives 500–5000 millimeters of rain each year on average. It may thrive in a broad range of soil types, but it does best in alluvial sandy loam that is somewhat deep and has adequate drainage. In addition to this, it may be found on shallow soils, such as rock fissures.

2.6 *Parts Used*

Bark, leaves, shoots, leaf juice, dried fruit, fresh fruit, seed, and latex are some of the diverse components of *F. religiosa* that are utilized in traditional medical practices as well as for various pharmacological activities. *F. religiosa* is also employed in certain scientific research. Numerous Ayurvedic preparations, such as Nyagrodhadi churna, Nalpamaradi tailam, Chandanasavam, and Saribadyasavam, all include the barks of the *F. religiosa* tree, which are considered an essential component (Simha and Laxminarayana, 2007; Warrier, 1993).

3. Microscopic Characters of *Ficus religiosa*

The exterior characteristics of the bark of the *F. religiosa* tree demonstrated that the bark was separated between an inner secondary phloem layer and an outer thick periderm layer. Phellem and phelloderm are two distinct layers that may be found below the periderm. The phellem zone has a thickness of 360 millimeters, and its cross-section is wavy and irregular. Phellem cells are organized into thin tangential membranous layers, and the older layers exfoliate in the form of thin membranes. This arrangement gives phellem cells their layered structure. The phellem layer is the most superficial of the plant's three primary cell layers. The phelloderm zone is rather extensive and easily distinguishable. The cells of the phelloderm differentiate into lignified sclereids. The secondary phloem separated between an inner non-collapsed zone that was narrow and an outside collapsed zone that was vast and open. The non-collapsed zone is made up of axial parenchyma, gelatinous fibers, and radial files of sieve tube components. Outer collapsed phloem is characterized by dilated rays, crushed and destroyed sieve tube members, thick-walled, fibers are lignified, and an abundance of parenchyma cells which are filled. In the outer secondary phloem zone, laticifers may be found in rather high numbers. Phloem rays may be either uniseriate or multiseriate depending on their

Table 4 Macroscopic Characteristics of *Ficus relgiosa.*

Morphological features	*Observation*	*Reference(s)*
Tree Height	Upto 5000 ft.	
Trunk size (diameter)	Upto 1 meter	(Singh and Jaiswal, 2014)
Tree Shape	Irregular shaped; widespread branches	
Leaf Shape	Heart-shaped; glossy; dark green	(Makhija *et al.*, 2010)
Leaf Arrangements	5-7 alternating pattern	
Size	7.5-10 cm	
Petiole	Slender petioles	

location. Multiseriate rays are always homocellular, but uniseriate rays might be homocellular or heterocellular (Babu et al., 2010; Makhija et al., 2010).

Table 5 Microscopic Characteristics of *F. religiosa*

Physiognomies	*Reference(s)*
Starch grains	(Babu *et al.*, 2010; Makhija *et al.*, 2010)
Crystals of Calcium Oxalate	
Oil Globules	
Xylem fibers	
Cork cells	
Phloem fibers	
Vascular bundles	
Parenchyma cells	
Biseriate Medullary Rays (BMR)	

4. Geographical Distribution of *Ficus religiosa*

There are many intriguing stories about the Peepal tree that have been passed down through the generations, even though no one is certain of the tree's origins. The peepal was perhaps the first tree to be portrayed in any form in India. The worship of the peepal is shown on a seal that was found at Mohenjo-Daro, which was one of the towns that were inhabited during the Indus Valley Civilization. The notion that the peepal tree was revered by Hindus even in those earlier periods is supported by excavations that have been carried out. Peepal is indigenous to the region spanning from India to southeast Asia, and it may be grown and found wild up to an altitude of 5000 feet. Many tropical (Figure 1 and Table 6) regions throughout the globe are suitable for growing the *Ficus religiosa* tree. In the United States, it is cultivated in the southern parts of the states of California, Florida, and Hawaii, as well as Homestead and Miami. It doesn't matter where the tree came from; it needs a lot of area to develop, and the soil has to be somewhat deep so that its roots may spread out across a wide area. It is a huge tree, reaching heights of around 20 meters, and has a crown that is fully grown. It requires a subtropical environment with hot summers and frost during the monsoon season to thrive, although it is adaptable to a broad range of soil types.

Around 500 species of Ficus may be found in the Asian–Australasian area, making it the richest and most diversified. On the other hand, around 110 and 130 species of Ficus exist in Africa and the Neotropics area, respectively, making for a less diverse population of the genus. A little less than half of the species of Ficus are monoecious, while the other half are dioecious in function (Singh *et al.*, 2011).

5. *Ficus religiosa* Products Available in the Market

Candies, arjun tea, and neotea arasa (Table 7) have all used *F. religiosa* as a component at one point or another in their production. Arjun tea is useful in treating

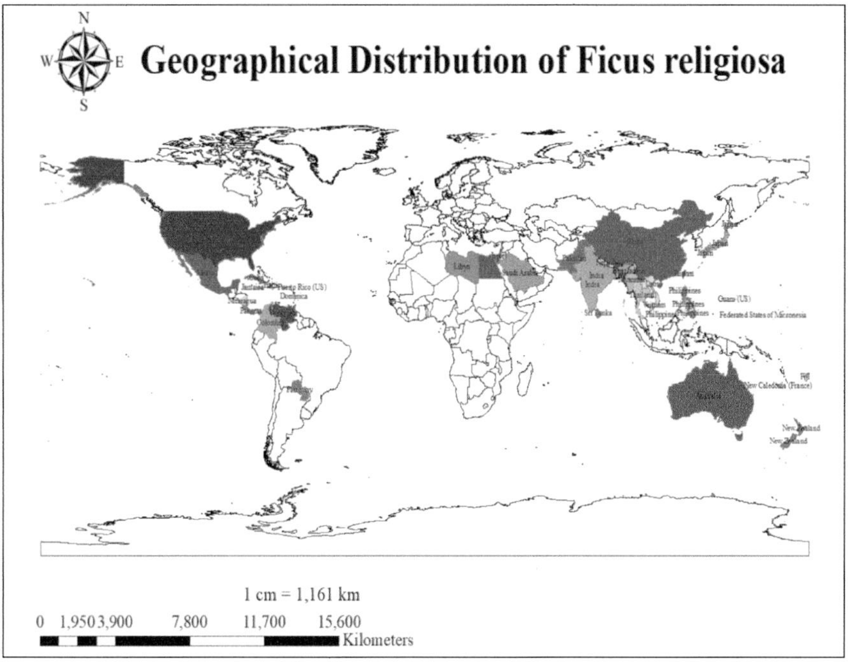

Fig. 1 Geographical distribution of *Ficus religiosa*

Table 6 Geographical Distribution of *F. religiosa*

Country/ Region	*Status*	*Reference(s)*
Chad	Native	(Orwa, 2009)
Egypt	Planted	(Flora of Pakistan)
Libya	Introduced	(Flora of Pakistan)
Madagascar	Planted	(CABI)
Bangladesh	Native	(Flora of Pakistan)
Bhutan	Present	(Mesoamericana, 2014)
China	Present	(Flora of China Editorial Committee. (2014). Flora of China. St. Louis)
Fuji	Planted	(CABI, Undated. Compendium record. Wallingford, UK: CABI)
Yunan	Native	(Rojo JP, 1999)
Hong Kong	Planted	PIER (2014)
India	Native	(Rojo JP, 1999)
Jharkhand	Present	(Kumar A., 2013)
Karnataka	Native	(CABI)
Kerala	Present	(CABI)
Madhya Pradesh	Native	(Firdousi, 2018)

Contd.

Table 6 *Contd.*

Country/ Region	*Status*	*Reference(s)*
Uttar Pradesh	Native	(Singh *et al.*, 2009)
West Bengal	Native	(CABI)
Israel	Introduced	(Seebens, 2017)
Japan	Present	(CABI)
Laos	Native	(CABI)
Myanmar	Present	(Kress WJ, 2003)
Nepal	Native	(Rojo JP, 1999)
Pakistan	Native	(Flora of Pakistan)
Philippines	Introduced	(DA, 1995)
Saudi Arabia	Present	(El-Sherbiny, 2011)
Singapore	Introduced	(Chong, 2009)
Sri Lanka	Introduced	(DJ, 2008)
Taiwan	Introduced	(Seebens, 2017)
Thailand	Native	(Rojo JP, 1999)
Vietnam	Native	(Rojo JP, 1999)
Columbia	Introduced	(Mesoamericana, 2014)
Paraguay	Introduced	(Mesoamericana, 2014)
Northern Mariana Islands	Introduced	(PIER, 2014)
New Caledonia	Introduced	(PIER, 2014)
New Zealand	Present	(James, 2011)
Guam	Introduced	(PIER, 2014)
Fiji	Introduced	(PIER, 2014)
Federated States of Micronesia	Present	(Wagner, 2014)
New South Wales	Naturalized	(Randall, 2012)
Australia	Present	CABI (Undated a)
Hawaii	Introduced	(Starr F, 2003)
Florida	Introduced	(Nadel, 1992)
California	Introduced	(Starr F, 2003)
United States	Introduced	(Orwa, 2009)
Puerto Rico	Introduced	(Liogier, 2000)
Panama	Introduced	(Mesoamericana, 2014)
Nicaragua	Introduced	(Mesoamericana, 2014)
Mexico	Introduced	(Mesoamericana, 2014)
Jamaica	Introduced	(Acevedo-Rodríguez, 2012)
Haiti	Introduced	(Acevedo-Rodríguez, 2012)
Dominican Republic	Introduced	(CABI)
Cuba	Introduced	(Acevedo-Rodríguez, 2012)
Costa Rica	Present	(Mesoamericana, 2014)

heart disease and decreasing cholesterol levels. It also includes the micronutrient coenzyme Q10 and functions as an antioxidant. Neotea arasa is useful against jaundice as well as heart disease due to the flavonoids and sterols it contains. Candies containing dry fruit powder are now being used as a therapy for asthma; alternatively, the powder may be taken after being mixed with water (Singh and Jaiswal, 2014).

Table 7 Herbal Products of *F. religiosa Available* in the Market

Name of Product	*Composition*	*Reference(s)*
Arjun tea	Bark of *F. religiosa*	(Amazon, 2018)
Neotea arasa ilia	Leaf Powder	(Amazon, 2018)
Nalpamaradi oil	Bark of *F. religiosa*	(Singh et al., 2018; Ayurtimes.com 2018)
Neotea arasam pala powder	Fruit of *F. religiosa*	(Amazon, 2018)
Neotea arasam pattai powder	Bark	(Amazon, 2018)
Nyagrodhadi churna	Stem bark of *F. religiosa*	(Anupama, 2014)
Oil pulling Concentrate	*F. religiosa*	(Drtungs.com 2018)
Sarasa syrup	*F. religiosa*	(Imispharma.com 2018)
Sarivadyasava	Bark of *F. religiosa*	(Ayurtimes.com 2018)
Slim honey	*F. religiosa*	(Indiamart, 2018)
Panchavalkadi tailam	Bark of *F. religiosa*	(Ayurmedinfo.com 2018)

6. Physiochemistry and Phytochemical Analysis

According to preliminary phytochemical analysis, flavonoids, tannins, saponins, cardiac glycosides steroids, and terpenoids were found in the *F. religiosa* barks (Anandjiwala *et al.*, 2008; Babu *et al.*, 2010). The *F. religiosa* barks included bergapten, lanosterol, bergaptol, -sitosterol, vitamin K1 and -sitosterol-d-glucoside (Varshney and Bhatnagar, 1965; Swami *et al.,* 1989; Swami and Bisht, 1996). Arginine, glycine, tyrosine, aspartic acid, methionine, serine, threonine, proline, tryptophan, leucine, valine, n-nonacosane, n-hentricontanen, n-octacosanol, and hexa-cosanol are all found in the leaves (Panda *et al.*, 1976; Behari *et al.*, 1984; Verma and Bhatia, 1986). The compounds undecane, asgaragine, tridecane, terpinene, tyrosine, thujene, tetradecane, pinene, limonene, dendrolasine, ylangene, bourbonene, alloaromadendrene, caryophyllene, copaene, and humulene are also identified in the fruit of Ficus religiosa, as detailed in Table 8.

7. Traditional and Contemporary Uses

Traditional Indian medicine has employed herbs to treat a variety of ailments (Table 9). The bark of *Ficus religiosa* has traditionally been used as an antimicrobial, antiprotozoal, antiparasitic, antiviral, astringent, antidiarrheal, and

Table 8 Phytochemicals present in different parts of *F. religiosa*

Phytochemical Constituents	*Parts*	*References*
Flavonoids	Bark	(Varshney and Bhatnagar 1965; Swami et al., 1989; Swami Bisht, 1996)
Tannins		
Terpenoids		
Cardiac glycosides steroids		
Saponins		
Bergapten		
-sitosterol-d-glucoside		
Vitamin K1		
-sitosterol		
Bergaptol		
Lanosterol		
Arginine	Leaves	(Panda et al., 1976; Behari et al.,1984; Verma and Bhatia, 1986)
Hexa-cosanol		
n-octacosanol		
n-nonacosane		
n-hentricontanen		
Leucine		
Proline		
Valine		
Tryptophan		
Threonine		
Serine		
Methionine		
Aspartic acid		
Tyrosine		
Glycine		
Myricetin	Fruit	(Ananda and Kunjani, 2000)
Quercetin		
Kaempefero		
Albuminoids	Seed	(Sheetal et al., 2008)
Phytosterolin		
β-sitosterol		

in the treatment of gonorrhea and ulcers, while the leaves have been used to cure skin problems. The leaves have antivenom action and help in the regulation of the menstrual cycle in females (Ghimire and Bastakoti, 2009). It has been utilized in the treatment of numerous disorders in Bangladesh, including cancer, inflammation, sexual disorders, and infectious diseases. The sensitive branches of this plant are used as a toothbrush in cases of severe fever. Laxatives may be made from fruits, latex can be used as a tonic, and fruit powder can be used to treat asthma (Chandrasekar et al., 2010). According to studies, the leaves of the *Ficus religiosa* tree are used to bind or knot the area where blood is oozing from wounds because the tree has constituents that have wound-healing properties and are also helpful for halting bleeding. A mixture of leaf powder, honey, and fennel seed helps with constipation. Another study revealed that the leaves of the tree are cooked, covered with ghee, and applied or tied to the affected area to treat and cure numerous skin infections, such as measles, epidemic parotitis, etc (Singh and Goel, 2009; Kumar et al., 2018).

The bark extract of *F. religiosa* has been used traditionally to treat patients with diarrhea, paralysis, diabetes, gonorrhea, and fractured bones. The antidote in the extract of the stem or bark may be used to prevent bleeding from cuts and wounds. Asthma and other respiratory disorders may be treated using *F. religiosa* fruit. Sexual syndromes, gastrointestinal problems, hematuria, cough, eye irritation, earache, toothache, back discomfort, headache, and blemished or itchy skin are all treated well with leaf extract of *F. religiosa* (Sirisha et al., 2010). According to Kaur et al. (2011), bark decoction of *Ficus religiosa* coupled with honey functions as a healing agent against numerous ailments such as ulcers, gonorrhea, and skin problems. Infections of the skin may be cured by applying a paste made from

Table 9 Traditional and Contemporary uses of *F. religiosa*

Parts of Plant	*Uses*	*References*
Bark	Astringent, cooling, aphrodisiac, antibacterial effective for the treatment of gonorrhoea, diarrhea and gastrohelcosis, anti-inflammatory, burns	(Warrier, 1993)
Bark Decoction	Cooling, gonorrhea, skin infections, scabies, hiccup, nausea	(Kapoor, 1990)
Leaves and young branch	Laxative, wound healing, skin ailments	(Kunwar et al., 2006)
Leaf Extract	Respiratory disorders, diarrhea, toothache, sexual disorders, hematuria, migraine, eye irritation, gastric complications, cough, scabies	(Kapoor, 1990; Warrier, 1993; Kunwar et al., 2006)
Leaf decoction	Anodyne for toothache	(Kunwar et al., 2006)
Fruit (Fresh or dry)	Asthma, laxative, fever, digestive Tuberculosis, hemorrhoids	(Khanom et al., 2000)
Seeds	Refrigerant, laxative	(Warrier, 1993)
Latex Extract	Neuralgic, anti-inflammatory, hemorrhages	(Warrier, 1993)

sensitive roots. Vaginal discharges and uterine problems have been successfully treated with the decoction. A powder form is suggested for use on burns. A mixture of root extract, salt, and brown sugar has diuretic effects.

In the past, its roots were used to cure painful, swollen joints by restoring a healthy uric acid level. Additionally, cold sores and gum disorders may be treated with roots. It has been operational in curing fluor albus in ladies. As the fruit matures, it can be used as an antidote to poison. You can cure diarrhea, and urinary issues, and improve fertility using fruit powder. The powder from these plants has also been shown to alleviate mental illness. It has been shown that eating fresh fruit may prevent cardiac syndromes and help with dehydration (Panchawat 2012; Gautam et al., 2014). *F. religiosa* has been historically used to manage nasal bleeding, hemorrhages, diabetes mellitus, constipation, fever, and jaundice (Sandeep et al., 2018). As a result of its medicinal properties, *Ficus religiosa* has been used to treat a broad range of conditions. In pharmaceutics, many elements of the *Ficus religiosa* tree have been employed, including the bark and its latex extract, fruit, leaves, aerial roots, and seeds, alone or in combination with other plants. Several Ayurvedic medicines rely on bark as their primary ingredient; for example, "Pancha Valkala Kashaya" is made by boiling and reducing "Pancha Valkaladi Tailum," which consists of "*Pancha Valkaladi*," "*Azadirachta indica*," "*Ficus religiosa*," "*Ficus benghalensis*" and other herbs (Panda, 2005; Singh and Panda, 2005).

Synergistic effect in terms of pharmaco-kinetics and pharmaco-dynamics is produced by the combination of traditional medicines, which lessens the aftereffects of treatment for infectious diseases by slowing the development of resistance to antibiotics and expanding the spectrum of antimicrobial activity (Aiyegoro and Okoh, 2009). By balancing Pitta and Kapha, *F. religiosa* is a recognized Ayurvedic medication that treats a wide range of conditions, including those caused by a disruption of the digestive system (such as hiccups, ulcers, mucositis, respiratory issues, joint and skin allergies, broken bones, and diabetes). When amalgamated with other herbs, *F. religiosa* functions as a drug vehicle (Anupaan), by boosting the assimilation, absorption, and circulation of medication into the body. For inflammatory ulcers, for instance, using a mixture of *F. religiosa* leaves and butter fat aids in the deep penetration of medication into the interior tissues at the afflicted location (Tiwari, 1998; Singh et al., 2011).

8. Pharmacological Actions of F. Religiosa against Different Diseases

The pharmacological actions of *Ficus religiosa* against various diseases are well-documented in the literature. These actions span a wide range of conditions, including cancer, fertility disorders, diabetes, nervous system disorders, digestive issues, and its antimicrobial and antiviral properties. A detailed summary of these therapeutic effects can be found in Table 10.

8.1 F. religiosa and Cancer

F. religiosa latex extract shows potential for the treatment of malignant tumors. The effects of using *F. religiosa* have been promising. Neroblastoma brain tissue cells, a human breast adenocarcinomic cell line, and colorectal cancer cell lines have all been exposed to its crude latex extract. Breast cancer and colorectal cancer cell lines treated with bark extract show a survival mechanism during G1 phase, whereas neroblastoma cell lines show a survival mechanism during G2 or metaphase. Inhibitory concentrations of latex extract ranged from 4.8 to 1.13 micrograms per milliliter across all cell lines evaluated. Toxic phytochemicals found in *Ficus religiosa* have shown promise in the therapy of cancerous tumors, according to this research (Saida, 2021). Extract of *F. religiosa* when combined with ethanol, to test against cytotoxic effects for brined shrimps. There is evidence that higher dosages enhance shrimp mortality. The LC50 and LC90 values were determined to be 2.7 and 4.62 g/ml, respectively. *F. religiosa* essential oil is pressed from the plant's leaves. In an *in vivo* test using brine shrimp, *F. religiosa* showed only a moderate impact (LC50 = 50 g/ml). 80–55% of cell death occurs at 100 g/ml indicating that it has anti-cytotoxic effect in a human breast cancer cell line (Rahman et al., 2014). The effects of *F. religiosa* aquaous and ethanolic bark extract on cervical cancer cell lines (SiHa and HeLa) were investigated. Both aqueous and ethanolic extracts continued to be beneficial in treating cervical cancer cell lines, with the former showing its efficacy even at low concentrations (Choudhari et al., 2011). The leaves of the *F. religiosa* tree are extracted. Researchers examined the effects of *F. religiosa* on human breast cancer cells throughout various dosage and time intervals. The extract had a modestly harmful effect on normal cells but showed permanent inhibition in malignant breast cells. Evidence at the cellular level showed that increased LEHDase activity in cancer cells caused a cell cycle arrest in the G1 and sub-G0 phases and damage to the mitochondrial membrane (Haneef et al., 2012).

8.2 F. religiosa and Fertility Issues

A variety of medicinal plants are reported to be used in the treatment of female infertility. A clinical experiment is performed to determine the efficacy of *F. religiosa* in the treatment of PCOS. The experimental animals in this study were rats, divided into seven groups. Four PCOS-induced groups were treated with leaf extract of *F. religiosa*. Estrous interims, body weight, ovarian and uterine weight, and Cyp19a1 and PPAR- levels were all examined in rats. Cyp19a1 and PPAR-γ were shown to be significantly regulated in rats treated with *Ficus religiosa* leaf extract, while those treated with Letrozole had an irregular erratic estrous cycle, variable organ weight, and hormonal levels. As a result, we may conclude that the putative component found in the leaves of *Ficus religiosa* may be beneficial in the treatment of PCOS in women (Suriyakalaa, 2020). An experimental study was conducted to analyze and correlate the efficacy of three antioxidants *i.e.*, *Ficus religiosa*, Sulforaphane, and Vitamin E to treat the cadmium-induced reproductive damage (Cd Blood and testicular tissues were tested to assess daily sperm production and the levels of cadmium, zinc, and testosterone). The test results showed a significant decline ($p<0.0001$) in testosterone, zinc, and daily sperm production due to cadmium exposure. A significant suppression can be observed in the groups treated with antioxidants. *Ficus religiosa* and sulforaphane are considered to be the most potent in reducing reproductive toxicity in comparison to vitamin E (Sarwat Jahan, 2013).

Forty-four diabetic and non-diabetic patients suffering from erectile dysfunction were divided into two groups for clinical testing of *F. religiosa*. Subjects were divided into a trial group and a placebo group. The trial group provided 10 grams of finely ground parts of *F. religiosa* provided dose proved to be useful in the treatment of ED and increased semen production in trial group patients (Virani et al., 2010).

8.3 *F. religosa* and Neurodegenerative Diseases

The efficacy of *F. religiosa* petroleum ether leaf extract against Parkinson's disease in rats was tested by administering the extract at three different doses: 100, 200, and 400 mg/kg. Besides monitoring locomotor activity, narcolepsy, and muscular stiffness, its influence on neurochemical markers such as MDA, CAT, SOD, and GSH was also evaluated using *in vivo* behavioral tracking. They were given 6 hydroxydopamine, which causes symptoms similar to Parkinson's disease, and haloperidol, which causes narcolepsy. The induced cataleptic state decreases significantly (p 0.001) at higher dosages (200, 400 mg/kg). Muscular dystrophy, lipid peroxidation, and the antioxidant enzymes superoxide dismutase, catalase, and glutathione were all significantly worsened in rats given 6 hydroxydopamine. Oil-ether extract of *F. religiosa* leaves (400 mg/kg) daily, resulted in a significant decrease in oxidative impairment and a significant increase in motor activities (Bhangale and Acharya, 2016). The bark extract of *F. religiosa* when treated with methanol is capable of restraining acetylcholinesterase enzyme. Thus, proven to help treat Alzheimer's disease (Panchawat, 2012).

The severity of PTZ-induced convulsions, depression, and cognitive impairment in mice was evaluated using a flavonoid-rich ethyl acetate crude extract of *F. religiosa*. The cognitive abilities of PTZ-induced mice were enhanced (dose-dependently) when a low dose of phenytoin (15 mg/kg) was combined with fractions (2.5, 5, and 10 mg/kg) of ethyl acetate crude extract (flavonoid rich) once daily for half a month. This was evidenced by a decrease in errors, an improvement in step-down latency as measured by the passive avoidance technique, an increase in mobility as measured by the tail test. A decrease in acetylcholinesterase activity, a decrease in catalase and glutathione levels, and an increase in thiobarbituric acid reactive compounds were identified in a biochemical analysis of brain tissue (Singh et al., 2014). To study the cognitive impairment and depression associated with PTZ (pentylenetetrazol), rats were given an extract of *F. religiosa* high in saponins at a dose of 1, 2, or 4 mg/kg subcutaneously every day for 15 days. It was shown to be protective against PTZ-associated depression in a tail suspension test on days 10 and 15, and a reduction in convulsions was seen in a tail suspension test on days 5, 10, and 15. The passive avoidance test showed that a saponin-rich extract did not ameliorate cognitive impairment in rat models (Singh et al., 2013). *Ficus religiosa* fruit extract was evaluated for its potential to prevent convulsions in animal models. All three of these methods—maximum electroshock (MES), pentylenetetrazol (PTZ), and picrotoxin—were used to cause convulsions. Cyproheptadine, a non-selective 5HT1/2 serotonin antagonist drug, was employed at a dose of 4 mg/kg to determine whether or not the protective effect of *Ficus religiosa* extract could be reversed. Pentobarbitone and its extract were also evaluated for their acute and neurological toxicity as well as their ability to induce and maintain sleep. There was no evidence of toxicity in the extract. The effectiveness of the extract in preventing seizures was evaluated by comparing it to the anti-epileptic drugs phenytoin and diazepam. In the PTZ-induced group, there was no hesitation in the outcomes. However, cyproheptadine pretreatment resulted in fewer convulsions after extract administration. This suggests that the extract's anticonvulsant effects are mediated in part by the serotonergic system (Singh and Goel, 2009).

Scopolamine-induced anterograde and retrograde amnesic rats were tested for the efficacy of methanolic extract of *F. religiosa*. A few behavioral modules i.e., passive avoidance techniques were utilized to assess memory and learning. 1 mg/kg dose of scopolamine was provided for the inducement of anterograde amnesia before training in an elevated plus maze and then induced before retrieval in case of retrograde amnesia. Dose-dependent amnesic rats; both retrograde and anterograde displayed a substantial improvement in memory by suppressing the effect of scopolamine when treated with 10, 50, and 100 mg/kg methanolic extract of *F. religiosa*. Those who were pre-treated with cyproheptadine resulted in the inversion of *Ficus* extract against amnesia (Kaur et al., 2010). Ethanolic leaf extract of *F. religiosa* test for memory improvement in two groups of albino rats i.e., Swiss and Wistar with the help of passive-avoidance technique. Extract-treated group results indicate an enhanced memory in scopolamine amnesic rats and inversion in the hypoxic condition due to the inducement of sodium nitrite (Devi et al., 2011).

8.4 *F. religiosa* and Digestive System Problems

A significant decline in ulceration and gastric exudation was observed when gastric ulcer-induced rats were treated with two different doses of leaf extract (250 mg/kg and 500mg/kg) of *F. religiosa* (Gregory et al., 2013). The bark extract of *Ficus religiosa* acts as an effective medicine when applied to various tests. Extracts given in different doses i.e., 100, 200, and 400 mg/kg body weight showed a significant ($P<0.05$) decline in ulcerative region acid secretions (Ali et al., 2011). Ulcer-induced Swiss Albino rats were tested for anti-ulcerative properties of *F. religiosa*. Ethanolic leaf extract treatment exhibits a significant decline ($p<0.001$) in the acidity of the stomach and a reduction in the ulceration area (Kumar et al., 2011; Al-Snafi, 2017).

8.5 *F. religiosa* and Diabetes

Orally taken aqueous extract of *ficus religiosa* (100 and 200 mg/kg) in experimental rats induced with type 2 diabetes (Table 10) which decreased the level of fasting blood glucose and regulated the oxidative stress. Resulting in the repression of malondialdehyde and reduced glutathione (Kirana et al., 2009). Rats were used which are STZ-induced diabetics to test the efficacy of different doses (25 mg/kg, 50 mg/kg, and 100 mg/kg) of *Ficus religiosa* aqueous extract in reducing blood glucose levels. All three doses showed hypoglycemic effects but more obvious changes can be observed in higher doses (100 and 50 mg/kg) as compared to 25mg/kg. Aqueous bark extract of *F. religiosa* displayed a substantial upsurge in body weight, serum insulin level, and the amount of glycogen stored in the liver and skeletal muscles. A notable decline in serum triglycerol and total cholesterol can be noticed as well. When treated with *Ficus religiosa*, the pancreas of STZ-induced rats showed an anti-lipid peroxidative effect comparable to that of glibenclamide (Pandit et al., 2010). Newborn rats weighing about 10–12 g induced with STZ (90 mg/kg i.p) were studied for diabetic cardiomyopathy. As a result, a noticeable increase can be seen in HbA1c, cardiac hypertrophy, TGF-β1, inflammatory marker (TNFα), malondialdehyde, LDH, CK-MB, BNP, and caspase-3 which reduced the superoxide dismutase, heart rate, and insulin levels. These newnatals were then treated with *F. religiosa* for eight weeks remarkably improved cardiac rate, diabetic condition, and oxidative stress (Sharma et al., 2014). After a month of treatment with 100 and 200 mg/kg administered oral dosages of *F. religiosa* aqueous extract, serum values for HbA1c, blood glucose after meal, and TNF- were measured in type 2 diabetes (STZ induced) rats. Values fell, however, the effects were more significant at larger doses, such as 200 mg/kg Serum (Kirana et al., 2011).

8.6 Anti-Aging Activity of *F. religiosa*

The bark, leaf, and adventitious root extract of *F. religiosa* was tested for anti-aging properties and wound healing via standardized methods. Results revealed the effectiveness of all extracts as anti-aging agents in cosmetic applications (Pandey et al., 2020).

8.7 Anticoagulant Activity of *F. religiosa*

Methanolic leaf extracts of *F. religiosa* and *F. benghalensis* were tested for anticoagulation in human plasma. Results demonstrated higher efficacy of *F. religiosa* extract than *F. benghalensis* with 17.7–26.7 seconds Prothrombin time and time for activated partial thromboplastin ranges from 47.7–72.3 seconds (Ambreen et al., 2021).

8.8 Antimicrobial Activity of *F. religiosa*

Evidence suggests that *Ficus religiosa* leaf extract is effective against a variety of microorganisms including *Escherichia coli, Shigella dysenteriae,*

Salmonella typhimurium, Salmonella typhi, Pseudomonas aeruginosa, Bacillus subtilis, Staphylococcus typh and *Staphylococcus aureus*. Chloroform extract of *Ficus religiosa* was effective against *O. chroococcum, B. cereus, S. faecalis, B. megaterium, K. pneumoniae, and S. lactis* (Kapile, 2022). Findings from tests using ethanol-treated fruit extract against *S. epidermidis* and *S. aureus* (gram-positive) and *P. vulgaris* and *Klebsiella pneumonia* (gram-negative) at doses of 15 and 30 mg/ml were mixed. When tested at a dosage of 15 mg/ml, an inhibitory mechanism was shown for *Staphylococcus epidermidis* and *Klebsiella pneumonia*, but a concentration of at least 24 mg/ml was necessary to inhibit *Staphylococcus aureus* and *Proteus vulgaris* (Kaur et al., 2014). Aqueous extract of *F. religiosa* leaves was used to formulate the nanoparticles of zinc oxide and titanium dioxide which exhibit significant larvicidal effects, especially for ZnO against a larval form of *A. stephensi* (Soni and Dhiman, 2020). Current research on *F. religiosa* leaf extract when treated with methanol also showed potent results by killing 70% of *Aedes aegypti* at early larval stages (Deepa et al., 2015). A variety of leaf extracts of *F. religiosa* i.e., Benzene, Acetone, Ethanol, and Hexane function as a repellent against *C. quinquefasciatus, A. stephensi,* and *A. aegypti*, due to the presence of tripnoids. Among all, ethanolic leaf extract was highly effective (4.0 mg/cm^2) (Pavela et al., 2019; Subash et al., 2019).

Candida albicans, a kind of yeast, has been said to be inhibited by an ethanolic extract of its leaves (Kapile, 2022). The effectiveness of a fine powder derived from *G. glabra* bark, *F. religiosa* stem, and *P. major* husk against Candida albicans was evaluated using the disc diffusion technique and compared to that of synthetic anti-fungal medications. Plant medicines outperformed synthetic anti-fungal by a wide margin (Sharma et al., 2016). The worm parasites may be eliminated with the help of *Ficus religiosa*. The effects of methanol on an extract of its bark are toxic to *Haemonchus contortus*. *Ascaridia galli* may be killed by a bark extract from the *F. religiosa* tree (Kapile, 2022). *F. religiosa* has shown impressive efficiency against earthworms (*Pheretima posthuma*) when the fruit is extracted into water (Sawarkar et al., 2011).

Table 10 Various pharmacological actions of *Ficus religiosa*

Pharmacological Actions	*Disorders*	*References*
Anti-Carcinogenic	Malignant tumors in the brain, human breast, and cervical region	(Saida, 2021; Rahman et al., 2014; Choudhari et al., 2011; Haneef et al., 2012)
Anti-Diabetic	Type 2 Diabetes	(Kirana et al., 2009; Kirana et al., 2011)
	Diabetic cardiomyopathy	(Sharma et al., 2014)
	Glycerol, Cholesterol levels in diabetic patients	(Pandit et al., 2010)

Contd.

Table 10 *Contd.*

Pharmacological Actions	*Disorders*	*References*
Fertility Disorders (Male/Female) Treatment	Reproductive damage	(Sarwat Jahan, 2013)
	Polycystic Ovary Syndrome	(Suriyakalaa, 2020)
	Erectile Dysfunction	(Virani et al., 2010)
Against Nervous disorder	Parkinson's disease	(Bhangale and Acharya, 2016)
	Alzheimer's disease	(Panchawat, 2012)
	Convulsions; Depression	(Singh et al., 2013; Singh and Goel, 2009)
	Amnesia	(Kaur et al., 2010)
	Cognitive impairment	(Devi et al., 2011)
Against Digestive disorders	Gastric ulceration	(Ali et al., 2011; Al-Snafi, 2017)
Anti-coagulant	Human blood plasma	(Ambreen et al., 2021)
Anti-aging & wound healing		(Pandey et al., 2020)
Anti-larval		(Soni and Dhiman, 2020; Deepa *et al.,* 2015; Pavela *et al.,* 2019; Subash et al., 2019)
Anti-microbial	Diseases caused by gram-positive and negative bacteria	(Kapile, 2022; Kaur et al., 2014)
Anti-fungal	Infection caused by *Candida albicans*	(Sharma et al., 2016)
Anti-helminthes	Ailments caused by earthworms and other helminthes	(Kapile, 2022; Sawarkar et al., 2011)
Anti-viral	HSV type 2, respiratory viruses; HIV 1 & HPV related malignancies	(Gosh et al., 2016; Cagno et al., 2015; Kusumoto et al., 1995; Choudhari et al., 2013)

8.9 Antiviral Activity of Ficus religiosa

The effectiveness of *F. religiosa* extracts against (HSVT-2) *Herpes Simplex Virus type-2* was tested. It was shown that bark extracts treated with both water and chloroform were efficient in blocking the activity of *Herpes Simplex Virus type 2*. Extracts treated with chloroform prevent the virus from entering cells, attaching to them, and replicating as effectively as those treated with water (Gosh et al., 2016). In vitro investigation showed that *F. religiosa* solvent-treated extracts are effective against RSV and HRV (respiratory viruses). The anti-HRV mechanism of methanolic bark extract was the most effective, while the anti-RSV mechanism of aqueous-treated bark extract was successful (Cagno et al., 2015). The bark extract of *F. religiosa* was tested for its ability to suppress the activity of HIV-1 proteases using both water and methanol as solvents (Kusumoto et al., 1995; Gosh et al., 2016). It has also been shown to be effective against HPV-related malignancies (Choudhari et al., 2013).

Table 11 Other Pharmacological Activities of *F. religiosa*

Actions	*Experiments*	*Reference(s)*
Broncho-constrictive	In guinea pigs stimulated with histamine and acetylcholine, an extract of fruits prepared using methanol at doses of 0.5, 1, and 2 milligrams per kilogram of body weight demonstrated significant benefits.	(Ahuja et al., 2011)
Anti-Asthmatic	Guinea pigs treated with an aqueous extract of the leaves at doses of 150 and 300 mg/kg of body weight exhibited anti-asthmatic activity.	(Kapoor, 2011)
Anti-Fertile	On the uterus of goats, an anti-fertility impact was seen when an extract of fruits that contained 1% methanol was administered.	(Goyal et al., 2013)
Anti-Inflammatory Properties	In golden Syrian hamsters with carrageen-induced inflammation, an ethanolic extract of bark at a dose of 100 mg/kg of body weight demonstrated anti-inflammatory benefits.	(Murugesan et al., 2012)
Wound Healing Properties	In rats of the Wistar Albino strain, the application of an ethanolic extract of leaves at a dose of 300 mg/kg of body weight promoted wound healing.	(Charde et al., 2017)
	In Wistar albino rats, topical use of ethanolic extracts of roots at a concentration of 10% promoted wound healing.	(Murti et al., 2011)
Anti-Ulcerative Properties	Antiulcer effects were shown in male albino Wistar rats when an ethanolic extract of bark at doses of 200 and 400 mg/kg of body weight was administered.	(Khan et al., 2011)

9. *F. religiosa* and its Toxicity

An extract of *F. religiosa* given at a level of 2000 mg/kg had no effects on albino mice (Gregory, 2013; Saha & Goswami, 2010). *F. religiosa* extract doses ranging from 50 to 2000 mg/kg also failed to significantly alter the reactions of the model individuals (Yadav, 2015). After administering an oral dosage of 5000 mg/kg of F. religiosa methanolic extract to rats for 2 days, there were no observed changes in the rats' autonomic reflexes (Parameswari et al., 2013). There was no mortality in a study where male Wistar rats were orally administered an extract of *F. religiosa* leaves at a dose of 4000 mg/kg (Bhangle and Acharya, 2016). According to the weight of the test subject, the median fatal dosage of a chemical or medication is deemed to be (Table 12) poorly deadly between 2000 and 5000 mg/kg, and non-lethal over 5000 mg/kg. Different fractions of *F. religiosa* are used in toxicological testing (Renata, 2019).

Table 12 Toxicity Levels of *F. religiosa*

Tested Dose Range	*Results*	*References*
50-2000 mg/kg	Non-Toxic	(Yadav, 2015)
4000 mg/kg	No Mortality	(Bhangle and Acharya, 2016)
5000 mg/kg	No signs for Acute Toxicity	(Parameswari et al., 2013)

Conclusion

The pharmacological potentials of the phytochemicals contained in the plant matrix of *Ficus* species have garnered great interest in the plant components of this genus. Its pharmacological actions may be due to the presence of Bioactive compounds flavonoids and terpenoids. The evaluated investigations validated the species' antibacterial, anti-diabetic, anti-inflammatory, and anticancer properties, providing scientific support for the use of the species in traditional treatments. Traditional use of this herb has been credited with relieving symptoms of urinary and urinogenital diseases. Antiviral compounds found in *F. religiosa* L. are effective against two respiratory viruses that have been linked to wheezing sickness and asthma aggravation. Acyclovir-resistant HSV-2 strains of genital herpes were also treated with F. religiosa because of the discovery of the bioactive metabolites as therapeutic natural chemicals. Furthermore, the overall conclusion of this review provides credence to the conventional use of the bark of *F. religiosa* for the cure of genital ulcers and sores its clinical potential has to be investigated further and Research is needed to determine its therapeutic efficacy and isolate its antiviral components.

References

Acevedo-Rodríguez, P. and Strong, M.T. (2012). Catalogue of seed plants of the West Indies.

Ahuja, D.B. (2011). Bronchospasm potentiating effect of methanolic extract of Ficus religiosa fruits in guinea pigs. *Journal of Ethnopharmacology*, 133(2), 324–328.

Aiyegoro, O.A. (2009). Use of bioactive plant products in combination with standard antibiotics: implications in antimicrobial chemotherapy. *Journal of Medicinal Plants Research*, 3(13), 1147–1152.

Al-Snafi, A.E. (2017). Pharmacology of Ficus religiosa-A review. *IOSR Journal of Pharmacy, 7*(3), 49–60.

Ambreen, S. (2019). Anticoagulant potential and total phenolic content of six species of the genus Ficus from Azad Kashmir, Pakistan. *Tropical Journal of Pharmaceutical Research* (18(6)), 1245–1251.

Anandjiwala, S., Bagul, M., Parabia, M. and Rajani, M. (2008). Evaluation of free radical scavenging activity of an ayurvedic formulation, Panchvalkala. *Indian Journal of Pharmaceutical Sciences, 70*(1), 31.

Anju Singh, A.S. and Monica Basu, M.B. (2009). Two new wood-decaying basidiomycetous fungi from India. *47*(1), 83–85.

Armstrong, K.J.E.J. Oo. B. (2004). A Checklist of the Trees, Shrubs, Herbs, and Climbers of Myanmar. W. John Kress, Robert A. DeFilipps, Ellen Farr & Daw Yin Yin Kyi. Washington, DC: Smithsonian Institution. Contributions from the United States National Herbarium, Vol. 45. 2003. 590pp. ISSN 0097 1618. Available free of charge while supplies last. *61*(2-3), 197–198.

Babu, K., Sabesan, G.S. and Rai, S. (2010). Comparative pharmacognostic studies on the barks of four Ficus species. *Turkish Journal of Botany, 34*(3), 215–224.

Behari, M., Rani, K., Matsumoto, T. and Shimizu, N. (1984). Isolation of active-principles from the leaves of Ficus religiosa. *Curr Agric, 8*(1-2), 73–76.

Berg, C.C., Corner, E.J.H. and Nooteboom, H. (2005). *Flora Malesiana. Series I, Seed plants. Volume 17, Part 2: Moraceae (Ficus)*: Nationaal Herbarium Nederland.

Bhangale, J.O. (2016). Anti-Parkinson activity of petroleum ether extract of Ficus religiosa (L.) leaves. *Advances in Pharmacological Sciences*.

CABI, U.J.C., Wallingford. (2020). CABI Compendium: Status inferred from regional distribution.

CABI, U.J.W., UK: CABI. CABI Compendium: Status as determined by CABI editor.

CABI, Undated a. CABI Compendium: Status inferred from regional distribution. Wallingford, UK: CABI.

Cagno, V., Civra, A., Kumar, R., Pradhan, S., Donalisio, M., Sinha, B.N.,... Lembo, D. (2015). Ficus religiosa L. bark extracts inhibit human rhinovirus and respiratory syncytial virus infection in vitro. *Journal of Ethnopharmacology, 176*, 252–257.

Cambridge, Massachusetts, USA: Missouri Botanical Garden and Harvard University Herbaria. http://www.tropicos.org/Project/Pakistan. (n.d.).

Chandrasekar, S., Bhanumathy, M., Pawar, A. and Somasundaram, T. (2010). Phytopharmacology of Ficus religiosa. *Pharmacognosy Reviews, 4*(8), 195.

Chantarasuwan, B., Berg, C.C. and van Welzen, P.C. (2013). A revision of Ficus subsection Urostigma (Moraceae). *Systematic Botany*, *38*(3), 653–686.

Charde, Rita M., Hemant J. Dhongade, Manoj S. Charde and A.V. Kasture. "Evaluation of antioxidant, wound healing and anti-inflammatory activity of ethanolic extract of leaves of Ficus religiosa." *Int J Pharm Sci Res* 19, no. 5 (2010): 73–82.

Chong, K.Y. (2009). A checklist of the total vascular plant flora of Singapore: native, naturalised and cultivated species. Raffles Museum of Biodiversity Research and Department of Biological Sciences.

Choudhari, A.S., Suryavanshi, S.A., Ingle, H. and Kaul-Ghanekar, R. (2011). Evaluating the antioxidant potential of aqueous and alcoholic extracts of Ficus religiosa using ORAC assay and assessing their cytotoxic activity in cervical cancer cell lines. *Biotechnol Bioinfo Bioeng*, 1(4), 443–450.

Choudhari, A.S., Suryavanshi, S.A. and Kaul-Ghanekar, R. (2013). The aqueous extract of Ficus religiosa induces cell cycle arrest in human cervical cancer cell lines SiHa (HPV-16 positive) and apoptosis in HeLa (HPV-18 positive). *PLoS One, 8*(7), e70127.

Committee, F.o.C.E. (2014). Flora of China Editorial Committee. (2014). Flora of China. St. Louis, Missouri and Cambridge, Massachusetts, USA: Missouri Botanical Garden and Harvard University Herbaria.

DA, M. (1995). A Pictorial Cyclopedia of Philippine Ornamental Plants. 214–217.

Datil, M. and Acevedo-Rodríguez, P. (2014). Ficus religiosa (sacred fig tree). *Forestry Compendium*(24168).

Deepa, J.G. (2015). Larvicidal activity of Indian medicinal plants on the dengue fever mosquito, Aedes aegypti Linnaeus. Int J Pure Appl Zool(3), 130–136.

Devi, W.B. (2011). Memory enhancing activities of Ficus religiosa leaves in rodents. *International Journal of Research in Ayurveda and Pharmacy* (IJRAP) 2(3), 834–838.

DJ, M. (2008). Mabberley's Plant-Book: A Portable Dictionary of Plants, their Classifications, and Uses., Cambridge, UK.

El-Sherbiny, A. A. (2011). Phytoparasitic nematodes associated with ornamental shrubs, trees and palms in Saudi Arabia, including new host records. *Pakistan Journal of Nematology* 29(2).

El-Sherbiny, A.J.P. J.o.N. (2011). Phytoparasitic nematodes associated with ornamental shrubs, trees and palms in Saudi Arabia, including new host records. *29*(2).

Ethnopharmacology, 133(1), 92–96.

Ficus religiosa using ORAC assay and assessing their cytotoxic activity in cervical cancer cell lines. *Biotechnol Bioinfo Bioeng* 1(4), 443–450.

Firdousi, S.J.F. and Fauna. (2018). First report of fungal disease at nursery of social forestry department, Jalgaon (MS) India. *24*(1), 88–90.

Flora of China Editorial Committee. (2014). Flora of China. St. Louis, M. a. (n.d.).

Flora of Pakistan, 2014. Flora of Pakistan/Pakistan Plant Database (PPD)., St. Louis, Missouri

Gautam, S., Meshram, A., Bhagyawant, S.S., Srivastava, N. Ficus religiosa-potential role in pharmaceuticals. *Int J Pharm Sci Res.* 2014; *5*(5):1616–1623.

Ghimire, K. and amp;. (2009). Ethnomedicinal knowledge and healthcare practices among the Tharus of Nawalparasi district in central Nepal. Forest Ecology and Management 257(10), 2066–2072.

Ghimire, K. and Bastakoti, R.R. (2009). Ethnomedicinal knowledge and healthcare practices among the Tharus of Nawalparasi district in central Nepal. *Forest Ecology and Management, 257*(10), 2066–2072.

Ghosh, M.C. (2016). Ficus religiosa L. bark extracts inhibit infection by herpes simplex virus type 2 *in vitro*. *Archives of Virology*(161), 3509–3514.

Gregory, M.D. (2013). Anti–ulcer activity of Ficus religiosa leaf ethanolic extract. *Asian Pacific Journal of Tropical Biomedicine* 3(7), 554–556.

Haneef, J.T. (2012). Bax translocation mediated mitochondrial apoptosis and caspase dependent photosensitizing effect of Ficus religiosa on cancer cells. *PLoS One* 7(7).

Hashmi, N.M. (2013). Nephroprotective effects of Ficus religiosa linn (peepal plant) stem bark against isoniazid and rifampicin induced nephrotoxicity in albino rabbits. *Pak Vet J* 33(3), 330–334.

Hesami, M. and Rahmati-Joneidabad, M. (2018). Morphological Characterization of Ficus religiosa Genotypes in Iran by Multivariate Analysis. *HortScience, 53*(7), 932–936.

Hesami, M., Daneshvar, M.H. and Yoosefzadeh-Najafabadi, M. (2018). Establishment of a protocol for *in vitro* seed germination and callus formation of Ficus religiosa L., an important medicinal plant. *Jundishapur Journal of Natural Pharmaceutical Products, 13*(4).

Huma Qureshi, H.Q., Muhammad Arshad, M.A. and Yamin Bibi, Y.B. (2014). Invasive flora of Pakistan: a critical analysis. *4*(1), 407–427.

Hunt, C. (2009). Mabberley's Plant-Book: A Portable Dictionary of Plants, their Classifications and Uses. *Reference Reviews, 23*(4), 35–36.

Jahan, S.K. (2014). Comparative analysis of antioxidants against cadmium induced reproductive toxicity in adult male rats. *Systems Biology in Reproductive Medicine* 60(1), 28–34.

James, T., Champion, P., Bullians, M. and Rahman, A. (2011). *Weed biosecurity breech through coco peat imports.* Paper presented at the Proceedings of the 23rd Asian-Pacific Weed Science Society Conference.

Jahan, S., Khan, M., Ahmed, S. and Ullah, H. (2014). Comparative analysis of antioxidants against cadmium induced reproductive toxicity in adult male rats. *Systems biology in reproductive medicine, 60*(1), 28–34.

Joshi, A.R. and Joshi, K. (2000). Indigenous knowledge and uses of medicinal plants by local communities of the Kali Gandaki Watershed Area, Nepal. *Journal of Ethnopharmacology, 73*(1–2), 175–183.

Kapile, C., Kulkarni, A., Pardeshi, P., Sayed, A. and Nehe, A. (2022). Ficus religiosa: A beneficial medicinal plant. *J Drug Deliv Ther*, 12(2-S), 210–218.

Kapoor, L.D. (1990). *Handbook of Ayurvedic medicinal plants*. CRC Press Inc.

Kapoor, M.J. (2011). Phytopharmacological evaluation and anti–asthmatic activity of Ficus religiosa leaves. *Asian Pacific Journal of Tropical Medicine* 4(8), 642–644.

Kariyawasam, A. (1995). *Buddhist ceremonies and rituals of Sri Lanka*: Buddhist Publication Society Kandy, Sri Lanka.

Kaur, A., Rana, A.C., Tiwari, V., Sharma, R. and Kumar, S. (2011). Review on ethanomedicinal and pharmacological properties of Ficus religiosa. *Journal of applied pharmaceutical science*, (Issue), 06-11.

Kaur, H.S. (2010). Anti-amnesic effect of Ficus religiosa in scopolamine-induced anterograde and retrograde amnesia. *Pharmaceutical Biology*, 48(2), 234–240.

Kaur, R.G. (2014). *In vitro* studies on antibiotic activity of Ficus religiosa fruits extract against human pathogenic bacteria. *Journal of Chemical and Pharmaceutical Research*, 6(11), 80–84.

Khan, M.S. (2011). Anti-ulcer activity of Ficus religiosa stem bark ethanolic extract in rats. *J Med Plants Res*, 5(3), 354–359.

Khanom, F.K. (2000). Tyrosinase inhibitory activity of Bangladeshi indigenous medicinal plants. Bioscience, *Biotechnology, and Biochemistry*, 64(9), 1967–1969.

Kirana, H.A. (2009). Aqueous extract of Ficus religiosa Linn. reduces oxidative stress in experimentally induced type 2 diabetic rats.

Kirana, H.J. (2011). The study of aqueous extract of Ficus religiosa Linn. on cytokine TNF-α in type 2 diabetic rats. *Pharmacognosy Research*, 3(1), 30.

Kress, W.J., DeFilipps, R.A., Farr, E. and Kyi, D.Y.Y. (2003). A checklist of the trees, shrubs, herbs, and climbers of Myanmar. *Contributions from the United States National Herbarium, 45*(1), 590.

Krishnamurthy, K. H. (2016). Peepul (Ficus religiosa, Linn.). *The Journal*, *23*(4).

Kumar, A.J.C.S. (2013). Fig wax scale, Ceroplastes rusci an emerging pest of Dalbergia sissoo and its parasitisation in India. *8*, 106–114.

Kumar, A.T. (2018). Ficus religiosa: A wholesome medicinal tree. *Journal of Pharmacognosy and Phytochemistry*, 7(4), 32–37.

Kumar, H.S. (2011). Anti-ulcer activity of Ficus religiosa L. leaves on experimental animals. *International Journal of Pharmaceutical Sciences Review and Research*, 11(2), 83–86.

Kunwar, R. a. (2006). *J. Ecol. Appl.*, 11, 85–97.(11), 85–97.

Kusumoto, I.T. (1995). Screening of various plant extracts used in ayurvedic medicine for inhibitory effects on human immunodeficiency virus type 1 (HIV-1) protease. *Phytotherapy Research*, 9(3), 180–184.

Liogier, A.H., Liogier, H.A. and Martorell, L. F. (2000). *Flora of Puerto Rico and adjacent islands: a systematic synopsis*: La Editorial, UPR.

Madulid, D.A. (1995). *Pictorial Cyclopedia of Philippine Ornamental Plants*: Bookmark.

Makhija, I.K., Sharma, I.P. and Khamar, D. (2010). Phytochemistry and Pharmacological properties of Ficus religiosa: an overview. *Annals of Biological Research, 1*(4), 171–180.

Mesoamericana, F. (2014). Flora Mesoamericana. St. Louis, Missouri, USA: Missouri Botanical Garden.

Murti, K.L. (2011). Exploration of healing promoting potentials of roots of Ficus religiosa. *Pharmacologia*, 2(12), 374–378.

Murugesan, P.R. (2012). Evaluation of Anti-inflammatory effects of Ficus religiosa Linn. Carrageenan induced acute inflammation. *Elixir Food Science* A, 51, 11124–11128.

Nadel, H.F. (1992). Escapees and accomplices: the naturalization of exotic Ficus and their associated faunas in Florida. *Florida Entomologist*, 29–38.

Nadel, H., Frank, J.H. and Knight Jr, R.J.F.E. (1992). Escapees and accomplices: the naturalization of exotic Ficus and their associated faunas in Florida. 29–38.

Orwa, C. (2009). Agroforestree Database: a tree reference and selection guide, version 4.0. http://www. worldagroforestry. org/sites/treedbs/treedatabases. asp.

Panchawat, S. (2012). Ficus religiosa Roxb.(Peepal): A phyto-pharmacological review. IJPS, 1(1), 435–446.

Panda, H. (2005). Herbs Cultivation and Medicinal Uses, second edition. 302–303.

Panda, S., Panda, N. and Sahue, B. (1976). Phytochemistry and pharmacological properties of Ficus religiosa: an overview. *Indian Veterinary Journal, 60*, 660–664.

Pandey, P.S. (2020). Enhanced anti-ageing and wound healing properties of Ficus religiosa L. bark, leaf and aerial root extract in human keratinocytes cell line (HaCaT). Vegetos, 33, 158–165.

Pandit, R.P. (2010). Antidiabetic effect of Ficus religiosa extract in streptozotocin-induced diabetic rats. Journal of Ethnopharmacology, 128(2), 462–466.

Parameswari, S.A., Chetty, C.M. and Chandrasekhar, K.B. (2013). Hepatoprotective activity of Ficus religiosa leaves against isoniazid+ rifampicin and paracetamol induced hepatotoxicity. Pharmacognosy research, *5*(4), 271.

Patil, M.S. (2011). Anticonvulsant activity of aqueous root extract of Ficus religiosa.

Pavela, R.M. (2019). Plant extracts for developing mosquito larvicides: From laboratory to the field, with insights on the modes of action. *Acta tropica*, 193, 236–271.

PIER. (2014). Pacific Islands Ecosystems at Risk., Honolulu, USA: HEAR, University of Hawaii. Retrieved from http://www.hear.org/pier/index.html

Pokharel, N. and Adhikari Pokharel, B. (2021). A Relationship Between Plants and Their Hindu and Buddha Cultures: An Analysis Ficus Religiosa (Pipal) *Transcultural Diplomacy and International Law in Heritage Conservation* (pp. 143–151): Springer.

Potential of Ficus religiosa (Peepal). Research Journal of Pharmacy and Technology, 12(11).

Rahman, M.K. (2014). Phytochemical, cytotoxic and antibacterial activity of two medicinal plants of Bangladesh. *Pharmacologyonline*, 1, 3–10.

Randall, R. (2012). A Global Compendium of Weeds. Perth, Australia: Department of Agriculture and Food Western Australia, 1124 pp.

Renata-Maria, V. (2019). Determination of acute and subacute toxicity of six plant extracts. *Medico Research Chronicles*, 6(2), 64–67.

Rifampicin and paracetamol induced hepatotoxicity. *Pharmacognosy Research*, 5(4), 271.

Rojo, J., Pitargue, F., Sosef, M.J.P.R.o.S.-E.A.N.M. and Plant, P. (1999). Ficus L. *1*, 277–289.

Sacred fig description. (n.d.). (EncyclopediaAmericana, Compiler) Retrieved from https://www.facebook.com/PharmaRiserss"Peepul"

Saida, L.T. (2021). Evaluation of chemo-preventive efficacy of Ficus religiosa latex extract by flow cytometry analysis and gene expression studies performed by RT-PCR in various cell lines. *Future Journal of Pharmaceutical Sciences*, 7, 1–21.

Sandeep, A.K. (2018). Ficus religiosa: A wholesome medicinal tree. *Journal of Pharmacognosy and Phytochemistry* 7(4), 32–37.

Sawarkar, H.A. (2011). *In Vitro* Anthelmintic Activity of Ficus benghalensis, Ficus carica and Ficus religosa: A Comparative Study. *International Journal of Pharmacy and Pharmaceutical Sciences* 3(Suppl 2), 152–153.

Seebens, H.B. (2017). No saturation in the accumulation of alien species worldwide. *Nature Communications*, 8(1), 14435.

Seebens, H., Blackburn, T.M., Dyer, E.E., Genovesi, P., Hulme, P.E., Jeschke, J.M.,... Arianoutsou, M.J.N.c. (2017). No saturation in the accumulation of alien species worldwide. *8*(1), 14435.

Sethuramani, A., Soundariya, R. and Dharani, V. (2021). Morpho-Playnological study on Ficus religiosa L. Leaves. *Journal of Pharmacognosy and Phytochemistry, 10*(2), 1358–1365.

Shah, M.R. (2014). Screening of phenolic content, antioxidant and *in vitro* eye irritation activities from Apiaceae family (dry seeds) for potential cosmetic applications. *International Journal of Pharmaceutical Sciences and Research*, 5(10), 4366.

Sharma, H.Y. (2016). Antifungal efficacy of three medicinal plants Glycyrrhiza glabra, Ficus religiosa, and Plantago major against oral Candida albicans: A comparative analysis. *Indian Journal of Dental Research* 27(4), 433.

Sharma, R.K. (2013). Anti-fertility activity of Ficus religiosa fruits extract on goat uterus *in vitro*. Int *J Drug Dev and Res*, 5(4), 330–335.

Sharma, R.K. (2014). Cardioprotective effects of Ficus religiosa in neonatal streptozotocin-induced diabetic cardiomyopathy in rats. *Biomedicine & Aging Pathology*, 4(1), 53–58.

Sharma, V., Mishra, S., Yesudas, R. and Rajput, R.S.A. Review on Ficus religiosa (Sacred Fig). [AD1]

Sheetal, A.B.M. (2008). Evaluation of free radicals scavenging activity of an Ayurvedic formulation, panchvankala. *Indian J Pharm Sci.* 70(1), 31–38.

Simha, K. and Laxminarayana, V. (2007). Standardization of Ayurvedic polyherbal formulation, Nyagrodhadi churna.

Singh, D.M. (2013). Effect of saponin fraction from Ficus religiosa on memory deficit, and behavioral and biochemical impairments in pentylenetetrazol kindled mice. Epilepsy & Behavior, 27(1), 206–211.

Singh, D. and Goel, R.K. (2009). Anticonvulsant effect of Ficus religiosa: role of serotonergic pathways. *Journal of Ethnopharmacology, 123*(2), 330–334.

Singh, D., Singh, B. and Goel, R.K. (2011). Traditional uses, phytochemistry and pharmacology of Ficus religiosa: A review. *Journal of Ethnopharmacology, 134*(3), 565–583.

Singh, M.P. (2005). Medicinal herbs with their formulations. Daya Books.

Singh, P.S. (2014). Ficus religiosa L. figs—a potential herbal adjuvant to phenytoin for improved management of epilepsy and associated behavioral comorbidities. *Epilepsy & Behavior*, 41,171–178.

Singh, S.A. (2018). Anticonvulsant effect of aqueous extract of aerial root of Ficus religiosa in animal models. *Journal of Drug Delivery and Therapeutics*, 8(1), 13–16.

Singh, S. and Jaiswal, S. (2014). Therapeutic properties of Ficus religiosa. *Int J Eng Res Gen Sci, 2*(5), 149–158.

Singh, T.G. (2020). Evaluation of renoprotective potential of Ficus religiosa in attenuation of diabetic nephropathy in rats. *Obesity Medicine*, 19, 100268.

Sirisha, N., Sreenivasulu, M., Sangeeta, K. and Chetty, C. M. (2010). Antioxidant properties of Ficus species-a review. *International Journal of Pharmtech Research, 2*(4), 2174–2182.

Soni, N. & amp;. (2020). Larvicidal and antibacterial activity of aqueous leaf extract of Peepal (Ficus religiosa) synthesized nanoparticles. *Parasite Epidemiology and Control*, 11.

Starr F, S.K. (2003). Ficus religiosa (Bo tree) Maui, Hawaii, USA: Haleakala Field Station, United States Geological Survey- Biological Resources Division.

Starr, F., Starr, K. and Loope, L. (2003). Ficus benjamina. United States Geological Survey-Biological Resources Division. Haleakala Field Station, Maui, Hawaii.

Subash, B.V. (2019). Laboratory efficacy of Ficus religiosa extracts against mosquito-borne disease. *Indian J. Appl. Res*, 9, 1–2.

Suriyakalaa, U.R. (2021). Upregulation of Cyp19a1 and PPAR-γ in ovarian steroidogenic pathway by Ficus religiosa: A potential cure for polycystic ovary syndrome. *Journal of Ethnopharmacology*, 267, 113540.

Swami, K.D. (1989). Chemical investigation of stem bark of Ficus-religiosa and Prosopis-spicigera. *Journal of the Indian Chemical Society*, 66(4), 288–289.

Swami, K.D. (1996). Constituents of Ficus religiosa and Ficus infectoria and their biological activity. *Journal of the Indian Chemical Society*, 73(11).

Swami, K. and Bisht, N. (1996). Constituents of Ficus religiosa and Ficus infectoria and their biological activity. *Journal of the Indian Chemical Society, 73*(11).

Swami, K.D., Malik, G.S. and Bisht, N.P.S. (1989). Chemical investigation of stem bark of Ficus-religiosa and Prosopis-spicigera. *Journal of the Indian Chemical Society, 66*(4), 288–289.

Tiwari, P., Ansari, V.A., Mahmood, T. and Ahsan, F. (2019). A review on taxonomical classification, phytochemical constituents and therapeutic potential of ficus religiosa (peepal). *Research Journal of Pharmacy and Technology, 12*(11), 5614–5620.

Tsai, L. and Yokoyama, J. (2019). Morphological and molecular evidence of nature hybridization between two related species, ficus religiosa and ficus rumphii in myanmar. *J. Myanmar Acad. Arts Sci.* Vol. XVII.:4

Varshney, I. and Bhatnagar, S. (1965). Saponins and sapogenins from Indian Ficus species. *Indian J. Pharm., 27*, 235.

Verma, R. and Bhatia, K. (1986). Chromatographic study of amino acids of the leaf protein concentrates of Ficus religiosa Linn and Mimusops elengi Linn. *Indian Journal of Pharmacy Practice, 23*, 231–232.

Virani, N.V. (2010). Clinical study on erectile dysfunction in diabetic and non-diabetic subjects and its management with Ficus relegiosa Linn. *Ayu*, 31(3), 272.

Wagner, W., Herbst, D., Tornabene, M., Weitzman, A. and Lorence, D. (2014). Flora of Micronesia website. Washington DC: Smithsonian Institution.

Warrier, P.K. (1993). *Indian medicinal plants: a compendium of 500 species* (Vol. 5): Orient Blackswan.

Wijaya, I.M.S. and Defiani, M.R. (2021). Diversity and distribution of figs (Ficus: Moraceae) in Gianyar District, Bali, Indonesia. *Biodiversitas Journal of Biological Diversity*, *22*(1).

YC, Y. (2015). Hepatoprotective effect of Ficus religiosa latex on cisplatin induced liver injury in Wistar rats. Revista Brasileira de Farmacognosia, 25, 278–283.

10 Antiviral Potential of *Tamarindus indica* L.: Exploring Nature's Shield

Parthasarathy Seethapathy[1*], Deepu Pandita[2*], Subbiah Sankaralingam[3] and Anu Pandita[4]

1. Introduction

The foundation of traditional medicine is centered on the utilization of medicinal plants. The utilization of plant preparations as sources of drugs has traditionally relied on anecdotal knowledge and superstitious beliefs that have been transmitted across generations (Balick and Cox, 2020). In recent times, there has been a notable surge in worldwide scholarly investigations about the field of medicinal plants (Gurib-Fakim, 2006). Medicinal plants have been extensively employed in various systems due to their capacity to potentially treat a diverse array of diseases (Alhazmi et al., 2021). Tamarind had probably undergone transcontinental dispersal to Asia at some point within the initial millennium BC. The evidence supports the existence of tamarind cultivation in Egypt as early as 400 BC. Additionally, references to tamarind can be found in the Indian Brahma Samhita Scriptures, which were written between 1200 and 200 BC (Singh and Niham, 2017). Theophrastus, a Greek philosopher and botanist, documented his observations on plants during the period of 370 to 287 BC. It is believed that his writings contain two descriptions of tamarind that originated from the region of East Africa. The utilization of tamarind as a medicinal plant dates back several centuries. Its fruits, considered the most valuable component, have frequently been documented for their therapeutic properties in various pharmacopoeias. However, there has been comparatively less research conducted on other components of plants. The leaves exhibit a well-documented hepatoprotective effect, which can be attributed to the abundance of polyhydroxylated compounds, many of which are flavonolic in nature. Both the seeds and the bark possess medicinal properties (Escalona-Arranz et al., 2010a).

The tamarind tree, *Tamarindus indica* Linn., family Fabaceae, subfamily Detarioideae is a tropical evergreen tree native to Africa, specifically Sudan and certain regions of the dry deciduous forests in Madagascar, and is classified as a

[1] Department of Plant Pathology, Amrita School of Agricultural Sciences, Amrita Vishwa Vidyapeetham, Coimbatore, Tamil Nadu, India.
[2] Government Department of School Education, Jammu, Jammu and Kashmir, India.
[3] PG and Research Department of Botany, Saraswathi Narayanan College, Madurai, 625022, India.
[4] Vatsalya Clinic, New Delhi, India.
* Corresponding author: spsarathyagri@gmail.com[1]; deepupandita@gmail.com[2]

DOI: 10.1201/9781003433583-10

tropical tree (Havinga et al., 2010). The etymology of the English word "tamarind" can be traced back to the Persian term "Tamar-i-hind," denoting the "date of India." This linguistic connection initially gave rise to the mistaken belief that the tamarind fruit was derived from an Indian palm. *Tamarindus indica* is a sizable, long-lasting tree that is frequently cultivated for its aesthetic appeal and ability to provide shade. It is also highly valued for its timber and the edible fruits it produces. The widespread cultivation of tamarind has been primarily observed in tropical and subtropical regions due to its diverse range of applications. The plant has become naturalized in various regions where it was initially cultivated, spreading to bushlands, pastures, coastline thickets, roadways, banks of rivers, and ruderal areas adjacent to villages. The prevalence of its cultivation is particularly notable in South India, specifically in the states of Tamil Nadu, Karnataka, and Andhra Pradesh (Sharma et al., 2013). In regions lacking a pronounced dry season, the tree exhibits the characteristic of being evergreen and has the potential to attain a vertical stature of 20 meters (66 feet) (Rao and Mathew, 2012). The species exhibits sensitivity to frost due to its tropical nature. This particular organism exhibits resilience in the face of arid soils and environments. The tree is extensively cultivated for its aesthetic appeal and the acidic fruits it produces, which are utilized in the production of beverages. Additionally, the fruits are a common ingredient in various decoctions that are employed as medicinal remedies. *T. indica* exhibits distinct variations, which can be classified into two main types: sweet and astringent. Tamarind, a type of fruit known for its acidic properties, is commonly found in numerous countries and thrives in sunny and warm environments. Although sweet-type varieties are not readily accessible. The fruit's distinct combination of sweetness and sourness makes it a unique culinary ingredient that is widely utilized in cooking. Furthermore, apart from its edible fruit, the various components of the plant, like roots, timber, bark, and leaves, exhibit nutritional and pharmaceutical attributes (Caluwé et al., 2009).

The laxative properties of tamarind pulp render it a beneficial remedy for fevers, while its infusion with water yields a highly invigorating beverage (De Caluwé et al., 2010). This substance is employed either in isolation or in conjunction with other cathartic agents for its laxative properties. When combined with other purgative medications, it diminishes the laxative effect. In the context of rural roads and highways, these trees serve the dual purpose of providing aesthetic value as well as offering shade. Tamarind is widely utilized in the culinary practices of Southern India (Nagendra and Mundoli, 2019). In recent times, the tamarind tree has experienced an increase in its prominence within the bonsai culture, particularly in Asian countries such as Indonesia and Taiwan (Lim and Lim, 2012). India annually produces approximately 0.3 million tonnes of tamarind, with the seed accounting for around 30–34% of the total fruit. This chapter has provided an examination of the health and nutritional aspects of tamarind. Tamarind exhibits various reported therapeutic actions, including anti-diabetic, hypolipidemic, hepatoprotective, anti-microbial as well as antiproliferative properties, among others. The plant in question is commonly utilized as a dietary vegetable by individuals residing in rural areas (Pal and Mukherjee, 2020).

2. Taxonomy and Scientific Classification

Tamarindus indica monospecific largest flowering tree, which is classified as a member of the Fabaceae family, specifically within the Caesalpinioideae subfamily. The classification of this entity is determined by its morphology, which indicates the presence of 5 to 8 distinct tribes. The species in question is a dicotyledonous organism with a diploid genome, characterized by a chromosome count of $2n = 24$. The Fabaceae family is recognized as one of the most extensive groups of angiosperms. This particular taxonomic group encompasses approximately 766 genera and 19,580 species, exhibiting a wide range of growth patterns across diverse climatic and habitat conditions. In the past, there existed a taxonomic differentiation between tamarinds originating from the West Indies and those originating from the East Indies. Tamarind trees in the West Indies were observed to yield pods that were three times longer than their width, containing a range of 1 to 4 seeds. On the other hand, tamarind trees in the East Indies were found to produce pods that were six times longer than their width, typically containing a higher quantity of seeds, ranging from 6 to 12 (Verheij and Coronel, 1991). A range of tamarind cultivars exist, exhibiting variations in both the color and sweetness of the fruit pulp.

The scientific classification of this species is as follows:

Domain	Eukaryota
Kingdom	Plantae
Phylum	Spermatophyta
Subphylum	Angiospermae
Class	Dicotyledonae
Order	Fabales
Family	Fabaceac
Subfamily	Detarioideae
Genus	*Tamarindus*
Species	*T. indica*

Binomial Name: *Tamarindus indica* L. 1753

Synonyms

Cavaraea Speg. 1916
Cavaraea elegans Speg. 1916
Tamarindus erythraeus Mattei. 1908
Tamarindus occidentalis Gaertn. 1791
Tamarindus officinalis Hook. 1851
Tamarindus somalensis Matteqi. 1908

Tamarindus umbrosa Salisb. 1796
Preferred Common Name: Tamarind
Vernacular name: Imli

International Common Names

English: Indian Tamarind, Kilytree, Tamarind
French: Tamarin, Tamarindier, Tamarinier
Spanish: Tamarindo
Chinese: Suan Dou

3. Botanical Description of the Plant

The tamarind tree is a large, long-living evergreen that typically grows untamed. The bark exhibits a scaly texture, ranging in color from gray to brown, and displays prominent fissures, resulting in a rough and scaly appearance. When the trunk and branches of the tree are injured, a crimson-colored resin is released. The longevity of the tree is notable, as it can bear fruit for a duration exceeding 50 years, while its lifespan can extend between 80 and 100 years (Parameswari et al., 2001). A mature tree can grow as tall as 30 meters (Devi and Boruah, 2020). Hard, dark-red heartwood and softer, yellowish sapwood make up tamarind wood (Devi and Boruah, 2020). The tamarind tree has a maximal crown height of 80 feet. The foliage is dense and forms an amorphous, vase-shaped outline around the tree crown. The tree flourishes in direct sunlight and has an inclination towards sandy, acidic, clayey, and loamy soils with a high resistance to xerotic conditions (wind-borne salt as in coastal regions) (Ranaivoson et al., 2015). This sizable evergreen tree with a stunning spreading crown is grown almost everywhere in India, except the Himalayas and western dry regions (Meena et al., 2022).

The evergreen leaves are pinnately compound (lobed) in shape and alternate in arrangement and have 10–18 pairs of opposite leaflets that are narrowly oblong. The leaflets are less than 5 cm in length, bright green in colour, elliptic-ovular in shape, and pinnate in venation (Sulieman et al., 2015). The opposite-sided leaflets of the pinnately compound (10–40 leaflet) leaves create a billowing effect in the wind (Kaur and Singh, 2016; Devi and Boruah, 2020). During extended periods of drought, the leaves can undergo abscission. The foliage of this tree exhibits a vibrant green coloration and possesses a visually delicate, feathery texture. The petiole and rachis are finely haired, and the midrib and net veining are more or less noticeable on both surfaces.

Pretty pale yellow or pinkish flowers grow on lax, small spikes that are about 2.5 cm wide. Flowers are elongated having 5 petals, 4 sepals, and 2-minute lower bracteoles that completely enclose the flower buds and fall off very early. The outer color of the four sepals, which are shed as the flower unfolds, frequently imparts a pink hue to the flower buds. The four sepals are unequal in length, ovate, and pink, cream, or pastel yellow. Out of the five petals, the posterior and lateral petals are enormous and showy, and their oblong shape slightly exceeds that of the calyx. The anterior petal segments have been reduced to filaments. The individual petals have an oblong shape and are white, cream, pale yellow, or pinkish with red or orange stripes. Flowers can be bisexual, dichogamous, or protogynous and highly cross-pollinated. The development of a flower blossom takes approximately 20 days after the first visible initiation. Racemes of flowers are produced, and the majority of them are yellow. The fruit is a brown indehiscent, 12 to 15-cm long legume that resembles a pod (pod-like fruits) and is sausage-shaped, curved or straight with rounded ends bearing a hard and brown shell. Fruit is a sub-cylindrical, indehiscent pod that is (5–10 cm × 2 cm), straight or curved, velvety, and rusty-brown. The pod shell is fragile and seeds are embedded in a gooey, edible pulp. It contains a soft, fleshy, juicy, acidic pulp that is thick and blackish-brown in colour and is packed with numerous hard-coated flattened, and glossy brown seeds. The irregularly shaped, three to ten, roughly 1.6 cm lengthy seeds are hard, shiny, and smooth (Rao and Mathew, 2012; Devi and Boruah, 2020). The fruit tastes sweet and sour. As Tamarindus ages, its branches droop from a single, central trunk and are often pruned to increase the density of the tree and assist in the harvesting of fruits. The leaflets fold at night. *Tamarindus* is a tropical species and frost vulnerable. The precise placement of *Tamarindus* viz-a-viz related genera will be made clear by the most recent morphological and molecular analyses and further research studies (Kaur and Singh, 2016).

Fig. 1 (a) Drawing of *T. indica* (Source: https://powo.science.kew.org) (b) Cultivated Tree (c) Foliage and Flowers (d) Flowers (e) Seed pods (f) Seeds.
(*Source:* https://tropical.theferns.info/viewtropical.php?id=Tamarindus+indica)

4. Geographical Distribution

The species is believed to have originated in arid Savannah regions of tropical Africa and the island of Madagascar. It is a tropical tree species that exhibits a remarkable ability to withstand temperatures as high as 47°C while displaying a pronounced vulnerability to frost. The precise origin of tamarind remains uncertain due to its extensive cultivation throughout ancient periods (Lim and Lim, 2012). The tamarind tree, which is indigenous to tropical Africa, has undergone extensive introduction and naturalization in over 60 countries worldwide. The species in uncertainty was introduced to tropical regions in the past, and it is now both cultivated and naturalized in numerous areas, including tropical Asia, the Arabian Peninsula, tropical America, the Caribbean, Australia, and several islands in the Pacific and Indian Oceans (Rivers and Mark, 2017). Tamarind is found in its natural habitat throughout Asia, spanning from Burma to Afghanistan, at altitudes of up to approximately 500 meters. These regions share comparable climatic characteristics, characterized by a combination of wet and semi-arid conditions commonly found in tropical areas. Additionally, it is observed in scattered regions within northern India.

The Indian subcontinent exhibits a continuous distribution pattern of the phenomenon under consideration, primarily observed in the southern and central regions. The predominant areas of production are primarily concentrated within the Asian countries of India, Bangladesh, Thailand, Sri Lanka, and Indonesia. Mexico and Costa Rica are considered to be the primary producers in the Americas. Despite its widespread use by local populations, tamarind is generally not commercially cultivated in Africa. Gambia, Senegal, Kenya, Tanzania, and Zambia are among the African nations that are involved in the production of diverse commodities (Diallo et al., 2007).

The species is indigenous to several countries in Africa, including Burkina Faso, Guinea, Guinea-Bissau, Ethiopia, Mali, Kenya, Madagascar, Mozambique, Uganda, Niger, Senegal, Sudan, Nigeria, Gambia, Chad, Eritrea, the Central African Republic, Tanzania and Zimbabwe. The extensive geographic range of tamarind can be partially attributed to human activities that facilitated introductions. The Indian subcontinent probably served as the initial point of dispersal for this phenomenon, subsequently spreading towards the eastern regions encompassing tropical and subtropical Asia, as well as the Pacific islands. The initial planting of a tamarind tree in the Hawaiian Islands occurred in the year 1797 (Kulkarni et al., 1993).

Crop cultivation has been introduced in various countries including the United States of America, Colombia, Myanmar, Cote d'Ivoire, Afghanistan, Brazil, Egypt, Ghana, Brunei, Indonesia, Liberia, Mexico, Iran, Guatemala, Haiti, Honduras, Pakistan, Sri Lanka, Jamaica, Laos, Malaysia, Nepal, Nicaragua, Mauritania, Australia, Puerto Rico, Vietnam, Bangladesh, Cambodia, China, Cuba, Dominican Republic, Panama, Papua New Guinea, Philippines, Thailand, Togo and Zambia (Gangaprasad et al., 2013). While the tree does not exhibit selectivity towards soil types, it thrives particularly well in deep alluvial soil. The tree exhibits sensitivity to frost, it demonstrates a preference for a warm climatic environment. The tamarind tree exhibits optimal growth in semi-tropical regions characterized by limited levels of precipitation. Tamarind thrives in regions characterized by high humidity and an average annual rainfall exceeding 1500 mm, but it is unable to withstand prolonged exposure to freezing temperatures. The plant exhibits a preference for moderately acidic, well-drained sandy soil with a pH range of 5 to 6. However, it is also capable of thriving in a wide variety of soil types, ranging from sandy to limestone, which may possess a rocky composition. Additionally, its adaptability to various types of soil and environmental circumstances, as well as its resistance to salt spray, powerful winds, and water deficiency, contribute to its widespread distribution (Hemalatha and Parameshwari, 2021).

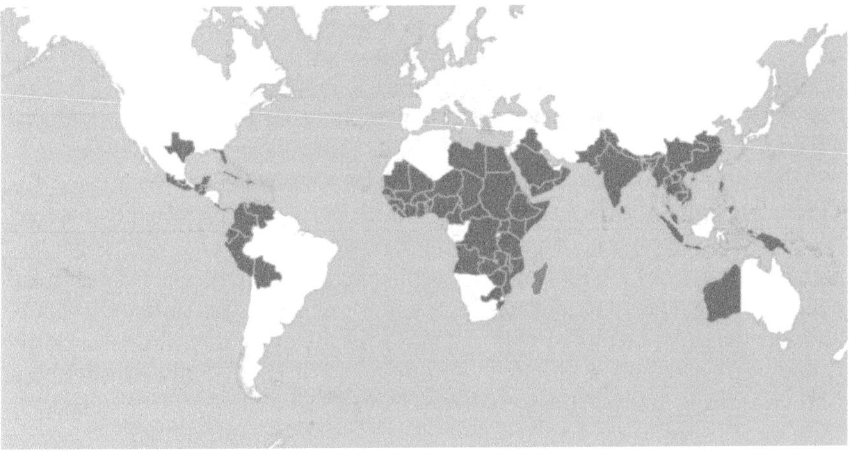

Fig. 2 The native range of *T. indica* L. is Comoros and Madagascar. (*Source:* https://powo.science.kew.org/)

5. Phytochemical Constituents and Metabolite Profiling

Tamarind fruits possess a diverse array of nutrients, including vitamins, electrolytes, mineral elements, and phytonutrients. According to the National Nutrient Database (USDA), the nutritional composition of tamarind per 100 grams includes, an energy content of 239 kcal, carbohydrate content of 62.5 g, protein content of 2.8 g, fat content of 0.6 g, no cholesterol and a dietary fiber content of 5.1 g. Tamarind fruits contain various vitamins, electrolytes, minerals, and phytonutrients in a 100 g serving. These vitamins include 14 μg folates, 1.94 mg niacin, 0.143 mg pantothenic acid, 0.066 mg pyridoxine, 0.428 mg thiamin, 30 IU vitamin A, 3.5 mg vitamin C, 0.1 mg vitamin E, and 2.8 μg vitamin K. The electrolytes found in tamarind fruits are sodium (28 mg) and potassium (628 mg). Additionally, tamarind fruits contain minerals such as calcium (74 mg), copper (0.86 mg), phosphorus (113 mg), iron (1.4-2.8 mg), magnesium (92 mg), selenium (1.3 μg), cadmium (0.33 mg) and zinc (8.52 mg) and carotene-β (18 μg) (Bhattacharya et al., 1994; Muzaffar et al., 2017). Also, the amino acid profile of tamarind seed includes methionine (113 mg/g), proline (287 mg/g), lysine (475 mg/g), aspartic acid (768 mg/g), arginine (450 mg/g), alanine (312 mg/g), leucine (531 mg/g), histidine (143 mg/g), isoleucine (313 mg/g), threonine (200 mg/g), tyrosine (287 mg/g), valine (306 mg/g), glycine (331 mg/g), serine (350 mg/g), phenyl alanine (318 mg/g), glutamic (1036 mg/g), and cysteine (106 mg/g) (Pal and Mukherjee, 2020).

Phytochemical and pharmaceutical analyses conducted on *T. indica* have demonstrated the presence of diverse essential elements, volatile compounds, phenolic substances, drugs, and lipid organic acids possessing anti-oxidant, anti-microbial, anti-bacterial, anti-snake venom, anti-diabetic and anti-inflammatory properties. The leaf structure of the plant is comprised of a total of thirteen discernible constituents, with linonene and benzyl-benzoate being the predominant compounds. The constituents found in the bark of roots encompass β-sitosterol, n-hexacosane, 21-oxobehenic acid, eicosanoic acid, octacosanyl ferulate, and (1)-pinitol. The seeds contain fatty acids, unsaponifiable matter originating from the oil of seeds and polyphenols. Moreover, the pulp is comprised of diverse organic acids (Azad, 2018).

The volatile components identified by Gas Chromatography (GC) and Gas Chromatography-Mass Spectrometry (GC-MS) in a study conducted by Sagrero-Nieves et al. (1994) reported supercritical fluid extraction of *T. indica* showing alpha copaene, alpha humulene, alpha pinene, alpha murolene, beta elemene, beta caryophyllene, beta pinene, and gamma cadinene. The tamarind phenolic components were identified with organic solvents extraction, followed by the isolation of metabolites through semi-preparative High Performance Liquid Chromatography (HPLC) (Sudjaroen et al., 2005). The structures of the compounds were determined using Liquid Chromatography–Electrospray-Ionisation-Mass Spectrometry (LC-ESI-MS), and Nano-Electrospray-Ionisation Mass Spectrometry (ESI-MS), as well as Nuclear Magnetic Resonance (NMR) spectroscopy. The results included the possibilities of oligomeric procyanidin tetramer (30.2%), procyanidin B2 (5.5%), procyanidin pentamer (17.6%), procyanidin hexamer (23.8%), procyanidin trimer (18.1%), and (2)-epicatechin (4.8%) from pericarp.

The fatty acid composition of sample includes myristic acid (present in small amounts), lauric acid (also present in small amounts), stearic acid (constituting 5.9% of the total fatty acids), palmitic acid (making up 14.8% of total fatty acids), oleic acid (comprising 27.0% of the total fatty acids), linolenic acid (accounting for 5.6% of the total fatty acids), linoleic acid (representing 7.5% of the total fatty acids), behenic acid (constituting 12.2% of the total fatty acids), arachidic acid (making up 4.5% of the total fatty acids), and lignoceric acid (comprising 22.3% of the total fatty acids) (Azad, 2018). The keto acid referred to in this context is alpha oxo-glutaric acid, which is also documented. Additional organic acid constituents include succinic acid, citric acid, stearic acid, galacturonic acid, lactic acid, and tartaric acid (Sulieman et al., 2015). The anti-oxidative components mentioned in the study conducted by Tsuda et al. (1994) include 4-dihydroxyacetophenone, methyl 3,4-dihydroxybenzoate, and 3,4-dihydroxy phenylacetate. Another compound with antioxidative properties is α,α-diphenyl-β-picrylhydrazyl mentioned in the study by Siddhuraju (2007).

The components with anti-microbial properties include alkaloids, tannins, sesquiterpenes, saponins, and phlobatamins (Doughari, 2006). Additionally, phenols and flavonoids have been found in the n-hexane, ethyl acetate, chloroform, and n-butanol extracts (Escalona-Arranz et al. 2013). The anti-bacterial components of tamarind analyzed by Bibekananda et al. (2014) included ciprofloxacin, gentamicin aqueous extracts, and hydroalcoholic extracts. The anti-snake venom compounds consist of nucleotidase enzyme, hyaluronidase, protease, and l-amino acid oxidase (Ushanandini et al., 2006). The anti-diabetic properties of the substance under investigation have been attributed to the presence of mucilage and pectin in the GLC of methyalated substances (Ibrahim et al., 1995). Additionally, Jain et al., (2007) identified (1)-pinitol as another component with antidiabetic and antiinflammatory effects.

Various tamarind ecological types have been classified and characterized based on their distinct sweet flavour profiles and other observable phenotypic traits. The primary constituents found in tamarind are anthranilate, linalool, limonene, and p-cymene. Nevertheless, the essential oils of this substance also contain compounds such as 6,10,14-trimethylpentadeca-5,9,13-trien2-one, longifolene, diphenyl-ether, and caryophyllene (Naeem et al., 2017). Seven hydrocarbons, including β-amyrin, β-sitosterol, and campesterol, were detected in tamarind seed oil (Arshad et al., 2019). Tartaric, acetic, and succinic acids, gum, tannins, sugar, pectin, alkaloids, sesquiterpenes, flavonoids, and glycosides have all been found in the aerial parts of this tree (Prasathkumar et al., 2021). Numerous active constituents were found in *T. indica* after phytochemical analysis was done on the plant, including phenolic compounds, L-(-) malic acid, arabinose, cardiac glycosides, galactose, glucose, mucilage and pectin, tartaric acid, xylose, and uronic acid (Kuru, 2014; Arshad et al., 2019).

Tamarind seed polysaccharide refers to a type of storage material found in the cell wall of tamarind seeds. It is obtained through the extraction process of tamarind seed powder. Tamarind seed polysaccharide is a complex polysaccharide with a high degree of branching, consisting of (1→4) glycosidic linkages. Furan derivatives and carboxylic acid made up the majority of the volatile components in the fruit pulp (Wong et al., 1998). The β-D-glucan backbone is characterized by the presence of side chains composed of α-D-xylopyranose and β-D-galactopyranosyl, which are linked to glucose residues through a (1→2)-α-D-xylopyranose linkage and a (1→6) linkage, respectively. The polysaccharide derived from tamarind seeds comprises glucose, galactose, and xylose (Joseph et al., 2012). The efficacy of this polysaccharide as a binding agent in weight granulation and immediate compression for various tablets was examined. The triterphenoids such as lupanone, lupeol, and some essential oil with benzyl benzoate and limonene are major compounds found in the tamarind leaves (Escalona-Arranz et al., 2010b). The primary fatty acids found in tamarind seeds were eicosanoic acid, oleic acid, palmitic acid, and linoleic acid (Idris et al., 2022). Proanthocyanidins in a variety of forms, including apigenin, epicatechin, catechin procyanidin B2, procyanidin dimer, eriodictyol, naringenin, procyanidin trimer, and taxifolin (Treml and Šmejkal, 2016).

6. Medicinal/Therapeutic Uses

The medicinal properties of the fruit are derived from its partially dried state. Tamarind seed powder is commonly utilized to manage abdominal discomfort, analgesic, anthelmintic, diarrhea, gastroenteritis, hyperglycemia, hypolipidemic, specific bacterial infections, parasitical infestations, healing of wounds, bowel movements, malaria, edoema, jaundice, ocular ailments, and ulcers. Moreover, it is used to manage elevated body temperature and respiratory infections. To alleviate pregnancy-induced nausea, certain women employ tamarind as a remedy. The tree contains a significant amount of essential amino acids and phytochemicals, making it a valuable resource. As a result, it has been documented to exhibit various beneficial properties such as anti-diabetic, anti-microbial, anti-venomic, anti-hyperlipidemic anti-oxidant, anti-malarial, anti-asthmatic, laxative and cardioprotective, and hepatoprotective activity. *T. indica* exhibits therapeutic properties for various medical conditions. It may also be regarded as cost-effective and readily accessible (Kuru, 2014; Rao et al., 2015; Azad, 2018).

A peptic ulcer is a distressing gastrointestinal ailment that impacts the stomach and duodenum. Previous studies have demonstrated that the seed extract of *T. indica* exhibits a protective effect against ulcer prototypes induced by alcohol, pylorus ligation, and ibuprofen in a dose-dependent manner. The potential treatment of ulcers using this approach presents a novel and innovative method. The protective properties of the *T. indica* seed are attributed to the presence of epicatechin, procyanidin, and polymeric tannins, which constitute the primary polyphenolic compounds in the seed. These substances function as a protective barrier against free radicals due to their anti-oxidant properties. Tannins exhibit the ability to impede ulcer development through the process of protein accumulation (Sumbul et al., 2011).

The content of tamarind fruit induces smooth muscle unwinding through the mechanism of calcium channel obstruction. Additionally, it elucidates the application of *T. indica* to cure diarrhoea (Mohammed and Yakasai, 2017). The ameliorative effect of *T. indica* extract of seeds has been demonstrated in cases of chemically induced acute kidney damage and renal cell carcinoma. The observed phenomenon can be elucidated by the anti-oxidative effect. Oxidative damage has been extensively linked to the development of cancer. However, it has been observed that certain polyphenol compounds present in *T. indica* seed extracts, such as 2-hydroxy-dihydroxyacetophenone, methyl 3,4-dihydroxybenzoate, 3,4-hydroxyphenylacetate, (–)-epicatechin, tannin, anthocyanins, and oligomeric proanthocyanidins, possess properties that induce antioxidant enzymes and inhibit cancer-related signal pathways (Farooqui et al., 2017).

Diabetes mellitus types 1 and 2 arise due to persistent inflammation, which leads to the impairment of the pancreatic β-cell islets (Soni and Singh, 2019). The dysregulation of insulin secretion significantly impacts both insulin receptor function and subsequent post-receptor events, leading to detrimental effects on various organs such as the liver, kidneys, and eyes. The protective effect of *T. indica* seed extract on pancreatic-cell islands can be attributed to its anti-inflammatory properties, capacity to regulate blood sugar, and ability to reverse damage to pancreatic tissue. The effects mentioned above, which exhibit similarities to the actions of insulin, are induced by an elevation in intracellular calcium ion concentrations within the pancreas and enhanced insulin activity in the plasma, rather than a reduction in absorption of glucose. Furthermore, it impacts the GLUT-2, GLUT-4, and SREBP-1c of the target tissue, specifically by elevating the mRNA level in the liver (Al-Kharousi et al., 2018).

6.1 Uses described in Pharmacopoeias and Traditional of Medicine

In Asia, Africa, and the Americas, tamarind-derived products are extensively employed as medicinal remedies. The tamarind tree leaves, fruits, and seeds have been broadly utilized in Indian Ayurvedic and traditional African medicine systems. The medicinal properties of tamarind are mentioned in ancient Sanskrit texts. The therapeutic benefits of tamarind fruits were extensively acknowledged in Europe after their introduction by Indian Arab traders (Rao and Mathew, 2012).

The ethnopharmacological significance of tamarind in Africa is noteworthy, as it possesses various medicinal properties (Havinga et al., 2010). These include its efficacy as an anti-helminthic agent for expelling worms and its anti-microbial, astringent properties, anti-septic, anti-viral, sunscreen, and properties (Panara et al., 2014). Additionally, tamarind has been utilized in Africa to facilitate wound healing in a diverse range of conditions such as nausea, vomiting during pregnancy, dysentery, eye dryness, chest pain, asthma, bacterial skin infections, colds, colic, boils, jaundice, keratitis, diabetes, diarrhea, cholesterol metabolism disorders, fever, gallbladder disorders, gastrointestinal disorders, leprosy, liver disorders, conjunctivitis, gingivitis, hemorrhoids, joint pain, chronic and acute constipation, sores, sore throat, eye inflammation, indigestion, saliva production, skin disinfection and sterilization, sprains and urinary swelling (Nahar et al., 2007; Joshi et al., 2023).

Preparations incorporating tamarind pulp, flowers, leaves, bark, and roots have been purported to possess various medicinal properties. These applications encompass the utilization of these substances as anti-inflammatory agents in North Africa, as herbal remedies in Burkina Faso, as remedies for leucorrhoea and skin ailments, as well as their traditional uses in India (Bhadoriya et al., 2011). In the northern region of Nigeria, the indigenous practice involves the preparation of a medicinal concoction by brewing the fresh stem bark and leaves in combination with potassium. This traditional remedy is commonly employed for treating various gastrointestinal disorders, general bodily discomfort, jaundice, yellow fever, and its purported blood tonic and skin cleansing properties (Komutarin et al., 2004). Tamarind is employed in the treatment of sore throats. Tamarind juice is consumed either rinsed or ingested to alleviate the ache and annoyance associated with sore throats (Narendra et al., 2009).

The prevalence of bacterial resistance to commonly used antibiotics is increasing, prompting the exploration of novel and practical therapeutic approaches due to the toxic effects of antibiotic usage. Medicinal plants are favoured due to their enhanced tolerability. *T. indica* has been identified as a potential antimicrobial agent. The anti-bacterial potential of *T. indica* extract has been observed in various bacterial strains, including *Escherichia coli*, *Bacillus subtilis*, *Salmonella paratyphi*, *Klebsiella pneumoniae*, *Burkholderia pseudomallei*, *Salmonella typhi* and *Staphylococcus aureus*. The presence of lupeol in this tree is associated with its antibacterial properties. The inhibitory effect of tamarind indica bark extract on *Salmonella typhi* was 25% in a chemical assay. The observed effects indicate lymphoproliferation, regardless of the presence or absence of phytohemagglutinin. This compound exerts its therapeutic effects on the level of parasitemia and is commonly employed in treating various infectious ailments, such as malaria. Additionally, it elicits a response from the immune system (Escalona-Arranz et al., 2010a). The methanolic extract of fruit pulp of *T. indica* was analyzed by GC-MS. The analysis revealed the presence of several bioactive compounds, including 5-Hydroxymethylfurfural, 1,6-anhydro-β-D-Glucopyranose, 5-methyl-Furancarboxaldehyde, Levoglucosenone, 3-O-Methyl-d-glucose, Triethylenediamine, 1-(2-furanyl)-1-Propcanone, Methyl 2-furoate, methyl ester-Hepta-2,4-dienoic acid. Various compounds, including 2-Heptanol, acetate, 3-Methyl-2-furoic acid, β-D-fructofuranosyl-α-D-Glucopyranoside, 5-[(5-methyl-2-fur-2-Furancarboxaldehyde, n-Hexadecanoic acid and cis-Vaccenic acid have been documented to possess multiple biological functions like anti-bacterial, anti-cancer, anti-fungal, antioxidant, anti-tubercular and other prophylactic properties (Fagbemi et al., 2022).

Based on available reports, it has been observed that the extract derived from the *T. indica* plant exhibits antiviral properties against various viruses, including the Tobacco Mosaic Virus (TMV), Cowpea Mosaic Virus (CMV), and Watermelon Mosaic Virus (WMV) (Perera et al., 2021). According to reports, it is purported that the substance also possesses anti-nematodal characteristics against *Bulinus truncatus* (Ross et al., 2003).

While it has been demonstrated that various components of *T. indica*, such as leaves and seeds, possess anti-inflammatory properties, it is essential to note that acetylsalicylic acid exhibits a more potent anti-inflammatory effect. Additionally, it possesses anti-inflammatory properties and exhibits inhibitory effects on the release of prostaglandins (PG) and nitric oxide (NO), similar to the pharmacological effects of diclofenac. Phospholipase A2 is expressed in reaction to lysosomal injury and facilitates the hydrolysis of phospholipids, which is essential for synthesizing inflammatory mediators. The prevention of cellular damage leads to the preservation of the cytoplasm and a decrease in the inflammatory response. The existence of flavonoids and polyphenols in *T. indica* has been associated with its anti-inflammatory and anti-nociceptive characteristics (Bhadoriya et al., 2012).

6.2 Traditional/Folklore Uses Described in Folk Medicine

The tamarind fruit is used as a poultice in Southeast Asia to treat feverish people by applying it to their foreheads. Due to the fruit's high concentrations of tartaric acid, malic acid, and potassium bitartrate, it has a laxative effect (Rösti et al., 2018). The world over, there is evidence of its use to treat constipation. It is used in traditional medicine for wound healing, respiratory issues, fever, malaria, dysentery, diarrhoea, and abdominal pain. Due to its laxative and aphrodisiac qualities, it is also widely used in tropical nations. Tamarind, called Imli in Hindi and Chincha in Ayurveda, is classified as a Kshara tree belonging to the Ksharastaka group, consisting of eight kshara plants. Chincha kshara is an autonomous medicinal substance that is additionally employed as a constituent in various formulations, including Agnisandeepano Rasa, Bhruhat Gudapippali, Gudapippali, Mahashankha Vati, and Shankha Vati, Shankhadravaka Rasa, among others. While it is commonly advised to use the entire plant for kshara preparation, suggesting that any part of the plant is appropriate for this purpose, the Rasashastra text Rasa Tarangini specifically recommends using Kasta (wood) as the suitable part for the preparation of Kshara (Kodlady et al., 2012).

6.3 Application Supported by Clinical Data

Tamarind seed extracts (TSE) contain a high concentration of triterpenes, procyanidins, polyphenolic bioactive substances, and polysaccharides. These compounds have been found to possess notable protective and curative properties against a range of individual pathophysiological conditions (Azad, 2014). The investigation of the acute toxicity of tamarind pulp extract revealed that no deaths occurred following the ingestion of 5000 mg/kg bw (body mass) and 3000 mg/kg bw of the extract (Lim et al., 2013). A research investigation was undertaken to examine the effects of *T. indica* leaf fluid extract on human blood cells. The findings indicated that when administered at concentrations ranging from 20 to 100 mg/ml, the extract did not induce noteworthy hemolysis. The protein denaturation ratio exhibited a significantly lower value when plant extracts were applied at concentrations ranging from 40 to 100 mg/ml compared to the control. This observation supports the assertion that plant extracts possess reduced toxicity. Furthermore, the excerpt demonstrated a defensive impact against oxidative damage induced by H_2O_2 in the membrane of human erythrocytes, using equivalent concentrations (Escalona-Arranz et al., 2013).

The clinical study provides evidence supporting the initial effectiveness of a food-based anti-inflammatory herbal mixture, NXT15906F6 (consisting of *T. indica* seeds and *Curcuma longa*), in reducing the clinical symptoms associated with knee osteoarthritis as early as day 5 of supplementation. The consumption of *T. indica* seeds and *C. longa* has been found to have a positive impact on knee pain reduction, as well as improvements in musculoskeletal function and joint mobility among participants. These improvements have been associated with enhanced physical performance and an overall enhancement in quality of life. Supplementation with NXT15906F6 has been observed to decrease the concentration of a marker associated with cartilage erosion, suggesting a potential involvement in the mitigation of degradation in joint tissues. Significantly, this trial serves to reaffirm the tolerability and safety of the botanical composition Tamarindus indica for human consumption (Prasad et al., 2023).

6.4 Pharmacological Activities Mainly Antiviral

The potential utility of tamarind seeds in the development of antiviral drugs for chikungunya is suggested by their antiviral abilities. The antiviral properties of the tamarind seed lectin have been substantiated by the research, employing a range of techniques such as luciferase-based assay, plaque elimination assay, and real-time PCR. The administration of Tamarind chitinase (chi)-like lectin (TCLL) exhibited a notable reduction in the infectivity of the chikungunya virus, amounting to a 64% decrease. Furthermore, there was an observed decrease of approximately 45% in the levels of viral RNA within the cells. Extensive research has been conducted on the utilization of lectin as an antiviral agent, encompassing various viruses such as HIV and HPV (Kaur et al., 2019). The research has also demonstrated that a lectin sourced from tamarind seeds exhibits an affinity for glycans, specifically sugar molecules containing N-acetylglucosamine (NAG), which are present in the viral capsule. Consequently, this interaction effectively impedes viral entry into host cells. The presence of N-acetylglucosamine (NAG) in the glycoproteins of the Chikungunya virus (CHIKV) suggests that the lectin derived from tamarind holds potential as an effective agent in combating this viral infection. The utilization of TCLL is being employed. Recently, there have been reports on the binding of Tamarind seed xyloglucan-specific endo-beta-1,4-glucanase to the glycan present on the envelope protein of Chikungunya virus, which is derived from the seeds of *T. indica*. The study assessed the practicality of employing an antigen capture test for the comprehensive detection of alphaviral infections. Sindbis virus (SINV), a member of the alphavirus genus, was utilized as a representative specimen, and the findings obtained were consistent with those observed for CHIKV (Choudhary et al., 2021).

The Tamarind trypsin inhibitor (TTI) has been observed to exhibit various advantageous effects in mitigating markers of inflammation such as tumour necrosis factor α (TNF-α) and leptin, as well as biochemical parameters including fasting glycemia, triglycerides, and very low-density lipoprotein (VLDL). Furthermore, TTI has demonstrated the ability to enhance pancreatic function and promote mucosal reliability in an obesity model. TTI can potentially impede the functioning of proteases that interact with SARS-CoV-2 infection, as well as the neutrophil action associated with the lung damage induced by the virus (Morais et al., 2021).

The study examined the potential of *T. indica* stem bark crude ethanol extract and column chromatographic fractions to inhibit virus replication and/or exhibit anti-Newcastle disease virus (NDV) activity, revealed that the crude extract of *T. indica* stems bark, at a concentration of 62.5 mg/mL, as well as its four chromotographic fractions (TiA, TiC, TiD), exhibited hemagglutination (HA) titre of 1:128 each. However, the TiB fraction demonstrated a slightly lower HA titre of 1:64. A concentration of 125 mg/mL resulted in a titre of 1:16 for both TiB and

TiD, while a titre of 1:8 was observed for TiA. Similarly, both the crude extract and TiC exhibited a 1:4 HA titre. Nevertheless, the minimum concentrations required for the inactivation of the virus were determined to be 0.24 mg/mL for the extract and 0.49 mg/mL for the fraction (Okoh et al., 2017).

The objective of this study was to utilize *in silico* methods to identify native and analogous peptides derived from the trypsin inhibitor extracted from tamarind seeds (TTIp) that have potential anti-SARS-CoV-2 activity. The investigation of the possibility of peptides to inhibit SARS-CoV-2 was conducted using molecular dynamics simulation. Specifically, the focus was on the interaction between these peptides and the binding sites of transmembrane serine protease 2 (TMPRSS2), which plays a crucial role in facilitating the entry of SARS-CoV-2 into the host cell. The optimal interaction potential energy (IPE) was observed between TMPRSS2 and a native peptide derived from trypsin cleavage, as well as its analogous peptide. Both peptides exhibited a significant number of hydrophobic residues, which is a prevalent physical-chemical characteristic shared by peptides that impede the entry of enveloped viruses, including SARS-CoV-2. This property is also observed in specific pharmaceutical compounds utilised for the treatment of COVID-19 (Luz et al., 2023).

7. Toxicity and Safety

There are several scientific investigations on animals conforming safe use of *T. indica*. A 2-year examination by Silva et al. (2009), reported no side effects in animals feeding on different doses of seed extract of *T. indica*. No changes were reported in the biochemistry of blood, urine, liver function test (LFT), and body weight in the animals fed on seed polysaccharide of *T. indica* for 28 days (Heimbach et al., 2013). Toxic effects were not reported but white blood cells and thrombocytes increased in number (Sharma et al., 2013). The seeds of *T. indica* contain tannins and other compounds making digestion challenging thus it is recommended to have boiled or water-imbibed seeds. The acidic content in seeds on long-term use can lead to the dental erosion. Ibuprofen and acetylsalicylic acid on ingestion with *T. indica* leads to an increase in their bio-availability and blood levels (Garba et al., 2003). The tamarind pulp extract has an LD50 (lethal dose 50) value exceeding 5000 mg/kg bw, indicating a high level of safety. This classification aligns with the recommendations of esteemed organizations like the World Health Organization (WHO) and the Organization for Economic and Cultural Development (OECD) (Abubakar et al., 2010).

Conclusion

The tamarind tree is a species that exhibits a wide range of applications, making it a versatile and multifunctional plant. It finds application in a wide range of domestic, medicinal, and industrial contexts. The pulp of this fruit is renowned for its exceptional flavour and taste, which can be described as a combination of sweetness and sourness. It is widely utilized in the preparation of various culinary creations and beverages. The various components of the tree, including the bark, root, foliage, fruit, and wood, collectively contribute to the sustenance of rural communities. The tamarind seed and other plant components exhibit antiviral, antibacterial, and antimicrobial characteristics and coagulation capabilities. These features have demonstrated potential in water treatment and extraction, offering an alternative to the challenges faced by developing countries.

References

Abubakar, M.G., Yerima, M.B., Zahriya, A.G. and Ukwuani, A.N. (2010). Acute toxicity and antifungal studies of ethanolic leaves, stem and pulp extract of Tamarindus indica. *Research Journal of Pharmaceutical, Biological and Chemical Sciences*, 1(4), pp.104–111.

Alhazmi, H.A., Najmi, A., Javed, S.A., Sultana, S., Al Bratty, M., Makeen, H.A., Meraya, A.M., Ahsan, W., Mohan, S., Taha, M.M. and Khalid, A. (2021). Medicinal plants and isolated molecules demonstrating immunomodulation activity as potential alternative therapies for viral diseases including COVID-19. *Frontiers in Immunology*, 12, p.637553.

Al-Kharousi, Z., Guizani, N., Al-Sadi, A. and Al Bulushi, I. (2018). Fresh fruit and vegetable bacteria: Diversity, antibiotic resistance and their possible contribution to gut microbiota. *Fruit and Vegetable Consumption and Health: New Research*, ed. W. Jongen (Hauppauge, NY: Nova Science Publishers, Inc), pp.39–66.

Arshad, M.S., Imran, M., Ahmed, A., Sohaib, M., Ullah, A., Nisa, M.U., Hina, G., Khalid, W. and Rehana, H. (2019). Tamarind: a diet-based strategy against lifestyle maladies. *Food Science & Nutrition*, 7(11), pp.3378–3390.

Azad, S. 2018. Tamarindo—Tamarindus indica. *In Exotic fruits* (pp. 403–412). Academic Press.

Balick, M.J. and Cox, P.A. 2020. Plants, people, and culture: the science of ethnobotany. *Garland Science.*

Bhadoriya, S.S., Ganeshpurkar, A., Narwaria, J., Rai, G. and Jain, A.P. (2011). Tamarindus indica: Extent of explored potential. *Pharmacognosy reviews*, 5(9), p.73.

Bhadoriya, S.S., Mishra, V., Raut, S., Ganeshpurkar, A. and Jain, S.K. (2012). Anti-inflammatory and antinociceptive activities of a hydroethanolic extract of Tamarindus indica leaves. *Scientia Pharmaceutica*, 80(3), pp.685–700.

Bhattacharya, S., Bal, S., Mukherjee, R.K. and Bhattacharya, S. (1994). Functional and nutritional properties of tamarind (Tamarindus indica) kernel protein. *Food Chemistry*, 49(1), pp.1–9.

Bibekananda, M., Dash, D.K. and Bhoi, B.B. (2014). Evaluation of *in vitro* antioxidant and antimicrobial properties of hydroalcoholic and aqueous seeds extracts of Tamarindus indica L. *World Journal of Pharmacy and Pharmaceutical Sciences (WJPPS)*, 3(1), pp.436–452.

Boland, D.J., Brooker, M.I.H., Chippendale, G.M., Hall, N., Hyland, B.P., Johnston, R.D., Kleinig, D.A., McDonald, M.W. and Turner, J.D. eds. (2006). *Forest trees of Australia*. CSIRO publishing.

Caluwé, E.D., Halamová, K. and Damme, P.V. (2009). Tamarind (Tamarindus indica L.): A Review of Traditional Uses. In Phytochemistry and Pharmacology. Publ: ACS Symposium Series, Bk Title, *African Natural Plant Products: New Discoveries and Challenges in Chemistry and Quality* (Vol. 1021, No. 5, pp. 85–110).

Choudhary, S., Neetu, N., Singh, V.A., Kumar, P., Chaudhary, M. and Tomar, S. (2021). Chikungunya virus titration, detection, and diagnosis using N-Acetylglucosamine (GlcNAc) specific lectin based virus capture assay. *Virus Research*, 302, p.198493.

De Caluwé, E., Halamouá, K. and Van Damme, P., (2010). Tamarindus indica L.–A review of traditional uses, phytochemistry and pharmacology. *Afrika focus*, 23(1), pp.53–83.

Devi, B. and Boruah, T. (2020). Tamarind (Tamarindus indica). *Antioxidants in Fruits: Properties and Health Benefits*, pp.317–332

Diallo, B.O., Joly, H.I., McKey, D., Hosaert-McKey, M. and Chevallier, M.H. (2007). Genetic diversity of Tamarindus indica populations: Any clues on the origin from its current distribution? *African Journal of Biotechnology*, 6(7).

Doughari, J.H. (2006). Antimicrobial activity of Tamarindus indica Linn. *Tropical Journal of Pharmaceutical Research*, 5(2), pp.597–603.

Escalona-Arranz, J.C., Garcia-Diaz, J., Perez-Rosés, R., De la Vega, J., Rodríguez-Amado, J. and Morris-Quevedo, H.J. (2014). Effect of Tamarindus indica L. leaves' fluid extract on human blood cells. *Natural Product Research*, 28(18), pp.1485–1488.

Escalona-Arranz, J.C., Péres-Roses, R., Urdaneta-Laffita, I., Camacho-Pozo, M.I., Rodríguez-Amado, J. and Licea-Jiménez, I. (2010a). Antimicrobial activity of extracts from Tamarindus indica L. leaves. *Pharmacognosy Magazine*, 6(23), p.242.

Escalona-Arranz, J.C., Pérez-Rosés, R., Jiménez, I.L., Rodríguez-Amado, J., Argota-Coello, H., Cañizares-Lay, J., Morris-Quevedo, H.J. and Sierra-González, G. (2010b). Chemical constituents of Tamarindus indica L. leaves. *Revista Cubana de Química*, 22(3), pp.65–71.

Escalona-Arranz, J.C., Perez-Roses, R., Urdaneta-Laffita, I., Morris-Quevedo, H., Camacho-Pozo, M.I., Rodriguez-Amado, J. and González, G.S. (2013). Role of polyphenols in the antimicrobial activity of ethanol Tamarindus indica L leaves fluid extract. *Boletín Latinoamericanoy del Caribe de Plantas Medicinales y Aromáticas*, 12(5), pp.516–522.

Fagbemi, K.O., Aina, D.A., Adeoye-Isijola, M.O., Naidoo, K.K., Coopoosamy, R.M. and Olajuyigbe, O.O. (2022). Bioactive compounds, antibacterial and antioxidant activities of methanol extract of Tamarindus indica Linn. *Scientific Reports*, 12(1), p.9432.

Farooqui, Z., Ahmed, F., Rizwan, S., Shahid, F., Khan, A.A. and Khan, F., (2017). Protective effect of Nigella sativa oil on cisplatin induced nephrotoxicity and oxidative damage in rat kidney. *Biomedicine & Pharmacotherapy*, 85, pp.7–15.

Gangaprasad, S., Rajkumar, R., Ravikumar, R.L., Savita, M.H., Krishnamurthy, K. and Hittalamani, S. (2013). Genetic diversity analysis in tamarind (Tamarindus indica L.). *Journal of Spices and Aromatic Crops*, 22(1).

Garba, M., Yakasai, I.A., Bakare, M.T. and Munir, H.Y. (2003). Effect of Tamarindus indica. L on the bioavailability of ibuprofen in healthy human volunteers. *European journal of Drug Metabolism and Pharmacokinetics*, 28, pp.179–184.

Gurib-Fakim, A., (2006). Medicinal plants: traditions of yesterday and drugs of tomorrow. *Molecular Aspects of Medicine*, 27(1), pp.1–93.

Havinga, R.M., Hartl, A. Putscher, J., Prehsler, S., Buchmann, C. and Vogl, C.R., (2010). Tamarindus indica L. (Fabaceae): patterns of use in traditional African medicine. *Journal of Ethnopharmacology*, 127(3), pp.573–588.

Heimbach, J.T., Egawa, H., Marone, P.A., Bauter, M.R. and Kennepohl, E., (2013). Tamarind seed polysaccharide: a 28-day dietary study in Sprague-Dawley rats. *International Journal of Toxicology*, 32(3), pp.198–208.

Hemalatha, C. and Parameshwari, S. (2021). The scope of tamarind (Tamarindus indica L.) kernel powder in diverse spheres: A review. *Materials Today: Proceedings*, 45, pp.8144–8148.

Ibrahim, N.A., El-Gengaihi, S., El-Hamidi, A. and Bashandy, S.A.E. (1994). Chemical and biological evaluation of Tamarindus indica L. growing in Sudan. *In Internat. Symposium on Medicinal and Aromatic Plants* 390 (pp. 51–58).

Idris, S.A., Rosli, N.R. and Aris, R.M.A.R. (2022). Supercritical carbon dioxide extraction of fatty acids compounds from tamarind seeds. *Materials Today: Proceedings*, 63, pp.S462–S466.

Jain, R., Jain, S., Sharma, A., Ito, H. and Hatano, T. (2007). Isolation of (+)-pinitol and other constituents from the root bark of Tamarindus indica Linn. *Journal of Natural Medicines*, 61, pp.355–356.

Joseph, J., Kanchalochana, S.N., Rajalakshmi, G., Hari, V. and Durai, R.D. (2012). Tamarind seed polysaccharide: A promising natural excipient for pharmaceuticals. *International Journal of Green Pharmacy* (IJGP), 6(4).

Joshi, A.V., Urmila, A., Swapnil, G. and Shahrukh, S. (2023). Review on Tamarindus indica. *Journal of Pharmacognosy and Phytochemistry*, 12(1), pp.303–308.

Kaur R., Mudgal R., Jose J., Kumar P., Tomar S. Glycan-dependent chikungunya viral infection divulged by antiviral activity of NAG specific chi-like lectin. *Virology*. (2019) Jan 2;526:91–98.

Kaur, M. and Singh, S. (2016). Physicochemical, morphological, pasting, and rheological properties of tamarind (Tamarindus indica L.) kernel starch. *International Journal of Food Properties*, 19(11), pp.2432–2442.

Kodlady, N., Patgiri, B.J., Harisha, C.R. and Shukla, V.J., (2012). Pharmacognostical and physicochemical analysis of Tamarindus indica Linn. stem. *Journal of Ayurveda and Integrative Medicine*, 3(1), p.6.

Komutarin, T., Azadi, S., Butterworth, L., Keil, D., Chitsomboon, B., Suttajit, M. and Meade, B.J. (2004). Extract of the seed coat of Tamarindus indica inhibits nitric oxide production by murine macrophages *in vitro* and *in vivo*. *Food and Chemical Toxicology*, 42(4), pp.649–658.

Kulkarni, R.S., Gangaprasad, S. and Swamy, G.S.K. (1993). Tamarind: economically an important minor forest produce. MFP News, 3(3).

Kuru, P. (2014). Tamarindus indica and its health related effects. *Asian Pacific Journal of Tropical Biomedicine*, 4(9), pp.676–681.

Lim, C.Y., Mat Junit, S., Abdulla, M.A. and Abdul Aziz, A. (2013). *In vivo* biochemical and gene expression analyses of the antioxidant activities and hypocholesterolaemic properties of Tamarindus indica fruit pulp extract. *Plos One*, 8(7), p.e70058.

Lim, T.K. and Lim, T.K. (2012). Tamarindus indica. *Edible Medicinal and Non-Medicinal Plants:* Volume 2, *Fruits*, pp.879–905.

Luz, A.B.S., de Medeiros, A.F., Bezerra, L.L., Lima, M.S.R., Pereira, A.S., e Silva, E.G.O., Passos, T.S., Monteiro, N.D.K.V. and de Araújo Morais, A.H. (2023). Prospecting native and analogous peptides with anti-SARS-CoV-2 potential derived from the trypsin inhibitor purified from tamarind seeds. *Arabian Journal of Chemistry*, 16(8), p.104886.

Mascarenhas, A.F. and Muralidharan, E.M. (1993). Clonal forestry with tropical hardwoods. In *Clonal Forestry II: Conservation and Application* (pp. 169–187). Berlin, Heidelberg: Springer Berlin Heidelberg.

Meena, J., Chandra, H. and Warkar, S.G. (2022). Carboxymethyl Tamarind Kernel Gum/ZnO-Biocomposite: As an Antifungal and Hazardous Metal Removal Agent. *Journal of New Materials for Electrochemical Systems*, 25(3).

Meena, V.S., Gora, J.S., Singh, A., Ram, C., Meena, N.K., Rouphael, Y., Basile, B. and Kumar, P. (2022). Underutilized fruit crops of Indian arid and semi-arid regions: Importance, conservation and utilization strategies. *Horticulturae*, 8(2), p.171.

Mohammed, K.G. and Yakasai, H.I. (2017). Tamarindus indica an appraisal on its uses in traditional medicine and its antinutritional componentes-Mini review". *Chronicles of Pharmaceutical Science*, 1(4), pp.199–203.

Morais, A.H.D.A., de Medeiros, A.F., Medeiros, I., de Lima, V.C., Luz, A.B., Maciel, B.L. and Passos, T.S. (2021). Tamarind (Tamarindus indica L.) seed a candidate protein source with potential for combating SARS-CoV-2 infection in obesity. *Drug Target Insights*, 15, p.5.

Muzaffar, K., Dar, B.N. and Kumar, P., (2017). Assessment of nutritional, physicochemical, antioxidant, structural and rheological properties of spray dried tamarind pulp powder. *Journal of Food Measurement and Characterization*, 11, pp.746–757.

Naeem, N., Nadeem, F., Azeem, M.W. and Dharmadasa, R.M. (2017). Tamarindus indica–A review of explored potentials. *International Journal of Chemical and Biochemical Sciences*, 12(2017), pp.98–106.

Nagendra, H. and Mundoli, S. (2019). *Cities and canopies: trees in Indian cities*. Penguin Random House India Private Limited.

Nahar, N., Choudhuri, M. and Alamgir, M. (2007). Preliminary in vitro evaluation of some traditional Ayurvedic antacids. *Advances in Traditional Medicine*, 7(4), pp.441–443.

Narendra, V., Gavatia, N.P., Bhaskar, G. and Mukul, T. (2009). Antioxidant potential of Tamarindus indica seed coat. *Journal of Pharmacy Research*, 2(11), pp.1705–1706.

Pal, D. and Mukherjee, S. (2020). Tamarind (Tamarindus indica) seeds in health and nutrition. *In Nuts and seeds in health and disease prevention* (pp. 171–182). Academic Press.

Panara, K., Harisha, C.R. and Shukla, V.J., (2014). Pharmacognostic and phytochemical evaluation of fruit pulp of Tamarindus Indica linn. *Int J Ayurvedic Med*, 5, pp.37–42.

Parameswari, K., Srimathi, P. and Maiarkodi, K., (2001). Standardisation of dormancy breaking treatment in tamarind (Tamarindus indica L.) seeds. *Legume Research-An International Journal*, 24(1), pp.60–62.

Pareek, O.P. and Sharma, S. (2017). *Systematic Pomology* (Vol. 1–2) (Set). Scientific Publishers.

Perera, W.P.R.T., Liyanage, J.A., Dissanayake, K.G.C., Gunathilaka, H., Weerakoon, W.M.T.D.N., Wanigasekara, D.N., Fernando, W.S.K., Rajapaksha, R.M.H., Liyanage, R.P. and Perera, B.T. (2021). Antiviral potential of selected medicinal herbs and their isolated natural products. *BioMed Research International*, 2021.

Prasad, N., Vinay, V. and Srivastava, A. (2023). Efficacy of a proprietary combination of Tamarindus indica seeds and Curcuma longa rhizome extracts in osteoarthritis: a clinical investigation. *Food & Nutrition Research*, 67.

Prasathkumar, M., Anisha, S., Dhrisya, C., Becky, R. and Sadhasivam, S. (2021). Therapeutic and pharmacological efficacy of selective Indian medicinal plants–a review. *Phytomedicine Plus*, 1(2), p.100029.

Ranaivoson, T., Brinkmann, K., Rakouth, B. and Buerkert, A. (2015). Distribution, biomass and local importance of tamarind trees in south-western Madagascar. *Global Ecology and Conservation*, 4, pp.14–25.

Rao, A.S., Kumar, A.A. and Ramana, M.V. (2015). Tamarind seed processing and by-products. *Agric. Eng. Int.: CIGR J*, 17, pp.200–204.

Rao, Y.S. and Mathew, K.M. (2012). Tamarind. In *Handbook of Herbs and Spices* (pp. 512–533). Woodhead Publishing.

Rivers, M.C. and Mark, J. (2017). Tamarindus indica. The IUCN Red List of Threatened Species. http://dx.doi.org/10.2305/IUCN.UK.2017-3.RLTS.T62020997A62020999.en.

Ross, I.A. (2003). Tamarindus indica. In Medicinal Plants of the World: Volume 1 Chemical Constituents, *Traditional and Modern Medicinal Uses* (pp. 455–463). Totowa, NJ: Humana Press.

Rösti, J., Schumann, M., Cléroux, M., Lorenzini, F., Zufferey, V. and Rienth, M. (2018). Effect of drying on tartaric acid and malic acid in Shiraz and Merlot berries. *Australian Journal of Grape and Wine Research*, 24(4), pp.421–429.

Sagrero-Nieves, L., Bartley, J.P. and Provis-Schwede, A. (1994). Supercritical fluid extraction of the volatile constituents from tamarind (Tamarindus indica L.). *Journal of Essential Oil Research*, 6(5), pp.547–548.

Sharma, S., Yadav, N., Pandey, A., Sharma, S. and Sharma, K.P. (2013). Antioxidant rich diet supplements (Spirulina and tamarind fruit pulp) mitigate hematological disorders in fluoride exposed mice. *Toxicological & Environmental Chemistry*, 95(10), pp.1739–1747.

Sharma, S.K., Singh, R.S. and Bhargava, R. (2013). Arid horticulture: an overview. *Annals of Arid Zone*, 52(3&4), 251–264.

Siddhuraju, P., (2007). Antioxidant activity of polyphenolic compounds extracted from defatted raw and dry heated Tamarindus indica seed coat. *LWT-Food Science and Technology*, 40(6), pp.982–990.

Silva, F.M.V., Leite, M.F., Spadaro, A.C.C., Uyemura, S.A., and Maistro, E.L. (2009). Assessment of the potential genotoxic risk of medicinal Tamarindus indica fruit pulp extract using *in vivo* assays. *Genetics and Molecular Research* 8(3), 1085–1092.

Singh, A.K. and Nigam, S.N. (2017). Ancient alien crop introductions integral to Indian agriculture: An overview. *In Proc. Indian Natn. Sci. Acad.* 83(3), 549–568.

Soni, N. and Singh, V.K., 2019. Traditional, nutraceutical and pharmacological approaches of Tamarindus indica (Imli). *European Journal of Biological Research,* 9(3), pp.141–154.

Sudjaroen, Y., Haubner, R., Würtele, G., Hull, W.E., Erben, G., Spiegelhalder, B., Changbumrung, S., Bartsch, H. and Owen, R.W. (2005). Isolation and structure elucidation of phenolic antioxidants from Tamarind (Tamarindus indica L.) seeds and pericarp. *Food and Chemical Toxicology*, 43(11), pp.1673–1682.

Sulieman, A.M.E., Alawad, S.M., Osman, M.A. and Abdelmageed, E.A. (2015). Physicochemical characteristics of local varieties of tamarind (Tamarindus indica L), Sudan. *International Journal of Plant Research*, 5(1), pp.13–18.

Sumbul, S., Ahmad, M.A., Mohd, A. and Mohd, A. (2011). Role of phenolic compounds in peptic ulcer: An overview. *Journal of Pharmacy and Bioallied Sciences*, 3(3), p.361.

Treml, J. and Šmejkal, K. (2016). Flavonoids as potent scavengers of hydroxyl radicals. *Comprehensive reviews in food science and food safety*, 15(4), pp.720–738.

Tsuda, T., Watanabe, M., Ohshima, K., Yamamoto, A., Kawakishi, S. and Osawa, T. (1994). Antioxidative components isolated from the seed of tamarind (Tamarindus indica L.). *Journal of Agricultural and Food Chemistry*, 42(12), pp.2671–2674.

Ushanandini, S., Nagaraju, S., Harish Kumar, K., Vedavathi, M., Machiah, D.K., Kemparaju, K., Vishwanath, B.S., Gowda, T.V. and Girish, K.S. (2006). The anti-snake venom properties of Tamarindus indica (leguminosae) seed extract. *Phytotherapy Research: An International Journal Devoted to Pharmacological and Toxicological Evaluation of Natural Product Derivatives*, 20(10), pp.851–858.

Vaughan, J. and Geissler, C. (2009). *The new Oxford book of food plants*. OUP Oxford.

Verheij, E.W.M. and Coronel, R.E. 1991. Plant Resources of South-East Asia. No. 2: *Edible Fruits and Nuts*.

Wong, K.C., Tan, C.P., Chow, C.H. and Chee, S.G. (1998). Volatile Constituents of the Fruit of Tamarindus indica L. *Journal of Essential Oil Research*, 10(2), pp.219–221.

11 *Cissampelos pareira* L.: The Herb with Antiviral Potential

Mahantesha B.N. Naika[1*] and Dhanush S. Sadhu[2]

1. Introduction

For all of mankind's basic needs, including food, medicine, housing, clothes, and so forth, nature has been the main source. Plants are the major foundations for traditional system of medicine that have been in existence for thousands of years and yet there is no dearth of novel medicines that mankind can utilise (Dar *et al.*, 2017). The use of medicinal plants dates back to 5000 years with major evidence dawning from ancient texts of India, Rome, Greece, Syria, Egypt and China (Samal, 2016). Ancient Indian texts such as *Rigveda, Atharvaveda, Charaka Samhita* and *Sushruta Samhita* speak about usage of various medicinal plants to treat several diseases (Samal, 2016). Considering the world's growing population, a major part of it continues to strive over medicinal plants and is markedly noticeable in developing countries wherein herbal medicine has had an unbroken lineage of traditional usage (WHO, 1998). As new civilizations dawn, plants are not only an indispensable part of our health care but also offer the best hope for establishing a base for safe future medicine (Hamburger and Hostettmann, 1991). Around 80 per cent of the health requirements of the world's population have been satiated by herbal medicines (Agisho *et al.*, 2014).

Infectious diseases have been noted out to be the prime reason for mortality and are increasing at a rapid pace in the last few years (Singh *et al.*, 2020). Viral diseases have taken the lead amongst such infectious diseases and have rung the alarm of concern across the world (Pathak and Baishya, 2013). Once having entered the body, they start to produce multiple copies of their genome and proteins by redirecting the body's metabolism (Sohail *et al.*, 2011). Medicinal plants are endowed with various bioactive compounds *viz.*, alkaloids, flavonoids, lignans, phenolic compounds and other such components which makes them a powerful elixir to tackle dreadful viral diseases (Pathak and Baishya, 2013; Ojo *et al.*, 2009; Jassim and Naji, 2003). In the present era, several clinical research have confirmed the antiviral potency of medicinal plants while recent scientific studies

[1] Centre for Biotechnology Research, Department of Biotechnology and Crop Improvement, College of Horticulture, Bagalkot, University of Horticultural Sciences, Bagalkot, Karnataka, India.

[2] Kittur Rani Channamma College of Horticulture, Arabhavi, University of Horticultural Sciences, Bagalkot, Karnataka, India.

* Corresponding author: mahantesha.bnn@uhsbagalkot.edu.in

DOI: 10.1201/9781003433583-11

have revealed the mechanisms governing the therapeutic abilities of different medicinal plants (Kumar Anbazhagan *et al.*, 2019) such as immuno-stimulatory properties (Webster *et al.*, 2006) and broad-spectrum antiviral activity of plant extracts (Pompei *et al.*, 1979). Moreover, emergences of new viral infections have paced out the survival of mankind (Mukhtar *et al.*, 2008) as many viral strains have developed resistance to conventional antiviral agents. Hence, it becomes important to focus on plant extracts exhibiting antiviral activity to overcome the posed challenge (Serkdjieva, 2003; Tolo *et al.*, 2006). This article highlights the antiviral potency of one such plant, *Cissampelos pareira* L.

Cissampelos pareira, referred to as ambastha or laghupatha in Ayurvedic texts, (Amresh *et al.*, 2007), is a perennial climber, having a widespread distribution across the tropical and subtropical regions of India which includes parts of Rajasthan, Himachal Pradesh, Punjab, West Bengal, Bihar and Tamil Nadu upto an altitude of 2000m (Singh and Nishteswar, 2013). Apart from India, the plant also flocks naturally in parts of East Africa, America, Australia, West Indies, Comoros, Mauritius, Seychelles and Madagascar (Mukerji and Bhandari, 1959; Semwal *et al.*, 2014).

2. Taxonomy

The epistemology of the generic name *Cissampelos* is from the Greek word Kissos (Ivy) and ampelos (vine) as it has similar growth patterns resembling ivy whereas the specific name *pareira* is a Portuguese term referring to roots of any wild vine (Sudhakaran, 2012).

It is a perennial climbing herb (**Figure 1**) coming under the family Menispermaceae sub-categorised within the Cocculeae tribe. Out of the total 37 species listed under the genus *Cissampelos*, only one species *i.e.*, *pareira*, is found to thrive on the Indian subcontinent (Mukerji and Bhandari, 1959).

Fig. 1 Leaves and Fruits of *Cissampelos pareira* L.
(Images adapted from https://en.wikipedia.org/wiki/Cissampelos_pareira#cite_note-2)

In its natural habitat, it is found to be supported over tree trunks with the help of a flexible and slender twining stem with alternately arranged simple, light green leaves. It is also known as velvet leaf because of its velvety pubescent lower surface. Green flowers, being unisexual and dioecious in nature, form the staminate umbels and pendulous pistillate spikes. Fruits are red orange hairy one seeded drupe (Sudhakaran, 2012). Seeds are curved like the fourth form of the moon, and this is a peculiar characteristic across the genera and hence popularly called as the moonseed family (Mennega *et al.*, 1983).

3. Ethnobotanical Uses

Cissampelos pareira finds its place in an ancient verse of Charaka Samhita with its formulations known to treat diarrhoea, fever, leprosy, asthma, heart related issues, worm infection and abdominal tumour (Charaka Samhita, Sutraprasthaana 4.12 & 4.16). Predominantly, it has been totalled for its use as febrifuge (jwarahara), proper lactation (stanyashodhana) and hormonal axis balance (Singh *et al.*, 2016). Ayurvedic formulations like Mahayogra guggulu, Agnimukh churn, Pathadi kwath amd Pusyanug churn are some of the well-known preparations of this herb (Rawat and Vashistha, 2011; Semwal *et al.*, 2014). The roots form an important ingredient in the medicinal preparations of the South American natives for treating a variety of women's ailments like menstural cramps, fibroid tumors and to ease childbirth (Rhodes, 1975) as well as it is known to possess blood purifying properties and thereby used to treat malaria, pneumonia, snake and dog bite (Maurya *et al.*, 2004; Katewa and Galav, 2006; Bora *et al.*, 2007; Khare, 2007; Samanta and Bhattacharya, 2011; Chaudhary *et al.*, 2018). Since the roots have also been deemed to possess hayatine methiodiodiole, it serves as an alternative for tubocurarine which is a muscle relaxing agent owing to its less toxicity and greater potentiality (Morita *et al.*, 1993). Leaves are also used to treat malarial fever in liquid form (Dwivedi *et al.*, 2019).

4. Phytochemical Constituents in *Cissampelos*

The genus *Cissampelos* has been gifted with a vast array of biologically active compounds known as bisbenzylisoquinoline alkaloids whose composition vary across the plant and the chemosphere diversity is acquainted substantially to genetic diversity of the species distributed across different geographical regions (Morita *et al.*, 1993).

The whole plant extract showed the presence of many phytochemical compounds out of which 54 phytomolecules have been isolated which includes mainly isoquinoline alkaloids (Kumari *et al.*, 2021). Cissamperine, a bisbenzylisoquinoline alkaloid, has been known for its antitumor activity (Morita *et al.*, 2003).

Table 1 Chemical Constituents of Roots, Stem and Leaves of *Cissampelos pariera.*

S. No.	*Plant Parts*	*Chemical Constituents*	*References*
1.	Root	**Alkaloids-** peolsine, hayatine, hayatinine, hayatidine, cissamine, berberine, reserpine	Anwer *et al.*, 1968; Sharma *et al.*, 2004
		Non-alkaloids- d-quercitol, thymol	Chowdury, 1972
2.	Leaves and Stem	**Alkaloids-** Cissamperine, Nuciferine, Bulbo-carpine, Corytuberine, Laudanosine, Magniflorine	Kupchan *et al.*, 1965; Ahmad et al., 1992
		Non-alkaloids- Cissampeloflavone, quercetin, galacturonic acid	Kupchan *et al.*, 1960; Vardhanabhuti and Ikeda, 2006; Rosario *et al.*, 1996

Along with antiviral potency, the herb has exhibited various properties summarised in Table 2 and in Figure 2.

Table 2 Therapeutic effects of *Cissampelos pareira.*

S. No.	*Therapeutic Effects*	*Mechanisms*	*References*
1.	Anti-inflammatory activity	An escalation of cAMP concentrations and diminishing cyclic nucleotide phosphodiesterase activity.	Feily and Namazi, 2011
2.	Anti-urolithic activity	Disrupts the formation of calcium oxalate calculi in the kidneys. The phytoconstituent berberine is key factor responsible for the curative activity.	Ramesh *et al.*, 2010
3.	Antiulcer activity	The root phytoconstituent quercitin responsible for protective effect against ulceration.	Nwafor and Okoye, 2005
4.	Antioxidant activity	Superoxide ion scavenging acivity and free radical DPPH stability.	Habila *et al.*, 2011
5.	Antidiabetic activity	Reduction of fasting blood glucose through a slackening gluconeogenesis and an escalated glucose metabolism governed by a rise in creatinine levels, serum lipids and liver glycogen.	Jannu *et al.*, 2011
6.	Cardiovascular activity	Induces isoproterenol dysfunction and regulation of intracellular Ca^{2+} through a spasmolytic activity mechanism.	Singh *et al.*, 2013
7.	Antimalarial activity	Isoliensinine, a phytoconstituents of the root, targets Plasmodium Trophozpite-to-schizont transition, characterised by a cell division arrest phenotype, through tans-membrane transport and mitotic division axis interactions.	Muema *et al.*, 2023

Contd.

Table 2 *Contd.*

S. No.	*Therapeutic Effects*	*Mechanisms*	*References*
8.	Antileukemic activity	Pareirubrines A and B involved in restricting the tumour growth.	Morita *et al.*, 1993
9.	Antivenom activity	Neutralises the proteolytic haemorrhagic effect of pit viper venom.	Camargo *et al.*, 2011

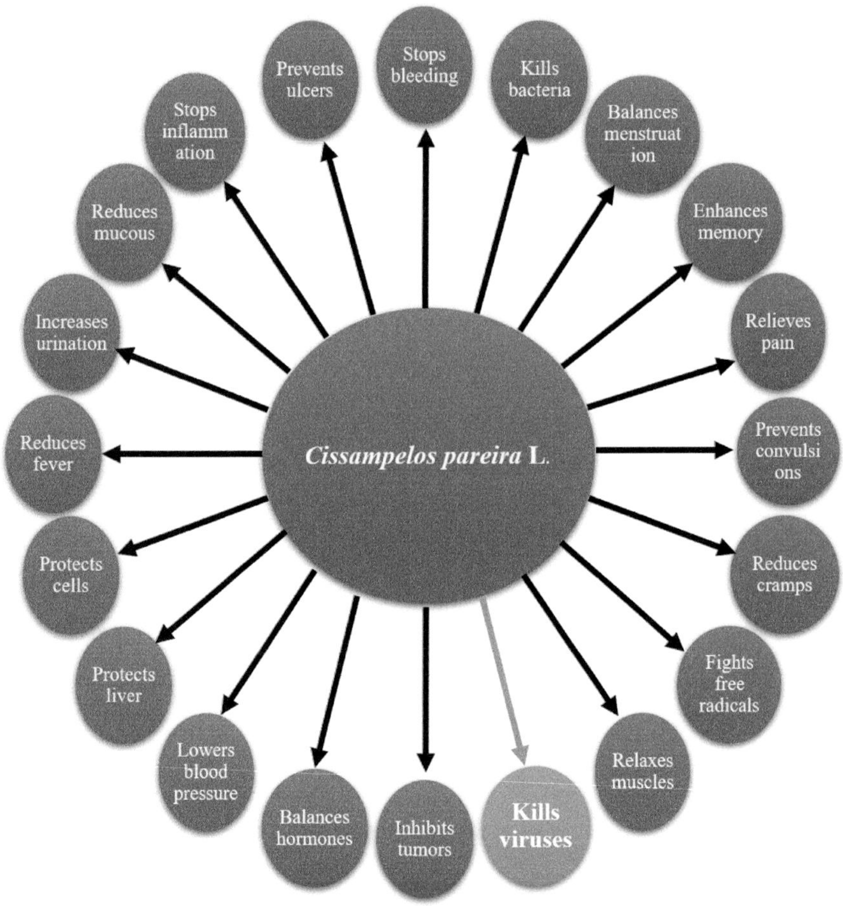

Fig. 2 Therapeutic activities of *Cissampelos pareira.*

Antiviral Potency of Cissampelos pareiera (Cipa)

Several studies have been conducted wherein extracts of *C. pareiera* have proved to exhibit antiviral activity. *Cipa* signatures have shown a positive corelation in knockdown signature of genes as well as protein inhibitors which play a key role in ribosomal protein processing and subsequent translation which form a part of transcription and translation processes in viruses (Haider *et al.*, 2021).

Anti-SARS-COV-2 activity: The whole plant extracts (aqueous and alcoholic) of *Cipa* exhibited anti-viral activity over SARS-COV-2 with an inhibition of 60% viral activity and 98% inhibition using hydroalcoholic whole plant extract. A further look into the single molecule constituents of *Cipa* showed a synergistic effect towards the viral inhibition with pareirarine having the most prominent inhibitory activity amongst the constituents. Additionally, the silico ligand binding assays were very clear in indicating the molecular therapeutic mechanism of *Cipa* bioactive compounds against the tested SARS-COV-2 drug targets (Haider, 2022).

Anti-dengue activity: The cell-based assays pointed out that the alcoholic extract of *Cipa* was found to exhibit a pan-DENV inhibitory activity against all four DENVs. Moreover, the extract when tested against paracetamol treated lines had a significant IC_{50} value in a time and dose dependent manner as well as maintaining a stable platelet count and a null discerning effect on erythrocyte viability, proving it to be clinically safe (Sood *et al.,* 2015).

Using a positron emission tomography (PET) CT biomarker, when *Cipa* plant powder along with Triphala (Amla Indian gooseberry extract) were administered to patients, it had a profuse impact on all four DENV lines (Ameen *et al.*, 2018).

Anti-cancer activity: There have been many predictions regarding priority compound families and related molecular, biological and fuctional pathways for COVID-19 from various sources which overlap with Cipa transcription signatures (Amresh *et al.*, 2007; Haider *et al.*, 2021; Singh *et al.*, 2013).

The Gene Ontology (GO) and Kyoto Encyclopedia of Genes and Genomes (KEGG) pathway enriched analysis carried out to trace the biological pathways and functions to discern a perceptible link between MCF7 cell lines treated with COVID-19 and *Cipa* at the gene expression level laid a substantial basis in discerning the upregulation and downregulation of genes governing the associated pathways (Guven, 2021).

Transcriptome studies conducted on MCF7 cell lines treated with *Cipa* extract using Affymetrix arrays expounded the downregulation of estrogen response signatures modulatory affiliated by estrogen receptor α (ERα) and gene knockdown experiments showcased the antiviral activity of *Cipa* in MCF7 cells infected with dengue though estrogen receptor 1 as the modulator (Haider, 2021).

Conclusion

C. pareira has been acknowledged amongst the important medicinal herbs in Ayurveda and other ancient medicinal texts to treat various ailments like malaria, inflammation, fever and diarrhoea. Presently, various *in vitro* and i*n vivo* studies, employing crude extracts of either stem or roots, have confirmed various biological properties and bridged the gap by confirming the traditional claims, giving a scientific base to it. Phytochemical studies backed up by qualitative chemical profiling, has proved *C. pareira* to be a rich source of flavonoids and isoquinoline alkaloids which constitute its active ingredient. The studies conducted on its antiviral potency is indeed a great insight for the herb to be used in tackling novel viral diseases. Although much progress has been made, there is always a scope for further research. This draws attention to the fact that in most of the experiments conducted, crude extracts of the herb have been used whereas many of the present-day pharmaceutical experiments focus on isolated phytomolecules which needs to be addressed. Furthermore, the exact mechanism of active ingredients still requires much more focus. Moreover, this chapter will provide the scientific community with understanding the antiviral potency of *C. pareira* which will aid in formulating further studies.

References

Agisho, H., Osie, M. and Lambore, T. (2104). Traditional medicinal plants utilization, management and threats in Hadiya Zone, Ethiopia. *Journal of Medicinal Plants*, 2(2): 94–108.

Ahmad, R., Malik, M.A. and Zia-ul-Haq. M. (1992). Alkaloids of Cissampelos pareira. *Fitoterapia* 63: 282–285.

Ameen, M.D., Vetriraj, M.D., Girish-Venkat, M.D., Andrew-John, M.D. and Samson, B.A.M.S. (2018). Natural antiviral therapy (Cissampelos pareira mix) efficacy against dengue virus monitoring by PET CT as biomarker. *Radiology Update*, 2(3): 101–111.

Amresh, G., Kant, R., Rao, V., Singh, P.N. (2007). Chemomodulatory influence of Cissampelos pareira (L.) Hirsuta on gastric cancer and antioxidant system in experimental animals. *Acta Pharmaceutica Sciencia*, 49: 71–83.

Anwer, F., Popli, S.P., Srivastava, R.M. and Khare, M.P. (1968). Studies in medicinal plants. 3. Protoberberine alkaloids from the roots of C. pareira Linn. *Experientia*, 24: 999.

Bora, U., Sahu, A., Saikia, A.P., Ryakala, V.K. and Goswami, P. (2007). Medicinal plants used by the people of Northeast India for curing malaria. *Phytotherapy Research*, 21: 800–804.

Camargo, Francisco Torres, Ana Ricciardi, Gabriela Ricciardi, Armando Dellacassa, Eduardo. (2011). SDS PAGE: a useful tool for preliminary screening of antisnake activity of plant extracts. *Boletin Latinoamericano y del Caribe de Plantas Medicinales y Aromaticas*. 10. 429–434.

Chaudhary, D., Paudel, S., Rana, R.M., Paudel, S. and Giri, P.M. (2018). Review on medicinal plants used for treatment of dog bite, *World Journal of Pharmaceutical Sciences*, 7: 498–510.

Chowdury, A.R. (1972). Chemical investigation on C. pareira Linn. *Science and Culture*, 38: 358–359.

Dar, R.A., Shahnawaz, M. and Qazi, P.H. (2017). Natural product medicines: A literature update. *Journal of Phytopharmacology*, 6(6): 349–351.

Dwivedi, M.K., Shyam, B.S., Lal, M., Singh, P.K., Sharma, N.K., Shukla, R. (2019). Geospatial mapping of antimalarial plants used by the ethnic groups of Anuppur district (Madhya Pradesh, India). *Indian Journal of Traditional Knowledge*, 18: 261–271.

Feily A., Namazi M.R. (2009) Cissampelos sympodialis Eichl (Menispermaceae) leaf extract as a possible novel and safe treatment for psoriasis. *Sao Paulo Medical Journal*, 127: 241–242.

Guven, E. (2021). Gene expression analysis of MCF7 cell lines of breast cancer treated with herbal extract of Cissampelos pareira revealed association with viral diseases. *Gene Reports*, 23: 101169.

Habila J.D., Bello I.A., Dzikwe A.A., Ladan Z., Sabiu M. (2011) Comparative evaluation of phytochemicals, antioxidant and antimicrobial activity of four medicinal plants native to Northern Nigeria. *Australian Journal of Basic and Applied Sciences*, 5: 537–543.

Haider, M., Anand, V., Enayathullah, M.G., Parekh, Y., Ram, S., Kumari, S., Anmol-Panda, G., Shukla, M., Dholakia, D., Ray, A., Bhattacharyya, S., Sharma, U., Bokara, K.K., Prasher, B. and Mukerji, M. (2022). Anti-SARS-CoV-2 potential of Cissampelos pareira L. identified by connectivity map-based analysis and in vitro studies. *BMC Complementary Medicine and Therapies*, 22(1): 114.

Haider, M., Dholakia, D., Panwar, A., Garg, P., Anand, V., Gheware, A., Singhal, K., Singh, D., Burse, S.A., Enayathullah, M.G., Parekh, Y., Ram, S., Kumari, S., Kumar, A., Ray, A., Medigeshi, G.R., Bokara, K.K., Sharma, U., Prasher, B., Mukerji, M. (2021). Traditional use of Cissampelos pareira L. for hormone disorder and fever provides molecular links of ESR1 modulation to viral inhibition. *BioRxiv*, 1–17.

Hamburger, M. and Hostettmann, K. (1991). Bioactivity in plants: the link between phytochemistry and medicine. *Phytochemistry*, 30: 3864–3874.

Jannu, V., Sai-Vishal, D., Ranjith-Babu, V., Harisha, B., Reddy R.C.S.D. (2011). Antidiabetic activity of hydro-alcoholic extract of C. pareira linn. leaves in streptozotocin induced diabetic rats. *International Journal of Pharmacy and Technology* 3: 3601–3611.

Jassim, S.A and Naji, M.A. (2003). Novel antiviral agents: a medicinal plant perspective. *Journal of Applied Microbiology*, 95(3): 412–427.

Katewa, S.S. and Galav, P.K. (2006). Additions to the traditional folk herbal medicines from Shekhawati region of Rajasthan. Indian Journal of Traditional Knowledge, 5: 494–500

Khare, C.P. (2007) *Indian Medicinal Plants* Springer-Verlag Berlin, vol 52–53.

Kumar-Anbazhagan, G., Palaniyandi, S. and Joseph, B. (2019). Antiviral Plant Extracts. *IntechOpen*. Chapter 4.

Kumari, S., Anmol, Bhatt, V., Suresh, P.S. and Sharma, U. (2021). Cissampelos pareira L.: A review of its traditional uses, phytochemistry and pharmacology. *Journal of Ethnopharmacology*, 274: 113850.

Kupchan S.M., Patel A.C. and Fujita E. (1965). Tumor inhibitors VI. Cissamperine, new cytotoxic alkaloid from Cissampelos pareira Var. hirsuta. Cytotoxicity of Bisbenzylisoquinoline alkaloids. *Journal of Pharmaceutical Sciences*, 4: 580–583.

Kupchan S.M., Yokoyama N. and Beal J.L. (1960) Menispermaceae alkaloids. I. The alkaloids of C. pareira Linn. and the origin of radix pareirae brave. *Journal of the American Pharmacists Asoociation*, 49: 727–731.

Maurya, R., Srivastava, S., Kulshreshta, D.K. and Gupta, C.M. (2004). Traditional remedies for fertility regulation. *Current Medicinal Chemistry*, 11: 1431–1450.

Mennega, A.M.V. (1982). Stem structure of the New World Menispermaceae. *Journal of Arnold Arboretum*. 63:145–171.

Mintah, S.O., Asafo-Agyei,T., Archer, M., Junior, P. A., Boamah, D., Kumadoh, D., Appiah, A., Ocloo, A., Boakye, Y. D. and Agyare, C. (2019). Medicinal plants for treatment of prevalent diseases, *Pharmacognosy-MedicinalPlants*,

Morita, H., Matsumoto, K., Takeya, K. and Itokawa H. (1993). Conformation of tropolone ring in antileukemic tropoloisoquinoline alkaloids. *Chemical and Pharmaceutical Bulletin*, 41(8):1478–1480.

Muema, J.M., Mutunga, J.M., Obonyo, M.A., Getahun, M.N., Mwakubambanya, R.S., Akala H.M., Cheruiyot, A.C., Yeda, R.A., Juma, D.W., Andagalu, B., Johnson, J.L., Roth, A.L. and Bargul, J.L. (2023). Isoliensinine from Cissampelos pariera rhizomes exhibits potential gametocytocidal and anti-malarial activities against Plasmodium falciparum clinical isolates. *Malaria Journal*, 22(1): 161.

Mukerji, B. and Bhandari, P.R. (1959). Cissampelos pareira l. source of a new curariform drug. *Planta Medica*, 7: 250–259.

Mukhtar, M., Arshad, M., Ahmad, M., Pomerantz, R. J., Wigdahl, B. and Parveen, Z. (2008). Antiviral potentials of medicinal plants. *Virus Research*, 131(2): 111–120.

Nwafor S.V. and Okoye C.F. (2005). Antiulcer properties of the ethanol root extract of Cissampelos mucronata. *Pharmaceutical Biology*, 43: 396–403.

Ojo, O.O., Oluyege, J.O. and Famurewa, O. (2009). Antiviral properties of two nigerian plants. *Academic Journals*, 3(7): 157–159.

Pathak, K and Baishya, K.A. (2013) Potential of some Indian medicinal herbs as antiviral agents. *American Journal of PharmTech Research*, 3(5): 11–17.

Pompei, R., Flore, O., Marccialis, M. A., Pani, A. and Loddo, B. (1979). Glycyrrhizic acid inhibits virus growth and inactivates virus particles. *Nature*, 281(5733): 689–690.

Ramesh, C., Nandakumar, K. and Rajesh, R. (2010). Antiurolithic activity of heart wood extract of Cedrus deodara in rats. *Journal of Complementary and Integrative Medicine*, 7: 11.

Rawat, R. and Vashistha, D.P. (2011). Common herbal plant in Uttarakhand, used in the popular medicinal preparation in Ayurveda. *International Journal of Pharmacognosy and Phytochemical Research*, 3: 64–73.

Rhodes, D.G. (1975). A revision of the genus Cissampelos. *Phytologia*, 30: 415–484.

Rosario S.L., da Silva A.J. and Parente J.P. (1996). Alkamides from Cissampelos glaberrima. *Planta Medica*, 62: 376–377.

Samal, J. (2016). Medicinal plants and related developments in India: A peep into 5-year plans of India. *Indian Journal of Health Sciences and Biomedical Research*, 9(1): 14-9 to 14-19.

Samanta, J. and Bhattacharya, S. (2011). Cissampelos pareira: a promising antifertility agent. *International Journal of Research in Ayurveda and Pharmacy*, 2: 439–442.

Semwal, D.K., Semwal, R.B., Vermaak, I. and Viljoen, A. (2014). From arrow poison to herbal medicine- the ethnobotanical, phytochemical and pharmacological significance of Cissampelos (Menispermaceae). *Journal of Ethnopharmacology*, 155: 1011–1028.

Serkedjieva, J. (2003). Influenza virus variants with reduced susceptibility to inhibition by a polyphenol extract from Geranium sanguineum L. *Pharmazie*, 58(1): 53–57.

Sharma, P.K., Chauhan, N.S. and Lal, B. (2004). Observations on the traditional phytotherapy among the inhabitants of Parvati valley in western Himalaya, India. *Journal of Ethnopharmacology*, 92: 167–176.

Singh, A., Mishra, A., Chaudhary, R. and Kumar V. (2020). Role of Herbal Plants in Prevention and Treatment of Parasitic Diseases. *Journal of Scientific Research*, 64(1): 50–58.

Singh, B.K., Pillai, K.K., Kohli, K., Haque, S.E. (2013). Effect of Cissampelos pareira root extract on isoproterenol-induced cardiac dysfunction. *Journal of Natural Medicines*, 67: 51–60.

Singh, S. and Nishteswar, K. (2013). Review on Cissampelos Pareira & Cyclea peltata (Patha dwaya) phyto-pharmacological perspectives. *International Journal of Ayurvedic Medicine*, 4: 282–289.

Singh, S., Nishteswar, K. and Patel, B. (2016). Therapeutic efficacy of Patha (Cissampelos pareira Linn.) - A review through classical texts of Ayurveda. *Journal of Ayurveda and Integrative Medicine* 1: 92–110.

Sohail, M.N., Rasul, F., Karim, A., Kanwal, U. and Attitalla, I.H. (2011). Plant as a Source of Natural Antiviral Agents. *Asian Journal of Animal and VeterinaryAdvances*, 6(12): 1125–1152.

Sood, R., Raut, R., Tyagi, P., Pareek, P.K., Barman, T.K., Singhal, S, et al. (2015). Cissampelos pareira Linn: Natural source of potent antiviral activity against all four dengue virus serotypes. *PLOS Neglected Tropical Diseases*, 9(12): e0004255.

Sudhakaran, M.V. (2012). Histo-morphological, fluorescent and powder microscopic characterization of Cissampelos pareira Linn. *Pharmacognosy Journal*, 4: 34.

Tolo, F.M., Rukunga, G.M., Muli, F.W., Njagi, E.N., Njue, W., Kumon, K., Mungai, G.M., Muthaura, C.N., Muli, J.M., Keter, L.K., Oishi, E. and KofiTsekpo, M.W. (2006). Anti-viral activity of the extracts a Kenya medicinal plant Carissa edulis against herpes simplex virus. *Journal of Ethnopharmacology*, 104(1/2): 92–99.

Vardhanabhuti B. and Ikeda, S. (2006). Isolation and characterization of hydrocolloids from monoi (Cissampelos pareira) leaves. *Food Hydrocolloids*, 20: 885–891.

Wang, W., Heng-Ghang, W. and Zhi-Duan C. (2007). Phylogeny and morphological evolution of tribe Menispermeae (Menispermaceae) inferred from chloroplast and nuclear sequences. *Perspectives in Plant Ecology Evolution and Systematics*, 8: 141–154.

Webster, D., Taschereau, P., Lee, T.D., Jurgens, T. (2006). Immunostimulant properties of Heracleum maximum Bartr. *Journal of Ethnopharmacology*, 106(3): 360–363.

WHO (1998). *Regulatory situation of herbal medicines. A worldwide review*. Pp 1-5. Geneva, Switzerland.

12 Secondary Metabolites of *Catharanthus roseus* (L.) G. Don and its Antiviral Properties

Neha Sharma[1], Manisha Sharma[1] and Shuchi Kaushik[2*]

1. A Brief Overview of *Catharanthus roseus* and its Secondary Metabolites

Catharanthus roseus (L.) G. Don belongs to the Apocynaceae family. It is a perennial medicinal herb. The phytochemical examination of *C. roseus* revealed the presence of alkaloids, saponin, terpenes, and the absence of steroids (Sindambiwe *et al.*, 1998; Nakamura *et al.*,1999; Bourne *et al.*, 1999; Wahab *et al.*, 2020). The antiviral properties of alkaloids like perivine, pericalline, periformyline, leurocristine, leurosivine, vincaleukoblastine, perividine, vindolinine, and carosine derived from *C. roseus* have been observed against several viruses, including herpes simplex virus type 1, vaccinia virus, poliovirus, and dengue virus type 2 (DENV-2). Leurocristine is effective against influenza, polio, meningococcal, and vaccinia viruses. Periformyline suppresses viruses of the poliovirus family, while vincaleucoblastine has extracellular virucidal activity against influenza, poliovirus, and vaccinia viruses. Perivine shows polio extracellular virucidal activity and is active against vaccinia (Ogasawara *et al.*, 2002; Fei *et al.*, 2003; Zhang *et al.*, 2003; Shyu *et al.*, 2006, Ozcelik *et al.*, 2011; Noor Zarina and Nazlina, 2020; Sharma *et al.*, 2022). Before the actual extraction method, several drying techniques are used to achieve a higher yield of bioactive components from *C. roseus*. It has been shown that drying the raw plant material considerably enhances the yield of chemicals during the extraction procedure. The desiccated *C. roseus* sample is prepared using a variety of processes, including freeze-drying and thermal drying. (Pham *et al.*, 2017). After that, the desired chemicals can be obtained using a variety of extraction methods. Vincristine, vinblastine, catharanthine, vindoline, and other compounds can be extracted using typical methods such as immersing the sample in solvents and leaving the mixture to settle overnight. In comparison to traditional extraction methods, ultrasonic extraction techniques are more sophisticated and produce a higher yield in less time. Alternative extraction techniques for *C. roseus* materials include microwave-assisted extraction, supercritical extraction, and enzyme-assisted extraction.

[1] Amity Institute of Biotechnology, Amity University Madhya Pradesh, India.
[2] State Forensic Science Laboratory, Madhya Pradesh, India.
* Corresponding author: shuchi.kaushik2@gmail.com

DOI: 10.1201/9781003433583-12

However, regardless of how well the extraction technique worked, the amount of vinca alkaloids recovered was heavily influenced by the amount of the component present in the plant tissue. Because *C. roseus* contains small amounts of vinca alkaloids, this is a cause for worry. Numerous studies are underway to facilitate enhanced yields of chemicals from *C. roseus* for bio-medicinal applications. There are two typical techniques for raising the output of vinca alkaloids:

A. In-vitro culture is experimented with to increase vinca alkaloids yield.
B. Enzyme-based synthesis of vinca alkaloids, and potential upregulation of transcript via recombinant DNA technology (Collu *et al.*, 2001).

2. A Summary of Viral Behavior

Viruses are obligatory intracellular parasites of host cells that hack host metabolic processes in addition to those that are specifically coded for by the viral genome. The paucity of serological indications and the wide variation in the window period of viral infections make prophylactic measures extremely difficult. Medication begins only after the symptoms are diagnosed clinically. To be successful, an antiviral drug must disrupt the molecular functions that viruses code for while having minimal impact on cellular metabolism. Unfortunately, a lot of naturally occurring substances and synthetic medicines that block the replication of viruses also block the molecular activities of tissues that are both infected and uninfected.

Therefore, there is always a high demand for a drug with specialized inhibitory functions, that won't obstruct any cellular metabolic pathway or biochemical flux. Numerous biochemical and synthetic molecules that are promising candidates for antiviral activity were tested in cultured cells under *in vitro* conditions. The *in vivo* investigations on virus-infected animal models did not substantially translate the *in-silico* results. It quickly became clear to researchers that getting the medication into the infected tissue's cells is extremely difficult, particularly when the virus has inflamed the tissues.

Antiviral medications should therefore be combined with anti-inflammatory medications to reduce the negative impacts of these drugs on body tissues. An untapped, abundant supply of antiviral agents comes from plants. Despite the limited number of studies on plants for antiviral agents, researchers have noted the occurrence of antiviral agents in plants at surprisingly high frequency. Typically, antiviral activity has been found in 20-30% of plants from tropical or temperate areas. It has been demonstrated that numerous compounds isolated from medicinal plants with a variety of chemical structures have antiviral action (Perez, 2003).

The MTT assay's cytotoxicity test against Vero cells revealed that the CC50 value for the crude preparation of *C. roseus* was 0.5 mg/mL. The *C. roseus* extract contains a phytochemical compound with potential antiviral action that is not cytotoxic to cells. Plaque reduction tests against HSV-1 revealed that the extract of *C. roseus* had 36, 20, and 4.7 as its selective indices (SI = CC50/EC50) in post-treatment, pre-treatment, and virucidal assays, respectively. These findings showed that the *C. roseus* extract contained several phytochemical compounds with potential antiviral action that was not cytotoxic to cells (Wahab *et al.*, 2020).

3. Potential Infectious Properties of Alkaloids

Alkaloids belong to a diverse category of chemical compounds occurring naturally and makes up the vast majority of secondary metabolites found in plants. Plant alkaloids can exist in crystalline, amorphous, flavorless, or non-volatile forms (Rosales *et al.*, 2012; Kukula-Koch *et al.*, 2017).

From the perspective of synthesis, a limited number of amino acid components, like lysine, phenylalanine, tyrosine, tryptophan, tyrosine, and ornithine, play crucial roles in the synthesis of most alkaloids. Various alkaloids are produced as a consequence of the conversion of these precursors into adaptable central intermediates. Plant species are regarded as alkaloid suppliers when they have an alkaloid content of more than 0.001%. The plant families Apocynaceae, Papaveraceae, Asteraceae, Amaryllidaceae, Fabaceae, Rubiaceae, Solanaceae, and Rutaceae have been identified as having potential pharmaceutical applications (Yang and Stöckigt, 2010). Alkaloids, which have a rich history in human medicine, are widely utilized in treating various illnesses, including neurological diseases, metabolic diseases, noncommunicable diseases, and infectious diseases (Kishore *et al.*, 2009; Hussaine *et al.*, 2018; Dey *et al.*, 2019).

Based on their chemical structure, alkaloids are primarily categorized into two main classes: indole alkaloids and isoquinoline alkaloids, each with a molecular mass of approximately 4000. Other notable groups of alkaloid phytochemicals include steroidal alkaloids, tropane, pyridine, and pyrrolizidine. Additionally, alkaloids can be classified based on their plant sources. For example, opiates are found in papaver plants of the Papaveraceae family, while cinchona plants of the Rubiaceae genus are known for their cinchona alkaloids. Alkaloids sourced from cactus, Rauvolfia, Catharanthus, Strychnos, and Ergot plants also contribute to the diverse array of natural alkaloids (Dey *et al.*, 2020).

Alkaloid phytochemicals may aid in the therapy of viral infections. On the other hand, the efficacy of currently available antiviral drugs is constrained by their variable toxicity towards patients and limited antiviral activity. Alkaloids, classified as phytochemicals, fulfill diverse biological and physiological roles and possess the capability to function independently as medications (Patel *et al.*, 2021). DNA and RNA, as well as protein envelopes in some instances, serve as the genetic building blocks of viruses. They depend on their environment and their hosts' metabolism to grow and flourish. They proliferate throughout the organism by taking advantage of and controlling the host's cellular machinery (Badshah *et al.*, 2020). Numerous alkaloid compounds can inhibit and interact with viruses. To prevent entry into host cells, a mechanism involves obstructing viral fusion or attachment by targeting the viral receptors present on the surface of the host cell. Additionally, these antiviral drugs can obstruct viral protein synthesis, viral protein assembly, and viral DNA or RNA synthesis (Kaur *et al.*, 2013; Fikatas *et al.*, 2021). By focusing on transport systems, these can prevent the spread of secondary viral infections to other healthy cells in a host body.

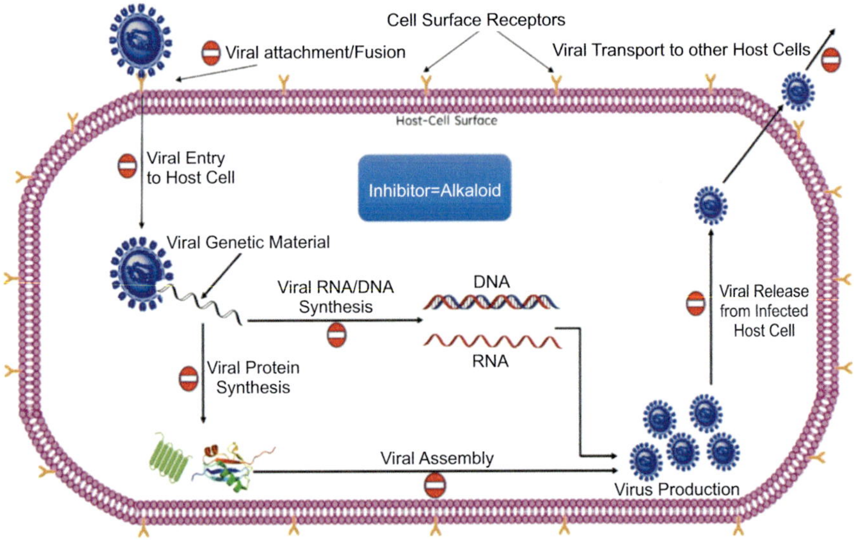

Fig. 1 (Adopted from Badshah *et al.* (2021). A representation of the different ways that alkaloids fight viruses that are highlighted: The antiviral activity of alkaloids encompasses a spectrum of actions, including targeting and inhibiting viral transport into host cells, hindering viral entry and fusion to host cell, viral DNA and RNA synthesis, viral protein synthesis, and interfering with viral glycoprotein. These varied mechanisms collectively contribute to the effectiveness of alkaloids against viruses.

4. Effectiveness of Catharanthus Roseus against Various Pathogens

4.1 Herpes Simplex Virus Type-1 (HSV-I)

HSV is divided into the HSV-I and HSV-II serovars. Common symptoms of HSV-I infections include pustules and bulges in the cells of the mouth and eyes. In more severe cases, HSV-I infections can lead to serious complications such as hearing-impaired, amaurosis, and potentially viral encephalitis. Contrarily, HSV-2 infections can lead to minor genital sores as well as significantly increase the risk of acquiring and transmitting human immunodeficiency virus, as well as other opportunistic illnesses. Cold sores, the common cold, and orolabial illness are all triggered by HSV-1. It causes gingivostomatitis, genital herpes, and keratitis labialis to infiltrate the mucosal epithelium. Sometimes it was described as more harmful, resulting in an immunocompromised person (Kolokotronis *et al.*, 2006, Kimberlin *et al.*, 2007, Madavaraju *et al.*, 2021).

Although acyclovir and similar nucleoside analogues were effective treatments, drug resistance has made them less effective, particularly for patients with compromised immune systems (Jiang, 2016). Alternative therapies have therefore been recorded to decrease the resistant properties (Reusser, 1996). The presence of alkaloids, saponin, and terpenoids in the methanolic extract of *C. roseus* may be the cause of the anti-HSV-1 action. Through plaque reduction assays, the *in-vitro* anti-HSV-1 activities at the various stages of viral infection were observed (Sindambiwe *et al.*, 1998; Nakamura *et al.*, 1999; Bourne *et al.*, 1999; Wahab *et al.*, 2020).

Table 1 *C. roseus* extract anti-HSV-1 action *in-vitro* at various concentrations. (*Source:* Wahab et al., 2020).

S. No.	*Phases of therapy*	*Concentration of C. roseus extract*	*Exposure time*	*Observation*
1	Pre-treatment (prophylactic activity)	0.125 mg/mL	24 hours before infection	100% plaque reduction
2	Post-treatment	0.250 mg/mL	48 hours post-adsorption	75% plaque reduction
3	Direct treatment	0.016 mg/mL	Introduce directly into the solution of a cell-free virus to assess its direct virucidal effects.	Greater than 50% decrease in plaque

4.2 Dengue Type 2 Virus

The dengue virus (DENV) belongs to the flavivirus genus. It is transmitted to humans through the bite of Aedes aegypti and Aedes albopictus mosquitoes, resulting in dengue fever. The genome of DENV consists of an 11-kilobyte positive-sense RNA strand encoding both structural and nonstructural proteins. Structural proteins include the capsid, envelope, and membrane precursor. Four DENV serovars (DENV1-4), all of which have the potential to be fatal, cause immunopathologic disease (southeast Asian hemorrhagic fever). It has a 10.7 kb genetic size and is a +ve sense ssRNA virus. It belongs to the Flaviviridae subfamily of Flavivirus. Methanol was used to create crude leaf extract of *C. roseus*. An *in-vitro* cytotoxicity assay utilizing the MTT assay was employed to determine the concentration of the extract required to induce 50% cell mortality CC50(cytotoxic concentration) (Dwivedi *et al.*, 2017, Hsiung *et al.*, 2017; Ramos-Castaeda *et al.*, 2017).

The foci forming unit reduction assay, which included pretreatment, post-treatment, and Virucidal efficacy testing, was used to demonstrate the antiviral activity. The results of a cytotoxicity assay on Vero cells show that the extract of *C. roseus* had a CC50 of 0.13 mg/mL. The extract of *C. roseus* had an EC50 of 0.025 mg/mL, or 50% Effective Concentration. The extract of *C. roseus* had a 5.2 selectivity score against DENV 2. The antiviral test employed three treatments: virucidal, pretreatment, and posttreatment. The outcomes showed that post-therapy was superior to pretreatment and virucidal in preventing viral replication. The findings of this study suggested that *C. roseus* extract has excellent potential as a potential naturally derived antiviral medication (Kuhn *et al.*, 2002; Mukhopadhyay *et al.*, 2005; Noor Zarina *et al.*, 2020).

4.3 Immunization and Polio

The *in-vitro* effectiveness of 36 alkaloids extracted from *Catharanthus roseus* or *C. lanceus* against type III poliovirus and the vaccinia virus was examined. Nine of these compounds had antiviral properties, with pericalline having the best results (Farnsworth *et al.*, 1968).

4.4 Coronavirus

Numerous endeavors have been undertaken to unearth a drug or vaccine with efficacy against the virus responsible for the 2019 Coronavirus Disease (COVID-19), namely Severe Acute Respiratory Syndrome Coronavirus-2 (SARS-CoV-2). The potential of medicinal herbs and their isolated phytonutrient in addressing a range of illnesses, including viral diseases, suggests promising therapeutic avenues for COVID-19 (Fakhri *et al.*, 2002; Mohammadi *et al.*, 2019; Ianevski *et al.*, 2020, Sharma *et al.*, 2020). The effectiveness of medicinal plants and the different compounds they contain in fighting the Severe Acute Respiratory Syndrome Coronavirus (SARS-CoV) is the subject of multiple ongoing studies. Alkaloids are a common class of phytochemicals found in a variety of plant families, including Papaveraceae, Amaryllidaceae, Apocynaceae, Asteraceae, and Solanaceae They possess the potential for exhibiting diverse biological activities and pharmacological effects, primarily attributed to the presence of one or more nitrogen atoms within their chemical structure (Cordell *et al.*, 2001; Dall'Acqua, 2012). Given the urgent need for effective treatment of SARS-CoV-2, the viral cause of COVID-19, exploring the pharmacological properties of alkaloids becomes particularly pertinent. One of the most common naturally occurring substances obtained from plants that have strong antiviral effects is alkaloids. As a result, these phytochemicals have offered potential leads for the development of COVID-19 treatment options. Alkaloids have shown potential anti-SARS-CoV properties by blocking pathogenesis-associated targets of the Coronaviridae family, which are essential for the viral life cycle. Recent research has demonstrated that alkaloids interact with structural coronavirus proteins, such as the spike (S) glycoprotein and nucleocapsid (N) on the virus surface, as well as nonstructural angiotensin-converting enzyme 2 (ACE2) in the cell membrane, and as a result, inhibit the coronavirus replication enzymes RdRp and 3-chymotrypsin-like protease (3CLpro) (Wink *et al.*, 2020; Pandeya *et al.*, 2020; Mani *et al.*, 2020).

4.5 Influenza A viruses

The influenza A viruses, belonging to the Orthomyxoviridae family, are characterized by their single-stranded, negative-sense RNA structure. Each year, seasonal influenza A strains cause significant economic losses and morbidity throughout the globe. Flu A (IAV) typically infects tracheal and bronchial epithelial cells, resulting in localized cellular damage and an acute inflammatory reaction in the recipient (Feng *et al.*, 2021).

5. Discussion

Phytochemical analyses reveal that extracts from *C. roseus* contain a rich array of specialized metabolites, such as alkaloids, saponins, and terpenoids. These compounds are traditionally utilized in treating various chronic human ailments (Abd Manaf *et al.*, 2010). Various secondary metabolites found in plants, including flavonoids, saponins, lignans, tannins, alkaloids, and thiophenes, have demonstrated significant antiviral effects against the replication of influenza virus, Hepatitis C Virus (HCV), and the coronavirus associated with Severe Acute Respiratory Syndrome (SARS-CoV) (Jassim and Naji, 2003; Ikeda *et al.*, 2005; Wen *et al.*, 2007; Naithani *et al.*, 2008; Lee *et al.*, 2012; Moradi *et al.*, 2017). The host cell environment is a necessity for the existence of viruses. Divergent invasion tactics also present a significant problem, in addition to genetic variation. The unlimited variety of chemical components present in medicinal plants can be employed to counteract genetic and invasion divergence, preventing the multiplication of DNA and RNA viruses. Ethnopharmacology enhances the likelihood of identifying a novel medication candidate (Naithani *et al.*, 2010). The cytotoxicity evaluation assay was used to measure the cytotoxic concentration required to kill 50% of the cell population or CC50 number. The *C. roseus* extract's CC50 value was established at 0.5 mg/mL. The increased SI (selective index value) results found with *C. roseus* extract could be attributed to the presence of unique phytochemical components. The abundance of active substances in *C. roseus* methanol extract, such as alkaloids (Nakamura *et al.*, 1999), saponins (Sindambiwe *et al.*, 1998), and terpenes, can be linked to the anti-HSV-1 activity (Bourne *et al.*, 1999).

Conclusion

The primary active ingredients in the traditional medicinal system, or ayurveda, are herbal compounds. A rich source of various kinds of alkaloids and phenolics with anticancer, antidiabetic, anti-Alzheimer's, antibacterial, antifungal, and antiviral activities is *C. roseus*, also known as Madagascar periwinkle or Cape periwinkle. In conclusion, *C. roseus* secondary metabolites, particularly alkaloids have demonstrated significant antiviral potential against various viruses. The plant's bioactive compounds, such as vinblastine, vincristine, catharanthine, and vindoline, exhibit mechanisms of action that include: Inhibition of viral replication, Interference with viral enzymes, and Modulation of immune responses

These findings suggest that *C. roseus*-derived secondary metabolites could serve as: Natural antiviral agents, adjunct therapies to enhance conventional antiviral treatments, and leads to developing novel antiviral drugs

For *in-vitro* antiviral action against simplex herpes virus (type I), vaccinia, polio type III viruses, and anti-dengue activity, various alkaloids and phenolics of *C. roseus* or *C. lanceus* were tested. The antiviral qualities of *C. roseus* are the main topic of this study. It has been discovered that a broad range of active phytochemicals have therapeutic uses against a variety of genetically and functionally diverse viruses.

These substances' antioxidant properties, scavenging abilities, inhibition of DNA and RNA synthesis, or blocking of viral reproduction, among other things, can be used to describe how they work as antivirals. A significant number of phytochemicals have shown promise as antiviral agents in numerous epidemiological and experimental studies.

A variety of *in-vitro* and *in-vivo* studies using various biological assays have been used to identify several potential leads, particularly in the last ten years. This study shows the complex interactions between various viral types and *C. roseus* extract on virus replication in Vero cells. The extract had negligible preventive effects and sporadic antiviral effects. Additional research is needed to identify the underlying mechanisms that add to its antiviral activity to clarify the mechanisms of inhibition. *Catharanthus roseus* offers a promising natural resource for combating viral infections, highlighting the importance of exploring plant-based solutions in the pursuit of effective antiviral therapies.

Author Contributions: All authors contributed to writing and editing the manuscript.

Funding: The present study received no any extramural funding support from any organization.

Conflicts of Interest: The authors declare no conflict of interest.

References

Abd Manaf, A., Alitheen, N.B., Mashitoh, A.R. and Shuhaimi, M. (2010). Cytotoxic effect of damnacanthal, nordamnacanthal, zerumbone and betulinic acid isolated from Malaysian plant sources. *International Food Research Journal. 17*(3), 711–719.

Babbar, N. (2015). An introduction to alkaloids and their applications in pharmaceutical chemistry. *The Pharma Innovation Journal*, *4*(10), 74–75.

Badshah, S.L., Faisal, S., Muhammad, A., Poulson, B.G., Emwas, A.H. and Jaremko, M. (2021). Antiviral activities of flavonoids. *Biomedicine & Pharmacotherapy*, *140*, 111596.

Badshah, S.L., Ullah, A. and Syed, S. (2020). The role of zinc-finger antiviral proteins in immunity against viruses. *Molecular Genetics, Microbiology and Virology*, *35*, 78–84.

Bourne, K.Z., Bourne, N., Reising, S.F. and Stanberry, L.R. (1999). Plant products as topical microbicide candidates: assessment of *in vitro* and *in vivo* activity against herpes simplex virus type 2. *Antiviral Research*, *42*(3), 219–226.

Collu, G., Unver, N., Peltenburg-Looman, A.M., van der Heijden, R., Verpoorte, R. and Memelink, J. (2001). Geraniol 10-hydroxylase1, a cytochrome P450 enzyme involved in terpenoid indole alkaloid biosynthesis. *FEBS letters*, *508*(2), 215–220.

Cordell, G.A., Quinn-Beattie, M.L. and Farnsworth, N.R. (2001). The potential of alkaloids in drug discovery. *Phytotherapy Research: An International Journal Devoted to Pharmacological and Toxicological Evaluation of Natural Product Derivatives*, *15*(3), 183–205.

Dall'Acqua, S. (2013). Plant-derived acetylcholinesterase inhibitory alkaloids for the treatment of Alzheimer's disease. *Botanics: Targets and Therapy*, *3*, 19–28.

Dey, P., Kundu, A., Chakraborty, H.J., Kar, B., Choi, W.S., Lee, B.M.,... and Kim, H. S. (2019). Therapeutic value of steroidal alkaloids in cancer: Current trends and future perspectives. *International Journal of Cancer*, *145*(7), 1731–1744.

Dey, P., Kundu, A., Kumar, A., Gupta, M., Lee, B.M., Bhakta, T. and Kim, H.S. (2020). Analysis of alkaloids (indole alkaloids, isoquinoline alkaloids, tropane alkaloids). In *Recent Advances in Natural Products Analysis* (pp. 505–567).

Dwivedi, V.D., Tripathi, I.P., Tripathi, R.C., Bharadwaj, S. and Mishra, S.K. (2017). Genomics, proteomics and evolution of dengue virus. *Briefings in functional genomics*, *16*(4), 217–227.

Faisal, S., Badshah, S.L., Kubra, B., Emwas, A.H. and Jaremko, M. (2023). Alkaloids as potential antivirals. A comprehensive review. *Natural Products and Bioprospecting*, *13*(1), 4.

Fakhri, S., Nouri, Z., Moradi, S.Z. and Farzaei, M.H. (2020). Astaxanthin, COVID-19 and immune response: Focus on oxidative stress, apoptosis and autophagy. *Phytotherapy Research*, *34*(11), 2790.

Farnsworth, N.R., Svoboda, G.H. and Blomster, R.N. (1968). Antiviral activity of selected Catharanthus alkaloids. *Journal of Pharmaceutical Sciences*, *57*(12), 2174–2175.

Fei, X.F., Wang, B.X., Li, T.J., Tashiro, S.I., Minami, M., Xing, D.J. and Ikejima, T. (2003). Evodiamine, a constituent of Evodiae Fructus, induces anti-proliferating effects in tumor cells. *Cancer Science*, *94*(1), 92–98.

Feng, X., Sureda, A., Jafari, S., Memariani, Z., Tewari, D., Annunziata, G.,... and Shen, A. Z. (2019). Berberine in cardiovascular and metabolic diseases: from mechanisms to therapeutics. *Theranostics*, *9*(7), 1923.

Feng, Y.L., Le, X.H. and Wu, S.P. (2021). Pathogenesis and pathological changes of avian influenza in human. *Avian Influenza in Human*, 29–40.

Fikatas, A., Vervaeke, P., Meyen, E., Llor, N., Ordeix, S., Boonen, I.,... and Schols, D. (2021). A novel series of indole alkaloid derivatives inhibit dengue and zika virus infection by interference with the viral replication complex. *Antimicrobial Agents and Chemotherapy*, *65*(8), e02349–20.

Hussain, G., Rasul, A., Anwar, H., Aziz, N., Razzaq, A., Wei, W.,... and Li, X. (2018). Role of plant derived alkaloids and their mechanism in neurodegenerative disorders. *International Journal of Biological Sciences*, *14*(3), 341.

Ianevski, A., Yao, R., Fenstad, M. H., Biza, S., Zusinaite, E., Reisberg, T.,... and Kainov, D. E. (2020). Potential antiviral options against SARS-CoV-2 infection. *Viruses*, *12*(6), 642.

Ikeda, T., Yokomizo, K., Okawa, M., Tsuchihashi, R., Kinjo, J., Nohara, T. and Uyeda, M. (2005). Anti-herpes virus type 1 activity of oleanane-type triterpenoids. *Biological and Pharmaceutical Bulletin*, *28*(9), 1779–1781.

Jassim, S.A. and Naji, M.A. (2003). Novel antiviral agents: a medicinal plant perspective. *Journal of Applied Microbiology*, *95*(3), 412–427.

Jiang, Y.C., Feng, H., Lin, Y.C. and Guo, X.R. (2016). New strategies against drug resistance to herpes simplex virus. *International Journal of Oral Science*, *8*(1), 1–6.

Kaur, P., Thiruchelvan, M., Lee, R.C.H., Chen, H., Chen, K.C., Ng, M.L. and Chu, J.J.H. (2013). Inhibition of chikungunya virus replication by harringtonine, a novel antiviral that suppresses viral protein expression. *Antimicrobial Agents and Chemotherapy*, *57*(1), 155–167.

Kimberlin, D.W. and Whitley, R.J. (2007). Antiviral therapy of HSV-1 and-2. *Human herpesviruses: biology, therapy, and immunoprophylaxis*.

Kishore, N., Mishra, B.B., Tripathi, V. and Tiwari, V.K. (2009). Alkaloids as potential anti-tubercular agents. *Fitoterapia*, *80*(3), 149–163.

Kolokotronis, A. and Doumas, S. (2006). Herpes simplex virus infection, with particular reference to the progression and complications of primary herpetic gingivostomatitis. *Clinical Microbiology and Infection*, *12*(3), 202–211.

Kuhn, R.J., Zhang, W., Rossmann, M.G., Pletnev, S.V., Corver, J., Lenches, E.,... and Strauss, J.H. (2002). Structure of dengue virus: implications for flavivirus organization, maturation, and fusion. *Cell*, *108*(5), 717–725.

Kukula-Koch W.A., Widelski J. Alkaloids. In: Badal S., Delgoda R., (2017). Editors. *Pharmacognosy.* Academic Press; Boston, MA, USA. pp. 163–198.

Kurek, J. (Ed.). (2019). *Alkaloids: their importance in nature and human life*. BoD–Books on Demand.

Lee, J., Lim, S., Kang, S.M., Min, S., Son, K., Lee, H.S.,... and Hwang, S.B. (2012). Saponin inhibits hepatitis C virus propagation by up-regulating suppressor of cytokine signaling 2. *PloS one*, *7*(6), e39366.

Madavaraju, K., Koganti, R., Volety, I., Yadavalli, T. and Shukla, D. (2021). Herpes simplex virus cell entry mechanisms: An update. *Frontiers in Cellular and Infection Microbiology*, *10*, 617578.

Mani, J.S., Johnson, J.B., Steel, J.C., Broszczak, D.A., Neilsen, P.M., Walsh, K.B. and Naiker, M. (2020). Natural product-derived phytochemicals as potential agents against coronaviruses: A review. *Virus Research*, *284*, 197989.

Mohammadi Pour, P., Fakhri, S., Asgary, S., Farzaei, M.H. and Echeverría, J. (2019). The signaling pathways, and therapeutic targets of antiviral agents: focusing on the antiviral approaches and clinical perspectives of anthocyanins in the management of viral diseases. *Frontiers in Pharmacology*, *10*, 1207.

Moradi, M.T., Karimi, A. and Lorigooini, Z. (2018). Alkaloids as the natural anti-influenza virus agents: a systematic review. *Toxin reviews*, *37*(1), 11–18.

Mukhopadhyay, S., Kuhn, R.J. and Rossmann, M. G. (2005). A structural perspective of the flavivirus life cycle. *Nature Reviews Microbiology*, *3*(1), 13–22.

Naithani, R., Huma, L.C., Holland, L.E., Shukla, D., McCormick, D.L., Mehta, R.G. and Moriarty, R.M. (2008). Antiviral activity of phytochemicals: a comprehensive review. *Mini Reviews in Medicinal Chemistry*, *8*(11), 1106–1133.

Naithani, R., Mehta, R.G., Shukla, D., Chandersekera, S.N. and Moriarty, R.M. (2010). Antiviral activity of phytochemicals: a current perspective. *Dietary Components and Immune Function*, 421–468.

Nakamura, N., Hattori, M., Kurokawa, M., Shiraki, K., Kashiwaba, N. and Ono, M. (1999). Anti-herpes simplex virus activity of alkaloids isolated from Stephania cepharantha. *Biological and Pharmaceutical Bulletin*, *22*(3), 268–274.
Nakamura, N., Hattori, M., Kurokawa, M., Shiraki, K., Kashiwaba, N. and Ono, M. (1999). Anti-herpes simplex virus activity of alkaloids isolated from Stephania cepharantha. *Biological and Pharmaceutical Bulletin*, *22*(3), 268–274.
Noor Zarina, A.W. and Nazlina, I. (2020). Efficacy of Catharanthus roseus extract against Dengue Virus Type 2 Infection *in vitro*. *Indian Journal of Public Health Research & Development*, *11*(1), 1314–1319.

Ogasawara, M., Matsunaga, T., Takahashi, S., Saiki, I. and Suzuki, H. (2002). Anti-invasive and metastatic activities of evodiamine. *Biological and Pharmaceutical Bulletin*, *25*(11), 1491–1493.

Özçelik, B., Kartal, M. and Orhan, I. (2011). Cytotoxicity, antiviral and antimicrobial activities of alkaloids, flavonoids, and phenolic acids. *Pharmaceutical Biology*, *49*(4), 396–402.

Pandeya, K.B., Ganeshpurkar, A. and Mishra, M. K. (2020). Natural RNA dependent RNA polymerase inhibitors: Molecular docking studies of some biologically active alkaloids of Argemone mexicana. *Medical Hypotheses*, *144*, 109905.

Patel, A., Vanecha, R., Patel, J., Patel, D., Shah, U. and Bambharoliya, T. (2022). Development of natural bioactive alkaloids: anticancer perspective. *Mini Reviews in Medicinal Chemistry*, *22*(2), 200–212.

Perez, R.M. (2003). Antiviral activity of compounds isolated from plants. *Pharmaceutical Biology*, *41*(2), 107–157.

Pham, H.N.T., Sakoff, J.A., Van Vuong, Q., Bowyer, M.C. and Scarlett, C.J. (2018). Screening phytochemical content, antioxidant, antimicrobial and cytotoxic activities of Catharanthus roseus (L.) G. Don stem extract and its fractions. *Biocatalysis and Agricultural Biotechnology*, *16*, 405–411.

Ramos-Castaneda, J., Barreto Dos Santos, F., Martinez-Vega, R., Galvão de Araujo, J.M., Joint, G. and Sarti, E. (2017). Dengue in Latin America: systematic review of molecular epidemiological trends. *PLoS neglected tropical diseases*, *11*(1), e0005224.

Reusser, P. (1996). Herpesvirus resistance to antiviral drugs: a review of the mechanisms, clinical importance and therapeutic options. *Journal of Hospital Infection*, *33*(4), 235–248.

Rosales, P.F., Bordin, G.S., Gower, A.E. and Moura, S. (2020). Indole alkaloids: 2012 until now, highlighting the new chemical structures and biological activities. *Fitoterapia*, *143*, 104558.

Sharma A., Tiwari, S., Deb, M.K. and Marty, J.L. (2020). "Severe acute respiratory syndrome coronavirus-2 (SARS-CoV-2): a global pandemic and treatment strategies," *International Journal of Antimicrobial Agents*, vol. 56, no. 2, Article ID 106054.

Sharma, N., Singh, I., Ajee, R.S., Kaushik, S. (2022). *Catharanthus Roseus*: A Source of Anticancer Phytomedicines. In: Kole, C. (eds) The Catharanthus Genome. Compendium of Plant Genomes. Springer, Cham. https://doi.org/10.1007/978-3-030-89269-2_2.

Shyu, K.G., Lin, S., Lee, C.C., Chen, E., Lin, L.C., Wang, B.W. and Tsai, S.C. (2006). Evodiamine inhibits *in vitro* angiogenesis: Implication for antitumorgenicity. *Life Sciences*, *78*(19), 2234–2243.

Sindambiwe, J.B., Calomme, M., Geerts, S., Pieters, L., Vlietinck, A.J. and Vanden Berghe, D.A. (1998). Evaluation of biological activities of triterpenoid saponins from Maesa lanceolata. *Journal of Natural Products*, *61*(5), 585–590.

Wahab, N.Z.A., Azizul, A., Ibrahim, N. (2020). Phytochemistry, cytotoxicity and antiviral activity of Catharanthus roseus. *Iranian Journal of Microbiology*, *12*(5), 460.

Wen, C.C., Kuo, Y.H., Jan, J.T., Liang, P.H., Wang, S.Y., Liu, H.G.,... and Yang, N.S. (2007). Specific plant terpenoids and lignoids possess potent antiviral activities against severe acute respiratory syndrome coronavirus. *Journal of Medicinal Chemistry*, *50*(17), 4087–4095.

Wink, M. (2020). Potential of DNA intercalating alkaloids and other plant secondary metabolites against SARS-CoV-2 causing COVID-19. *Diversity*, *12*(5), 175.

Yang, L. and Stöckigt, J. (2010). Trends for diverse production strategies of plant medicinal alkaloids. *Natural product reports*, *27*(10), 1469–1479.

Zhang, Y., Wu, L.J., Tashiro, S.I., Onodera, S. and Ikejima, T. (2003). Intracellular regulation of evodiamine-induced A375-S2 cell death. *Biological and Pharmaceutical Bulletin*, *26*(11), 1543–1547.

13 *Centella asiatica* (L.) Urban-Antiviral Properties

Rushikesh R Mali[1] and Tabassum Khan[2*]

1. Introduction

Centella asiatica (Linn.) (*C. asiatica*) has been used as a remedy in mild and chronic ailments for thousands of years, mainly relying on experiences and folk medicines as it continues to get significant attention. Recently, there has been an enormous emphasis on the study of plants, and a substantial form of data has developed to demonstrate the immense potential of medicinal plants utilised in diverse traditional systems of medicine. *C. asiatica*, or Mandukaparni, is a prostrate, hardly fragrant, stoloniferous perennial plant often found growing up to 500–600 meters above sea level in crop fields and other waste areas. *C. asiatica* has been utilized as medicine not only in the Indian subcontinent but also in China, Sri Lanka, Nepal, Madagascar, and other tropical parts of the globe (Gohil et al., 2010). It is largely recognized as a "Food of Brain" in the Indian subcontinent since it is one of the most important herbs for healing skin issues, repairing wounds, and renewing the nerves and brain cells. This chapter reviews scientific classification and taxonomy, botanical attributes, geographic distribution, and its pharmacological activity with a special focus on its antiviral activity.

2. Taxonomy and Scientific Classification

C. asiatica clonal herbaceous plant species is a member of the kingdom Plantae's *Apiceae* family (usually called the *Umbelliferae*). The taxonomical classification is indicated here (Biswas et al., 2021; Brinkhaus et al., 2000; Uddin et al., 2017).

Kingdom	:	Plantae
Subkingdom	:	Embryophyta
Division	:	Tracheophyta
Class	:	Magnoliopsida
Order	:	Apiales
Family	:	Apiaceae

[1] Department of Quality Assurance, SVKM's Dr. Bhanuben Nanavati College of Pharmacy, Mumbai, India.
[2] Department of Pharmaceutical Chemistry & Quality Assurance, SVKM's Dr. Bhanuben Nanavati College of Pharmacy, Mumbai, India.
* Corresponding author: tabassum.khan@bncp.ac.in

DOI: 10.1201/9781003433583-13

Subfamily : Hydrocotyle
Genus : Centella
Species : asiatica

Synonyms

C. asiatica (L.) is known by several common names like "Gotu kola" or "Indian pennywort," "water pennywort" or "marsh pennywort" in many other regions of the world such as Europe and America, "Mandukaparni" in Sanskrit/Aayurveda, "brahmi" in Hindi/Unani (Belwal et al., 2019) .

3. Botanical Description of the Plant

C. asiatica (L.) is a perennial herbaceous plant that grows up to 15–17 cm tall, it is prostrate, stoloniferous, and has a mild or no scent (Fig. 1). However, some bigger varieties may reach a height of 25 cm. It mainly grows around water and has a glabrous, striated, and nodal-rooting stem. It has small green leaves arranged alternatively in clusters at nodes (Chandrika and Kumara, 2015). The leaves have long petioles, orbicular-reniform in shape, about 2–7 cm in length and 1–5 cm in breadth with crenate edges and glabrous on both sides. The flowers are white-purple-to-pink coloured borne in fascicled umbels, and flowering is from April to June. It produces rectangular, globular, and somewhat thickened pericarps that are roughly 4–5 cm long, and it yields tiny, spherical fruits grown throughout the growing season (Vasantharuba et al., 2012).

Fig. 1 Centella asiatica.
(*Source:* Botanical Garden SVKM's Dr. Bhanuben Nanavati College of Pharmacy)

4. Geographical Distribution

C. asiatica is mainly found in tropical and subtropical swampy regions at 200–2100 metres above sea level (Samant and Pant, 2006), particularly in warm-environmental areas like China, India, Sri Lanka, Madagascar, South Africa,

Malesia, Eran, Northern Australia, and eastern Europe (Gohil et al., 2010), in most of the tropical or wet pantropical regions, including rice fields, and rocky, higher elevations, with about 20 species linked to *C asiatica*.

5. Phytochemical Constituents and Metabolite Profiling

C. asiatica is a rich source of terpenoids (asiaticoside, centelloside, madecassoside, and brahmoside), amino acids (alanine, serine, aspartate, and glutamate), phenols (kaempferol and quercetin), and carbohydrates (glucose, mesoinositol, and centellose), among several others, that have found broader application as both health and food supplements. Saponins (triterpenoids) are the principal active components. Asiaticoside is a trisaccharide with three aglycone linkages: madecassoside, madasiatic acid, and asiatic acid. (Singh and Rastogi, 1969; Belwal et al., 2019). These triterpene saponins and their sapogenins primarily contribute to tissue regeneration and vascular repairs by preventing the synthesis of collagen at the injury site. Other substances identified include brahmoside and brahminoside, with CNS and uterorelaxant effects, although clinical research data is still needed to validate these effects. The crude extracts of isothankuniside and thankuniside glycosides were found to show antifertility action in mice model (Heidari et al., 2012), centelloside and its derivatives are useful in the treatment of venous hypertension (Gohil et al., 2010). The other constituents reported in *C. asiatica* include tannins, essential oils (Apiol, α-humulene, α-pinene, α-copaene, α-caryophyllene, β-chariophylen, trans-β-pharnesen, β-elemene, and germachrene D) (Joshi et al., 2007), phytosterols (campesterol, sitosterol, stigmasterol), mucilages, resins, aminoacids (alanine, serine, aminobutyrate, aspartate, glutamate, lysine, and treonine), fatty acids (linoleic acid, linolenic, oleic, palmitic and stearic acids), flavonoids (cheratin and kemferol derivatives), an alkaloid (hydrochotine) and bitter component, vallerine (Singh & Rastogi, 1968; Sahu et al., 1989; Inamdar et al., 1996; Schaneberg et al., 2003).

6. Medicinal/Therapeutic Uses

C. asiatica is a moderate adaptogen used medicinally in India for hundreds of years; it is frequently found in skincare anti-aging products. Gotu kola/*C. asiatica* is a particularly effective herb reported to delay the onset of ageing, according to Charaka, who is frequently regarded as the founder of the Indian Traditional System of Medicine (Ayurveda). It is included in the top ten herbs with anti-aging benefits, and its antioxidative characteristics may contribute to this (Chaudhary et al., 2010). Some of the traditional, folk, pharmacological uses of *C. asiatica* along with a special emphasis on plant antiviral applications are explained in this section .

6.1 Uses described in Pharmacopoeias and in Traditional Systems of Medicine

C. asiatica is regarded as an ethnomedicinal plant extensively utilised as a medicine in several ancient civilizations, including Ayurveda, Unani, and folk remedies. The literature reflects significant contributions made by a small number of German authors who were among the first to systematically document and compile information on *C. asiatica*. These authors also provided well-documented evidence of the plant's long-standing therapeutic use, particularly in Southeast Asia and India. They reported that *C. asiatica* was mentioned in approximately 500 AD, in the texts of codified Sanskrit medicine (Ayurveda). Additionally, they stated that *C. asiatica* is the medicinal plant referred to as "manduk-parni" in Sanskrit, referenced in the "Sushruta Samhita," an ancient Indian literature on medicine, written around 1200 BC. It was formally incorporated into the Indian Pharmacopoeia in the 19th century, where it was advised in the treatment of wound healing, skin problems, amenorrhea, diarrhoea, fever, lupus, varicose ulcers, eczema, and psoriasis (Brinkhaus et al., 2000). The use of *C. asiatica* as medicine is documented; it was used to cure leprosy on the island of Mauritius and by traditional healers in Java and in other Indonesian islands, and Madagascar. According to Indian Ayurveda, this plant possesses multifunctional roles and has been utilised for thousands of years in Ayurveda traditional medicine to treat both moderate and chronic ailments (Gohil et al., 2010). The aerial portions of *C. asiatica* have been used in Sri Lanka and Indian Ayurvedic traditional medicine to treat skin conditions, syphilis, rheumatism, mental sickness, epilepsy, hysteria, and in the treatment of dehydration (Matsuda et al., 2001). Most of the biological actions of *C. asiatica* have been ascribed to the presence of pentacyclic triterpene compounds mainly asiatic acid, madecassic acid, and triterpene saponin-asiaticoside, madecassoside. It is also a good source of several micronutrients, iron, phosphorus, sodium, vitamin C, vitamin A, carotene, and dietary fibres.

Additionally, this plant is reported to flourish in tropical areas of Oceania, Madagascar, Central America, and equatorial Africa. Close relatives of *C. asiatica* were utilised in leprosy therapy in the 19th century due to the range of the plant's herbal qualities, and this was recognised by American eclectics who were knowledgeable about the different types of medicinal features related to the plants used in leprosy treatment. French authorities originally authorised *C. asiatica* as a drug in the 1880s. Since that time, *C. asiatica* formulations have been used to treat several conditions that were previously listed as well as those that were discussed in the section "Clinical Applications". Dr. Boiteau used it for long-term leprosy therapy, which he underwent for several years, and was a contributing factor in his eventual recovery in 1852. Another noteworthy occurrence of *C. asiatica* at the Centennial International Exhibition of Melbourne in 1888 was the Brisbane group of physicians' presentation of extracted juice as a medication. Based on all the traditional uses *C. asiatica* and *C. asiatica* extracts are also included in the Polish and European Pharmacopoeia. When this herb is a dried, fragmented plant made up of aerial parts of *C. asiatica*, it has at least 6.0% of all triterpenoid derivatives expressed as asiaticoside. *C. asiatica* is used in dermatological disorders as an anti-inflammatory for eczema, and to heal small wounds, scratches, burns, and hypertrophic wounds. It is used in traditional Asian medicine to treat venous insufficiency as a diuretic, antibacterial, antiviral, anti-

activities of *C. asiatica* are given in Figure 2. Although all these multifunctional features of *C. asiatica* have been documented in the literature, this chapter primarily focuses on the health advantages of *C. asiatica*, especially its antiviral activity.

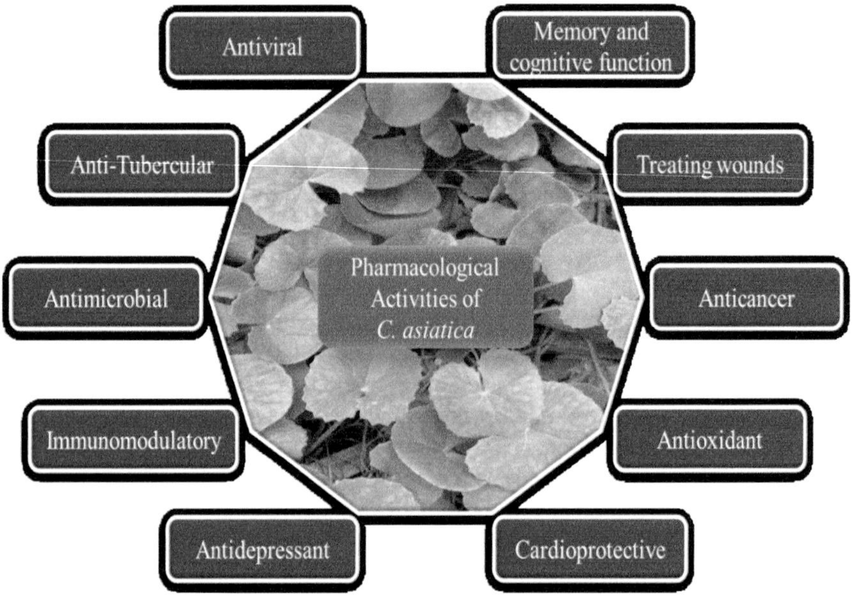

Fig. 2 Pharmacological activities of *C. asiatica.*

The crude water extract of *C. asiatica* showed anti-herpes simplex action (Yoosook et al., 2000). Plate inhibition assays of *C. asiatica* have demonstrated its anti-HSV-1 and 2 inhibitory activities with a median effective dosage (ED50) of 362 μg/ml for HSV-1 and 298 μg/ml for HSV-2, respectively. Additionally, the methanol and aqueous extracts of *C. asiatica* were found to have excellent virucidal, preventative, and distinct antiviral attachment activities with pertinent assays against the pseudorabies virus (PrV), as well as being the most effective antiviral attachment agents among other tested plant extracts with percentage cell viability up to 60%. Although there are fewer findings on *C. asiatica's* antiviral abilities, there are several articles on the plant's other pharmacological traits. Some include antibacterial, antifungal, antiulcer, antidiabetic, and anti-inflammatory action (Gohil et al., 2010). Whereas, a *C. asiatica's* ethanol extract (0.005–0.1 mg ml^{-1}) showed significant virucidal activity against the vesicular stomatitis virus (VSV) based on the study done by Ali et al. (1996).

Many researchers are working to create plant-based medicine as there is a rising demand to create antiviral drugs. In one such work, Khan et al. (2021) used a bioinformatics technique to investigate the phytochemicals present in the *C. asiatica* plant as inhibitors against the main protease (M-pro), a crucial regulatory enzyme of the SARS-CoV-2 lifecycle. A total of 107 phytochemicals from *C. asiatica* were subjected to molecular docking using computer-aided drug design methods (PyRx 0.8), and based on the results, the top nine phytochemicals that showed the greatest promise were identified. Those phytochemicals were the inhibitors of M-pro had a higher affinity for binding and lower side effects. During their study, they found a few amino acids, including Lys5, Met6, Thr111, Gln127, Leu282, Glu288, Glu290, Phe291, Asp295, Arg298, and Gln299, at the active sites of the M-pro, which had an area of 27.045 and a volume of 5.047. Nine compounds exhibited good binding energy ranging from −9 Kcal/mol to −9.5 Kcal/mol when molecular docked with the major SARS-CoV-2 viral protease (PDB ID: 6LU7). However this result still needs to get experimental confirmation with a strong potential for translation and drug development as an effective therapy against SARSCoV-2 infections (Khan et al., 2021).

In a recent study, Maduray et al. (2022) used aqueous leaf, stolon, and root extracts to synthesize silver nanoparticles (AgNPs). An extensive study on AgNPs' anti-HIV efficacy was conducted utilising cells infected with MT-4 (Maduray et al., 2022). Utilizing *C. asiatica* leaf (L), stolon (S), and root (R) extracts, they created AgNPs utilising Green one-pot synthesis procedures. Then spectroscopic and microscopic methods were used for the evaluation of produced AgNPs. Then MTT assay was performed to evaluate the cytotoxic activity of the produced AgNPs' and anti-HIV activity was assessed. The result showed that L-AgNPs, S-AgNPs, and R-AgNPs with spherical shapes and average particle sizes of 16 nm, 17 nm, and 26 nm, respectively, were successfully synthesised. For L-AgNPs, S-AgNPs, and R-AgNPs, the protection offered by synthetic AgNPs to HIV-infected MT-4 cells was ≤10% at dosages ranging from 11 to 1375 μg/mL, 9-1125 μg/mL, and 2.52-315 μg/mL respectively, with minimal toxicity. This work supports the possibility that AgNPs made using *C. asiatica* extracts could be effective in HIV warranting more focussed research (Maduray et al., 2022).

Komaikul et al. (2022) determined the effectiveness of neutral electrolyzed water (NEW) in inactivating HCoV-OC43 and SARS-CoV-2 on the surfaces of plastic and *C. asiatica*. They used plaque assay and high-performance liquid chromatography (HPLC) to determine the quantities of infectious HCoVOC43 and triterpenoid generated by *C. asiatica*. Their findings showed that HCoV-OC43 persistence on *C. asiatica* leaves was equivalent to that on inert polystyrene. On *C. asiatica* leaves, HCoV-OC43 remained stable for around 24 hours when maintained and covered at room temperature with high humidity (>90% RH). Both infectious HCoV-OC43 and SARS-CoV-2 in liquid were found to be successfully inactivated by NEW with 197 ppm of available chlorine concentration (ACC) (3.68 and 4.34 log reduction, respectively), and dried HCoV-OC43 on the surfaces of *C. asiatica* leaves were also inactivated (2.31 log reduction). It was also found that the asiaticoside levels were considerably greater after 5 minutes of soaking *C. asiatica* leaves in NEW containing 205 ppm of ACC or water when compared to the unsoaked group, which had much lower values (29.96 ± 0.78 mg/g dry weight). Based on these results, it is clear that NEW could be a possibility for inactivation of coronavirus, even though the *C. asiatica* leaves that have been infected by it offer a risk for transmission (Komaikul et al., 2022). In another study by Latha *et al.* anti-SARS-CoV-2 properties of phytochemicals derived from several plant species, including *C. asiatica*, were screened *in silico* and their safety and effectiveness were assessed *in-vitro* and *in-vivo*. For *in-silico* screening of phytoconstituents, molecular docking was used, and subsequently, the hits were removed, separated, and purified. In addition to *C. asiatica*, ethanolic extracts from other plants were also tested. Remdesivir was used as a positive control for testing the *in vitro* security and anti-SARS-CoV-2 activity of plant methanol extracts in VeroE6 cells. Wistar male and female rats were used in the investigation of acute and subacute toxicity. The results determined as percent cell viability showed that *C. asiatica* has higher viability after 24 and 48 hours of treatment. Furthermore, they also found that both the E and N genes of *C. asiatica* demonstrated potent *in-vitro* anti-SARS-CoV-2 activity. To test for acute and subacute cytotoxicity in Wistar male and female rats, the extract of *C. asiatica* was mixed alongside *Andrographis paniculate* and *Withania somnifera* based on *in-vitro* results and found that there were no major changes in the haematological, biochemical, and histopathological criteria. This established that the test plant extracts were safe and active *in vitro* and *in vivo* (Latha et al., 2022).

6.5 Mode of Action of Antiviral Compounds

Large amounts of pentacyclic triterpenoid saponins, or centelloids, are present in *C. asiatica*. Asiaticoside, centelloside, madecassoside, brahmoside, brahminoside, thankuniside, sceffoleoside, centellose, asiatic-, brahmic-, centellic-, and madecassic acids are some of the terpenoids of this group. The isoprenoid pathway synthesizes the secondary plant metabolites known as triterpene saponins, which are hydrophobic triterpenoid structures (aglycones) comprising hydrophilic sugar chains (glycone). Due to their anti-microbial activity and protective effect against pathogen infections, these triterpenoids function as phytoanticipins in plants. Due to the extensive range of pharmacological activity linked to these secondary metabolites, preparations of *C. asiatica* are employed in both conventional and alternative medicine. Also, there have been claims that the triterpenes have a variety of medicinal uses, including antiviral utility (Darshani et al., 2022). There are reports demonstrating the activity of asiatic acid and its derivatives on the inhibition of human enterovirus 71 replication (Zhao et al., 2014). In their study, Yoosook et al. (2000) mentioned that the anti-HSV-1 and 2 activity of *C. asiatica* is because of the asiaticoside present in the plant (Yoosook et al., 2000). Similarly, Komaikul *et al.* noted an increase in the concentration of asiaticoside when the *C. asiatica* leaves were soaked in the NEW containing water (Komaikul *et al.*, 2022). This indicates that asiaticoside can be responsible for the antiviral activity of *C. asiatica*. In a study, Huang *et al.* (2022) elucidated the possible mechanism of asiaticoside in treating COVID-19 using network pharmacology and molecular docking method, which explains the control of several immunological or inflammatory targets and signalling cascades. They discovered 45 distinct main targets in their investigation that asiaticoside may activate. According to a Kyoto Encyclopaedia of Genes and Genomes (KEGG) enrichment analysis, asiaticoside may inhibit the inflammatory and immune-related signalling pathways involved in COVID-19, such as interleukin-17 (IL-17) signalling, T helper 17 (Th17) cell differentiation, MAPK, and the PI3K-Akt signalling pathway. More research is needed to put these beliefs to comprehensive evidence-based support. When taken as a whole, these results imply that asiaticoside could have a lung protective effect (Huang et al., 2022; He et al., 2023). Also, reports are indicating the anti-microbial activity of *C. asiatica,* which is mainly due to the inhibition of NF-κB in LPS-induced macrophages, these triterpene compounds reduce the production of IL-6, TNF-, IL-1, COX-2, and iNOS (Yun et al., 2008). In addition to these, other signalling pathways involved in the antimicrobial mechanism of action for *C. asiatica* include modulation of GSK-3β, MAPK, PI3K/Akt, TGf-β1 /Smad Notch, FAK/Akt pathways (Biswas et al., 2021).

6.6 Future of Research and Drug Delivery

Numerous research findings on *C. asiatica's* pharmacology highlight this herb's totipotency and suitability in both *in-vitro* and *in-vivo* research studies. Centellosides of asiaticoside, asiatic acid, madecassoside, and madecassic acid are the key therapeutic ingredients (Maya et al., 2021). These substances, like herbal extracts and triterpenoids, have more calming effects on multisystem diseases like type 2 diabetes mellitus, gestational diabetes, sleep deprivation, hyperlipidaemia, atopic dermatitis, liver injury, wound healing, baldness, gastric mucosal injury, drug-induced liver toxicity, breast cancer, gastric ulcers, oral sub mucous fibrosis, leukaemia, and migraine. Asiatic acid, actively reduces symptoms of Parkinson's disease, cognitive decline, renovascular hypertension, myocardial ischemia/ reperfusion injury, Alzheimer's disease, atherosclerosis, acute pancreatitis, colon carcinogenesis, ovarian carcinoma, Periodontitis, sepsis, pelvic inflammatory disease, cancer, antimicrobial, antiviral, endometriosis, and pulmonary fibrosis. *C. asiatica* is currently over utilized due to its diverse healthcare applications, and as a result, it has been listed as a seriously threatened plant species by the International Union for Conservation of Nature (IUCN). Strategies for continuous, large-scale, invitro-grown production to suit the increasing demands of the herbal sector are needed. More emphasis on clinical studies to explore the potential is warranted, hence future research should focus on establishing evidence of the effectiveness of secondary metabolites, with translational potential. For a more comprehensive understanding of *C. asiatica's* variety, genetic characterization in the form of genetic fingerprinting is advocated (Perera et al., 2021).

7. Toxicity and Safety

The safety of *C. asiatica* is ensured by the long-standing knowledge of its usage in traditional medicine and cuisines across the world. However, before using it in pharmacological therapy, further safety information will be needed. Based on the reported survey, which includes scant information on oral administration in human studies and *in vitro* human cell lines, the Cosmetic Ingredient Review (CIR) expert panel's evaluation report from 2015 has validated the safety of various extracts of *C. asiatica* in the current usage practises (Bergfeld et al., 2015). According to the European Medicines Agency's evaluation report on *C. asiatica* from 2010, there are no reported adverse events from pharmacovigilance data, and the tolerance of oral *C. asiatica* preparations has also been satisfactory. However, sporadic reports of nausea and gastrointestinal symptoms following oral administration of *C. asiatica* preparations have been made. The long-term use of medicines using *C. asiatica* refined extracts in Europe, which have been approved and marketed for more than 30 years, is also highlighted in this chapter.

Research and knowledge of possible drug interactions are essential in addition to safety data to assure the safe consumption of *C. asiatica* as a food, a herbal supplement, or a medicine while taking other drugs. There are a limited number of studies on *C. asiatica's* drug interactions. According to these studies, using an *in vitro* recombinant human Cytochrome P450 assay, the two main triterpenes present in *C. asiatica*, asiaticoside, and madecassoside, may interact with other medications by decreasing the activity of human cytochrome P450 (Winitthana et al., 2011). Drug interactions should be viewed because CYP enzymes, mostly found in the liver, regulate drug metabolization, making their inhibition or induction essential. Asiaticoside and madecassoside inhibited two of the six CYP isoforms that this study looked at (CYP2C19 and CYP3A4). The inhibitory effects of two triterpenes were, however, somewhat moderate when compared to the IC50 values for other known selective inhibitors. As a result, asiaticoside and madecassoside may interact with medications metabolized by CYP2C19 and CYP3A4.

Additionally, it is noted in the literature that *C. asiatica* supplements may have the potential to affect the liver, making it unsafe to use them with other treatments that may potentially do so. However, there is very little evidence to back up this assertion. Additionally, it is asserted that combining *C. asiatica* with other diuretics may have a positive synergistic effect. Furthermore, due to its propensity to act as a sedative, it is claimed that *C. asiatica* may potentiate anti-anxiety drugs (Sabaragamuwa et al., 2018).

Conclusion

C. asiatica is a perennial herbaceous plant that goes by a variety of names, including "Gotu cola," "Indian pennywort," "water pennywort," and "marsh pennywort." It is a well-known medicinal plant from the Apiaceae family and thrives in marshy, tropical, and subtropical areas close to water. It has reportedly been used in a variety of traditional medical systems, including Ayurveda, Unani, Chinese, and Sri Lankan medicinal systems, to treat conditions like wound healing, skin issues, amenorrhea, diarrhoea, fever, lupus, varicose ulcers, eczema, and psoriasis, along with the these to treat the brain and mental health. It also has origins in folk medicine, where it was used as a treatment for several ailments like asthma, skin diseases, ulcers, and physical aches as well as to enhance memory and as a nervine tonic, among other things. Triterpenoids (asiaticoside, centelloside, madecassoside, and brahmoside), amino acids (alanine, serine, aspartate, and glutamate), phenols (kaempferol and quercetin), and carbohydrates (glucose, mesoinositol, and centellose, among others, found in both dietary supplements and food products) are the main active components of *C. asiatica*. The four primary active triterpenoids that are said to be involved in their activity are asiaticoside, madecassoside, asiatic acid, and madecassic acid. Numerous *in vitro* and *in vivo* studies have revealed that *C. asiatica* and its triterpenoids have a wide spectrum of pharmacological and therapeutic activities. *C. asiatica* can be useful to treat number of pharmacological conditions, including memory and cognitive function, wound healing, cytotoxic and anticancer, antioxidant, cardioprotective, radioprotective, antidepressant, immunomodulatory, antibacterial and antifungal, anti-protozoal, actions that fight leprosy and tuberculosis, and even diabetes. Along with all of these pharmacological effects, it has attractive antiviral activity by obstructing several relevant signalling pathways in the viral life cycle. The available toxicity data on *C. asiatica* strongly indicates that there is very little toxicity or side effects in its use, which opens up the potential for future research and development of this plant in viral and other therapeutics areas.

References

Ahmad, F.B. and Ismail, G. (2003). Medicinal plants used by kadazandusun communities around crocker range. *ASEAN Review of Biodiversity and Environmental Conservation (ARBEC)*. http://www.arbec.com.my/pdf/art1janmar03.pdf.

Alfarra, H.Y. and Omar, M. (2013). Centella asiatica: from folk remedy to the medicinal biotechnology–a state revision. *International Journal of Biosciences (IJB)*, *3*(6), 49–67. https://doi.org/10.12692/ijb/3.6.49–67.

Ali, A.M., Mukram Mackeen, M., El-Sharkawy, S.H., Hamid, J.A., Ahmad, F.B.H. and Lajis, N.H. (1996). Antiviral and Cytotoxic Activities of Some Plants Used in Malaysian Indigenous Medicine. *Pertanika J. Trop. Agric. Sci*, *19*(3), 129–136. https://www.researchgate.net/publication/236960016_Antiviral_and_Cytotoxic_Activities_of_Some_Plants_Used_in_Malaysian_Indigenous_Medicine.

Belwal, T., Andola, H.C., Atanassova, M.S., Joshi, B., Suyal, R., Thakur, S., Bisht, A., Jantwal, A., Bhatt, I.D. and Rawal, R.S. (2019). Gotu Kola (Centella asiatica). In *Nonvitamin and Nonmineral Nutritional Supplements* (pp. 265–275). Elsevier. https://doi.org/10.1016/B978-0-12-812491-8.00038-2.

Bergfeld, W.F., Donald, F.A.C.P., Belsito, V., Hill, R.A., Klaassen, C.D., Liebler, D.C., Marks, J.G., Shank, R.C., Slaga, T.J. and Snyder, P.W. (2015). *Safety Assessment of Centella asiatica-derived Ingredients as Used in Cosmetics*. https://www.cir-safety.org/sites/default/files/centel062015FR.pdf.

Biswas, D., Mandal, S., Chatterjee Saha, S., Tudu, C.K., Nandy, S., Batiha, G.E.S., Shekhawat, M.S., Pandey, D.K. and Dey, A. (2021). Ethnobotany, phytochemistry, pharmacology, and toxicity of Centella asiatica (L.) Urban: A comprehensive review. In *Phytotherapy Research* (Vol. 35, Issue 12, pp. 6624–6654). John Wiley and Sons Ltd. https://doi.org/10.1002/ptr.7248.

Brinkhaus, B., Lindner, M., Schuppan, D. and Hahn, E.G. (2000). Chemical, pharmacological and clinical profile of the East Asian medical plant Centella aslatica. *Phytomedicine*, *7*(5), 427–448. https://doi.org/10.1016/S0944-7113(00)80065-3.

Chandrika, U.G. and Prasad Kumara, P.A.A.S. (2015). Gotu Kola (Centella asiatica). In *Advances in Food and Nutrition Research* (Vol. 76, pp. 125–157). Academic Press Inc. https://doi.org/10.1016/bs.afnr.2015.08.001.

Chaudhary, S.A., Gadhvi, K.V. and Chaudhary, A.B. (2010). Comprehensive review on world herb trade and most utilized medicinal plant. *International Journal of Applied Biology and Pharmaceutical Technology*, *1*(2), 510–517. https://api.semanticscholar.org/CorpusID:56460232.

Darshani, P., Sen Sarma, S., Srivastava, A.K., Baishya, R. and Kumar, D. (2022). Anti-viral triterpenes: a review. *Phytochemistry Reviews*, *21*(6), 1761–1842. https://doi.org/10.1007/s11101-022-09808-1.

Ding, B., Niu, W., Wang, S., Zhang, F., Wang, H., Chen, X., Chen, S., Ma, S., Kang, W., Wang, M., Li, L., Xiao, W., Guo, Z. and Wang, Y. (2022). *Centella asiatica* (L.) Urb. attenuates cardiac hypertrophy and improves heart function through multi-level mechanisms revealed by systems pharmacology. *Journal of Ethnopharmacology*, *291*, 115106. https://doi.org/10.1016/j.jep.2022.115106.

Duke, J.A. (2002). *Handbook of Medicinal Herbs* (2nd Edition). CRC Press. https://doi.org/10.1201/9781420040463.

Gohil, K., Patel, J. and Gajjar, A. (2010). Pharmacological review on *Centella asiatica*: A potential herbal cure-all. *Indian Journal of Pharmaceutical Sciences*, *72*(5), 546. https://doi.org/10.4103/0250-474X.78519.

Hamidpour, R. (2015). Medicinal Property of Gotu kola (*Centella asiatica*) from the Selection of Traditional Applications to the Novel Phytotherapy. *Archives in Cancer Research*, *3*(4). https://doi.org/10.21767/2254-6081.100042.

He, Z., Hu, Y., Niu, Z., Zhong, K., Liu, T., Yang, M., Ji, L. and Hu, W. (2023). A review of pharmacokinetic and pharmacological properties of asiaticoside, a major active constituent of *Centella asiatica* (L.) Urb. In *Journal of Ethnopharmacology* (Vol. 302). Elsevier Ireland Ltd. https://doi.org/10.1016/j.jep.2022.115865.

Heidari, M., Heidari-Vala, H., Sadeghi, M.R. and Akhondi, M.M. (2012). The inductive effects of *Centella asiatica* on rat spermatogenic cell apoptosis *in vivo*. *Journal of Natural Medicines*, *66*(2), 271–278. https://doi.org/10.1007/s11418-011-0578-y.

Huang, J., Zhou, X., Gong, Y., Chen, J., Yang, Y. and Liu, K. (2022). Network pharmacology and molecular docking analysis reveals the mechanism of asiaticoside on COVID-19. *Annals of Translational Medicine*, *10*(4), 174–174. https://doi.org/10.21037/atm-22–51.

Inamdar, P.K., Yeole, R.D., Ghogare, A.B. and de Souza, N.J. (1996). Determination of biologically active constituents in *Centella asiatica*. *Journal of Chromatography A*, *742*(1–2), 127–130. https://doi.org/10.1016/0021-9673(96)00237-3.

Jantwal, A., Durgapal, S., Upadhyay, J., Rana, M., Tariq, M., Dhariwal, A. and Joshi, T. (2021). *Centella asiatica*. In *Naturally Occurring Chemicals Against Alzheimer's Disease* (pp. 257–269). Elsevier. https://doi.org/10.1016/B978-0-12-819212-2.00021-9.

Jayashree, G., Kurup Muraleedhara, G., Sudarslal, S. and Jacob, V.B. (2003). Anti-oxidant activity of *Centella asiatica* on lymphoma-bearing mice. *Fitoterapia*, *74*(5), 431–434. https://doi.org/10.1016/S0367-326X(03)00121-7.

Joshi, V.P., Kumar, N., Singh, B. and Chamoli, R.P. (2007). Chemical Composition of the Essential Oil of *Centella asiatica* (L.) Urb. from Western Himalaya. *Natural Product Communications*, *2*(5), 1934578X0700200. https://doi.org/10.1177/1934578X0700200515.

Joy, J. and Nair, C.K.K. (2009). Protection of DNA and membranes from gamma-radiation induced damages by <I>Centella asiatica</I> *Journal of Pharmacy and Pharmacology*, *61*(7), 941–947. https://doi.org/10.1211/jpp/61.07.0014.

Khan, M.A., Al, M.A., Khan, M., Rahman, S., Ahsan, A., Sanjana, J.M., Mahmud, S., Mahfuz, B., Sharfaraz, A., Khan, M.I., Al Emran, A., Ahmed, F. and *Moni*, M.A. (2021). *Bioinformatics approaches to investigate the phytochemicals of Centella asiatica against the main protease of SARS-CoV-2*. 1–30. https://doi.org/https://doi.org/10.21203/rs.3.rs-1099054/v1.

Komaikul, J., Ruangdachsuwan, S., Juntarapornchai, S., Wanlayaporn, D., Ketsuwan, K., Masrinoul, P., Yoksan, S., Puthavathana, P. and Kitisripanya, T. (2022). Effectiveness of neutral electrolyzed water in inactivating HCoV-OC43 and SARS-CoV-2 on the surfaces of plastic and the medicinal plant *Centella asiatica* (L.) urban. *Heliyon*, *8*(8), e10294. https://doi.org/10.1016/j.heliyon.2022.e10294.

Latha, D., Hrishikesh, D., Shiban, G., Chandrashekar, C. and Bharath, B. (2022). *In silico*, *in vitro* screening of plant extracts for anti-SARS-CoV-2 activity and evaluation of their acute and sub-acute toxicity. *Phytomedicine Plus*, *2*(2), 100233. https://doi.org/10.1016/j.phyplu.2022.100233.

Latif Khan, M. (2009). Ethnomedicinal plants of Manas National Park. In *Northeast India Article in Indian Journal of Traditional Knowledge*. https://www.researchgate.net/publication/236623479.

Maduray, K., Moodley, R., Ramdhani, S. and Parboosing, R. (2022). The anti-HIV activity of biogenic silver nanoparticles synthesized from *Centella asiatica* extracts. *Journal of Herbal Medicine*, *35*, 100592. https://doi.org/10.1016/j.hermed.2022.100592.

Matsuda, H., MorikaWA, T., Ueda, H. and Yoshikawa, M. (2001). Medicinal Foodstuffs. XXVII. Saponin Constituents of Gotu Kola (2): Structures of New Ursane- and Oleanane-Type Triterpene Oligoglycosides, Centellasaponins B, C, and D, from *Centella asiatica* Cultivated in Sri Lanka. *Chemical and Pharmaceutical Bulletin*, *49*(10), 1368–1371. https://doi.org/10.1248/cpb.49.1368.

Maya, P., Mahayasih, W., Harizal, Herman and Ahmad, I. (2021). *In silico* identification of natural products from *Centella asiatica* as severe acute respiratory syndromecoronavirus 2 main protease inhibitor. *Journal of Advanced Pharmaceutical Technology and Research*, *12*(3), 261–266. https://doi.org/10.4103/japtr.JAPTR_297_20.

Mitra, S.S., Mukherjee, A., Nongdam, P., Pandey, D.K. and Dey, A. (2021). Endophytes producing active constituents in *Centella asiatica* with a special emphasis on asiaticoside and madecassoside: A review update. In *Volatiles and Metabolites of Microbes* (pp. 409–427). Elsevier. https://doi.org/10.1016/B978-0-12-824523-1.00013-4.

Mohamed, S., Hambali, Z., Samah, B.A., Dev, R. and Dev, O. (2009). Comparison on Cognitive Effects of *Centella Asiatica* in Healthy Middle Age Female and Male Volunteers. In *European Journal of Scientific Research* (Vol. 31, Issue 4). http://www.eurojournals.com/ejsr.htm.

Narayanan, M.K.R. and Kumar, N. (2007). Gendered knowledge and changing trends in utilization of wild edible greens in Western Ghats, India. *Indian Journal of Traditional Knowledge*, *6*, 204–216. https://api.semanticscholar.org/CorpusID:68617612.

Oyenihi, A.B., Ahiante, B.O., Oyenihi, O.R. and Masola, B. (2020). Centella asiatica: its potential for the treatment of diabetes. In *Diabetes: Oxidative Stress and Dietary Antioxidants* (pp. 213–222). Elsevier. https://doi.org/10.1016/B978-0-12-815776-3.00021-8.

Perera, W.P.R.T., Liyanage, J.A., Dissanayake, K.G.C., Gunathilaka, H., Weerakoon, W.M.T.D.N., Wanigasekara, D.N., Fernando, W.S.K., Rajapaksha, R.M.H., Liyanage, R.P. and Perera, B.T. (2021). Antiviral Potential of Selected Medicinal Herbs and Their Isolated Natural Products. *BioMed Research International*, *2021*. https://doi.org/10.1155/2021/7872406.

Polash, S.A., Saha, T., Hossain, M.S. and Sarker, S.R. (2017). Phytochemical contents, antioxidant and antibacterial activity of the ethanolic extracts of *Centella asiatica*(L.) Urb.leaf and stem. *Jahangirnagar University Journal of Biological Sciences*, *6*(1), 51–57. https://doi.org/10.3329/jujbs.v6i1.33731.

Poovizhi T*, Krishnaprabha V. and Uthira L. (2022). Perspective study on Pharmacological and Therapeutic potential of *Centella asiatica*. *Medicon Medical Sciences*. https://doi.org/10.55162/MCMS.02.032.

Prakash, V., Jaiswal, N. and Srivastava, M. (2017). A review on medicinal properties of *Centella asiatica*. In *Asian Journal of Pharmaceutical and Clinical Research* (Vol. 10, Issue 10, pp. 69–74). Innovare Academics Sciences Pvt. Ltd. https://doi.org/10.22159/ajpcr.2017.v10i10.20760.

Punturee, K., Wild, C.P., Kasinrerk, W. and Vinitketkumnuen, U. (2005). Immunomodulatory activities of *Centella asiatica* and Rhinacanthus nasutus extracts. *Asian Pacific Journal of Cancer Prevention : APJCP*, *6*(3), 396–400. http://www.ncbi.nlm.nih.gov/pubmed/16236006.

Rahmatullah, M., Ferdausi, D., Mollik, A., Jahan, R., Chowdhury, M. and Haque, W. (2010). A survey of medicinal plants used by kavirajes of chalna area, Khulna district, Bangladesh. *African Journal of Traditional, Complementary and Alternative Medicines*, *7*(2), 91–97. https://doi.org/10.4314/ajtcam.v7i2.50859.

Rajendran, K., Balaji, P. and Basu, J. (2008). Medicinal plants and their utilization by villagers in southern districts of Tamil Nadu. *Indian Journal of Traditional Knowledge*, *7*(3), 417–420. https://api.semanticscholar.org/CorpusID:196168510.

Sabaragamuwa, R., Perera, C.O. and Fedrizzi, B. (2018). *Centella asiatica* (Gotu kola) as a neuroprotectant and its potential role in healthy ageing. *Trends in Food Science & Technology*, *79*, 88–97. https://doi.org/10.1016/j.tifs.2018.07.024.

Sahu, N.P., Roy, S.K. and Mahato, S.B. (1989). Spectroscopic determination of structures of triterpenoid trisaccharides from *Centella asiatica. Phytochemistry*, *28*(10), 2852–2854. https://doi.org/10.1016/S0031-9422(00)98106-8.

Samant, S.S. and Pant, S. (2006). Diversity, distribution pattern and conservation status of the plants used in liver diseases/ailments in Indian Himalayan Region. *Journal of Mountain Science*, *3*(1), 28–47. https://doi.org/10.1007/s11629-006-0028-6.

Schaneberg, B.T., Mikell, J.R., Bedir, E. and Khan, I.A. (2003). An improved HPLC method for quantitative determination of six triterpenes in *Centella asiatica* extracts and commercial products. *Die Pharmazie*, *58*(6), 381–384. http://www.ncbi.nlm.nih.gov/pubmed/12856998.

Singh, B. and Rastogi, R.P. (1968). Chemical examination of *Centella asiatica* linn—III. *Phytochemistry*, *7*(8), 1385–1393. https://doi.org/10.1016/S0031-9422(00)85642-3.

Singh, B. and Rastogi, R.P. (1969). A reinvestigation of the triterpenes of *Centella asiatica. Phytochemistry*, *8*(5), 917–921. https://doi.org/10.1016/S0031-9422(00)85884-7.

Singh, S., Gautam, A., Sharma, A. and Batra, A. (2010). *Centella asiatica* (L.): a plant with immense medicinal potential but threatened. *International Journal of Pharmaceutical Sciences Review and Research,* 4(2):9–17.

Tiwari, S., Singh, S., Patwardhan, K., Gehlot, S. and Gambhir, I.S. (2008). Effect of *Centella asiatica* on mild cognitive impairment (MCI) and other common age-related clinical problems. In *Article in Digest Journal of Nanomaterials and Biostructures,* 3(4). https://www.researchgate.net/publication/201910063.

Uddin, S.M., Alam, K.M., Hoq, O.M. and Nahar Nuri, Z.N. (2017). The therapeutic use of *Centella asiatica. Article in International Journal of Chemistry*, (1): 21–26.

Vasantharuba, S., Sundaram, S.P., Thangaiah, A., Seevaratnam, V., Banumathi, P., Premalatha, M.R. and Arumugam, T. (2012). Functional properties of *Centella asiatica* (L.): A review. *International Journal of Pharmacy and Pharmaceutical Sciences*, *4*, 8–14. https://www.researchgate.net/publication/286360480.

Wang, L., Guo, T., Guo, Y. and Xu, Y. (2020). Asiaticoside produces an antidepressant-like effect in a chronic unpredictable mild stress model of depression in mice, involving reversion of inflammation and the PKA/pCREB/BDNF signaling pathway. *Molecular Medicine Reports*, *22*(3), 2364–2372. https://doi.org/10.3892/mmr.2020.11305.

Wattanathorn, J., Mator, L., Muchimapura, S., Tongun, T., Pasuriwong, O., Piyawatkul, N., Yimtae, K., Sripanidkulchai, B. and Singkhoraard, J. (2008). Positive modulation of cognition and mood in the healthy elderly volunteer following the administration of *Centella asiatica. Journal of Ethnopharmacology*, *116*(2), 325–332. https://doi.org/10.1016/j.jep.2007.11.038.

Winitthana, T., Niwattisaiwong, N., Patarapanich, C., Tantisira, M.H. and Lawanprasert, S. (2011). *In vitro* inhibitory effects of asiaticoside and madecassoside on human cytochrome P450. *Toxicology in Vitro*, *25*(4), 890–896. https://doi.org/10.1016/j.tiv.2011.02.010.

Wu, T., Geng, J., Guo, W., Gao, J. and Zhu, X. (2017). Asiatic acid inhibits lung cancer cell growth *in vitro* and *in vivo* by destroying mitochondria. *Acta Pharmaceutica Sinica B*, *7*(1), 65–72. https://doi.org/10.1016/j.apsb.2016.04.003.

Yoosook, C., Bunyapraphatsara, N., Boonyakiat, Y. and Kantasuk, C. (2000). Anti-herpes simplex virus activities of crude water extracts of Thai Medicinal Plants. *Phytomedicine*, *6*(6), 411–419. https://doi.org/10.1016/S0944-7113(00)80068-9.

Yun, K.J., Kim, J.Y., Kim, J. bin., Lee, K.W., Jeong, S.Y., Park, H.J., Jung, H.J., Cho, Y.W., Yun, K. and Lee, K.T. (2008). Inhibition of LPS-induced NO and PGE2 production by asiatic acid via NF-κB inactivation in RAW 264.7 macrophages: Possible involvement of the IKK and MAPK pathways. *International Immunopharmacology*, *8*(3), 431–441. https://doi.org/10.1016/j.intimp.2007.11.003.

Zahara, K. (2014). Clinical and therapeutic benefits of *Centella asiatica*. *Pure and Applied Biology*, *3*(4), 152–159. https://doi.org/10.19045/bspab.2014.34004.

Zhao, C., Xu, J., Zhang, Y., Zhao, L. and Feng, B. (2014). Inhibition of Human Enterovirus 71 Replication by Pentacyclic Triterpenes and Their Novel Synthetic Derivatives. *Chemical and Pharmaceutical Bulletin*, *62*(8), 764–771. https://doi.org/10.1248/cpb.c14-00088.

14 Unlocking the Antiviral Power of *Piper longum* L.: Nature's Potential Against Viruses

Parthasarathy Seethapathy[1*], Deepu Pandita[2*], Subbiah Sankaralingam[3] and Anu Pandita[4]

1. Introduction

Approximately 20,000 plant varieties are used in conventional medical practices around the world. These plants are also seen as possible ways to find novel medications. *Piper longum* (L.) (1753), a slender aromatic climber best known as the pippali or long pepper, is a flowering vine in the Piperaceae family. Pepper comes from the Sanskrit word pippali for long pepper. The long pepper also known as the Indian, Javanese, or Indonesian long pepper is grown for its fruit which after drying finds use as a spice (Kumar et al., 2011). It is native to South Asia and has been used for a long time in the traditional Ayurvedic medicine system for diverse purposes (Biswas et al., 2022). Piperine alkaloids in fruits contribute to the pungency and hot flavor of this plant and are the principal constituent accounting for 3–5% of P. longum (Yadav et al., 2020). *P. nigrum* produces pepper of green, white, and black colour with an analogous but mostly calefacient flavour, and is a close relative of the long pepper plant. The *P. retrofractum* long pepper species native to Indonesia and Java is a good rejuvenator in Ayurvedic medicine and alleviates muscular pain and inflammation on topical application (Salleh and Ahmad, 2020).

Traditional uses embrace the whole long pepper and parts like fruit, but no comprehensive information on its application has been compiled. *P. longum* is affordable, broadly accessible and efficacious against numerous maladies, including cancer, depression, obesity, diabetes, snakebite, inflammation, scorpion-sting, and liver damage (Yadav et al., 2020). The purpose of this chapter is to emphasize modern progress in the materia medica and pharmacognosy mainly the anti-viral research on the *P. longum*, as well as enlighten scholars about this plant to encourage further investigation. In ancient Indian Ayurvedic texts, its

[1] Department of Plant Pathology, Amrita School of Agricultural Sciences, Amrita Vishwa Vidyapeetham, Coimbatore, Tamil Nadu, India.
[2] Government Department of School Education, Jammu, Jammu and Kashmir, India.
[3] PG and Research Department of Botany, Saraswathi Narayanan College, Madurai, India.
[4] Vatsalya Clinic, Krishna Nagar, New Delhi, India.
* Corresponding author: spsarathyagri@gmail.com[1]; deepupandita@gmail.com[2]

DOI: 10.1201/9781003433583-14

medicinal and culinary applications are described in detail. In the sixth or fifth century BC, it arrived in Greece, although Hippocrates referred to it as a medicine rather than a spice. Before the Columbian exchange, long pepper was an essential and long-familiar spice among the Greeks and Romans. Although Theophrastus differentiated the two in his first botany work, the ancient history of black pepper and long pepper is frequently intermingled and confounded. Pliny incorrectly presumed that the dried long pepper and black pepper originated from identical plants, even though the Romans frequently referred to both as piper. In Europe, round peppercorns or black pepper started to vie with the long pepper in the 12th century and by the 14th century, it had supplanted it (Kumar et al., 2011). Long pepper is indigenous to the subtropical and tropical regions of the globe, which includes the Indian subcontinent, the Middle East, Sri Lanka, and the Americas. It is believed that Roman emperors valued it much more than black pepper because of its commercial worth and economic significance (Singh, 2018).

The Age of Discovery was fueled by the search for more affordable and reliable sources of black pepper. Only after discovering the American continents and the chilli pepper, which the Spanish called pimiento, did the popularity of the long pepper decline. Chili peppers, which, when desiccated, resemble long peppers in shape and flavor, were simpler to cultivate in several European-friendly regions. Long pepper is a very uncommon flavor in European recipes; however, it continues to be utilized in Indian and Nepalese vegetable and pickled vegetables, some North African seasoning blends, and Indonesian and Malaysian food preparation. It is generally known as pippali and is regularly seen in Indian food stores (Barboza et al., 2022). The underground roots and dried aerial parts of the female plants are referred to as the economic components. These parts are frequently used for their many fragrant and healthy qualities. The long pepper possesses a pleasant and aromatic scent, yet its taste is characterized by a sharp and intense spiciness that persists and induces a numbing sensation, contradicting its seemingly innocuous fragrance (Babu et al., 2006). Recently, there has been a revitalization of interest in long pepper due to its unique flavor profile. As a result, it has gained some popularity among food enthusiasts, chefs, and specialty markets. However, its distribution is still more limited compared to other spices.

2. Taxonomy and Scientific Classification

There are 4000 species in the Piperaceae family, which are split into 5 groups. The two largest genera in this family are Piper with 2,000 species and Peperomia with 1,600 species. Piper types are very important for business, the economy, and medicine. From an economic perspective, *P. nigrum* and *P. longum* are two of the most important species in the Piperaceae family (Yadav et al., 2020). Piperaceae is a large and diverse family that includes known species such as *Piper betle* (used as stimulant), *P. nigrum* (used as black and white pepper), *P. ornatum* (used as a decorative plant), *P. retrofractum* (used as a herb), and *P. sarmentosum* (used as indigenous vegetable and therapeutic plant). The piperaceae is one of the most significant angiospermic groups in Asia, with 1050 species distributed among three taxa. Casimir de Candolle compiled a monograph of the entire family for *'Prodromus systematis naturalis regni vegetabilis'* in 1869 (Suwanphakdee et al., 2016). The taxonomy of long pepper is given below:

Domain	Eukaryota
Kingdom	Plantae
Phylum	Spermatophyta
Subphylum	Angiospermae
Class	Dicotyledonae
Order	Piperales
Family	Piperaceae
Genus	*Piper*
Species	*longum*

Botanical name: *Piper longum* L.

Preferred Common Name: Indian long pepper

Other Scientific Names: *Piper latifolium* Hunter (1809); *Chavica roxburghii* Miq. (1844)

Synonyms

P. sarmentosum Wall.
C. sarmentosa Miq.
C. longa H.Karst.
P. retrofractum Vahl
P. turbinarium Noronha.
P. peepuloides Wall
C. peepuloides Miq.
P. chaba Hunt

Vernacular names

English : Indian long pepper, Java pepper, Bengal pepper, Jaborandi pepper, Balinese pepper
Burmese : Peik-khyen, Peikchin
French : Poivre long
Dutch : Langwerpige peper
German : Bengalischer Pfeffer, Langer Pfeffer, Jaborandi-Pfeffer
Hungarian : Bengáli bors
Hindi : Pipar, piplamool
Italian : Pepe lungo
Swedish : Långpeppar
Nepalese : Pipal, Piplamol, Popal
Portuguese : Pimenta-longa

Turkish	: Dar biberi
Unani	: Pipal
Sanskrit	: Magadhi, Chapala, Kati, Chanchala, Granthika, Kola, Kana, Krishna, Tiktatandula, Korangi, Krishnapippali, Pippali, Vaidehi
Kannada	: Lippali, Hippali, Thippili
Bengáli	: Pipl Amor
Urdu	: Pippal
Malaya	: Magadhi, Pit poot, Thippili, Pippali, Thippili
Gujarati	: Pipli
Telugu	: Pippallu
Marathi	: Pimpli
Tamil	: Pippili, thippili, kandan lippilli,sirumulam

3. Botanical Description of the Plant

P. longum is a trailing, creeping, slender, compact , aromatic, flowering, dioecious ascending herbaceous vine with perennial woody rootstocks and node-bearing stems with thickened nodes. *P. longum* grows in India, Myanmar, Indonesia, Nepal, Malaysia, Sri Lanka, and Vietnam and has chromosome numbers varying from $2n = 24$ to $2n = 96$ (Kumar et al., 2011; Dey et al., 2020). The stems are medicinal, fibrous, and engineered for capacity. The plant in challenge is a perennial creeping under-shrub that exhibits a spreading growth habit at ground level. The stems exhibit a creeping growth habit, characterized by jointed segments that bear erect branches responsible for fruit production. Additionally, the young shoots of this plant are covered in a layer of fine hairs. The branches exhibit dimorphism, with vegetative branches characterized by their creeping nature and ability to root at nodes. Conversely, fertile splits are erect in posture, featuring swollen nodes and lacking adventitious roots. The petiole is lobed, variable in length, longest in lower leaves, upper leaves almost sessile and amplexicaul, and can reach a maximum length of 10 cm. The leaves are ovate, cordate with broadened round lobes at the base, alternate, simple, alternate in arrangement, petiolate or sessile, clearly dimorphic, spreading, not glossy, gland-dotted, membranous, devoid of stipules, and of various blade diameters. The lower leaves measure 5 to 7 cm in length, while the upper leaves measure 2 to 3 cm (Kumar et al., 2011; Dey et al., 2020).

The spikes are cylindrical and possess a peduncle. The male spikes are longer and slender, while the female spikes measure 1.3–2.5 cm in length and have a diameter of 4–5 mm. The fruits are ovoid and are embedded within a fleshy spike. As they ripen, the spike changes color from green to black. The flowering process occurs continuously throughout the year, and the flowers of this particular species exhibit dioecism. The spike structure of the inflorescence is made up of unisexual, tiny, densely packed flowers, and gives it its structural identity. These flowers give rise to compact clusters of gray berries. These flowers eventually give rise to small grey-berry–clusters. The female spikes exhibit a characteristic of having short, thick stalks that range in length from 1.5–2.5 cm and have a thickness of 0.5–0.7 cm. The male spikes exhibit a slender and elongated morphology, characterized by longer stalks measuring between 2.5 and 7.5 cm. The female spikes bear circular bracts and three stigmas and are truncated than male spikes (1.25–2 cm). Flowers exist as singular shafts. The fruits are oblong, blunt, and bluish-green in colour and develop into 2.5-3.5 cm lengthy and 5 mm thickened fleshy spikes (Jiang et al., 2013; Yadav et al., 2020).

Pippalimula is the root radix and mature female spikes are accumulated and desiccated as commercialized pippali. Piplamul is classified into three grades I, II, and III. Grade I has thick roots and underground stems, whereas grades II and III have thin roots, stems, or broken fragments. The commercial drug is nearly completely composed of cylindrical, straight or slightly curved, transversely cut pieces (5–25 mm in length; 2–7 mm in diameter). Few have well-defined, swollen inter-nodes with numerous scars of leaves and rootlets. The surface colour is grimy light brown (Gajurel et al., 2021). Fig. 1 shows the drawing of *P. longum* (a), leaves, inflorescence, and fruits (b-c), and dried catkins (d) of long pepper.

Fig. 1 (a) Drawing of *Piper longum* (*Source:* http://plantillustrations.org/Zenker, J.C., Merkantilische Waarenkunde (1831-1835) Merkent. Waarenk. (1831-1835), (b-c) Long pepper's Leaves, Inflorescence, and Fruits (d) Dried long pepper Catkins. (*Source:* https://efloraofindia.com/2011/03/21/piper-longum/)

4. Geographical Distribution

P. longum, commonly referred to as the long pepper is a vine-like shrub indigenous to the Indian subcontinent, Middle East, Sri Lanka, and America and the Indo-Malaya region. Long pepper also grows wild in tropical evergreen rainforests of Singapore, the Philippines, Indonesia, Timor, India, Middle-East, Nepal, Malaysia, Sri Lanka, and America as well as the Indian subcontinent. The plant exhibits natural growth patterns in the Indian tropical rainforests (Sivarajan and Balachandran, 1994; Singh et al., 2022).

The primary source of Indian long pepper is predominantly obtained from indigenous plant species. However, a few other places, including the Khasi Hills, Kerala, Karnataka, the lower elevations of West Bengal, Maharashtra, Eastern Uttar Pradesh, Madhya Pradesh, and Tamil Nadu, also cultivate this crop. The species can be observed in its natural habitat within the forested regions of Andhra Pradesh, as well as the Andaman and Nicobar Islands. The plant is indigenous to the Indian evergreen forests and is under cultivation in Assam, Andhra Pradesh, and Tamil Nadu zones. The long pepper flourishes in ideal tropical forest circumstances characterized by mild shade, high humidity, ample soil moisture, and adequate drainage. The ideal soil for pepper is friable clay loam, that is rich in nutrients. The Gol Tipalli variety from West Bengal, the Asli and Suvali varieties from Assam, the Pipal Nan Sori variety from Maharashtra, and the Vishwam variety developed by the Kerala Agricultural University and the Central Institute of Medicinal and Aromatic Plants (CIMAP) are widely cultivated varieties in India (Nair, 2023). Fig. 2 represents the worldwide distribution of *P. longum*.

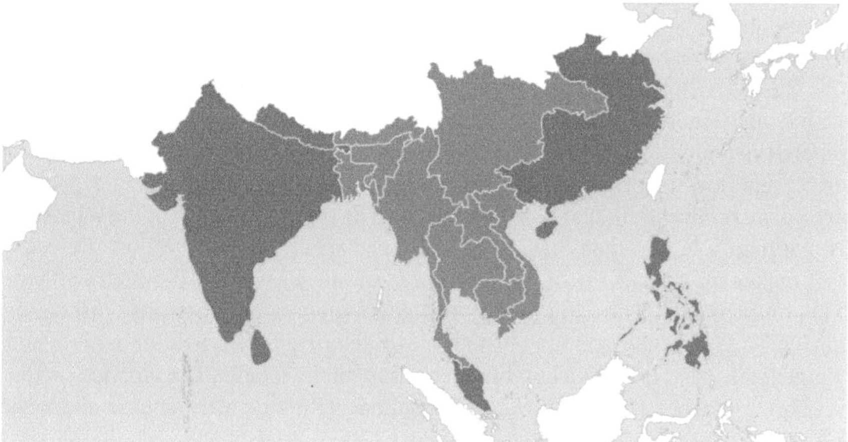

Fig. 2 The native range of *P. longum* is E. Himalaya to China (S. Yunnan) and Indo-China. (*Source:* https://powo.science.kew.org/)

5. Phytochemical Constituents and Metabolite Profiling

The desiccated female flowers (pippali) and rootstock (pippalimula) are marketable plant portions used to cure a wide range of diseases, including the typical flu, high temperatures, indigestion, bronchitis, asthma, leprosy, and cancer. The plant contains over 50 alkaloids and numerous essential oil compounds. The alkaloids, resin, volatile oil, and terpenoids occur in fruits. Steam distillation of the long pepper's desiccated spikes yields essential oil (0.7%–0.8%). Numerous bioactive metabolites like alkaloids, flavonoids, esters and steroids discovered in the extracts of plants and essential oils from the fruits and roots have been reported to be analgesic, anti-amoebic, anti-angiogenic, anti-arthritic, anti-asthmatic, anti-cancer, anti-depressant, anti-diarrheic, anti-dysenteric, anti-microbial, anti-fungal, anti-viral, anti-obesity, anti-inflammatory, anti-platelet, analgesic, antioxidant, anti-hyperlipidaemic, anti-hyperglycaemic, anti-parasitic, anti-snake-venom agents, immunomodulatory and radioprotective (Sunila and Kuttan, 2004; Kumar et al., 2011; Sharma et al., 2014). Many pharmacological properties of long pepper were credited to its anti-oxidative and anti-inflammatory nature, plus its capability to regulate an array of signaling pathways and catalysts (Biswas et al., 2020). Long pepper contains a high concentration of mineral deposits, vitamin supplements, and rich nutrients. Long pepper seeds (100 g) contain 39.5 g of carbohydrate, 12.2 g of protein, and 5.8 g of fibre and relatively high mineral concentrations such as fixed oil (6.6 g), volatile oil (1.5 g), and piperine (4.5 g). The below-ground portion of the root system and stem exhibited the highest concentration of piperine. The piperine concentration in fruits exhibits an increase as they progress in maturity, with levels rising from 0.53% at 14–16 days to 0.9% at 40–45 days. Brachystamide-B, a dimer of desmethoxypiplartine, N-isobutyl deca piperine, retrofractamide A, pergumidiene, piperlongumine, piperlonguminine, tetrahydropiperlongumine and trimethoxy cinnamoyl-piperidine have also been reported in the plant roots.

Piperine is the most prevalent of the fruit alkaloids and related compounds, which also include asarinine, pipernonaline, methyl piperine, pellitorine, piperettine, piperlongumine, piperundecalidine, and piperlonguminine. When dried, piperine, the principal alkaloid and active constituent of the long pepper, is present at a concentration of 3–5% by weight. The piperine content of the substance is marginally higher than that found in black pepper. A bland, yellow-green oil that makes up around 1% of the material is produced when it is water-distilled. This oil possesses a specific gravity of 0.86 and emits an aroma reminiscent of ginger. The substance exhibits a distinctive olfactory characteristic and a bitter and pungent taste that induces a numbing sensation on the tongue. The piperine (0.15–0.18%), piplartine (0.13–0.20%), and residual yellow crystal pungent alkaloids have been observed. According to Pillai et al. (2000), triacontane, reducing sugars, dihydrostigmasterol and glycosides occur in plants. Piperine exhibits a strong attraction towards lipids and fatty compounds, thereby promoting the formation of micelles. It also induces the active transportation of amino acids through the stimulation of gamma-glutamyl transpeptidase and alters the epithelial cell wall. Based on these findings, it is postulated that the consumption of piperine in moderate amounts may serve as a thermonutrient (Wadt and da Silva, 2016).

Ultimately, the cellular interactions of some of the putative druglike phytochemicals with protein targets implicated in neurological disorders were found as, aristolcatoms, alkaloids and their derivatives, benxodioxoles, benzene, and substituted derivatives, carboxylic acids and derivatives, cinnamic acids and derivatives, diarylheptanoids, furanoid lignans, isoflavonoids, naphthalenes, organic nitrogen compounds, oxanes, phenanthrenes and derivatives, phenol ethers, phenol lipids, phenylpropanoic acids, polycyclic hydrocarbons, piperidines, pyridines, pteridines and derivatives, oxanes, naphthalenes , isoflavanoides, organooxygen compounds, organic nitrogen compounds, pyridines and derivatives, saturated hydrocarbons, and steroid and unsaturated hydrocarbons (Choudhary and Singh, 2018). Figure 3 shows the structures (generated by ChemSpider) of some of the active phytocompounds of *P. longum.*

6. Medicinal/Therapeutic Uses of Long Pepper

Long pepper finds utilization in improving appetite and digestion and curing indigestion, intestinal gas, stomachache, constipation, gonorrhea, diarrhea, heartburn, tongue paralysis, chronic malaria, cholera, viral hepatitis, spleen diseases, respiratory infections, lung problems including bronchitis, asthma and cough and tumors (Bhavan et al., 2013; Prasad and Tyagi, 2016).

Long pepper in Ayurveda finds use as a "Rasayana" to cure respiratory disorder and also forms an essential component in digestive formulations and as an ingredient of "Trikatu" (three pungent herbs of black pepper, long pepper, and ginger) in numerous Ayurvedic formulations. *In-vivo* investigations suggest that Trikatu enhances the bio-availability of medicaments administered along with it. According to Ayurveda, the root is an emmenagogue, carminative, liver tonic, stomachic, abortifacient, and aphrodisiac. The fruits have diuretic, haematinic, digestive, and general tonic value, useful in snakebites, liver inflammation, joint pains, scorpion stings, and night blindness. Singly long pepper is used to correct digestive perturbations and secondary respiratory disorders (Johri and Zutshi, 1992; Kaushik, 2018; Kaushik et al., 2020; Kalamol et al., 2023).

6.1 Uses Described in Pharmacopoeias and Traditional Medicine

There is inadequate evidence to assist the use of Indian long pepper to treat gastrointestinal and respiratory disorders, arthritis, menstrual problems, and numerous other conditions. Additionally, Indian long pepper is utilized to improve the body's ability to absorb other drugs. (Poddar et al., 2020). Indian long pepper is most commonly used in the form of ayurvedic medicine for curing chronic bronchitis, stomachache, constipation, chronic malaria, asthma, diarrhea, cholera, gonorrhea, paralysis of the tongue, viral hepatitis, respiratory pathological processes, cough, bronchitis, spleen disorders, and deadly tumors (Bezerra et al., 2013; Ramamoorthy et al., 2019). The Ayurvedic formulations mentioned in Ayurvedic Pharmacopoeia of India (The Ayurvedic Pharmacopoeia of India Parts II and IV, Ministry of Health and Family Welfare, Department of AYUSH, Government of India) are represented in Fig. 4.

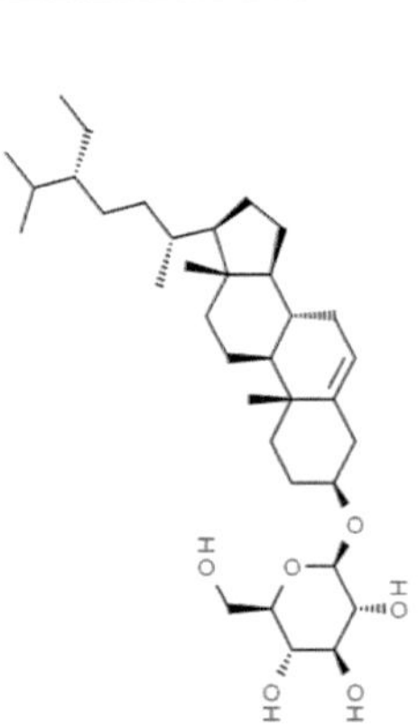
Daucosterol

Pipercide

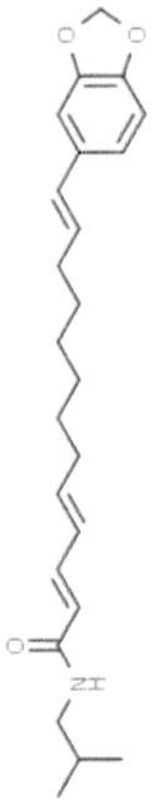
Guineensine

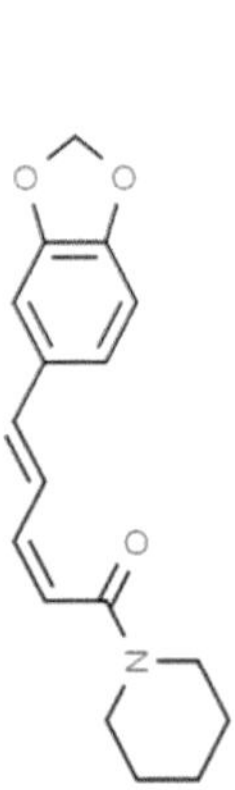
Isopiperine

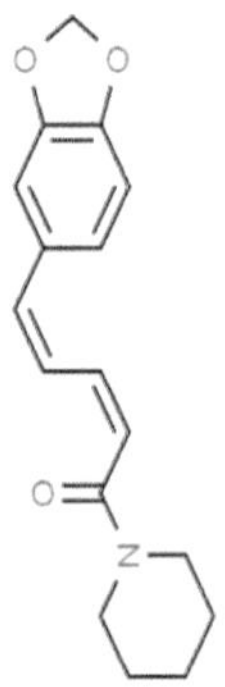
Chavicine

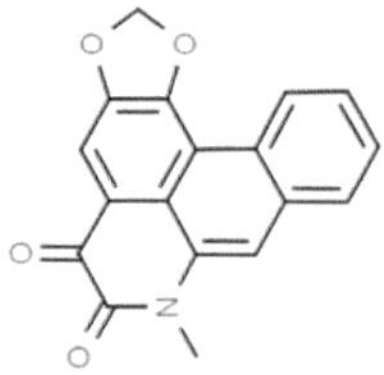
Cepharadione A

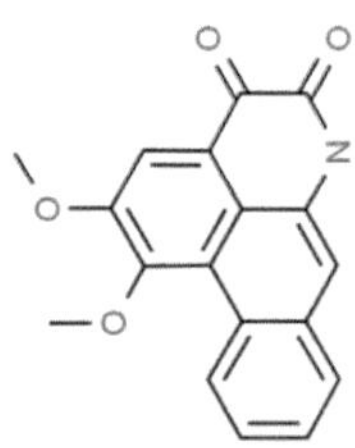
Cepharadione B

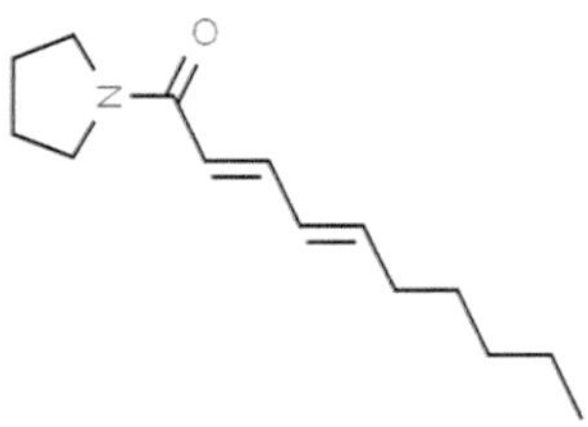
Sarmentine

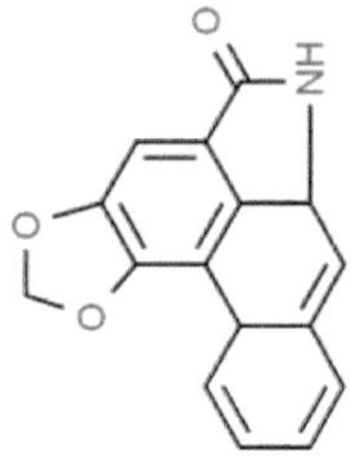
Cepharanone A

Fig. 3 *Contd.*

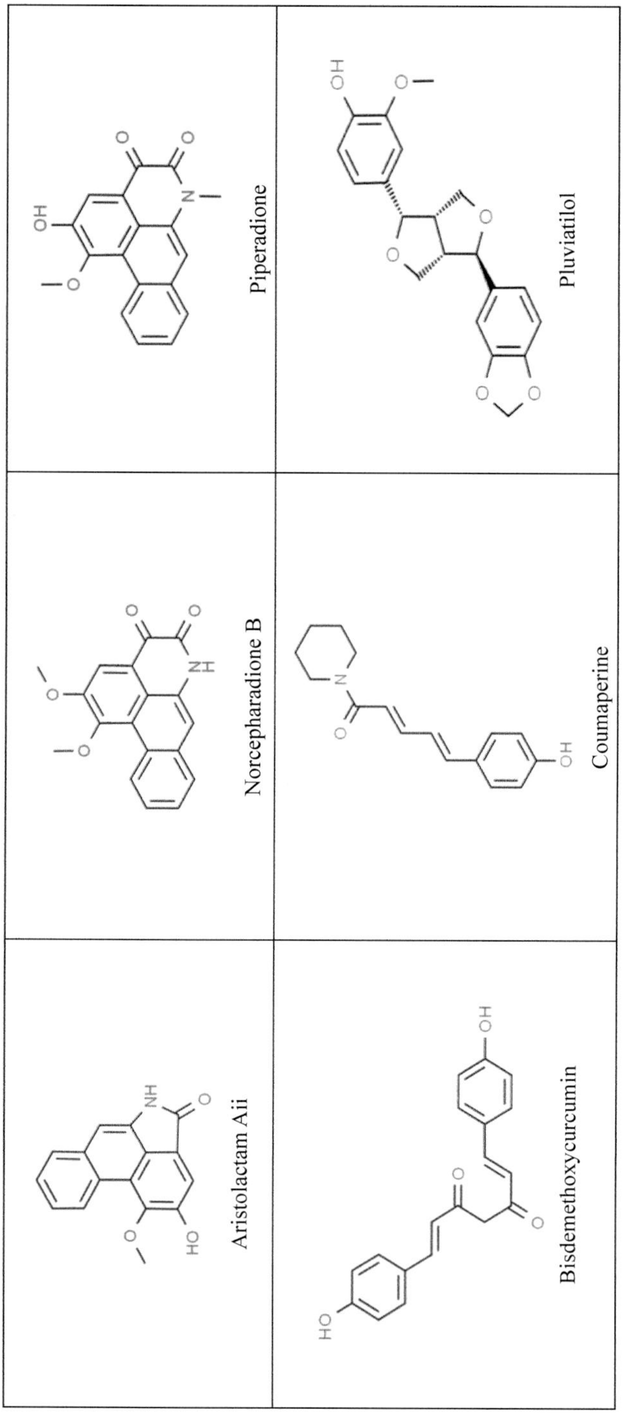
Aristolactam Aii
Norcepharadione B
Piperadione
Bisdemethoxycurcumin
Coumaperine
Pluviatilol

Fig. 3 *Contd.*

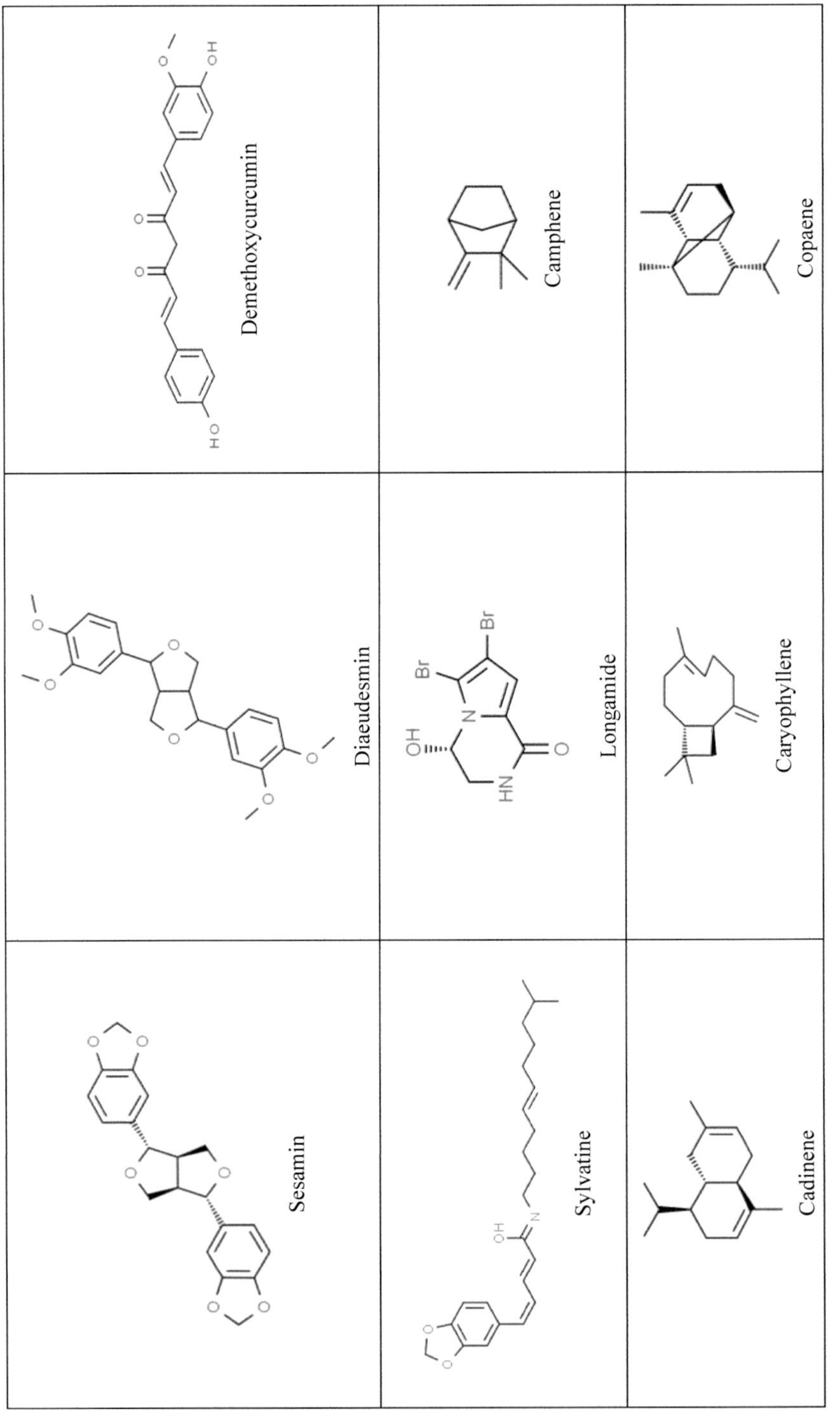

Sesamin
Diaeudesmin
Demethoxycurcumin
Sylvatine
Longamide
Camphene
Cadinene
Caryophyllene
Copaene

Fig. 3 *Contd.*

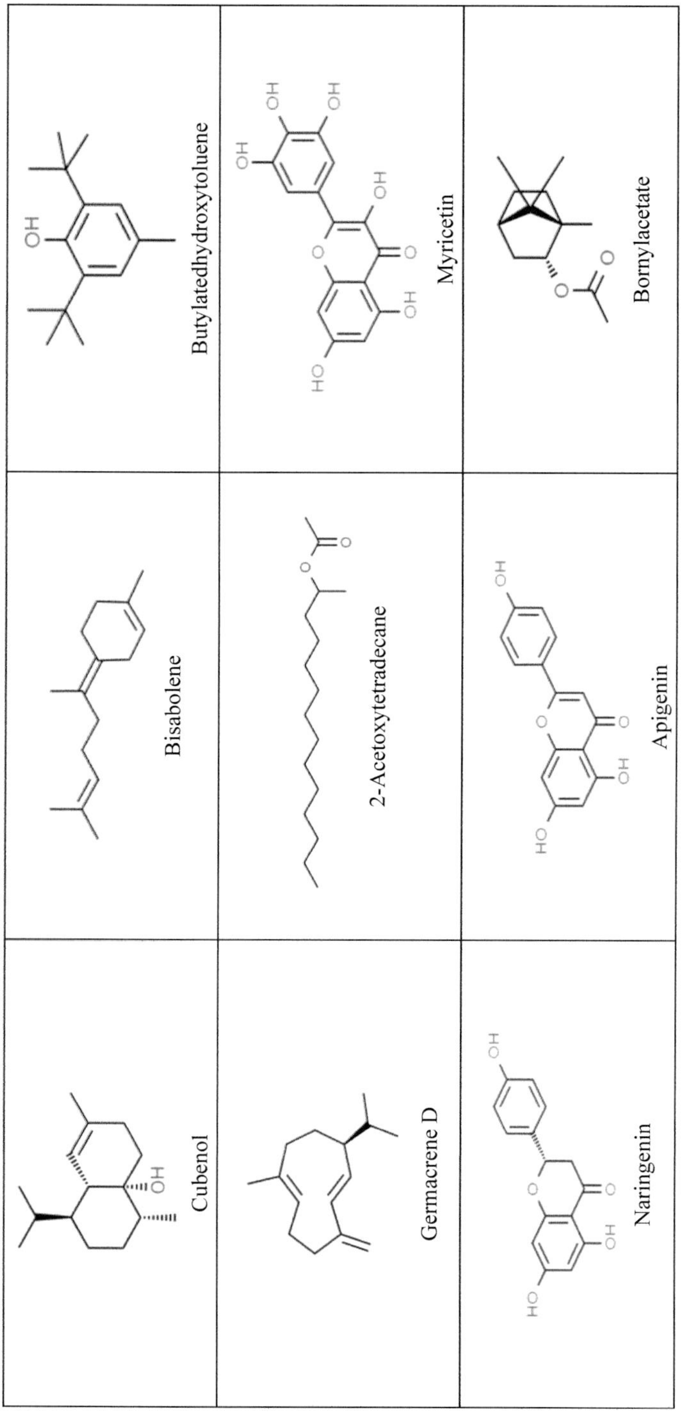

Fig. 3 Structure of some active phyto-compounds of *P. longum* (generated in www.ChemSpider.com).

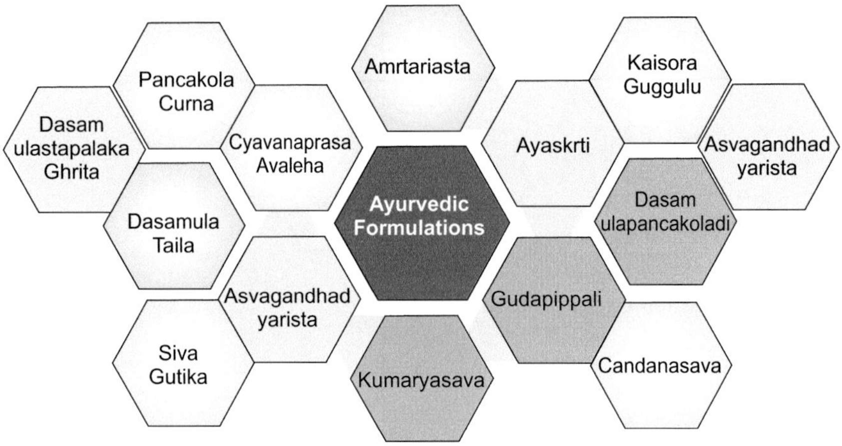

Fig. 4 Ayurvedic formulations of *P. longum.*

The popular Ayurvedic (churna) medicines made from *P. longum* and other Piper species (Hazra et al., 2019) are represented in **Fig. 5**.

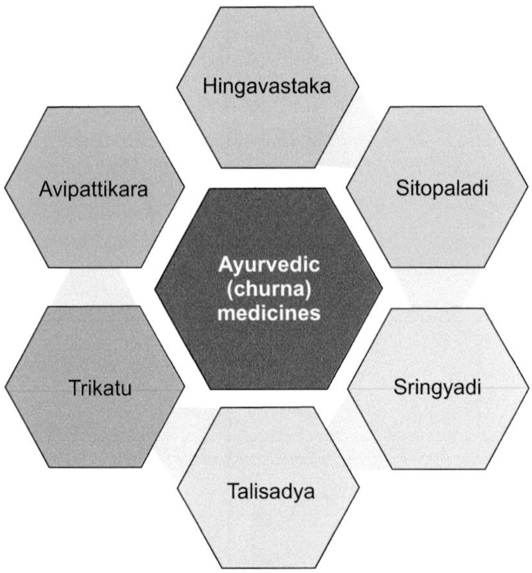

Fig. 5 Ayurvedic (churna) medicines made from *P. longum* and other Piper species.

6.2 Application Supported by Clinical Data

The rise of male hormonal birth control is predicated by a reduction in sperm concentration unaffected by sexual desire and sexual ability. The combined use of depot medroxy progesterone acetate (DMPA) and extract of Javanese long pepper (JLP) at 0.94 mg and 1.88 mg reduces spermatozoa percentage. It is unclear if this combination impacts blood chemistry, hemophilia, or body weight (Yurnadi et al., 2012). Occasionally excruciating peripheral coldness may be linked to reduced peripheral blood flow. Long pepper extraction has been shown to enhance blood circulation, which may be a beneficial treatment for peripheral coldness. The purpose of this randomized, double-blind, placebo-controlled, crossover trial was to evaluate effects of long pepper extract on the peripheral chill in healthy women. Assessment of peripheral blood flow and skin temperature after exposure to frigid water was the primary endpoint. Blood flow and skin temperature were quantitatively substantially higher after cold water stress in women who consumed long pepper extract than those who consumed a placebo. This suggests that long pepper extract could be used to prevent and treat peripheral coldness in women (Yoshino, 2018). The *P. longum* ethanol extract (10 mg/dose/animal) and piperine (1.14 mg/dose/animal) inhibits the development of solid tumors and increase the lifespan of rodents induced with Dalton's lymphoma ascites cells. At 250g/mL, piperine proved cytotoxic against Ehrlich ascites carcinoma cells and Dalton's lymphoma ascites (Anuradha et al., 2004). The antioxidant potential of *P. longum* against free radical-induced cellular damage is promising. Antioxidant activity is demonstrated by extracting petroleum ether of root and piperine from roots decreasing lipid peroxide levels and maintaining glutathione content (Natarajan et al., 2006). By the use of morphological, biochemical, and histopathological parameters, the hepatoprotective effect of plant fruit extract against carbon tetrachloride (CCl4) induced acute, chronic, reversible, and irreversible liver injury was evaluated in rats. The fruit extract promotes regeneration by inhibiting fibrosis but provides little defense against acute injury or cirrhosis (Koul and Kapil, 1993). The piperine defends against hepatotoxicity induced by tertiary carbon tetrachloride and butyl hydroperoxide by decreasing lipid peroxidation in *in vivo* and *in vitro* conditions (Christina et al., 2006).

6.3 Pharmacological Activities Mainly Antiviral

It was discovered that long pepper has significant anti-Hepatitis B Virus (HBV) activity. The Hepatitis B virus has circular and double-stranded DNA and is a member of the Hepadnavirus family. This virus causes acute and chronic liver infections. During the assessment search for Hepatitis B virus inhibitory compounds derived from long pepper ethanol-based extract, it was discovered to have superior anti-HBV activity in *in vitro* conditions. The combination of bioassay-guided fractionation and repeated purification led to the isolation of four novel compounds, including two new glycosides longumosides A and B, and two new amide alkaloids erythroside A (1) and erythroside B (1).-1-[1-oxo-9(3,4-methylenedioxyphenyl)-8,9-dihydroxy-2E-nonenyl]-piperidine, threo-1-[1-oxo-9(3,4-methylenedioxyphenyl)-8,9-dihydroxy-2E-nonenyl]-piperidine and two compounds, 3,4-dihydroxy-2-piperidinone and 5,6-dihydro-2(1H)-pyridinone derived from a natural source for the first time. In addition, using the Hep G 2.2.15 cell line, the anti-HBV activity of compounds 2–6 and the known compound Piperine were evaluated. Piperine exhibited extraordinary HBV inhibitory activity, inhibiting the release of hepatitis B virus surface antigen (HBsAg) and Hepatitis B e-antigen (HBeAg) (Jiang et al., 2013).

Hema and Mohan (2020) reported that the methanolic and chloroform extracts from *P. longum* seeds exhibited potent antiviral activity against vesicular stomatitis virus (VSV) and human para influenza virus (HPIV) on the HeLa cell lines. The fruits of *P. longum* are an essential component of the sura kudineer, a potential polyherbal siddha anti-viral drug (Priya and Kumari, 2017). Singh et al. (2021) evaluated the antiviral properties of *P. longum* against ribosomal inactivating proteins by suppressing the proteins in viral infections. The therapeutic qualities of focaccia bread may be improved by adding botanical components such as long pepper and *Plectranthus amboinicus*. Borage is employed as a medicinal agent to treat communicable ailments such as viral influenza and cough, owing to its antiviral, antibacterial, and antipyretic properties. Similarly, *P. longum* exhibits antibacterial effects and possesses immunomodulatory properties. Commonly referred to as Indian long pepper, Piper longum finds extensive usage in Siddha medicine for the management of cough and respiratory disorders. The upper respiratory infection primarily affects symptoms such as cough, sinusitis, sore throat, and pharyngitis. In addition, it can lead to the development of lung inflammation. It is suggested that this herb could be incorporated into one's regular dietary intake (Moyeenudin and Thiruchelvi, 2021). The identification of piperine as a potential drug candidate was achieved using computational docking simulation. In this study, we have chosen eight crucial proteins that play significant roles in the structure and function of Dengue fever and the Ebola viruses. These proteins were identified as potential targets for *in silico* docking with piperine, a bioactive compound. The profile of piperine in terms of absorption, distribution, metabolism, and excretion indicated that it exhibited a favorable safety profile and possessed notable characteristics resembling those of pharmaceutical drugs. The results of molecular dynamic modeling and binding free energy calculation demonstrated that piperine exhibited inhibitory effects on the Methyltransferase of the Dengue virus and the VP35 Interferon Inhibitory Domain (IID) of the Ebola virus, when compared to the commercial antiviral Ribavirin. Moreover, a comprehensive statistical analysis employing multivariate and clustering methodologies demonstrated that piperine exhibited a higher degree of attraction towards viral proteins compared to Ribavirin (Nag and Chowdhury, 2020).

A systematic *in silico* investigation was done on the papain-like protease of the COVID-19 virus and the chemical elements of long pepper. The pharmacokinetic properties of 32 plant-based constituents of long pepper were screened to aid in the selection of the most active constituents. Aristololactam, Fargesin, Lignans Machilin F, Piperundecalidine, l-asarinin, and Pluviatilolare were used for molecular docking after selection. The inhibitor-receptor complex is also subjected to molecular dynamic (MD) analysis, which suggests substantial inhibition and a firm complex of I-Asarinin with PLpro. The docking scores and modeling outcomes indicate that I-Asarinin is a potential drug candidate against Coronavirus disease (COVID-19) (Lakhera et al., 2021). The results of another study indicate that the addition of guduchi and pippali in Ayurvedic formulation has demonstrated a reduction in the duration of hospitalization and an improvement in the time taken for recovery. The Ayurvedic science add-on group exhibited improved levels of general well-being and activity during the 3-month follow-up period after leaving (Kataria et al., 2022). Long pepper phytocompounds on human receptor Angiotensin Converting Enzyme 2 (ACE2), and Trans-Membrane Protease Serine 2 (TMPRSS2) proteins have been studied. Vicenin 2, rosmarinic acid, and orientin were the most effective phytoconstituents in terms of molecular interaction and drug-like properties against host receptor proteins namely ACE2 and TMPRSS2. Using phytomolecules to activate ACE2 and TMPRSS2 against Severe-Acute-Respiratory-Syndrome-Related Coronavirus-2 (SARS-CoV2) allows for the logical design of future anti-COVID medications (Jindal and Rani, 2023).

6.4 Future of Research and Drug Delivery

P. longum holds a reservoir of alkaloids and related phyto-compounds, the most copious of which is piperine, alongwith brachystelma, iperonaline, dehydro pipernonaline piperidine, piperettine, brachystine, methyl piperine, pellitorine, piperidine, asinine, piperderidine, brachystelma-A, piperlongumine, piperine, refractomide A, piperlongumine, tetrahydro piper longumine, pipercide, longamide, pregumidiene, tetrahydropiperine, trimethoxy cinnamoyl-piperidine, and piperlongumine (Zaveri et al., 2010). With this array of molecules, researchers are exploring the use of synthetic biology and nanoparticle therapy to improve drug stability, enhance bio-availability, and target specific cells or tissues in the body.

7. Toxicity and Safety

Studies on the long-term use of high doses of piperine are lacking and, in general, are acknowledged to be safe and sound only in average doses. The fruits show contraceptive action in *in-vitro* models, hence, consumption of fruits during the female pregnancy and lactation needs to be prevented. Piperine interferes with the enzymatic drug biotransformation, inhibits hepatic UDP-glucuronyltransferase and aryl hydrocarbon hydroxylase, and alters the pharmacokinetic parameters of phenytoin and barbiturates (Meghwal and Goswami, 2013). The investigations on the single oral dose in 3g/kg body weight and 90 days of chronic toxicity laboratory animals, disclosed no harmful consequences. Investigations of the plant-isolated compounds on mice under *in-vitro* conditions reported median lethal dose (LD50) values of piperlongumine, piperine, and piperlonguminine (Kumar et al., 2011). Therefore, acute toxicity studies showed no morbidity or mortality when 3g/kg to 5g/kg is administered to laboratory models during pharmacologic studies, nevertheless, during pregnancy and lactation, *P. longum* fruits must be used carefully due to potential interactions. Indian long pepper may enhance the amount of propranolol, phenytoin, and theophylline absorbed by the body (Puri et al., 2021). Consuming an overdose of long pepper with these compounds may increase harmful effects and side effects. Nanoparticles can be loaded with drugs and designed to release them in a controlled manner, to improve efficacy and reduce ramifications. The targeted drug delivery systems can deliver medicines directly to the affected tissues or cells, minimizing damage to the healthy cells and reducing after-effects. Implantable devices, such as drug-eluting stents and microchips, are being developed to provide long-term and targeted drug delivery. These devices can be placed directly in the human body and programmed to release medications at specific times or in response to certain situations. They offer the potential for continuous drug administration and improved patient compliance. Also, Clustered regularly interspaced palindromic repeats (CRISPR/Cas) have opened up novel prospects to treat genetic disorders. Researchers are exploring ways to deliver gene-editing tools and therapeutic genes to target cells using various delivery systems, including viral vectors and nanoparticles.

Conclusion

Piper longum has noteworthy effects and consequences against human diseases including microbial infections, viral diseases, cancer and tumors, diabetes, indigestion, intestinal gas, stomachache, constipation, gonorrhea, diarrhea, heartburn, tongue paralysis, inflammation, depression, obesity, chronic malaria, cholera, viral hepatitis, spleen diseases, respiratory infections, lung problems including bronchitis, asthma and cough and hepatotoxicity. Moreover, plant is nontoxic, free from adverse effects but safe and effective in several disorders and no death has been reported with high doses of plant extracts. Hence, with the knowledge domain concentrated in this chapter, scientific exploration of various pharmacologically interconnected pathways can be taken by scientists. Although previous studies revealed a wide range of biological effects for piperine bioactive constituents. As a result, few in-vitro studies were conducted on animal models, however, many lacked proper experimental settings of doses and controls. As a result, forthcoming research is required to interpret the mechanism of bioactive constituents and their effects on antiviral properties in animal models and humans using the proper experimental protocols so that we can protect human health from a variety of viral diseases.

References

Anuradha, V., Srinivas, P.V. and Madhusudana Rao, J. (2004). Isolation and synthesis of isodihydropiperlonguminine. *Natural product research*, 18(3), pp. 247–251.

Babu, K.N., Divakaran, M., Ravindran, P.N. and Peter, K.V. (2006). Long pepper. In *Handbook of herbs and spices* (pp. 420–437). Woodhead Publishing.

Barboza, G.E., García, C.C., de Bem Bianchetti, L., Romero, M.V. and Scaldaferro, M. (2022). Monograph of wild and cultivated chili peppers (Capsicum L., Solanaceae). PhytoKeys, 200, p.1.

Bezerra, D.P., Pessoa, C., De Moraes, M.O., Saker-Neto, N., Silveira, E.R. and Costa-Lotufo, L.V. (2013). Overview of the therapeutic potential of piplartine (piperlongumine). *European Journal of Pharmaceutical Sciences*, 48(3), pp.453–463.

Bhavan, P.S., Saranya, C., Manickam, N., Muralisankar, T., Radhakrishnan, S. and Srinivasan, V. (2013). Effects of Piper longum, Piper nigram and Zingiber officinale on survival, growth, activities of digestive enzymes and contents of total protein, vitamins and minerals in the freshwater prawn Macrobrachium rosenbergii. *Elixir Biotechnology*, 58, pp.14824–14828.

Biswas, P., Ghorai, M., Mishra, T., Gopalakrishnan, A.V., Roy, D., Mane, A.B., Mundhra, A., Das, N., Mohture, V.M., Patil, M.T. and Rahman, M.H. (2022). Piper longum L.: A comprehensive review on traditional uses, phytochemistry, pharmacology, and health-promoting activities. *Phytotherapy Research*, 36(12), pp.4425–4476.

Choudhary, N. and Singh, V. (2018). A census of P. longum's phytochemicals and their network pharmacological evaluation for identifying novel drug-like molecules against various diseases, with a special focus on neurological disorders. *PLoS One*, 13(1), p.e0191006.

Christina, A.J.M., Saraswathy, G.R., Robert, S.H., Kothai, R., Chidambaranathan, N., Nalini, G. and Therasal, R.L. 2006. Inhibition of CCl4-induced liver fibrosis by Piper longum Linn.?. *Phytomedicine*, 13(3), pp.196–198.

Dey, P., Goyary, D., Chattopadhyay, P., Kishor, S., Karmakar, S. and Verma, A., 2020. Evaluation of larvicidal activity of Piper longum leaf against the dengue vector, Aedes aegypti, malarial vector, Anopheles stephensi and filariasis vector, Culex quinquefasciatus. *South African Journal of Botany*, 132, pp.482–490.

Gajurel, P.R., Kashung, S., Nopi, S., Panmei, R. and Singh, B. 2021. Can the Ayurvedic pippali plant (Piper longum L.) be a good option for livelihood and socio-economic development for Indian farmers? *Current Science* (00113891), 120(10).

Hazra, A. K., Chakraborty, B., Mitra, A. and Sur, T. K., 2019. A rapid HPTLC method to estimate piperine in Ayurvedic formulations. *Journal of Ayurveda and Integrative Medicine*, 10(4): 248–254.

Hema, R.N. and Mohan, S. 2020. A Scientific Review on Siddha Polyherbal formulation-Mudakkaruththaan Ilakam. *Journal of Research in Biomedical Sciences*, 3(4), pp.101–109.

Jiang, Z.Y., Liu, W.F., Zhang, X.M., Luo, J., Ma, Y.B. and Chen, J.J. 2013. Anti-HBV active constituents from Piper longum. *Bioorganic & medicinal chemistry letters*, 23(7), pp.2123-2127.

Jindal, D. and Rani, V., 2023. *In silico* studies of phytoconstituents from Piper longum and Ocimum sanctum as ACE2 and TMRSS2 inhibitors: strategies to combat COVID-19. *Applied Biochemistry and Biotechnology*, 195(4), pp.2618-2635.

Johri, R.K. and Zutshi, U. (1992). An Ayurvedic formulation 'Trikatu' and its constituents. *Journal of Ethnopharmacology*, 37(2), 85–91. https://doi.org/10.1016/0378-8741(92)90067-2

Kalamol, M.K., Mohanan, A. and Venkatesh, R.N. 2022. Medicated Porridge (Karkkidaka Kanji) as a Healthy Seasonal Food—A Review. *In Biology and Life Sciences Forum* (Vol. 18, No. 1, p. 46). MDPI.

Kataria, S., Sharma, P., Ram, J.P., Deswal, V., Singh, M., Rana, R., Singhal, R., Tripathi, A., Kumar, K. and Trehan, N. 2022. A pilot clinical study of an add-on Ayurvedic formulation containing Tinospora cordifolia and Piper longum in mild to moderate COVID-19. *Journal of Ayurveda and Integrative Medicine*, 13(2), p.100454.

Kaushik, R., 2018. Trikatu-A combination of three bioavailability enhancers. *International Journal of Green Pharmacy* (IJGP), 12(03).

Kaushik, R., Jain, J. and Khan, A.D. 2020. Trikatu: Transforming Food into Medicines. *Innovations in Food Technology: Current Perspectives and Future Goals*, 501–508.

Koul, I.B. and Kapil, A. 1993. Evaluation of the liver protective potential of piperine, an active principle of black and long peppers. *Planta medica*, 59(05), pp.413–417.

Kumar, S., Kamboj, J. and Sharma, S. 2011. Overview for various aspects of the health benefits of Piper longum linn. fruit. *Journal of Acupuncture and Meridian Studies*, 4(2), 134–140.

Lakhera, S., Devlal, K., Ghosh, A. and Rana, M. 2021. *In silico* investigation of phytoconstituents of medicinal herb 'Piper Longum'against SARS-CoV-2 by molecular docking and molecular dynamics analysis. *Results in Chemistry*, 3, p.100199.

Meghwal, M. and Goswami, T.K. 2013. Piper nigrum and piperine: an update. *Phytotherapy Research*, 27(8), pp.1121–1130.

Moyeenudin, H.M. and Thiruchelvi, R. 2021. The Antiviral and Antibacterial properties of Plectranthus amboinicus and Piper longum with the Addition to Focaccia Bread Nutritional Value and Sensory Evaluation. *Research Journal of Pharmacy and Technology*, 14(9), pp.4951–4956.

Nag, A. and Chowdhury, R.R. 2020. Piperine, an alkaloid of black pepper seeds can effectively inhibit the antiviral enzymes of Dengue and Ebola viruses, an *in silico* molecular docking study. *Virusdisease*, 31, pp.308–315.

Nair, K.P., 2023. Horse-Radish, Long Pepper and Galangal. *In A Compendium of Unique and Rare Spices: Global Economic Potential* (pp. 101-112). Cham: Springer Nature Switzerland.

Natarajan, K.S., Narasimhan, M., Shanmugasundaram, K.R. and Shanmugasundaram, E.R.B. 2006. Antioxidant activity of a salt–spice–herbal mixture against free radical induction. *Journal of Ethnopharmacology*, 105(1-2), pp.76–83.

Poddar, S., Sarkar, T., Choudhury, S., Chatterjee, S. and Ghosh, P. 2020. Indian traditional medicinal plants: A concise review. *International Journal of Botany Studies*, 5(5), pp.174–190.

Prasad, S. and K Tyagi, A. 2016. Historical spice as a future drug: therapeutic potential of piperlongumine. *Current Pharmaceutical Design*, 22(27), pp.4151–4159.

Priya, N.C. and Kumari, P.S., 2017. Antiviral activities and cytotoxicity assay of seed extracts of Piper longum and Piper nigrum on human cell lines. *International Journal of Pharmaceutical Sciences Review and Research*, 44(1), pp.197–202.

Puri, I., Pargotra, P.P. and Magotra, A. 2021. Role of AYUSH during COVID-19 crisis. *Journal of Ayurveda and Integrated Medical Sciences*, 6(5), pp.283–287.

Ramamoorthy, R., Muthalagu, M., Andra, S., Ravichandran, B. and Narayanasamy, M. 2019. Investigation on antimicrobial, antioxidant and cytotoxicity properties of triple bark extract formulated using traditional medicinal plants. *SN Applied Sciences*, 1, pp.1–7.

Salleh, W.M.N.H.W. and Ahmad, F. 2020. Phytopharmacological investigations of Piper retrofractum Vahl.–a review. *Agriculturae Conspectus Scientificus*, 85(3), pp.193–202.

Sharma, A.K., Kumar, S., Chashoo, G., Saxena, A.K. and Pandey, A.K. 2014. Cell cycle inhibitory activity of Piper longum against A549 cell line and its protective effect against metal-induced toxicity in rats.

Singh, A. 2018. Critical review on various ethonomedicinal and pharmacological aspects of piper longum linn. (long pepper or pippali). *International Journal of Innovative Pharmaceutical Sciences and Research*, 6(1), pp.48–60.

Singh, I., Krishna, R., Israr, J. and Kumar, A. 2023. Determination of piperine in ethyl acetate extract of piper longum root and evaluation of antioxidant, antibacterial activity. *Biochemical & Cellular Archives*, 23(1) 189–194.

Singh, R., Goel, S., Bourgeade, P., Aleya, L. and Tewari, D. 2021. Ayurveda Rasayana as antivirals and immunomodulators: potential applications in COVID-19. *Environmental Science and Pollution Research*, pp.1–27.

Singh, R., Rasheed, H., Ahmed, S., Singh, H. and Sharma, A. 2022. Most Modern Approach to the Phytochemical Evaluation and Use of Pepper Species in Ayurvedic Formulations. *In Chemistry, Biological Activities and Therapeutic Applications of Medicinal Plants in Ayurveda* (pp. 148–175). Royal Society of Chemistry.

Sivarajan, V.V. and Balachandran, I. (1994). *Ayurvedic drugs and their plant sources*. New Delhi, India: Oxford and IBH publishing.

Sunila, E.S. and Kuttan, G. 2004. Immunomodulatory and antitumor activity of Piper longum Linn. and piperine. *Journal of ethnopharmacology*, 90(2-3), pp.339–346.

Suwanphakdee, C., Simpson, D.A., Hodkinson, T.R. and Chantaranothai, P. 2016. Taxonomic notes on the genus Piper (Piperaceae). *Nordic Journal of Botany*, 34(5), pp.605–618.

Wadt, P.G.S. and da Silva, L.M. (2016). Sufficiency range and foliar position for evaluation of long pepper nutritional status. *Científica (Jaboticabal)*, 44(3), pp.431-438.

Yadav, V., Krishnan, A. and Vohora, D., (2020). A systematic review on Piper longum L.: Bridging traditional knowledge and pharmacological evidence for future translational research. *Journal of Ethnopharmacology*, 247, p.112255.

Yoshino, S. (2018). Effects of Long Pepper Extract on Peripheral Coldness in Japanese Healthy Women—A Randomized Double—blind Placebo—controlled Crossover Trial—. Pieronline, 46(2), pp.219–225.

Yurnadi, Y., Asmida, Y., Suryandari, D.A. and Moeloek, N. 2012. Combination of depot medroxy progesterone acetate and javanese long pepper extract on body weight, hematology, and blood biochemistry as a safe contraception model. *Makara Journal of Science*, pp.155–162.

Zaveri, M., Khandhar, A., Patel, S. and Patel, A. 2010. Chemistry and pharmacology of Piper longum L. *International Journal of Pharmaceutical Sciences Review and Research*, 5(1), pp.67–76.

scaber is widespread in tropical America, Asia, and Africa (Peer et al., 1980; Poli et al., 1992). It has long been valued as a medicinal herb in many countries in Southeast Asia, Latin America, and Africa (Hammer et al., 1993). Subtropical or tropical wet mountain woods are its native habitat. Of the thirty-two species in the genus, only the *E. scaber* is reported to flourish in India (Geetha et al., 2012). The plant is widely spread in the forests of Chhattisgarh, Assam, and Andaman and Nicobar Islands. The plant is abundantly found across the Western Ghats. Its common names include elephant's foot and prickly-leaved elephant foot.

5. A General Profile of Phytochemicals of *Elephantopus scaber*

The plant is a rich source of bioactive compounds, widely categorized as alkaloids, flavonoids, quinones, steroids, phenols, glycosides, terpenoids, and essential oils (Ahmed et al., 2009).

5.1 Phenolic Acids and Flavonoids

In nature, phenolic acids may be found free or bound in the form of esters and glycosides. Phenolics include more complex chemicals like flavonoids and tannins as well as small-molecular weight molecules such as coumarins, phenylpropanoids, and benzoic acid derivatives **(Fig. 2)**. The plant's ethanolic fraction contains a

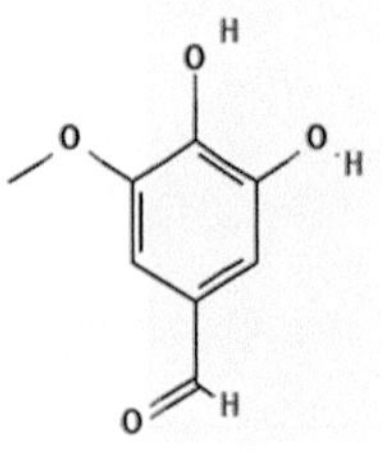

3,4-dihydroxy benzaldehyde

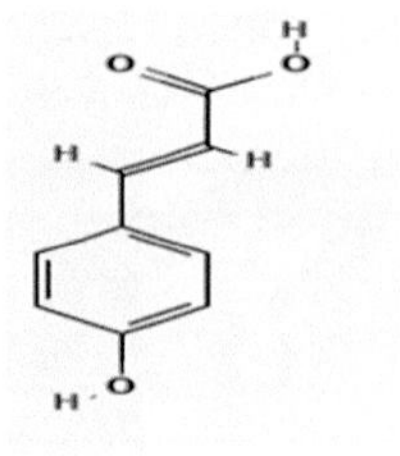

p-coumaric acid

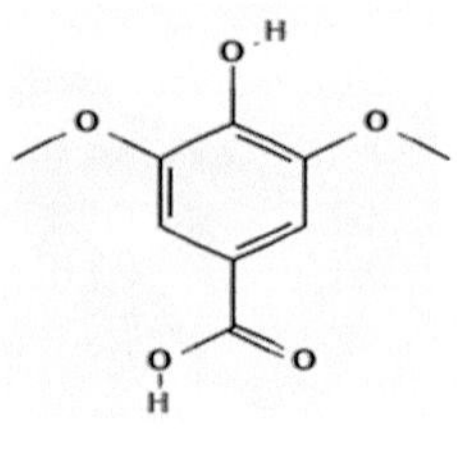

syringic acid

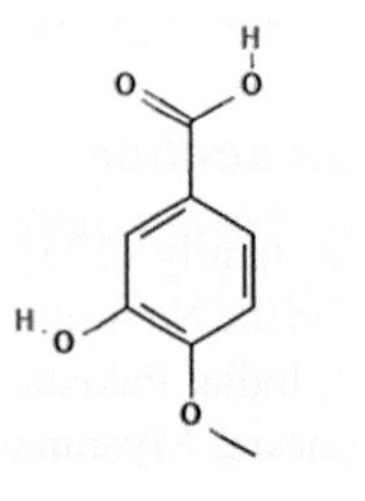

isovanillic acid

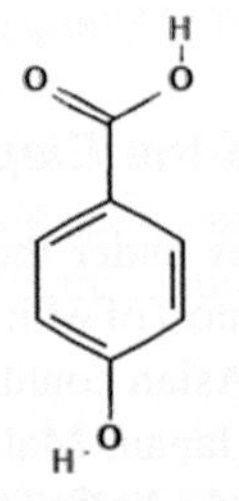

p-hydroxybenzoic acid

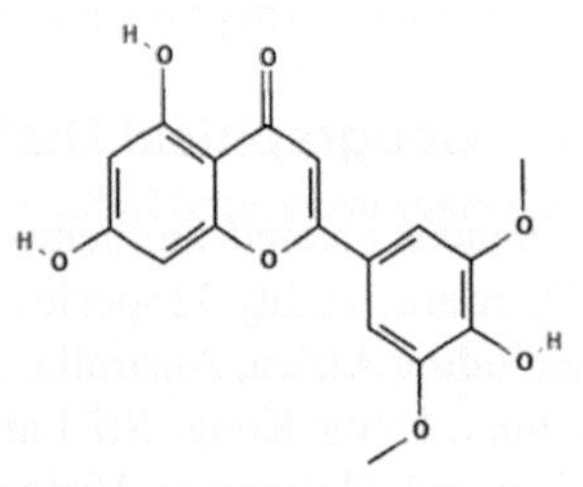

Tricin

Fig. 2 Bioactive Phenolic acids present in *E. scaber.* (*Source:* PubChem)

variety of phenolic chemicals. Lutein and its flavonoid glycosides, luteolin-4-O—D glucoside, and luteolin-7-Oglucuronide 6”-methyl ester were found in methanolic extracts of *Elephantopus scaber* aerial parts (**Figure 3**) (Zhang, 2011; Hisham et al., 1992).

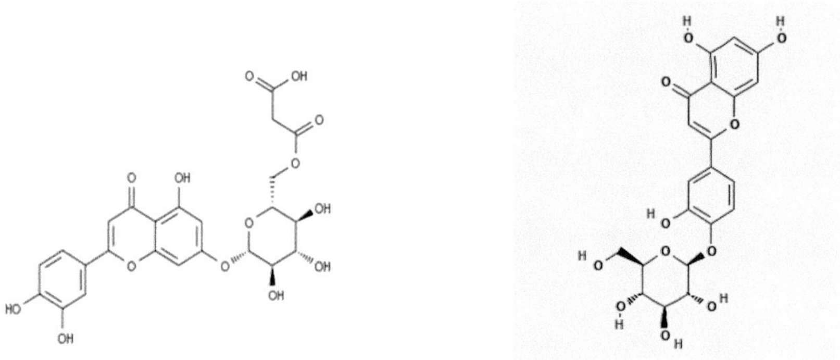

luteolin-7-Oglucuronide 6"-methyl ester | luteolin-4-O-β-D glucoside

Fig. 3 Bioactive flavonoids prssent *E. scaber.* (*Source:* PubChem)

5.2 *Terpenoids*

Terpenoids (or isoprenoids) are naturally occurring molecules derived from the 5-carbon compound isoprene and its derivatives known as terpenes and diterpenes. All terpenoids are formed by combining one or more isoprene molecules. Terpenes have several diverse functions in mediating favorable and antagonistic relationships between species and safeguarding a wide range of animals, plants, and microorganisms against predators, pathogens, and competitors. Sesquiterpene lactones, a family of naturally occurring terpenoids, are 15-carbon chemicals mostly found in the *E. scaber* (**Fig. 4**). They have a bicyclic, monocyclic, or linear structure. These lactones have a variety of pharmacological properties including antibacterial, antiviral, anti-inflammatory, and anticancer. The sesquiterpene lactones have been isolated from different extracts such as ethanol extracts, chloroform extracts, aqueous extracts, or acetone extracts of the whole plant or part of the plant. The isolation procedures used a variety of techniques to purify sesquiterpene lactones such as Column chromatography, high-performance liquid chromatography, thin-layer chromatography, and other gradient elution techniques. Structural stereochemistry of isolated sesquiterpene lactones has been elucidated by spectral tools such as Nuclear Magnetic Resonance (NMR), Mass Spectrometry, X-ray crystallography, and infrared spectra (IR) (Huang et al., 2010; Geetha et al., 2012; Xu et al., 2006; Su et al., 2009). Table 1 represents the vital sesquiterpene lactones found in the *E. scaber*.

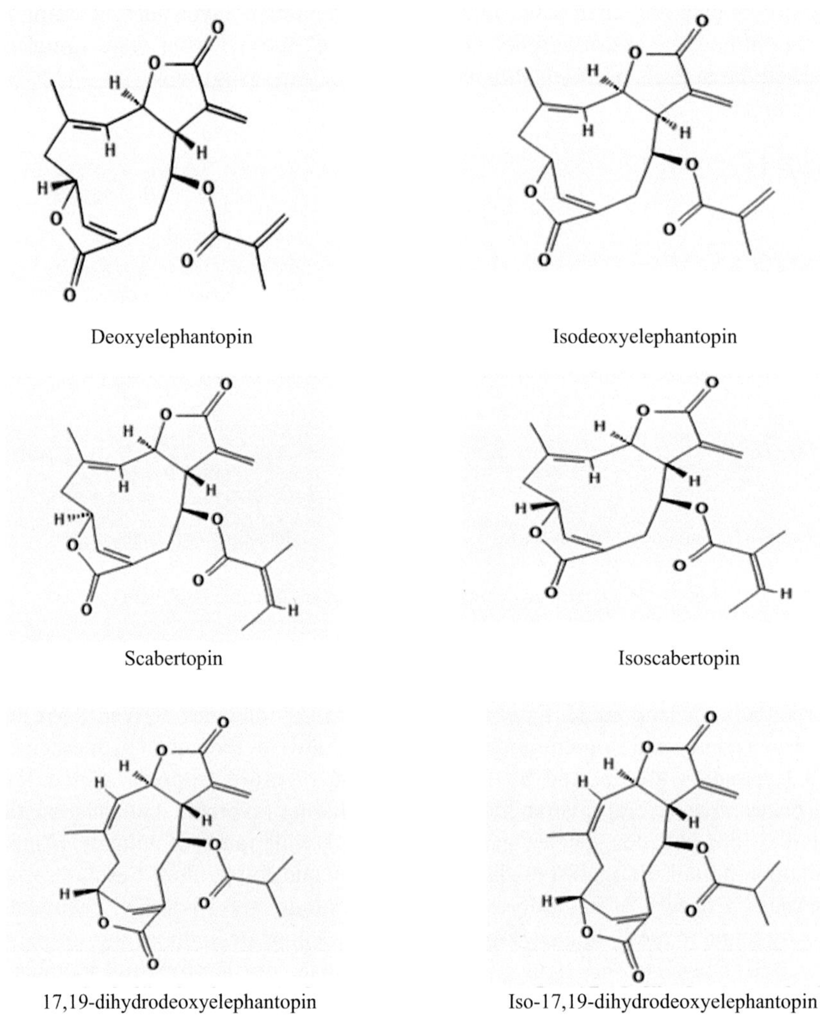

Fig. 4 Major bioactive sesquiterpenoids in *E. scaber.* (*Source:* PubChem)

5.3 Essential Oils

A concentrated hydrophobic liquid called an essential oil is made up of plant-derived volatile fragrance components. Terpenes, especially monoterpenes and sesquiterpenes, make up the majority of the molecules in essential oils. To extract the essential oils, the entire *E. scaber* plant was steam-distilled with Clevenger-type equipment, yielding a transparent, yellowish essential oil (**Fig. 5**). Further, analytical procedures have helped to identify 21 different components in this essential oil (Table 1) (Wang et al., 2004).

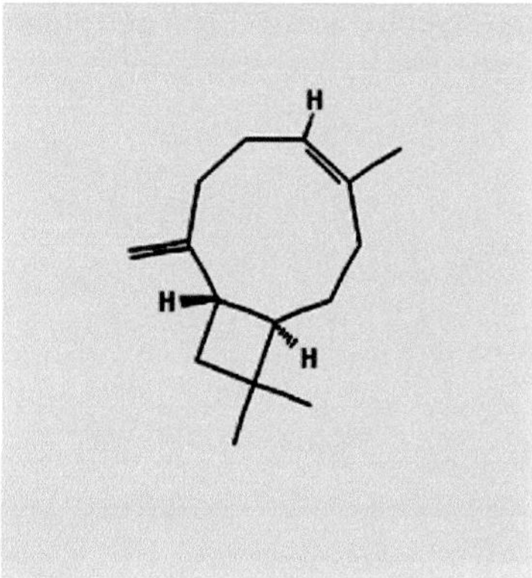

Isocaryophyllene

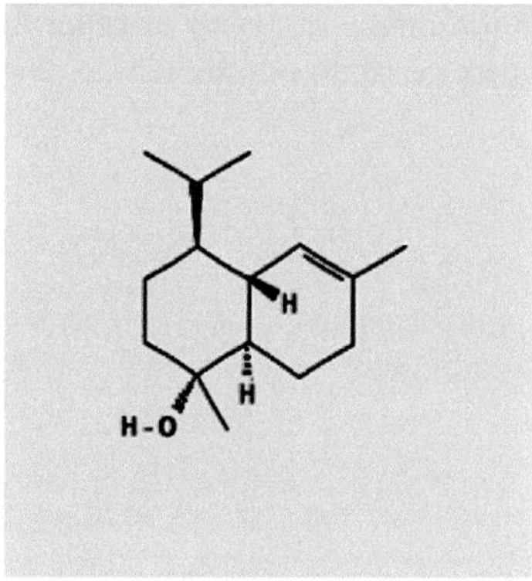

Cadinol

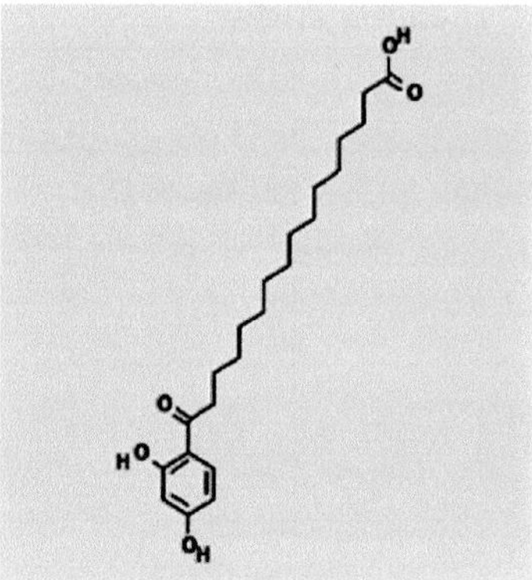

Hexadecanoic acid

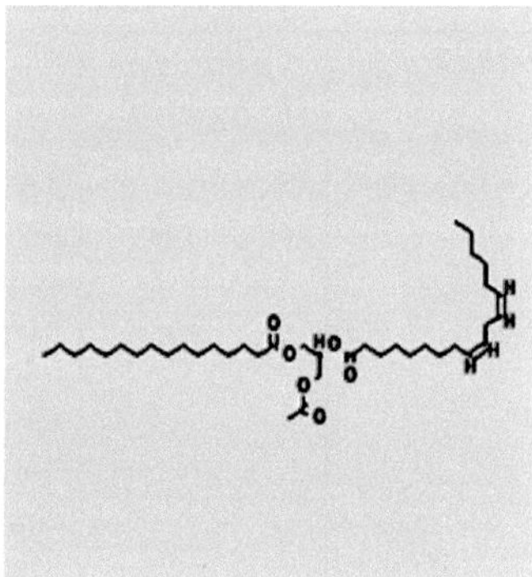

Octadecadienoic acid

Fig. 5 Major essential oils and hydrocarbons present in *E. scaber* (Source: PubChem)

5.4 Hydrocarbons

Cluster analysis of the oils extracted from *E. scaber* leaf has helped to classify them into two categories, one high in octadecadienoic acid and hexadecenoic acid and the other high in the five alkane homologs. Further analysis of dried plant leaf samples with dichloromethane produced 24 different compounds. Among them, 20 compounds were detected and identified by utilizing mass spectrometry (Wang et al., 2005). The major hydrocarbons are mentioned in Table 1.

5.5 Steroids and other Constituents

The extraction of *Elephantopus scaber* whole plant with ethanol, mild petroleum, and methanol, has helped to identify the presence of stigmasterol and lupeol.

Additionally, a group of other compounds were detected using light petroleum ether extraction of the whole plant. (**Fig. 6**).

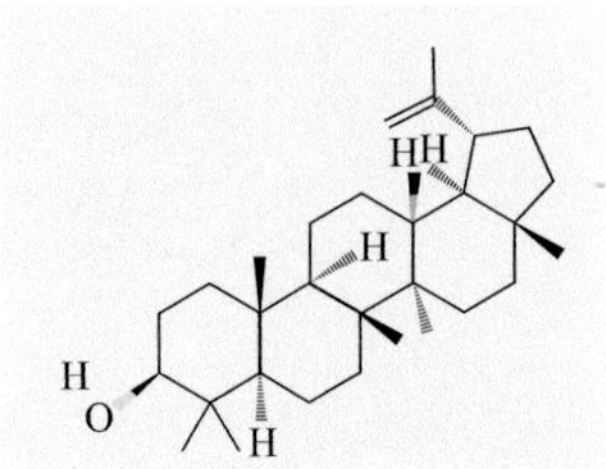

Lupeol

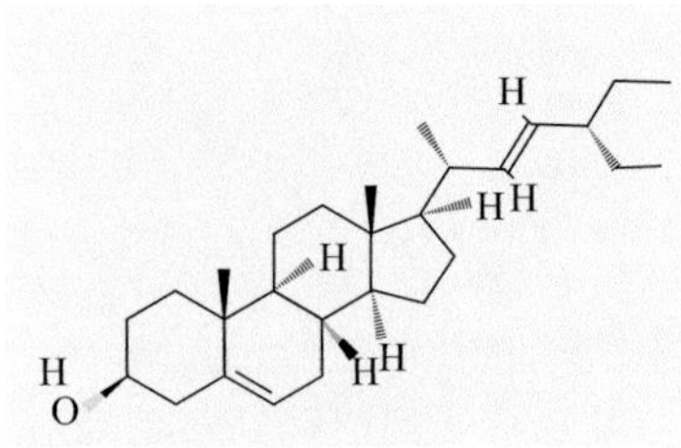

Stigmasterol

Fig. 6 Bioactive sterols present in *E.scaber.* (*Source:* PubChem)

Table 1 Pharmacological profile of *Elephantopus scraber.*

Plant Part	*Bioactive Compounds*	*Sub-Compounds*	*References*
Whole plant	Phenolic compounds	3,4-dihydroxy benzaldehyde p-coumaric acid syringic acid, isovanillic acid p-hydroxybenzoic acid ferulic acid, vanillic acid 3-methoxy-4-hydroxyl cinnamic aldehyde, tricin, syringic acid 2-hydroxybenzolate acid	Zhang, 2011 Hisham et al., 1992
Aerial part	Flavanoids	luteolin-7-Oglucuronide 6”-methyl ester luteolin-4-O-β-D glucoside	Chang. et al., 2012
	Terpenoids	Deoxyelephantopin Isodeoxyelephantopin Scabertopin Isoscabertopin Elescaberin 17,19-dihydrodeoxyelephantopin Iso-17,19 dihydrodeoxyelephantopin 11,13-dihydrodeoxy-elephantopin	Huang et al., 2010 Geetha et al., 2012 Xu et al., 2006 Su et al., 2009 Liang et al., 2008 Than et al., 2005 Desilva et al.,1982 Chang et al., 2011

Contd.

Table 1 *Contd.*

Plant Part	*Bioactive Compounds*	*Sub-Compounds*	*References*
Whole plant	Essential oils	Cyclosativene Copaene Isopropyl dimethyl hexahydronaphthalene Trimethyl dimethylenedecahydronaphthalene Zingiberene B-caryophyllene Caryophyllene Dimethyl-6-(4- methyl-3-pentenyl)-2-norpinene B-sesquiphellandrene Isocaryophyllene A-santalol Ledol A-bisabolol Caryophyllene oxide B-bisabolol Isopropyl dimethyl tetrahydronaphthalenol Cadinol Hexahydrofarnesyl acetone Hexadecanoic acid Phytol Octadecadienoic acid	Wang et al., 2004
Leaves	Hydrocarbons	Hexadecanoic acid (8.19–39.22%); Octadecadienoic acid (trace –29.22%); Alkane homologues, i.e., n-tetradecane, n-pentadecane, n-hexadecane, n-heptadecane and n-octadecane	Wang et al., 2005
Whole plant	Steroids	Lupeol Lupeol acetate Ursolic acid Stigmasterol	Chang et al., 2011 Than et al., 2005; Su et al., 2009

6. Therapeutic Uses of *Elephantopus scaber*

E. scaber has been traditionally used in many nations to cure several illnesses. Different sections of the plant have been shown to possess essential pharmacological constituents. Antibacterial, antioxidant, hepatoprotective, antidiabetic, analgesic, anti-inflammatory, antiasthmatic, wound healing, and antiplatelet activities are among the various pharmacological functions and actions of the plant that have been shown in a number of research. These pharmacological properties of *the E. scaber* are discussed below.

6.1 Use of Elephantopus scaber in traditional systems of Medicine

In many countries, the *E. scaber* is used as a traditional medicine to cure or manage various human diseases (**Fig. 7**).

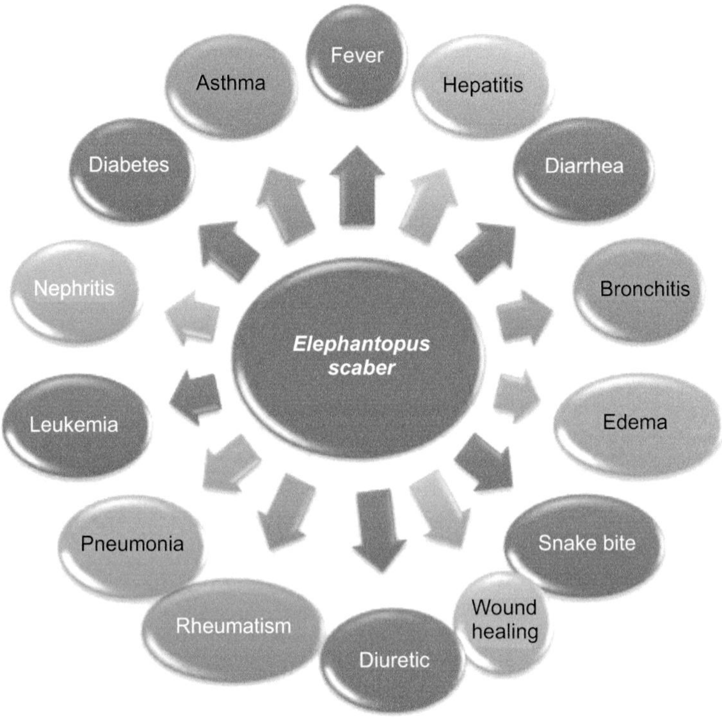

Fig. 7 Uses of *E. scaber* in various Pathologies.
(*Source:* Generated in Microsoft Power Point)

In Brazil, the entire plant is used to induce urination, lower fever, remove bladder stones, and fight cough, bronchitis, and asthma. Root decoction has been used in Malaysia to hasten abdominal contractions and reduce inflammation during delivery (Inta et al., 2008). In Thailand, people use *E. scaber* as a galactagogue, tonic, and remedy for chapped lips. In Madagascar, the plant's aerial portion has also been utilized as an antipyretic. Taiwanese use whole plant extract to cure hepatitis; in Mauritius, people use it to treat diarrhoea, urinary issues, and acne; and Nigerians use a hot water leaf extract to treat arthritis (Sudip et al., 2018). In India, the whole *E. scaber* plant is used as a toothbrush to get rid of toothaches. Furthermore, the entire plant has been said to be useful in treating several ailments, including snake bites, rheumatism, leukemia, diabetes, pneumonia, and filariasis. The aqueous extract of the leaves is administered topically to cure dermatitis and ulcers. The leaf is also crushed and combined with salt to cure diarrhoea (Achuta et al., 2010). Table 2 below summarizes the various uses of different parts of the plant in folklore medicine across the world.

Table 2 Folklore uses of the plant, *E. scaber* in different parts of the world.

S. No.	*Plant Part*	*Country*	*Folklore Uses*	*References*
1	Whole Plant Decoction	Columbia, Brazil, Southern China, India, Hong Kong, Taiwan	Bronchitis, fever, pnuemonia arthralgia, toothache, insomnia, bladder stones diuresis, snake bites, bacterial infection, leukaemia, rheumatism, hepatitis, nephritis, galactogogue, fliariasis, chapped lips, flatulence, diabetes, scabies, edema	Achuta et al., 2010; Peer and Metzger 1980; Kabiru and Por, 2013
2	Root Decoction	Malaysia, Mauritius, Thailand	Accelerate contraction of the abdominal area and prevent inflammation after childbirth, as a tonic, urinary problems, pimples, bloody stool	Inta et al., 2008
3	Aerial Decoction (Leaves)	Madagascar, Nigeria, India	Eczema, ulcers, asthma, dysentery, arthritis	Daniel et al., 2006; Jasmine and Daisy, 2008; Achuta et al., 2010

6.2 Potential Therapeutic Effect of Elephantopus scaber

The potential therapeutic effects of *E. scaber* in various human diseases supported by different clinical investigations are discussed below.

6.2.1 Antiasthmatic Activity

E. scaber leaf extracts were tested in guinea pigs for preliminary phytochemical screening and their ability to treat asthma. Animals were given acetylcholine and histamine to cause tracheal constriction and bronchospasm. It was discovered that the plant's ethanolic extraction considerably reduced the bronchospasm and trachea constriction in a concentration-dependent manner and further inhibited the degranulation of mast cells. It is presumed that flavonoids or steroids found in the plant may be responsible for the antiasthmatic properties found in the ethanolic extract of the *E. scaber* leaf because the flavonoids and steroids are believed to have anticholinergic, antihistaminic, and mast-cell stabilizing properties (Sagar and Sahoo, 2012).

6.2.2 Anticancer Activity

These studies suggest *E. scaber* extracts may be effective in treating colon cancer. Using mammalian cell lines, Chan et al. (2015) evaluated the *E. scaber* extract and its fractions for anticancer activities. The proliferation of HT-29 and HCT116 cells was shown to be considerably decreased by the plant extract and further caused apoptosis in these cells. The best results in the HCT116 cell line were shown by the ethyl acetate fraction of the *E. scaber*. As per these mechanistic investigations, the plant extracts induced apoptosis by upregulating proapoptotic proteins, lowering the potential of the mitochondrial membrane, cleaving poly (ADP-ribose) polymerase (PARP), and activating caspase-3. The participation of the mitochondrial signalling system was also shown by a weakened mitochondrial membrane potential and an excess of reactive oxygen species. Additional studies showed that plant fractions increased intracellular reactive oxygen species (ROS), which prompted a hike in the expression of p53 protein. The treatment with plant fractions also caused Bax to oligomerize, depolarize the membrane potential of mitochondria, and trigger a cascade of caspases in a time-dependent manner. N-acetyl-l-cysteine (NAC), which reduces intracellular ROS levels, protects the mitochondrial membrane's integrity, and prevents cell death (**Fig. 8**). The caspase cascade causes the PARP cleavage, promoting DNA fragmentation and finally leading to apoptosis (Chan et al., 2015).

6.2.3 Antioxidant and Anti-inflammatory Activity

Researchers have examined the processes underlying the *E. scaber's* anti-neuroinflammatory and antioxidant effects. It has been shown that treatment of lipopolysaccharide (LPS)-induced microglia cells (BV-2) with *E. scaber* leaf extract decreased the expression of anti-inflammatory molecules such as interleukin-1 (IL-1), Nuclear factor beta (NF-B) and tumour necrosis factor (TNF). Additionally, it was shown that leaf extract inhibited the inflammation in BV-2 cells via NF-B signalling, and oxidative stress (induced by LPS) was reduced by activating the multifunctional regulatory factors such as HO-1 and Nrf2 and enhanced antioxidants mediated free radical neutralisation. The toxicological tests have further indicated that the leaf extract is safe for use in medicine with no adverse effects seen at a dose of 2000 mg/kg body weight of mice (Chan et al., 2017).

6.2.4 Antidiabetic Activity

The effects of the *E. scaber* crude extract and its acetone fractions on diabetes were investigated in streptozotocin (STZ) treated diabetic rats. The crude extract showed a significant hypoglycaemic effect in these mice when administered orally

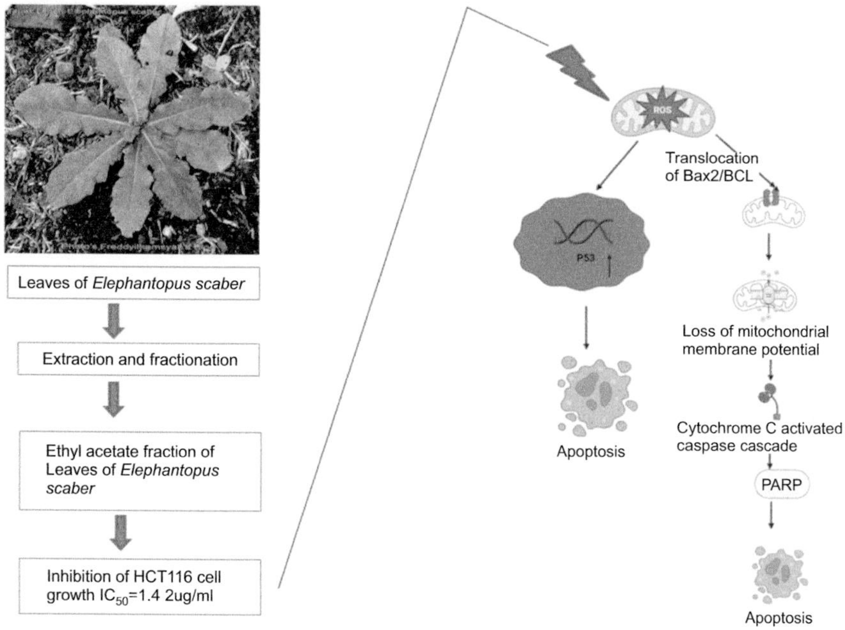

Fig. 8 Molecular mechanism by which *E. scaber* promotes cell growth inhibition and cell death by apoptosis (*Source:* Generated in BioRender)

at a dose of 150 mg/kg. There was also an improvement in the declining insulin levels in these mice (Daisy et al., 2009). On the other hand, the acetone fractions obtained from the *E. scaber* showed a time-dependent reduction in blood glucose levels. The initial phytochemical analysis of *E. scaber* fractions suggested that terpenoids may be responsible for this hypoglycaemic effect. In a separate study, rats with alloxan-induced diabetes were given an oral dose of the plant's aqueous extract derived from its leaves and roots, which resulted in a significant reduction in blood sugar, glycated haemoglobin, and enzymes associated with gluconeogenesis. According to a report, 2,623 triptolides and terpenoids have been extracted from the whole plant. These compounds are believed to play a role in the anti-diabetic properties of the *E. scaber* (Rao et al., 2007; Modilal and Daisy, 2011).

6.2.5 Hepatoprotective Activity

The hepatoprotective properties of *E. scaber* were studied in mice. The mice whose liver damage was induced by ethanol were given the plant extract and investigated further for the ability of *E. scaber* extracts to repair liver damage. S ignificant healing was seen in these animals. This finding substantiates the historical use of the plant as a liver tonic (Wan et al., 2012).

In another study, it was found that administering methanol extract of *E. scaber* to rats with carbon tetrachloride-induced liver damage resulted in the reversal of that condition (Sheeba et al., 2012). Additionally, the aqueous extract of the plant has shown hepatoprotective effects in rats with previously induced liver damage caused by lipopolysaccharide treatment. Further, it is believed that the *E. scaber* protects against LPS-induced acute hepatic damage in SD mice by an antioxidant action, suppression of the p38 MAP kinase pathway, and cyclooxygenase (COX-2) production (Hung et al., 2011).

6.3 Antimicrobial Activity

Lupeol, a pharmacologically active triterpenoid derived from the *E. scaber* has been shown via molecular docking studies to inhibit the activity of autolysin. The autolysin enzymes are involved in peptidoglycan turnover by enzymatic hydrolysis of the cell wall polymer network to generate sites for the insertion of new cell wall building blocks (Dubée et al., 2011). Dysregulation of cell wall autolysin leads to bacterial cell lysis (Darnell et al., 2019). Staphylococcus aureus is killed when the triterpenoid molecule makes strong contact with the autolysin's active site residues and deactivates its enzymatic activity (Vergese et al., 2011). In an investigation, the disc diffusion method was used to examine the antibacterial activity of *E. scaber* Linn's methanolic extract. Petri dishes with 10 ml of Muller Hinton agar media were chosen with a 24-hour bacterial strain culture. Two sterile filter paper discs (6mm) containing 100 g of leaf extract each disc were placed on the medium's surface. The methanolic extract of *E. scaber* leaves possessed significant antibacterial activity against *Staphylococcus aureus* (29 mm), *E. coli* (27 mm), *Pseudomonas aeruginosa* (23 mm), *Bacillus subtilis* (29 mm) and *P. vulgaris* (25 mm) at 100 μg disc-1 as determined by the measurement of the zone of inhibition shown around the discs (Perumal et al., 2004). The extract also inhibited the growth of *Bacillus megaterium*, *Xanthomonas campestris,* and *E. coli* (Ammani et al., 2011). As compared to chloroform and petroleum ether extracts, methanol extract was found to be more efficient against bacteria such as *Staphylococcus aureus*, *Salmonella paratyphi* A, *Klebsiella pneumonia*, *Pseudomonas aeruginosa*, *Salmonella sonnei*, *E. coli*, and *Salmonella typhimurium* (Jenny et al., 2004). The strong antibacterial activity appears to back up this plant's historic uses as a treatment for infections and other ailments.

Some studies have investigated the impact of aqueous and methanolic extract of *E. scaber* on pathogenic fungi *Aspergillus niger*, *Aspergillus flavus, Rhizopus indicus*, and *Mucor indicus.* These studies have indicated the antifungal activity in methanolic extract increased in a dose-dependent manner and was highest against *Mucor indicus* (32 mm) and lowest against *Rhizopus indicus* (14 mm). The inhibition zone is measured in diameter units by the disc diffusion method (Kamalakannan et al., 2012).

6.4 Wound Healing Property

The wound healing property of Deoxyelephantopin was tested in comparison to the ethanolic extract of *E. scaber*. The results have demonstrated that

Deoxyelephantopin is more effective in wound healing. It promotes cellular proliferation, formation of granulation in tissue, synthesis of collagen, and increases the rate of wound contraction (Singh, 2012). This may be because deoxyelephantopin contains an active moiety called methylene lactone. An aqueous extract of E. scaber is taken orally to treat wounds. (Basha et al., 2013).

6.5 Antiviral Pharmacological Activities of Elephantopus scaber

E. scaber is a rich source of various phytochemicals believed to contribute to the antiviral properties of the plant. Many biological assays have been carried out to establish the antiviral properties of the plant. These assays include reduction assay, neutralization assay, cytopathic assay, and various other inhibition assays. Several researchers have investigated the different extracts of *E. scaber in vitro* to test their antiviral properties against the various types of viral infections. Some of these antiviral properties of the plant are discussed below.

6.5.1 Anti-Herpes Simplex Virus (anti-HSV) Activity of Elephantopus scaber

The herpes simplex virus (HSV) causes painful blisters and ulcers in humans. The virus primarily spreads by skin-to-skin contact. This virus has two types, HSV-1 and HSV-2 (WHO report 2015). The HSV-1 causes sores in the mouth or around the lips known as fever blisters or cold sores. HSV-2 typically causes vaginal sores. Both viral types can spread to the mouth, genitalia, and other body areas. HSV can spread easily through direct contact with the mucous membranes, saliva, or fluid from the open sores of an infected person (WHO report 2020). Even if the infected person doesn't have visible sores or other obvious symptoms of infection, HSV can still be transmitted. Herpes cannot survive for very long outside of the body. HSV infection is being treated with antiviral medication e.g. acyclovir (ACV). The drug specifically inhibits viral DNA polymerase (Crumpacker et al., 1982). The cost of ACV therapy is high and its prolonged use may result in the development of HSV strains that are resistant to the drug. Such long-term therapy can cause mutations in the HSV-thymidine kinase or DNA polymerase genes, which are particularly concerning as they can lead to opportunistic infections in people with weakened immune systems (Elion et al., 1993) However, the search for new antiviral drugs has been simplified by testing plant extracts for their effectiveness against HSV (Crumpacker et al., 1982). The extracts of the *E. scaber* were investigated for antiviral activity. The study showed that an ethanol extract of *E. scaber* had anti-herpes simplex virus action on African green monkeys (GMK). HSV-1 strain F and HSV-2 strain G were cultivated on GMK cells to cause infection. A plaque titration experiment was used to determine the toxicity of the ethanol extract of roots and leaves in GMK cells. The ethanol extract's cytotoxicity was evaluated to have a 50% dosage of 4.97 mg/mL, corresponding therapeutic indices (TI) of 0.35 for HSV-1 and 0.54 for HSV-2 (Raenu and Yingmanee, 2015).

To ascertain the method of antiviral activity of *E. scaber*, viral adsorption and viral replication assays were carried out. The therapeutic index (TI; CD50/ED50) of the ethanol extract against HSV-1 was 14.14 g/mL, and that for HSV-2 was 9.12 g/mL and 0.54 g/mL (Raenu and Yingmanee, 2015). This shows that *E. scaber* extract is effective during viral adsorption and that its inhibitory activity is greater against HSV-2 than against HSV-1 (Raenu and Yingmanee, 2015). The mode of HSV inhibition by the *E. scaber* is depicted in **Fig. 9**.

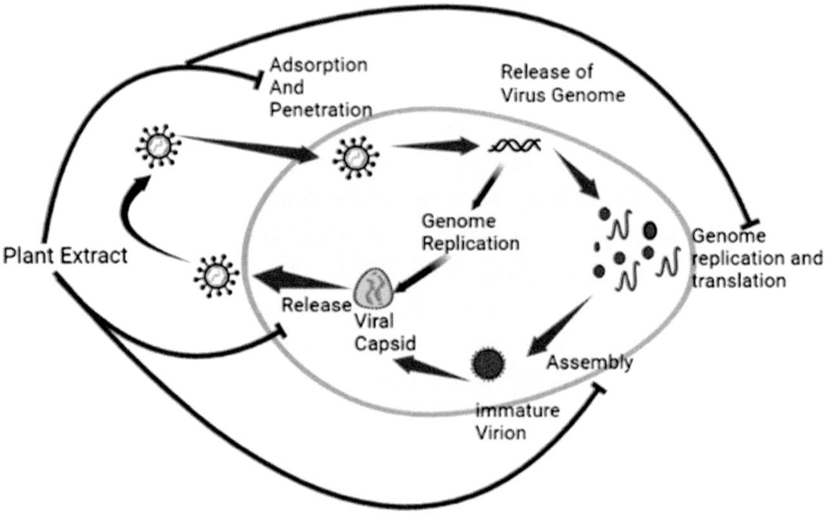

Fig. 9 Model depicting the Inhibitory Mechanism of *E. scaber* extract on Viral Growth. (*Source:* Generated in BioRender)

6.5.2 Anti-HIV Activity of Elephantopus scaber

The anti-HIV properties of the water-soluble extracts of the leaves and roots of E. scaber were examined. The initial screening of plant extracts for their ability to inhibit HIV-1 reverse transcriptase was conducted at a concentration of 200 μg/ml. The inhibition assay for HIV-1 reverse transcriptase with the plant extract was carried out using the radiometric method (Wiwat and Kwantrairat, 2013). It was found that plant extract exhibited a high inhibitory effect on HIV-1 reverse transcriptase with more than 90% inhibitory response. Compared to the ethanolic extract, the aqueous extract of the plant proved to be more effective in inhibiting the viral enzyme. The maximal inhibitory concentration was 69.9 g/ml in the leaves and 107.57 g/ml in the roots. The inhibitory enzyme mode observed wasn't non-competitive as determined by Lineweaver Burk plots (Wiwat and Kwantrairat, 2013). Further, to identify the factor that inhibits the HIV reverse transcriptase, the aqueous fraction of *E. scaber* was subjected to ammonium sulphate precipitation by using varying amounts of salt. Protein separation was accomplished using SDS-PAGE and gel elution (**Fig. 10**).

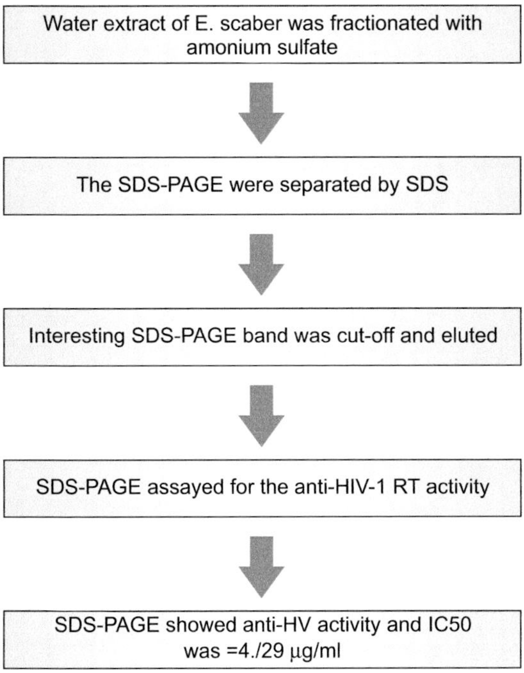

Fig. 10 Schematic of isolation of antiviral protein from *Elephantopus scaber.* (*Source:* Generated in Microsoft Power Point)

The purified proteins were tested against anti-HIV-1 reverse transcriptase activity, which revealed that one of the purified proteins has anti-HIV activity at IC50 = 4.29 g/ml. The protein has been further characterized whose isoelectric point (pi) and the molecular weight was determined and was found to be 4.65 and 34.5 kDa respectively (Wiwat and Kwantrairat, 2013). Further, the proposed mechanism of the mode of action of *E. scaber* extracts on HIV is depicted in **Fig. 11**.

6.5.3 Anti-human Respiratory Syncytial Virus (anti-hRSV) Activity of Elephantopus scaber

Human respiratory syncytial virus (hRSV) is a virus with a single-stranded, negative-sense RNA that primarily targets the lower respiratory tract in infants and young children. Unfortunately, the immune response to RSV infection is not effective in providing lasting protection, resulting in persistent RSV infections throughout adulthood (Huntley et al., 2002). There is currently no cure for RSV infection. RSV in high-risk individuals may only be treated and prevented with

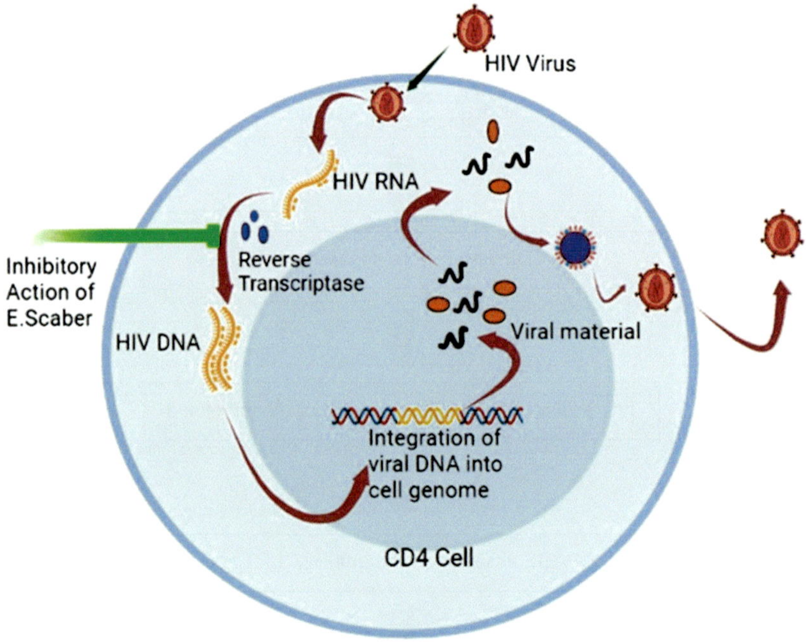

Fig. 11 Mechanism of action of *Elaphantopus scaber* on HIV Growth. (*Source:* Generated in BioRender)

the help of two currently authorized medications: ribavirin and RSV-specific neutralizing antibody (RSVIG) preparations that have been effective in treating RSV infections, but their administration is neither simple nor affordable (Wyde et al., 1993).

The use of ribavirin is limited by concerns regarding its efficacy and safety, as well as the complicated and time-consuming aerosol inhalation regimen required for administration. RSVIG usage can provide passive protection against RSV infections. However, there are high expenditures involved, and RSVIG's therapeutic effectiveness has not yet been shown (Kneyber et al., 2000). Therefore, it is crucial to develop novel, effective, and safe therapeutic options to address HSV and RSV infections. Medical herbs may be useful sources for the identification of powerful natural antiviral medicines since they have historically been used in the battle against infectious illnesses.

Twenty-one herbs, including *E. scaber*, were examined by Li et al. (2004) for anti-RSV screening tests. All the herbs that were tested showed moderate to strong anti-RSV effectiveness. The aqueous extract of *E. scaber*, in particular, was found to possess anti-RSV properties, with IC50 values below 50 g/ml. These findings may explain why these herbs have been traditionally employed in the treatment of respiratory viral infections.

The effect of polyphenolic components in the potent herbs was assessed for their anti-RSV activity. Plant extracts were separated using polyamide chromatography, revealing that many aqueous plant extracts lost their anti-RSV activity after treatment with a polyamide column. However, the processed extracts from *E. scaber* still maintained anti-RSV effectiveness, with IC50 values of 37.5. The findings revealed that the herbs' anti-RSV effect may potentially be attributed to certain other types of ingredients than polyphenolic chemicals (Li et al., 2004). A bioassay-guided isolation of *E. scaber* was carried out in an investigation carried out by Geng et al. (2011), yielding five dicaffeoyl derivatives, including a novel chemical, 1a,2b-O-dicaffeoylcyclopentan-3b-ol from its rhizomes (**Fig. 12**). All of the dicaffeoyl compounds had strong anti-RSV efficacy *in vitro*. Based on the findings, the novel compound's structure was elucidated through spectroscopic analyses. A cytopathic effect reduction assay, *in vitro* antiviral activity of these dicaffeoyl derivatives against the respiratory syncytial virus (RSV) was carried out. The results showed stronger anti-RSV activity than the positive control drug, ribavirin (Geng et al., 2011).

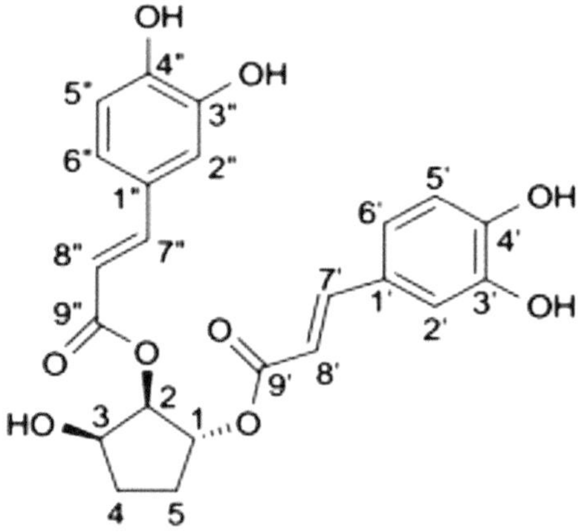

Fig. 12 Structure of 1a,2b-O-dicaffeoylcyclopentan-3b-ol. (*Source:* PubChem)

7. Toxicity and Safety of *Elephantopus scaber*

Several studies were conducted to evaluate the potential toxicity of *E. scaber.* According to Poli et al. (1992) and Sankar et al. (2001), an oral dose of 6 g/Kg of aqueous and ethanolic extract of the whole plant of *E. scaber* was given mice models which did not result in any adverse effects or fatalities. In contrast, when the same extracts with different dosage concentrations, the highest being 3mg/kg were administered through the intraperitoneal route, it induced different pathologies such as myopathy, athetosis, impaired balance and coordination, and even led to the death of mice. Thus the result indicated that the LD50 values of the extracts, whether administered intravenously or orally, were higher than 2 and 6 g/kg, respectively, indicating a low level of acute toxicity (Poli et al., 1992; Sankar et al., 2001).

Singh et al. (2005) reported that the isolated chemical compound, deoxyelephantopin, had a maximum tolerable dose of 40 mg/kg body weight of rat, while the aqueous and ethanol extracts of *E. scaber* had a maximum tolerated dose of 3 g/kg body weight (Singh et al., 2005).

Other studies have evaluated the toxicity of *E. scaber* using various plant extracts, and they have consistently shown that the plant is non-toxic even at significantly higher concentrations ranging from 2000 mg to 6000 mg (Daisy and Jasmine, 2009; Battu et al., 2012; Sagar and Sahoo, 2012).

In one more investigation, the toxicity of ethanolic extract of *E. scaber* leaves was assessed using mice models. The experimental group of mice was orally given a confined dose of 5000 mg/kg of body weight. No mortality was observed in the experimental group even after two weeks of treatment. Additionally, no significant difference in body weight was observed between the experimental group (treated with plant extract) and the control group (untreated). These results suggest that *E. scaber* extract may be considered safe and non-toxic for oral ingestion up to 5000 mg/kg body weight in mice (Ho et al., 2012).

8. Conclusion and Future Prospectives

The renewed interest in herbal remedies can be traced back to the long-standing tradition of using plants as a source of medicine. This interest has been fueled by the emergence of new diseases, the ineffectiveness of current medications, and the adverse effects of synthetic pharmaceuticals. The ethnomedicinal claims of a small herb namely *E. scaber* which almost grows in all parts of the world have been scientifically demonstrated by researchers through various in-depth studies. The primary pool of sesquiterpenoids, a subclass of terpenoids, is one of the several phytochemicals that provide the plant its therapeutic qualities. These phytochemicals from the different parts of *E. scaber* have been examined for therapeutic potential against several diseases. This chapter attempted to offer insights into traditional usage, including its antiviral properties in particular as well as its antitumor, antimicrobial, antihepatotoxic, antidiabetic, antioxidant, anti-inflammatory, and antiasthmatic effects.

Further research is needed to confirm and determine *E. scaber's* therapeutic significance in different human disorders. It is anticipated that this plant may provide new lead compounds with antiviral and anticancer properties. Due to numerous scientific articles that continually emphasize its biochemical makeup and therapeutic potential, *E. scaber* cultivation may eventually become a cash crop for the regions where it flourishes. Further synthesis is also required to reduce dependence on the natural source and produce medications from isolated chemicals from the plant on a large scale. Finding plant-derived compounds for novel drug development may be greatly aided by combining high-throughput technologies and traditional treatments. Given the wide variety and complexities of pharmacological compounds and phytochemicals found in *E. scaber*, an aggressive, persistent strategy is required to identify unidentified bioactive components with high antiviral properties.

References

Ahmad, A., Abass, F.M., Sufia, H. and Lim, H.K. (2009). Extraction, Saperation and identification of Chemical ingredients of Elephantopus Scaber L. using Factorial Design of experiment. *International Journal of Chemistry*. 1(1), pp. 36–49.

Aslam, S.M., Ahmad, M.S. and Mamat, A.S. (2016). Phytochemical Evaluation of Polyherbal Formulation of Clinacanthus nutans and Elephantopus scaber to Identify Flavonoids. *Pharmacognosy Journal*, 8, pp. 534–541

Basha, S.K., Anjaneyulu, E. and Sudarsanam, G. (2013). Herbal plants for wound healing among Sugali tribes of Yerramalai forest of Kurnool District, Andhra Prdesh, India, *Indian J. Applied Pure Biol*, 28, pp. 19–25.

Battu, G.R., Rao, Y.V. and Dasari, VSP. (2012). Antihepatotoxic effect of Elephantopus scaber L. on carbon tetrachloride-induced hepatotoxicity in rats. *Recent Res Sci Technol*, 4, pp. 21–24.

Chan, C.K., Tan, L.T., Andy, S.N., Kamarudin, M.N.A., Goh, B.H. and Kadir, H.A. (2017). Anti-neuroinflammatory Activity of Elephantopus scaber L. via Activation of Nrf2/HO-1 Signaling and Inhibition of p38 MAPK Pathway in LPS-Induced Microglia BV-2 Cells. *Frontiers in Pharmacology*, 8, 397.

Chan, C.K., Supriady, H., Goh, B.H. and Kadir, H.A. (2015). Elephantopus scaber induces apoptosis through ROS-dependent mitochondrial signaling pathway in HCT116 human colorectal carcinoma cells. *J. Ethnopharmacol*.168, pp. 291–304.

Chang, C.L., Shen, C.C., Ni, C.L. and Chen, C.C. (2011). A new sesquiterpene from Elephantopus scaber. *HungKuang J.* 65, pp. 49–56.

Cragg, G.A. and Newman, D.J. 2005. Biodiversity: A continuing source of novel drug leads. *Pure Appl Chem*, 77, pp. 7–24.

Crumpacker, C.S., Schnipper, L.E., Marlowe S.I., et al. (1982). Resistance to antiviral drugs of herpes simplex virus isolated from a patient treated with acyclovir. NEJM , 306, pp. 343–346.

Daisy, P. and Priya, C.E. (2010). Hypolipidemic and Renal Functionality Potentials of the Hexane Extract Fractions of Elephantopus scaber Linn. *International Journal of Biomedical Science*, 6, pp. 241–245.

Daisy, P., Jasmine, R., Ignacimuthu, S. and Murugan, E. (2009). A novel steroid from Elephantopus scaber L. an ethnomedicinal plant with antidiabetic activity. *Phytomedicine*, 16, pp. 252–257.

Daisy, P., Priya, C.E. and Lily. (2011). A study on the regenerative potential of root extracts of Elephantopus scaber L. *African Journal of Pharmacy and Pharmacology* 5, pp. 1832–1837.

Daniel, M. (2006). Medicinal plants: chemistry and properties. Science Publishers Enfield, New Hampshire, pp. 85–86.

Darnell, R.L., Knottenbelt, M.K., Todd Rose, F.O., Monk, I.R., Stinear, T.P. and Cook, G.M. (2019). Genomewide Profiling of the Enterococcus faecalis Transcriptional Response to Teixobactin Reveals CroRS as an Essential Regulator of Antimicrobial Tolerance, mSphere. 4, pp. 3.

Desilva, L.B.D., Herath, W.H.M.W., Jennings, R.C., Mahendran, M. and Wannigama, G.E. (1982). A new sesquiterpene lactone from Elephantopus scaber. *Phytochemistry*, 21, pp. 1173–1175.

Dubée, V., Chau, F., Arthur, M., Garry, L., Benadda, S., Mesnage, S., Lefort, A. and Fantin, B. (2011). The *in vitro* contribution of autolysins to bacterial killing elicited by amoxicillin increases with inoculum size in Enterococcus faecalis. *Antimicrobial agents and chemotherapy*, 55, pp. 910–912.

Elion, G.B. (1993). Acyclovir discovery, mechanism of action and selectivity. *J Med Virol.*41 Suppl, 1, pp. 2–6

Farnsworth, N.R. (1985). Medicinal plants in therapy. *Bull World Health Organ*, 63, pp. 965–981.

Ganga, R.B., Rao, Y.V. and Dasar, V.S.P. (2012). *Recent Res. Sci. Technol.* 4: 21–24.

Geetha, B.S., Nair, M.S., Latha, P.G. and Remani, P. (2012). Sesquiterpene lactones isolated from Elephantopus scaber l. inhibits human lymphocyte proliferation and the growth of tumour cell lines and induces apoptosis *in vitro. J. Biomed. Biotechnol*, pp. 1–8.

Geng, H.W., Zhang, X.L., Wang, G.C., Yang, X.X., Wu, X., Wang, Y.F., Li, Y.L. (2011). Antiviral dicaffeoyl derivatives from Elephantopus scaber. *Journal of Asian Natural Products Research*, 13(7), pp. 665–669.

Hammer, M.L.A. and Johns, E.A. (1993). Tapping an Amazonian plethora: four medicinal plants of Marajo Island, Para (Brazilourna). *J. Ethnopharm.* 40.

Hiradeve, S.M., Rangari, V.D. (2014). Elephantopus scaber Linn. A review on its activity of on alcohol induced liver damage in mice. *Evid Based Complement Altern Med*, pp. 1–8

Hisham, A., Pieters, I., Claeys, M., Dommisse, R., Berghe, D.V. and Vlietinck. (1992). *Planta Med.* 58, pp.474–475

Hisham, A., Pieters, I., Claeys, M., Dommisse, R., Berghe, D. V. and Vlietinck, A. (1992). Guaianolide Glucosides from Elephantopus scaber. *Planta medica*, 58(5), pp. 474–475.

Ho, W.Y., Yeap, S.K., Ho, C.L., Rahim, R.A. and Alitheen, N.B. (2012). An ethnobotanical study of medicinal plants of Rewa district Madhya Pradesh. *Indian J Traditional Knowledge*, 26, pp. 191–202.

Huang, C.C., Lo, C.P., Chiu, C.Y. and Shyur, L.F. (2010). Deoxyelephantopin, a novel multifunctional agent, suppresses mammary tumour growth and lung metastasis and double survival time in mice. *Br. J. Pharmacol*, 159(19), pp. 856–871.

Hung, H.F., Hou, C.W., Chen, Y.L., Lin, C.C., Fu, H.W., Wang, J.S. and Jeng, K.C. (2011). Elephantopus scaber inhibits lipopolysaccharide-induced liver injury by suppression of signaling pathways in rats, *The Am J Chi Med.* 39, pp. 705–717.

Huntley, C.C., Weiss, W.J., Gazumyan, A., et al. (2002). RFI-641, a potent respiratory syncytial virus inhibitor. Antimicrob Agents, Chemother 46, pp. 841–847.

Inta, A., Shengji, P., Balslev, H., Wangpakapattanawong, P. and Trisonthi, C. (2008). A comparative study on medicinal plants used in Akha's traditional medicine in China and Thailand, cultural coherence or ecological divergence, *J Ethnopharmacol.* 116, pp. 508–517.

Jasmine, R. and Daisy, P., Selvakumar, B.N. (2007)(b). A novel terpenoid from Elephantopus scaber with antibacterial activity against beta lactamase-producing clinical isolates. *Res. J. Microbiol.* 2, pp. 770–775.

Jasmine, R., Daisy, P. (2007)(a). Effect of crude extract and fraction from Elephantopus scaber on hyperglycemia in Streptozotocin-diabetic rats. I*nternat.J.Biol. Chem.* 1(2), pp. 117–121.

Jasmine, R. and Selvakumar, B.N. (2011). Synergistic action of two Indian medicinal plants of clinical isolates of vancomycin resistant Enterococci species. *Pharmacologyonline.* 3, pp. 1192–1198.

Jenny, A.D., Saha, S., Paul, M., Dutta, M.Z. and Nath, A.K. (2012). Antibacterial activity of aerial part of extract of Elephantopus scaber Linn.*Pharm. Res*, 2, pp. 38–41.

Kabiru, A. and Por, L.Y. (2013). Elephantopus species: traditional uses, pharmacological actions and chemical composition. *Advance Life Sci. Technol.* 15, pp. 6–13.

Kamalakannan, P., Kavitha R., Elamathi, R., Deepa, T. and Sridhar, S. (2012). Study of the phytochemical and antimicrobial potential of methanol and aqueous extracts of aerial parts of Elephantopus scaber L *Int. J. Curr. Pharm. Res*, 4, pp. 18–21.

Kiritikar, K.D. and Basu, B.D. (1991). *Indian Medicinal Plants*, 2nd ed. International Book Distributors, Deharadun, pp. 1328–1329

Kneyber, M.C.J., Moll, H.A. and Groot, R.D. (2000). Treatment and prevention of respiratory syncytial virus infection. *Eur J Pediatr*, 159, pp. 399–411.

Kumar, S.S., Perumal, P. and Suresh, B. (2004). Antibacterial studies on leaf extract of Elephantopus Scaber Linn. *Ancient Science of Life*, pp. 23:6–8.

Liang, Q.L., Min, Z.D. and Tang, Y.P. (2008). A new elemanolide sesquiterpene lactone from Elephantopus scaber. *J. Asian Nat. Prod. Res.* 10, pp. 403–407.

Modilal, M.R.D. and Daisy, P. (2011) Hypoglycemic effects of Elephantopus scaber in alloxan-induced diabetic rats, *Ind J Novel Drug Delivery*, 3, pp. 98–103.

Peer, L.M. and Metzger, J. (1980). Medicinal Plants of pus Rohr. *In: Flora Reipublicae Popularis Sinicae To-East and Southeast Asia: Attributed Properties Uses*. The MIT Press, London, pp. 43–47.

Poli, A., Nicolau, M., Simoes, C.M.O., Nicolau, R.M.R. and Zanin, M. (1992). Preliminary pharmacologic evaluation of crude whole plant extracts of Elephantopus scaber. Part I: *in vivo studies. J Ethnopharmacol*, 37, pp. 71–76.

Raenu Yand Yingmanee T. (2015). Inhibitory effect of Rosmarinus offi cinalis L., Senna alata (L.) Roxb., Acalypha indica L. and Elephantopus scaber L. against herpes simplex viruses The International Conference on Herbal and Traditional Medicine.

Rao, M.U., Sreenivasulu, M., Chengaiah, B. and Reddy, K.J. (2010). Herbal medicines for diabetes mellitus: a review. *Int J Pharm Tech Res*, 2, pp. 1883–1892.

Sagar, R. and Sahoo, H.B. (2012). Evaluation of antiasthmatic activity of ethanolic extract of Elephantopus scaber L. leaves. *Indian J. Pharmacol*, 44(3), p. 398.

Sala, E., Guasch, L., Iwaszkiewicz, J., Mulero, M., Salvado, M.J. (2011). Identification of human IKK-2 inhibitors of natural origin (Part II): *In silico* prediction of IKK-2 inhibitors in natural extracts with known anti-inflammatory activity. *Eur J Med Chem*, 46, pp. 6098–6103.

Sankar, V., Kalirajan, R., Sales, S.V., Raghuraman, S. (2001). Antiinflammatory activity of Elephantopus scaber in albino rats. *Indian J Pharm Sci*, pp. 523–525.

Sheeba, K.O., Wills, P.J., Latha, B.K., Rajalekshmy, R. and Latha, M. (2012). Antioxidant and antihepatotoxic efficacy of methanolic extract of Elephantopus scaber Linn in Wistar rats, *Asian Pacific J Trop Dis*, 2, pp. 904–908.

Shinwari, Z.K. (2010). Medicinal plants research in Pakistan. *J Med Plants Res* 4, pp. 161–176.

Singh, A.P. (2005). Promising phytochemicals from Indian medicinal plants. *Ethnobot. Leaflets*, 9, pp. 15–23.

Singh, S.D.J, Krishna, V., Mankani, K.L., Manjunatha, B.K., Vidya and S.M., Manohara, Y.N. (2005). Wound healing activity of the leaf extracts and deoxyelephantopin isolated from Elephantopus scaber Linn. *Indian J Pharmacol*. 37, pp. 238–242.

Slikkerveer, L. (2006). The challenge of non-experimental validation of mac plants, towards a multivariate model of transcultural utilization of medicinal, aromatic and cosmetic plants. In: medicinal and aromatic plants: agricultural, commercial, ecological, legal, pharmacological and social aspects,Springer. 17, pp. 1–28.

Su, M., Wu, X., Chung, H.Y., Li, Y., Ye, W. (2009). Antiproliferative activities of five Chinese medicinal herbs and active compounds in Elephantopus scaber. *Nat. Prod. Commun*. 4 (8), pp. 1025–1030.

Sudip K. M., Haraprasad, P., Ishan, P. and Sankhadip. (2018). Biological Potential of Elephantopus scaber Linn. *Int. J. Pharm. Sci. Rev. Res.*, 50(2), pp: 130–134.

Sunilbabu, K., Ammani, K., Bobbarala, V. and Usha, Y. (2011). *In vitro* Antibacterial Screening of Five Local Medicinal Plants of Araku Environment Drug Invention Today, 3, pp. 14–15.

Than, N.N., Fotso, S., Sevvana, M., Sheldrick, G.M., Fiebig, H.H., Kelter, G., Laatsch, H. (2005). Sesquiterpene lactones from Elephantopus scaber. Z. *Naturforsch*. 60, pp. 200–204.

Wan, Y.H., Swee, K.Y., Ho, C.L., Raha, A.R. and Alitheen, N.B. (2012). Hepatoprotective activity of Elephantopus scaber on alcohol-induced liver damage in mice, Evidence-Based Com Alt Med 1–2.

Wang, L., Jian, S., Nan, P., Liu, J. and Zhong, Y., (2004). Chemical composition of the essential oil of Elephantopus scaber from southern China. Z. *Naturforsch*. 59, pp. 327–329.

Wang, L., Jian, S., Nan, P., Liu, J. and Zhong, Y. (2005). Chemotypical variability of leaf oils in Elephantopus scaber from 12 locations in China. *Chem. Nat. Compd.* 41, pp. 491–493.

Wiwat C. and Kwantrairat S. (2013). HIV-1 Reverse Transcriptase Inhibitors from Thai Medicinal Plants and Elephantopus scaber Linn; *Mahidol University Journal of Pharmaceutical Sciences*, 40 (3), pp. 35–44.

Wright, C.I., Buren, L.V., Kroner, C.I. and Koning, M.M.G. (2007). Herbal medicines as diuretics: a review of the scientific evidence. *J. Ethnopharmacol.* 114, pp. 1–31.

Wyde, P.R., Ambrose, M.W., Meyerson, L.R. and Gilbert, B.E. (1993). The antiviral activity of SP-303, a natural polyphenolic polymer, against respiratory syncytial and parainfluenza type 3 viruses in cotton rats. *Antiviral Res* 20, pp. 145–154.

Xu, G., Liang, Q., Gong, Z., Yu, W., He, S. and Xi, L. (2006). Antitumor activities of the four sesquiterpene lactones from Elephantopus scaber L. *Exp. Oncol.* 28, pp. 106–109.

Yaolan L., Linda S.M., Ooi , H.W., Paul, P.H. and Vincent E.C. (2004). Antiviral Activities of Medicinal Herbs Traditionally Used in Southern Mainland China. *Phytother. Res.* 18, pp. 718–722.

Zhang H.B., Kong, L.J., Hang, Q.L., Jiang, J.H. and Min, Z.D. (2011). *Chinese J. Exp. Traditional Med. Formulae.* 3, pp. 101–103.

16 Antiviral Bioactive Compounds of *Terminalia chebula* Ritz. (Myrobalan) and their Antiviral Mechanisms

Noman Ahmad[1], Hamaad Raza Ahmad[1], Abrar Ahmad[2], Rimsha Hassan[2], Hanzla Siddique[2], Ghousia Liaquat[3], Ghulam Murtaza[4], Tahira Jameel[3], Muhammad Imran[3], Mubeen Fatima[5], Muhammad Nadeem[3] and Muhammad Ashar Ayub[3*]

1. Taxonomy and Scientific Classification

Table 1 outlines the scientific classification of *Terminalia chebula* Ritz (Myrobalan) and offers insights into its taxonomy. *Terminalia chebula Ritz.* is a medium to large deciduous tree of the Combretaceae family. Combretaceae is often known as the white mangrove family. Mangrove ecosystems support three genera: Conocarpus, Laguncularia, and Lumnitzera (mangals). *Terminalia chebula* is a flowering plant in the order Myrtales. The family has around 530 species of trees, shrubs, and lianas divided into ten genera. Combretaceae are found across the subtropics and tropics (Koriem, 2014). Some species of this family, like *Terminalia ivorensis*, yield good building lumber (Samuel *et al.,* 2012). The leadwood tree, *Combretum imberbe*, is also a member of this family. The widely cultivated *Quisqualis indica* has been assigned to the genus Combretum. The non-proteinogenic amino acid excitotoxin Quisqualic acid, a potent-amino-3-hydroxy-5-methyl-4-isoxazole propionic acid (AMPA) agonist, is detected in several Quisqualis species (Jordaan *et al.,* 2011; Mahajan *et al.,* 2017).

2. Synonyms

The Combretaceae family member Haritaki is known as the "Prince of Medicines" in Tibet. The medicinal plant Terminalia chebula Ritz. (Myrobalan) is widely distributed in India, Sri Lanka, and Myanmar, and its synonyms vary according to the Ayurvedic region (Table 2). Many scientists have also used additional terms for various Terminalia chebula species, as mentioned in Table 3. It forms a sizable

[1] Institute of Soil and Environmental Sciences, University of Agriculture Faisalabad, Pakistan.
[2] Department of Eastern Medicine and Surgery, Government College University Faisalabad, Pakistan.
[3] Institute of Agro-Industry and Environment, The Islamia University of Bahawalpur, Pakistan.
[4] Department of Food Science, Government College University, Faisalabad, Pakistan.
[5] Centre for Applied Molecular Biology (CAMB), University of the Punjab, Lahore, Pakistan.
* Corresponding author: ashar.ayub@iub.edu.pk

DOI: 10.1201/9781003433583-16

Table 1 Scientific classification of *Terminalia chebula Retz.* (Myrobalan).

Taxonomic Rank	*Synonyms/Common Terms*
Kingdom	Plantae - plantes, planta, vegetal, plants
Subkingdom	Viridiplantae: green plants
Infrakingdom	Streptophyta: land plants
Super-division	Embryophyta
Division	Tracheophyta: vascular plants, tracheophytes
Subdivision	Spermatophytina: spermatophytes, seed plants, phanérogames
Class	Magnoliopsida
Superorder	Rosanae
Order	Myrtales
Family	Combretaceae (mangrove family)
Genus	Terminalia L. (tropical almond)
Species	*Terminalia chebula* (Gaertn.) Retz. (myrobalan)

Table 2 Several vernacular names of *Terminalia chebula* (Myrobalan) in different languages.

Local Names	*Language*	*Country*	*Reference*
Anelekai, Arale, Alale, ಅಳಲೆ	Kannada	Karnataka in the southwestern region of India	(Basha and Code, 2017)
Bad amier Chébule, Myrobalan chébule	French	French	(Chattopadhyay and Bhattacharyya, 2007)
Black myrobalan, Chebulic myrobalan, Chebulic Myrobalam, Chebulic Myrombalam, Gallnut	English	United States	(Gowtham *et al.*, 2019)
Chebulische Myrobalane	German	Germany	(Basha and Code, 2017)
Halilah, Harr, Hareer, Halila Zard, Halila Siyah	Urdu	Pakistan, India	(Basha and Code, 2017)
Shilikha, Hilika, Hilikha, Shilikha, শিলিখা	Assamese	Assam state of India	(Chandil *et al.*, 2020)
Halela	Kashmiri	Kashmir	(Chandil *et al.*, 2020)
Halela, Harar, Haser	Punjabi	Indian Punjab & Pakistani Punjab	(Ashwini *et al.*, 2011)
Harad, Harra, हरीतक, हर्रा	Hindi	India	(Said *et al.*, 2012)
Haritaki	Sanskrit	Indo-European languages	(Chandil *et al.*, 2020)
Haritaki, Hora, হরীতকী	Bengali	Bangladesh	(Bulbul *et al.*, 2022)
He zi	Chinese	China	(Khan *et al.*, 2015)
Hirda, Haritaki, Harda, Hireda	Marathi	The Indian state of Maharashtra	(Gupta, 2012)
Hirdo, Himaja, Pilo-Harda, Hardo, હરડી, હરડે	Gujarati	The Indian state of Gujarat	(Bulbul *et al.*, 2022)

portion of all internationally recognized healthcare systems. Additionally, this medicinal herb has been discovered to be an inexpensive and valuable source of various phytochemical compounds that play a vital role in developing medications for several disorders. *Terminalia chebula Ritz.* frequently referred to as "black myrobalan" or "chebulic myrobalan", is an evergreen tree in the Combretaceae family and has been regarded as one of the most significant therapeutic plants in Ayurvedic medicine (Chandil *et al.,* 2020). Its ancient Hindi names are "Haritaki" and "Abbhaya" because they relate to the God Shiva's death and bravery (Ashwini *et al.,* 2011). Kumar et al. (2017) discovered that Haritaki is not mentioned in the Ayurvedic Pharmacopoeia of India (API). Still, there are repeated references to its synonyms, qualities, and hints in the Brahattrayi and Ashtanga Samgraha. Haritaki fruit is recommended for usage in Brahattrayi in the following preparations: Jvara, Prameha, Kushtha, Unmada, Krami Roga, Pandu, Grahani, Visha, Madatyaya, Apasmara, and Bhutabadha (Ikram *et al.,* 2021).

According to Madanapala Nighantu, it is known as Hartaki because it is native to the Himalayas, where Hara lives, and naturally Harita (greenish-yellow) in colour (Chandil *et al.,* 2020). In the Kannada language, "Har" means taking away as it eliminates illnesses (Basha and Code, 2017). It is best described as one of the most valuable herbal remedies in Ayurvedic, Siddha, Unani, and homeopathic medicine (Kaur, 2022). In addition to being antioxidant and antidiabetic, it is also antibacterial, antiviral, antifungal, anticancer, and has wound-healing properties (Chandil et al., 2020).

Table 3 Synonyms used for the different species of *Terminalia chebula Retz.*

Homotypic Synonyms	
Synonyms	*Source*
Buceras chebula (Retz.) Lyons	World Synonymic Checklist of Vascular Plants
Myrobalanus chebula (Retz.) Gaertn.	
Heterotypic Synonyms	
Combretum argyrophyllum K.Schum.	World Synonymic Checklist of Vascular Plants
Myrobalanifera citrina Houtt.	
Myrobalanus gangetica (Roxb.) Kostel.	
Myrobalanus tomentella Kuntze	World Checklist of Vascular Plants (WCVP) and International Plant Names Index
Terminalia aruta Buch. -Ham.	GBIF Backbone Taxonomy
Terminalia acuta Walp.	World Synonymic Checklist of Vascular Plants
Terminalia argyrophylla King & Prain	World Checklist of Vascular Plants (WCVP) and International Plant Names Index
Terminalia aruta Buch. -Ham. ex G. Don	World Synonymic Checklist of Vascular Plants
Terminalia gangetica Roxb.	
Terminalia glandulipetiolata De Wild.	The World Checklist of Vascular Plants (WCVP)
Terminalia reticulata Roth	World Synonymic Checklist of Vascular Plants
Terminalia tomentella Kurz	World Checklist of Vascular Plants (WCVP) and International Plant Names Index

3. Botanical Description of the Plant

Terminalia chebula Ritz. is a blooming evergreen tree of the Combretaceae family. It grows at an elevation ranging from 1500 to 2000 metres in predominantly clayey and shady soils (Mehra et al., 2012). Although most of their trees are medium height, some can reach heights of 40 to 80 feet (Nyabuto, 1994). It features a spherical canopy with expanding branches. The trunk is dark brown and has several longitudinal cracks (Loganathan, 2019).

The leaves are oblong and elliptic, with two large ducts at the apex of petioles and an acute lip, clavate at the base, glabrous above, and yellow pubescence beneath (Kar *et al.,* 2016). Its leaves are slightly wider than Indian blackberry leaves, four to eight inches long, pinnately whitish green, and flexible but rough to the touch (Haroon *et al.,* 2015). The leaves are not dense on the stems. Around February and March, its leaves fall off, and the tree looks slimy. New leaves emerge in the rainy season. The rainy season is also known as the time of the plantation. In the spring, they grow fresh, whitish-yellow leaves with a particular smell. Some people also like to plant myrobalan trees in their gardens due to their aesthetic value (Raju *et al.,* 2012; Dutta and Devi, 2015).

Flowers are monoecious, yellowish-white, and have a foul odour (Mitra and Zaman, 2020). They appear in short axillary spikes (Rai and Joshi, 2009). The flowering starts in May and June and fruit in December. The plant size ranges from 1–2 inches (Menpara and Chanda, 2014). On the outside, there are five grooves or ribs.

Fruits are drupe, obovoid, and five-ribbed (Saha *et al.,* 2020). The shape of the fruit is not the same. Some are long, and some are small. Fruits begin to develop in November and are harvested from January through April. When immature, the fruit is greenish, turning yellow to grey when ripe. Fruits that fall from the tree before ripening are very small and turn black when dry. The seed has a pale-yellow colour (Said and Naveed, 2012). The physical characteristics of the Myrobalan fruit are presented in Table 4, providing a clear description of its morphological traits. It is considered good if the fruit ripens on the tree itself (Kala, 2022). Shiny fruit is an indication of high-quality fruit. When raw fruit is placed in water, it sinks if it is not greasy and heavy, showing it is not very useful. The fruit of *Terminalia chebula Retz.* has been used for decades apart from the chemistry of *Terminalia chebula* compounds.

Types:

Terminalia chebula is of three types:

- **(a)** Small Myrobalan - the unripe fruit
- **(b)** Yellow Myrobalan - the ripening fruit stage occurs after the seed has matured.
- **(c)** Large Myrobalan - the fully ripened fruit (Chattopadhay and Bhàttacharyya, 2007)

Parts Used: From January through April, ripened fruits are picked by shaking the plants. These are dried in thin layers in the shade. Green fruits are also harvested

Table 4 Physical characteristics of *Terminalia chebula* Ritz.

Characteristic	*Observation*
Colour	Yellowish brown to brown
Odour	Strong, unpleasant odour
Taste	Fruit tastes sweet, sour, or bitter depending on the fruit type
Shape	Smooth ellipsoid/longitudinal wrinkle to ovoid drupes
Size	2.1–4.0 cm (0.83–1.76 in) long and 1.5–2.6 cm (0.48–0.96 in) broad
Solubility	The extract is water and oil-in-water formulation soluble
Extra Features	Fruits are drupes composed of three to six longitudinal ribs and are stiff and stony in texture. The seeds are pale yellow and 1.5 to 3.0 cm long.

and dried until they become black for medicinal purposes (Dodke and Pansare, 2017). Very young fruits are also valued in medicine.

Microscopical Character of *Terminalia chebula*: It shows the epicarp of several layers of corky cells, often containing a few starch grains. The mesocarp comprises round parenchymatous cells with brownish cell contents (Debnath *et al.,* 2017). There are numerous stellate crystals of calcium oxalate, but starch is absent, and parenchymatous cells have cellulose walls (Dickison, 2000; Nagadesi *et al.,* 2022). In semi-mature fruits, isolated vascular strands of vessels are intermingled with parenchymatous tissues, and vessel walls have pitted thickening (Shukla *et al.,* 2013). The powder shows pitted vessels and rounds of parenchymatous cells containing resins and oils of brownish colour (Roy *et al.,* 2015).

The macro and microscopic characteristics of the 'Behera' were revealed by microscopy slide preparation with a small number of sediments and a floral hydrate solution gently passed with distilled water. Under the microscope, a group of flask-shaped true streets with thick walls of sclerenchyma, fibre cells, Xylem fibres with luminescence, and stone cells with broad luminous prismatic crystals were also visible in the blistering fruit's additional two full sections (Singh and Sharma, 2010).

4. Geographical Distribution

Terminalia chebula Ritz. thrives in areas with temperatures ranging from 35 to 45 degrees Celsius and annual rainfall of 1000 to 3000 millimeters. It occurs in both dry and damp deciduous forests (Mahapatra et al., 2005). It can grow on different types of soils but grows best on loose and well-drained soils (Banerjee *et al.,* 2022). Myrobalan trees grow at elevations ranging from 300 to 900 metres in North India, including the Satpura hills of Madhya Pradesh and Maharashtra, as well as the Panchamahal area of Gujarat (Rangari, 2008). Myanmar and Sri Lanka also have it as detailed in Table 5. This tree is found more or less in every part of India, but it is abundant in North India, from Bengal, Assam, and Brahma, to Madras. It has trees in most of the lower parts of the Himalayas. It is also present in some areas of Punjab in Pakistan, its tree is not long-lived, and after some time, it dries up and falls (Dhillon *et al.*, 2014; Shaheen *et al.*, 2018). One of the significant Non-Timber Forest Product (NTFP) species is *Terminalia chebula ritz,* also known as chebulic Myrobalan (Hanumantha *et al.*, 2019).

Across Ayurveda, Sidda, traditional Chinese, and Unani medicine, the species referred to as the "King of medicine" is frequently used to cure a broad spectrum of disorders in Asia and Africa (Shaikh *et al.*, 2020; Kailash *et al.*, 2022). The Ayurvedic herb *Terminalia chebula* is a member of the Combretaceae family. The Middle East (Figure 1) and tropical nations like China, India, and Thailand are habitats for these plants. It is a wide, tropical, evergreen tree with thick, cracked, black bark (Gupta and Dubey, 2019). Since ancient times, the fruit of the tree has been utilized as a traditional medicine for the domestic treatment of various human maladies due to its wide range of health advantages. *Terminalia chebula* has long been a staple of Ayurveda, Unani, and homeopathic medicine and is now a modern medicine cynosure (Rana *et al.*, 2018). The inclusion of several phytochemicals, including antioxidants, anthocyanins, alkaloids, terpenes, flavonoids, and polysaccharides, may be responsible for the reported health benefits (Bag et al., 2013).

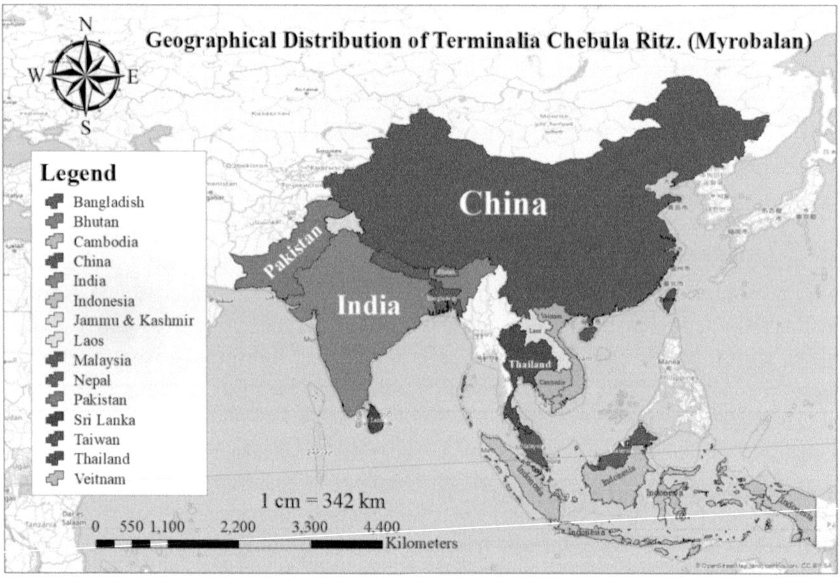

Fig. 1 Geographical distribution of *Terminalia chebula Ritz.* (Myrobalan).

5. Phytochemical Constituents and Metabolite Profiling

Terminalia chebula Ritz. fruits have been frequently documented to have significant levels of tannins and phenolic compounds (Jain, 2020). Myrobalan is also renowned for its rich content of ascorbic acid. Terflavin A, chebulagic acid, terchebulin, chebulinic acid, punicalagin, and corilagin are among the essential chemicals found in many tannins (hydrolyzable tannins) (Nigam *et al.*, 2020).

Table 5 Geographical distributions of *Terminalia chebula Ritz.* (Myrobalan)

S.No.	*Geographical Area*	*Distribution Pattern*	*Reference (s)*
1	Bangladesh	Rare in Bangladesh's districts of Moulvibazar, Chittagong, and Mymensingh.	(Akbar, 2020; Kabir *et al.*, 2014)
2	Bhutan	Bhutan's Goshing, Ngangla, and Phangkhar Gewogs deciduous forests.	(Purna and Tenzi, 2012; Phurpa *et al.*, 2017)
3	Cambodia	Protected areas in Cambodia's deciduous and agricultural forests include Pang Khang Lech, Veun Sai-Siem Pang Conservation Area (VSSPCA), Siem, and Virachey National Park.	(Philip, 1999; Loveridge *et al.*, 2017)
4	China	It is indigenous to West Yunnan and is grown in Fujian (Min), Guangdong, and Zhuang Autonomous Region.	(Flora of China Editorial Committee, 2018)
5	India	Cis-Himalaya, with Himalayan peaks rising to 1,500 meters.	(Bag *et al.*, 2013)
6	Indonesia	Cultivated in Riau Province and Sumatra of Indonesia.	(Mahyar *et al.*, 1991)
7	Laos	Cultivated in the Nakai-Nam Theun National Park in Nakai District, Khammouane Province, Laos, and the northern Thai dry deciduous woodlands and grasslands.	(Philip, 1999)
8	Malaysia	Grown in the rainforest of Sarawak's protected parkland.	(Nicholson, 1965; Rathinamoorthy and Thilagavathi, 2014)
9	Myanmar	Cultivated in Myanmar's deciduous forests of Bago, Kachin, Mandalay, and Yangon.	(Rathinamoorthy and Thilagavathi, 2014)
10	Nepal	Nepal's deciduous woodlands of the Rasuwa and Jhapa districts.	(Yadav *et al.*, 2010; Raut *et al.*, 2018; Saru *et al.*, 2020)
11	Pakistan	Cultivated in Northwest Khyber Pakhtunkhwa Province of Pakistan.	(Arshad *et al.*, 2010)
12	Sri Lanka	Their native environment is in the intermediate zones, and they are farmed in agroclimatic zones such as Wilpattu National Park in Anuradhapura and Gal Oya National Park in Badulla.	(Government of Sri Lanka Forest Department, 2009; Green, 1990; San *et al.*, 2013)
13	Taiwan	Cultivated in Chushan, Nantou, and Kaohsiung cities of Taiwan.	(Safavi *et al.*, 2015)
14	Thailand	Thailand's Chae-Son district of Lampang Savannas and dry deciduous woodlands.	(Philip, 1999; Wilart, 2020)
15	Vietnam	Planted in Vietnam's Buôn Ma Thuột and Đắk Lắk province amid dry deciduous and grassy woodlands.	(Philip, 1999; Nguyen, 2016)

Table 6 contains a comprehensive list of these chemicals (Huang *et al.,* 2018) devised a reverse-phase HPLC technique for determining Fourteen bioactive compounds in *Terminalia chebula Ritz*. (Im *et al.,* 2017) discovered 48 polyphenol components in a methanol extract of *Terminalia chebula Ritz*. Many methyl and ethyl esters of bioactive components have been documented to be artifacts produced during the fractionation and separation techniques (Ruan *et al.,* 2018). Myrobalan and tannin-rich species, including Punica granatum and Rheum sp., also contain these tannins and other secondary metabolites (Courtney and Cock, 2022; Mandeville and Cock, 2018; Cock and Courtney, 2021; Wright *et al.,* 2019). Several phytochemicals, including gallic acid, ellagic acid, and hydroxycinnamic acids, and their derivatives, have been described. (Table 6). Flavonoids found in fruits include quercetin, rutin, and methylation metabolites of quercetin (Rahaman *et al.,* 2022).

Table 6 The phytochemical composition of the fruits of *Terminalia chebula Ritz.*

S.No.	*Chemical Constituents*	*Nature of the Component*	*Reference*
1	1,2,3,4,6-penta-O-galloyl-ß-D-glucose (PGG)	Hydrolyzable Tannins	(Bae *et al.,* 2017)
2	1,6,-di-O-galloyl-D-glucose	Hydrolyzable Tannins	(Juang *et al.,* 2004)
3	3,4,6-tri-O-galloyl-D-glucose	Hydrolyzable Tannins	(Pellati *et al.,* 2013)
4	Casuarinin	Hydrolyzable Tannins	(Kesharwani *et al.,* 2017)
5	Chebublic acid	Hydrolysable Tannins	(Feng *et al.,* 2021)
6	Chebulagic acid	Hydrolyzable Tannins	(Han *et al.,* 2006)
7	Chebulanin	Hydrolyzable Tannins	(Cheng *et al.,* 2003)
8	Chebulinic acid	Hydrolyzable Tannins	(Mishra *et al.,* 2013)
9	Corilagin	Hydrolysable Tannins	(Li *et al.,* 2019)
10	Ellagic acid	Hydrolyzable Tannins	(Sarabhai *et al., 2013*)
11	Ethyl gallate	Hydrolyzable Tannins	(Karel *et al.,* 2004)
12	Flavonol glycosides	Phenolic Compound	(Venkatesan *et al.,* 2017)
13	Gallic acid	Hydrolyzable Tannins	(Kumar *et al.,* 2010)
14	Luteolin	Phenolic Compound	(Pfundstein *et al.,* 2010)
15	Neochebulinic	Hydrolyzable Tannins	(Juang *et al.,* 2004)
16	Punicalagin	Hydrolysable Tannins	(Nigam *et al.,* 2020)
17	Tannins	Hydrolyzable Tannins	(Chattopadhyay and Bhattacharyya, 2007)
18	Terchebulin	Hydrolyzable Tannins	(Miyasaki *et al.,* 2013)

6. Medicinal/Therapeutic Uses

Traditional medicines have long been regarded as a valuable and inexpensive source of specific plant metabolites, widely employed in producing medications to treat various ailments. A considerable majority of the world's population, particularly in underdeveloped nations, is reliant on herbal remedies. The application of Myrobalan fruits and plant-based products in medicine is becoming more prevalent since natural medications are inexpensive, have a biological basis, have better safe buffers, and have few or no adverse effects. *Terminalia chebula Retz.* refers to the Combretaceae family and is a member of the valuable therapeutic plants used in Ayurvedic, Siddha, Unani, and Homeopathic remedies. Because of its remarkable wound-healing capacity and the broad range of therapeutic characteristics, it is known as the "King of Medicines" in Tibet. It is ranked first in Ayurvedic pharmacopoeia (Mani *et al.,* 2021).

Terminalia chebula Ritz. is a moderate, suitable, and efficient laxative in traditional medicine. M oreover, its bioactive constituents are non-toxic and have a medicinal effect. An important component in Triphala's popular herbal cure for enlarged liver, stomach problems, and eye irritation is *Terminalia chebula Ritz.* (Riaz *et al.,* 2015; Srikantha *et al.,* 2018). Being a tonic and purifying the mind cleanses the mind from mental disorders and strengthens the senses, memory, and mind (Adams *et al.,* 2007). Treats melancholia by absorbing irregular cerebral secretions due to moisture absorption (Mobeen and Moazzam, 2022). It is used to strengthen the stomach (Akbar, 2020). It could treat diarrhoea by frying it in a small amount of oil (Singh *et al.,* 2002). It is the fact that the blood is purified and pure, that blood diseases are removed, and we are healthy. It is also used to treat depression and melancholy (Khan *et al.,* 2018). This research provides a birds-eye perspective of the biochemical and pharmacological activities of different preparations and extracted bioactive components of *Terminalia chebula Ritz.*

6.1 Uses described in Pharmacopoeias and Traditional Systems of Medicine

The Indian subcontinent's Ayurvedic Pharmacopoeia cites the following medicinal applications of Terminalia chebula Ritz fruits: Gulma (gastric glands), Hrdroga (cardiac disorder), Jirnajvara (severe fever), Visamajvara (intermittent fever), Kasa (coughing), Pandu (anaemia), Prameha (hyperglycemia), Sirorgoa (conditions above the collar), Sotha (inflammatory condition), Tamaka Swasa (bronchial asthma) (Indian Ayurvedic Pharmacopoeia, 2017). In the Indian fruit market, Myrobalan is abundantly available. Traditionally, they are employed as a general tonic and abrasive, treating scars and infections. The dried powder is used in dental treatment to treat stomatitis and gum blisters (Jose *et al.,* 2011). The myrobalan herb is an antidote to snake bites (Alagesaboopathi, 2013). Kadukkai is a Tamil Nadu tribal name for *Terminalia chebula Ritz.* It is utilized to cure multiple maladies, including temperature, congestion, diarrhoea, gastroenteritis, dermatitis, candidiasis, renal system issue, and skin infections. In China, the medicine is used to treat sore throats, coughs, and diarrhoea. In Tib, dried fruit is prescribed to treat ulcers and cholera (Kausar *et al.,* 2021).

Monograph or pharmacopoeia details of *Terminalia chebula* are included in "The Ayurvedic Pharmacopoeia of India" (Part-I and Volume-I). They are also mentioned in "The Unani Pharmacopoeia of India" in several preparations.

Terminalia chebula fruit is included in "The Siddha Pharmacopoeia of India." *Terminalia chebula Ritz.* fruit extract is also included in the American Pharmacopeial Convention's Herbal Remedies Handbook. *Terminalia chebula* has been assigned to the American Herbal Pharmacopoeia's handbook priority committee's list of possible concerns. According to the European Novel Food Directory Terminalia chebula fruit and bark were used in nutritional supplements in Europe before May 15, 1997 (Jain and Tiwari, 2020).

6.2 Traditional/Folklore Uses described in Folk Medicine

The Indian subcontinent is equipped with a diverse range of medicinal herbs. These herbs are frequently utilised by people from all walks of life, whether as folk cures or home remedies from the Siddha, Ayurvedic, and Unani systems. *Terminalia chebula Ritz.* (commonly famous as Chebulic Myrobalan or Black Myrobalan) is a South Asian evergreen shrub found from India, Pakistan, and Nepal east to Southwestern China, and south to Sri Lanka, Vietnam, and Malaysia (Meher *et al.*, 2018). The plants produce a fruit with a rough seed but a mushy pulp (Ashrafi *et al.*, 2018). The myrobalan seed, "harda," is widely used in Indian cuisine (Jain and Tiwari, 2020). The herb has a strong reputation in Ayurvedic and Tibetan medicine.

Terminalia chebula Ritz. has a long history of usage in ancient Indian medicine. *Terminalia chebula Ritz.* is known as 'Haritaki' in Sanskrit, and it has multiple meanings, including the removal of all ailments (Harayet), having an inherent greenish hue (Harita), and blooming in Lord Shiva's dwelling (Hara), the Himalayas. 'Haritaki' is called 'Abhaya' in Sanskrit, which refers to its fearlessness in tackling sickness. It is a Rasayana with the plain interpretation of Path (Ayana) of the Juice (rasa) or Elixir (turning to gold) that balances the Tridoshas through multiple energetic impulses in the valuable body characteristics of *Terminalia chebula Ritz.*, including treating fever (Jwaraghna), cough and flu clinical signs (Kashaghna), reducing ritual cleansing (Virechnopaga), slowing the ageing process (Vayasthapana), treating diarrhoea (Grahi), improving sight (Chakshushya), relieving pain (Shulahara), maintaining a healthy diet (Pachana), stimulating the desire to eat (Rochana), treating piles (Arshohara), treats skin dermatitis (Kusthahara), and reduces inflammation (Shothahara) (Bag et al., 2013).

6.3 Applications supported by Clinical Data

Terminalia chebula (Myrobalan) is a superfood that has become increasingly popular in recent years for its impressive nutritional benefits. This small fruit is packed with powerful antioxidants, vitamins, minerals, and other nutrients that can help improve your overall health and well-being. It has long been used in Ayurveda as a natural cure for various ailments, and current research has confirmed its potential to help maintain optimal health. Myrobalan is also known for its positive effects on the digestive and immune systems, and heart health. This article explores the amazing nutritional benefits of Myrobalan and how to unlock them for optimal health.

Anti-inflammatory properties: Myrobalan is rich in flavonoids and antioxidants, which can help reduce inflammation. Chronic inflammation, which occurs constantly, can lead to many health conditions, such as heart disease and diabetes.

Antioxidant properties: Myrobalan is also a powerful antioxidant. Antioxidants are beneficial nutrients found in a wide variety of foods. They help neutralize free radicals, harmful compounds that can cause damage to cells and DNA and increase the risk of disease.

Cardiovascular benefits: Research suggests that Myrobalan may provide cardiovascular benefits, such as reduced blood pressure and improved blood lipid profile. Many studies have found that daily fruit intake can help lower cholesterol and triglyceride levels, which are important indicators of heart health.

Digestive health: According to research, Myrobalan can help improve digestive health. It has been shown to have positive effects on the gut microbiome, which refers to the communities of microorganisms that live in our digestive system and significantly impact overall health.

Immune health: Myrobalan's anti-inflammatory and antioxidant properties may help maintain optimal immune health. It can also help support the growth of beneficial microbes in the gut and prevent the development of harmful microbial activity, which is responsible for infections and common digestive disorders.

Cancer prevention: Myrobalan contains high amounts of antioxidants, which can help prevent and fight cancer by neutralizing free radicals. Free radicals may cause mutations in healthy cells' DNA and lead to tumour growth.

Diabetes prevention: Daily consumption of Myrobalan may help prevent type 2 diabetes by improving metabolic health.

Weight management: Daily consumption of Myrobalan may help improve metabolic health and aid in weight loss.

Digestive health: Myrobalan has been studied for its potential to improve digestive health and protect against certain digestive disorders.

Eye health: Daily consumption of Myrobalan may help prevent eye diseases, such as macular degeneration, due to the high amounts of antioxidants in the fruit.

6.4 Pharmacological Activities: Mainly Antiviral

Terminalia chebula Ritz. extracts or separated functional elements have been found to possess a variety of pharmacological properties. Rich phenolic components, antioxidants or scavenging of free radical properties are among the most extensively researched capabilities (Dharmaratne *et al.*, 2018). *Terminalia chebula* extraction can shield antioxidant enzymes against reactive oxygen species (ROS) effects. The ethanolic extracts of Terminalia fruits can reduce lipid peroxidase concentrations and defend against photosensitization-induced oxidative stress by preventing the mechanism of Fe^{+2} driven lipid peroxidation, which may be effective in lowering Iron toxicity caused by light (Hussain *et al.,* 2021; Gupta, 2012; Sarkar *et al.,* 2012).

Oral administration of *Terminalia chebula* bark extracted in aqueous alcohol-based solvents results in positive histological changes in seminiferous tubules (Gupta, 2012). Myrobalan aqueous extracts suppress human sexual waypoints more effectively than other extracts (Khanna and Sharma, 2013).

Terminalia chebula alcoholic extract has been believed to produce bioactive molecules with anti-carcinogenic and antiproliferative properties that reduce cell survival, hinder cell growth, and cause cells to die in a dose-dependent manner (Saleem *et al.,* 2002). Chebulagic acid, a chemically modified biomolecule from Terminalia chebula alcoholic extract, has dual inhibitory potency against COX and 5-LOX. It also inhibits cell proliferation in HCT-15, COLO-205, DU-145, MDA-MB-231, and K562 cell lines. According to current research, triphala inhibits enzymatic activity (Reddy *et al.,* 2009; Sharma *et al.,* 2019).

Terminalia chebula Ritz has been shown to work as a drug to treat high blood pressure (Amalraj and Gopi, 2017). *Terminalia chebula* extract contributes to cardioprotective action by stabilising intracellular membranes and reducing hypoperfusion (Ashwin *et al.,* 2011). Suchalatha *et al.* (2007) found that using an alcoholic extract of *Terminalia chebula* (TCE) also helps protect against the metabolic changes caused by isoproterenol.

Upadhyay *et al.* (2014) looked into how phytochemicals in *Terminalia chebula Retz.* fruit stops cancer cells from growing. They found that the most effective compounds against cell growth were chebulinic acid, tannic acid, and ellagic acid. In a dose-dependent manner, ethanol extract of *Terminalia chebula* fruit downregulates and induces cell apoptosis in many mammary carcinoma lines such as MCF-7 and S115, osteogenic sarcoma cells, adenocarcinoma of the prostate cell, and a non-cancerous encapsulated human urinary tract cell line (Dodke and Pansare, 2017; Altememy *et al.,* 2022).

In the form of an ointment, ethanolic extracts from *Terminalia chebula* fruits help take care of the skin (Li *et al.,* 2011). Transdermal application of *Terminalia chebula* ethanol leaf extracts results in the substantially quicker repair of injured tissue owing to increased contraction rates and a shorter tissue regeneration time (Alam *et al.,* 2011; Yazarlu *et al.,* 2021). According to clinical studies, wounds treated with Myrobalan extract have higher protein levels, deoxyribonucleic acid, and collagen in the granulation tissue (the tissue that forms after a wound heals) (Prasad *et al.,* 2010; Shao *et al.,* 2019).

Rao *et al.* (2006) found that the fruits and nuts of *Terminalia chebula* lower blood sugar in a way that depends on how much you take in. Chebulagic acid-containing aqueous extracts of *Terminalia chebula*, taken at a dose of 200–250 ml/kg of body weight, decreased glucose levels and helped treat type II diabetes (Huang *et al.,* 2012; Ali, 2013).

Mehra *et al.* (2012) found that chebulagic acid from *Terminalia chebula* has the same effect on muscle spasms as papaverine. The plant may be used as an antidote to snake bites (Kadir *et al.,* 2015; Ancheril *et al.,* 2016). Haritaki promotes memories by supporting brain functions (Haghani *et al.,* 2022). It is an adjuvant herb for treating persistent fever (Jami *et al.,* 2014). Lopez et al. (2017) say that it is suitable for increasing physical activity and that long-term use can help people who are overweight lose weight. This plant has the potential to stop bleeding (Ballabh *et al.,* 2007). Fresh fruit eaten before meals hasten metabolism, but when consumed with meals, it elevates brainpower, nurtures the senses, and disinfects the digestive and urogenital systems (Bode, 2004). Haritaki improves digestion, controls intestinal function, and promotes nutrient absorption (Samy *et al.,* 2008; Mehra *et al.,* 2012). Ashwini *et al.* (2011) found that the plant has a sympathomimetic effect and helps people deal with anxiety. Chebulinic acid with brown dye powder, which is the primary colouring agent, helps the body get rid of toxins and extra fat (Schmidt & Brandys, 2016). *Terminalia chebula Ritz.* works well as a skin shine and complexion enhancer (Randhawa *et al.,* 2021). According to Yakaew *et al.* (2016), *Terminalia chebula* leaf extract is an effective ayurvedic skin protectant against photodamage (Variya *et al.,* 2016). Sheng *et al.* (2016) identified the bioactive fraction of *Terminalia chebula Ritz.* fruit aqueous extract and investigated its antidiarrheal effects.

Table 7 Numerous pharmacological activities associated with *Terminalia chebula Retz.*

Pharmacological activities	*Reference(s)*
Adaptogenic	(Akila *et al.,* 2022)
Antiallergic	(Jha and Sit, 2020)
Antianaphylactic	(Dusi and Saminathan, 2021)
Anti-arthritic	(Daram *et al.,* 2021)
Antibacterial	(Dharmaratne *et al.,* 2018)
Anti-carcinogenic	(Cock and Cheesman, 2021)
Anticaries	(Lee and Hwang, 2021)
Antidiabetic	(Alema *et al.,* 2020)
Antimycotic agents	(Venkatachalam *et al.,* 2020)
Antiedemic	(Kim *et al.,* 2022)
Antimutagenic	(Singamaneni *et al.,* 2020)
Antioxidant	(Jeong *et al.,* 2019)
Antiprotozoal	(Meher *et al.,* 2018)
Antispasmodic	(Gupta *et al.,* 2020)
Antiulcer	(Akter *et al.,*2019)
Antiviral	(Sarka *et al.,* 2022)
Cardioprotective	(Manikandan *et al.,* 2022)

Contd.

Table 7 *Contd.*

Pharmacological activities	*Reference(s)*
Chemopreventive	(Prasad *et al.*, 2020)
Cytoprotective	(Lin *et al.*, 2022)
Gastrointestinal motility	(Keerthana *et al.*, 2018)
Hepatoprotective	(Feng *et al.*, 2021)
Hypocholesterolemic	(Eltimamy *et al.*, 2022)
Hypolipidemic	(Kim *et al.*, 2022)
Immunomodulatory	(Patil *et al.*,2022)
Purgative	(Rajani and Kumar, 2021)
Radioprotective	(Bulbul *et al.*, 2022)
Renoprotective	(Asif *et al.*, 2019)
Wound healing	(Nigam *et al.*,2020)

6.5 Mode of Action of Antiviral Compounds

Several studies have shown that *Terminalia chebula Ritz.* inhibits viral infections caused by HSV-I (Herpes Simplex Virus-I) which is transmitted by oral route and causes infectious disease in or around the mouth, cytomegalovirus (CMV) which causes mononucleosis or liver diseases (hepatitis), and influenza A virus (human immunodeficiency virus type-1) (Oyuntsetseg *et al.*, 2014; Dhama *et al.*, 2018; Nigam *et al.*, 2020; Kesharwani *et al.*, 2017). Gallic acid and tannins in *Terminalia chebula Ritz.* fruits are proven to be HIV-1 integrase inhibitors. The galloyl component is crucial in stopping HIV-1 integrase from working *(Ahn et al.*, 2002; Ahn *et al.*, 2006). *Terminalia chebula Ritz.* outperforms human immunodeficiency virus-1 infection *in vivo* and *in vitro* when combined with acyclovir, a synthetic analogue of the purine nucleoside, as evidenced by a decrease in virus growth in rodent brains (Park *et al.*, 2005; Sharifi-Rad *et al.*, 2018). Punicalagin ($C_{48}H_{28}O_{30}$) and chebulic acid (CA), hydrolyzable tannins derived from dried *Terminalia chebula Ritz.* fruits were discovered to inhibit HSV-I entrance in pulmonary epithelial cells (A549) by aiming and neutralising the HSV-I viral genome, thereby preventing entry, attachment, cell-to-cell dispersal, and opportunistic infections (Lin *et al.*, 2011; Nigam *et al.*, 2020). A study that looked at *Terminalia chebula's* effect on cytomegalovirus using immunocompromised hosts found that *Terminalia chebula* significantly decreased the ability of human cytomegalovirus to replicate (Upadhyay *et al.*, 2014; Kannan, 2018). *Terminalia chebula Ritz.* has recently been studied for its ability to combat HSV-II (Herpes Simplex Virus-II) transmitted through sexual contact, also known as genital herpes infection. It has been shown that chebulinic and chebulic acids are derived from *Terminalia chebula Ritz.* extract possesses significantly higher immediate antiviral action and effectiveness against herpes simplex viral protein type 2, inhibiting viral adhesion and subsequent infection of human cells when compared to acyclovir, a typical prescription used to treat Herpes simplex virus infestation (Pandey *et al.,* 2020; Kesharwani *et al.,* 2017). The ethanolic extracts of *Terminalia chebula Ritz.* have been proven to have significant anti-hepatitis B viral (HBV) action by lowering the quantity of exogenous HBV virion genetic material, indicating that they might be used as a potent anti-HBV medication in the coming years (Kim *et al.,* 2001; Philips *et al.,* 2021). The antiviral activity of a *Terminalia chebula Ritz.* bark liquor preparation was evaluated by suppressing syncytium development and influencing cell membrane activation in baby hamster renal tissues infected with Newcastle disease virus (NDV), indicating that *Terminalia chebula Ritz.* extract can inhibit NDV-hemagglutinin neuramidase (HN) glycoprotein cell surface expression in a dose-dependent manner (Nigam *et al.,* 2020; Mahajan and Haswani, 2022).

6.6 Future of Research and Drug Delivery

One of the most versatile herbs, Terminalia chebula has a wide range of pharmacologic and therapeutic uses. This adaptable ayurvedic plant is a one-of-a-kind source of many chemical compounds with specific chemical configurations, despite exhibiting various pharmacological effects due to numerous bioactive metabolites. Comparatively, limited investigation has been conducted on the potential therapeutic uses of this herb against illnesses, notably multidrug-resistant pathogenic infections. As a result, substantial research is required to capitalise on their therapeutic potential in combating ailments, especially drug-resistant infections. As the world moves towards using medicinal herbs that are good for the environment and have traditional pharmaceutical properties, a pharmaceutical research framework should be set up to develop technologically advanced medicines from *Terminalia chebula* chemicals that are effective against a wide range of illnesses and to deal with the problem of chemoresistance after a thorough study of its therapeutic potential, mechanism of action, molecular structure, and pharmacokinetics.

Delivery strategies have immensely helped convert promising therapeutics into successful therapies. As medication delivery methods advance, medicine gets safer and more pleasant for patients. In the future, to increase the pharmacological/therapeutic activity efficiency, *Terminalia chebula* extract can be converted into pills. It can be diluted with any inert solution to enhance absorption and metabolic activity or mixed with oil-based and water-based extracts for superficial application to many skin diseases. Its diluted extract or nanoparticles can be used as eye drops etc. However, the traditional uses of these compounds must be supported by *in vivo* and laboratory investigations for mechanism-based pharmacological assessment.

7. Toxicity and Safety

Terminalia chebula Ritz. may be risk-free if taken orally for three to six months or less (Kapoor *et al.,* 2014). However, it should not be used without the supervision of a physician. Myrobalan can have a few side effects, such as gastrointestinal discomfort, cardiovascular disorders, and allergic reactions. It should not be consumed by pregnant or breastfeeding women or individuals taking certain medications, such as antihistamines, anti-epileptics, or antidepressants. Individuals with allergies to plants in the Euphorbia family, such as poinsettia and the genus Croton, should avoid using Myrobalan. Table 8 shows the acute and chronic toxicity of *Terminalia chebula*, as well as its safety outcomes. There is not yet enough information on the toxicity of *Terminalia bellerica* and *Terminalia chebula*. It's recommended to refrain from using it until more is understood.

Table 8 Toxicological information regarding the *Terminalia chebula Ritz.*

Test	*Results*	*References*
Acute Toxicity Test	LD50 $\leq$ 5000 $mgkg^{-1}$ (mice)	(Bulus *et al.*, 2011)
	LD50 $\leq$ 5000 $mgkg^{-1}$ (rat)	(Sireeratawong *et al.*, 2012)
Sub-acute Toxicity Test	LD50 $\geq$ 2753.44 $mgkg^{-1}$ (mice)	(Jain and Tiwari, 2020)
Chronic Toxicity Test	LD50 $\leq$ 1200* $mgkg^{-1}$ (rat)	(Akpovona *et al.*, 2016)

*1200 $mgkg^{-1}$ was the maximum dose tested up to 270 days

Conclusion

Terminalia chebula Ritz. is a nutrient-dense herb with various beneficial properties that have been utilized as a natural cure for multiple illnesses in Ayurvedic pharmacopoeia. Haritaki (*Chebulic Myrobalan*) can be enjoyed as freshly picked or dried fruit, bakery products including pastries, biscuits, buns, and crackers, served as a garnish over yoghurt or milkshakes or fresh juices. Recent studies have proved its ability to assist in maintaining a healthier lifestyle and boost cardiac, immunological, and gastrointestinal functioning. It also can potentially prevent and treat viral infections, malignancies, hyperglycemia (diabetes mellitus), and vision problems. *Terminalia chebula* fruit extract contains various active ingredients, including 1,2,3,4,6-penta-O-galloyl-ß-D-glucose (PGG), casuarinin, chebulinic acid, chebulagic acid, chebulanin, ellagic acid, ethyl gallate, gallic acid, luteolin, neochebulinic, and terchebulin compounds. *Terminalia chebula* preparation has been shown to inhibit HSV-II (Herpes Simplex Virus-II) infection by preventing viral entry into the host. It is also loaded with antioxidants, minerals, and vitamins that significantly bolster your health and lifestyle.

References

Adams, M., Gmünder, F. and Hamburger, M. (2007). Plants traditionally used in age related brain disorders—A survey of ethnobotanical literature. *Journal of Ethnopharmacology*, 113(3), 363–381.

Ahn, M.J., Yoon, K.D., Kim, C.Y., Kim, J.H., Shin, C.G. and Kim, J. (2006). Inhibitory activity on HIV-1 reverse transcriptase and integrase of a carmalol derivative from a brown Alga, Ishige okamurae. *Phytotherapy Research: An International Journal Devoted to Pharmacological and Toxicological Evaluation of Natural Product Derivatives*, 20(8), 711–713.

Ahn, M.J., Kim, C.Y., Lee, J.S., Kim, T.G., Kim, S.H., Lee, C.K., ... and Kim, J. (2002). Inhibition of HIV-1 integrase by galloyl glucoses from *Terminalia chebula* and flavonol glycoside gallates from Euphorbia pekinensis. *Planta Medica*, 68(05), 457–459.

Akbar, S. and Akbar, S. (2020). *Terminalia chebula Retz*. (Combretaceae) (Syn.: T. zelanica Van Heurck & Müll. Arg.). *Handbook of 200 Medicinal Plants: A Comprehensive Review of Their Traditional Medical Uses and Scientific Justifications*, 1779–1793.

Akila, B., Kiruthiga, G., Chaithanya, K.P., Mirunaleni, P., Kumar, P.N., Monika, T., ... and Aishwarya, A. (2022). Kadukkai-*Terminalia chebula Retz*. (Elixir of Rejuvenation)–A Gift of Siddha Medicine. *Journal of Pharmaceutical Research International*, 30–38.

Akpovona, A.E., Onoagbe, I.O., Eze, G.I. and Omonkhua, A.A. (2016). Acute and subchronic toxicity studies of ethanol extract of Terminalia macroptera stem bark in Wistar albino rats. *Journal of Medicine and Biomedical Research*, 15(1), 62–73.

Akter, S., Begum, T., Begum, R., Tonny, T.S., Yasmin, M., Shifa, S. and Afroze, F. (2019). Phytochemical analysis and investigation of antiinflammatory and anti-ulcer activity of Terminalia bellirica leaves extract. *Int J Pharmacogn*, 6, 54–65.

Alagesaboopathi, C. (2013). Ethnomedicinal plants used for the treatment of snake bites by Malayali tribal's and rural people in Salem district, Tamilnadu, India. *Int J Biosci*, 3(2), 42–53.

Alam, G., Singh, M.P. and Singh, A. (2011). Wound healing potential of some medicinal plants. *International Journal of Pharmaceutical Sciences Review and Research*, 9(1), 136–145.

Alema, N.M., Periasamy, G., Sibhat, G.G., Tekulu, G.H. and Hiben, M.G. (2020). Antidiabetic activity of extracts of Terminalia brownii Fresen. Stem bark in mice. *Journal of Experimental Pharmacology*, 12, 61.

Ali Asgar, M.D. (2013). Antidiabetic potential of phenolic compounds: A review. *International Journal of Food Properties*, 16(1), 91–103.

Altememy, D., Alizadeh, A. and Ghasemi, S. (2022). Current Medical Applications of *Terminalia Chebula*: An Overview. *Journal of Pharmaceutical Negative Results*, 13(4), 183–192.

Amalraj, A. and Gopi, S. (2017). Medicinal properties of Terminalia arjuna (Roxb.) Wight & Arn.: a review. *Journal of Traditional and Complementary Medicine*, 7(1), 65–78.

Ancheril, I.J., Sharanesh, T., Naveen, K. and Deepthi, V. (2016). Snake bite treatment in Prayoga samuccayam. *Ancient Science of Life*, 35(4), 245.

Ashrafi, N., Gharanjig, K., Hosseinnezhad, M., Khajeh Mehrizi, M., Imani, H. and Razani, N. (2018). Dyeing properties and color fabrics using natural dye and mordant. *Progress in Color, Colorants and Coatings*, 11(2), 79–83.

Ashwini, R., Gajalakshmi, S., Mythili, S. and Sathiavelu, A. (2011). *Terminalia chebula*-a pharmacological review. *J Pharm Res*, 4(9), 2884–2887.

Asif, M., Ahmed, M.W., Khair, S., Murugeswaran, R., Meena, R., Alam, M., ... and Negi, R.K. (2019). Pharmacognostical studies of Halela Siyah (*Terminalia chebula* Retz.): An important Unani medicinal plant. *Research Journal of Pharmacognosy and Phytochemistry*, 11(4), 205–211.

Bae, S., Kim, S.Y., Do, M.H., Lee, C.H. and Song, Y.J. (2017). 1, 2, 3, 4, 6-Penta-O-galloyl-ß-D-glucose, a bioactive compound in Elaeocarpus sylvestris extract, inhibits varicella-zoster virus replication. *Antiviral Research*, 144, 266–272.

Bag, A., Bhattacharyya, S.K. and Chattopadhyay, R.R. (2013). The development of *Terminalia chebula Retz.* (Combretaceae) in clinical research. *Asian Pacific Journal of Tropical Biomedicine*, 3(3), 244–252.

Ballabh, B., Chaurasia, O.P., Ahmed, Z. and Singh, S.B. (2008). Traditional medicinal plants of cold desert Ladakh—used against kidney and urinary disorders. *Journal of Ethnopharmacology*, 118(2), 331–339.

Banerjee, S., Khatri, P.K. and Banerjee, S.K. (2022). Soil and Vegetation in Pachmarhi Biosphere Reserve and Their Correlation. *Soil Health and Environmental Sustainability*, 87–105.

Basha, S.J. and Code, Q.R. (2017). A review on *Terminalia chebula*. *IJPR*, 7(10).

Bode, M. (2004). Ayurvedic and Unani health and beauty products: Reworking India's medical traditions (Doctoral dissertation, Universiteit van Amsterdam [Host]).

Bulbul, M.R.H., Chowdhury, M.N.U., Naima, T.A., Sami, S.A., Imtiaj, M.S., Huda, N. and Uddin, M.G. (2022). A comprehensive review on the diverse pharmacological perspectives of *Terminalia chebula Retz. Heliyon*, e10220.

Bulus, T., Atawodi, S.E. and Mamman, M. (2011). Acute toxicity effect of the aqueous extract of Terminalia avicennioides on white albino rats. *Science World Journal*, 6(2), 1–4.

Chandil, S., Pansare, T.A. and Bamoriya, H. (2020). Medicinal Importance of Traditional Herbal Plant Haritaki (Terminalia Chebula): An Overview.

Chattopadhyay, R.R. and Bhattacharyya, S.K. (2007). PHCOG REV: Plant Review. Terminalia chebula: An update. *Pharmacognosy Reviews* Jan-May, 2007; 1: 151, 157.

Cheng, H.Y., Lin, T.C., Yu, K.H., Yang, C.M. and Lin, C.C. (2003). Antioxidant and free radical scavenging activities of *Terminalia chebula. Biological and pharmaceutical bulletin*, 26(9), 1331–1335.

Cock I, Courtney R. (2021) Comparison of the Antibacterial Activity of Australian Terminalia Spp. Extracts Against Klebsiella Pneumoniae: Novel Treatment for Ankylosing Spondylitis.

Cock, I.E. and Cheesman, M. (2021). Plants of the genus Terminalia: Phytochemical and antioxidant profiles, proliferation, and cancer. *Cancer*, 495–503.

Combretum argyrophyllum K.Schum. in GBIF Secretariat (2022). GBIF Backbone Taxonomy. Checklist dataset https://doi.org/10.15468/39omei accessed via GBIF.org on 2022-12-16.

Courtney, R. and Cock, I.E. (2022) Comparison of the antibacterial activity of Australian Terminalia spp. extracts against Klebsiella pneumoniae: a potential treatment for ankylosing spondylitis. *Inflammopharmacology,* 30:207–223

Daram, P., Jitta, S.R., Shreedhara, C.S., Misra, C.S., Gourishetti, K. and Lobo, R. (2021). Investigation of antiinflammatory and anti-arthritic potentials of Terminalia catappa bark using *in vitro* assays and carrageenan-induced inflammation, complete Freund's adjuvant induced arthritis model in rats. *South African Journal of Botany*, 141, 313–321.

Debnath, M., Chaudhuri, S., Nanda, A., Biswas, M. and Haldar, P.K. (2017). Pharmacognostic and preliminary physicochemical study of vidangadilouham–an ayurvedic antidiabetic herbomineral preparation. *Journal of Ayurvedic and Herbal Medicine*, 3(1), 15–26.

Dhama K., Karthik K., Khandia R., et al., (2018) Medicinal and therapeutic potential of herbs and plant metabolites/extracts countering viral pathogens-current knowledge and future prospects. *Curr Drug Metab*, 19:236–263.

Dharmaratne, M., Manoraj, A., Thevanesam, V., Ekanayake, A., Kumar, N.S., Liyanapathirana, V., ... and Bandara, B.M. (2018). Terminalia bellirica fruit extracts: *in-vitro* antibacterial activity against selected multidrug-resistant bacteria, radical scavenging activity and cytotoxicity study on BHK-21 cells. *BMC Complementary and Alternative Medicine*, 18(1), 1–12.

Dhillon, H.S., Thakur, A.K. and Singh, K.J. (2014). Growth and propagation aspects of some medicinally important trees in Chandigarh, India: a review. *Journal of Medicinal Plants*, 2(5), 29–35.

Dickison, W.C. (2000). Integrative plant anatomy. Academic press.

Dodke, P.C. and Pansare, T.A. (2017). Ayurvedic and modern aspect of *Terminalia chebula Retz.* Haritaki an overview. *International Journal of Ayurvedic and Herbal Medicine*, 7(2), 2508–2517.

Dusi, S. and Saminathan, J. (2021). Crude Extract of *Terminalia Chebula* Fruit Effect on Gastrointestinal Disorders Using Different Animal Models. *International Journal of Pharmaceutical and Bio Medical Science*, 1(9), 192–196.

Dutta, G. and Devi, A. (2015). Phenology and population structure of six tree species in tropical forest of Assam, northeast India. *Tropical Ecology*, 56(3), 393–399.

Eltimamy, M., Elshamarka, M., Aboelsaad, M., Sayed, M. and Moawad, H. (2022). Effects of alcoholic extract of *Terminalia Chebula* dried fruit on blood biochemical profile in diabetic rats. *Journal of Diabetes & Metabolic Disorders*, 1–12.

Feng, X.H., Xu, H.Y., Wang, J.Y., Duan, S., Wang, Y.C. and Ma, C.M. (2021). *In vivo* hepatoprotective activity and the underlying mechanism of chebulinic acid from *Terminalia chebula* fruit. *Phytomedicine*, 83, 153479.

Feng, X.H., Xu, H.Y., Wang, J.Y., Duan, S., Wang, Y.C. and Ma, C.M. (2021). *In vivo* hepatoprotective activity and the underlying mechanism of chebulinic acid from *Terminalia chebula* fruit. *Phytomedicine*, 83, 153479.

Flora of China Editorial Committee. (2018). *Flora of China*.

Gowtham, D.S., Rao, P.P., Goutham, M., Priya, K.K. and Sivudu, K.V. (2019). Haritaki–A Wonder Drug.

Gupta, A., Kumar, R., Bhattacharyya, P., Bishayee, A. and Pandey, A.K. (2020). Terminalia bellirica (Gaertn.) roxb. (Bahera) in health and disease: A systematic and comprehensive review. *Phytomedicine*, 77, 153278.

Gupta, J. and Dubey, R.K. (2019). Indigenous trees–An overview. *Indian Horticulture*, 64(4).

Gupta, P.C. (2012). Biological and pharmacological properties of *Terminalia chebula Retz*. (Haritaki)-An overview. *Int J Pharm Pharm Sci*, 4(3), 62–68.

Haghani, M., Jaafari, M., Meftahi, G.H., Behzadnia, M.J., Bahari, Z. and Jangravi, Z. (2022). Analgesic effects of *Terminalia Chebula* is mediated by suppression of brain protein expression of NGF and NF-κB and oxidative markers following neuropathic pain in rats.

Han, Q., Song, J., Qiao, C., Wong, L. and Xu, H. (2006). Preparative isolation of hydrolyzable tannins chebulagic acid and chebulinic acid from *Terminalia chebula* by high-speed counter-current chromatography. *Journal of Separation Science*, 29(11), 1653–1657.

Hanumantha, M., Bisanalli, S.I., Patil, R.S. and Manjunatha, G.O. (2019). Utilization pattern of non-timber forest products (NTFPS) in Siddapura taluk of Uttara Kannada district of western Ghat region, Karnataka. *Journal of Pharmacognosy and Phytochemistry*, 8(3), 901–905.

Haroon, R., Jelani, S. and Arshad, F.K. (2015). Comparative Analysis of Antioxidant Profiles of Bark, Leaves, and Seeds of Syzigium cumini (Indian Blackberry). Editorial Board.

Hassler, M. (2021). Synonymic checklists of the vascular plants of the world. *Catalogue of Life Checklist* (Version 2021-08-06).

Huang T-Y., Chi L-M. and Chien K-Y. (2018) Size-exclusion chromatography using reverse-phase columns for protein separation. *J Chromatogr A,* 1571:201–212.

Huang, Y.N., Zhao, D.D., Gao, B., Zhong, K., Zhu, R.X., Zhang, Y. and Gao, H. (2012). Anti-hyperglycemic effect of chebulagic acid from the fruits of *Terminalia chebula* Retz. *International Journal of Molecular Sciences*, 13(5), 6320–6333.

Hussain, S.Z., Naseer, B., Qadri, T., Fatima, T. and Bhat, T.A. (2021). Anola (Emblica officinalis): Morphology, Taxonomy, Composition and Health Benefits. In Fruits Grown in Highland Regions of the Himalayas (pp. 193–206). Springer, Cham.

Ikram, A., Saeed, F., Munir, H., Sultan, M.T., Afzaal, M., Ahmed, A. and Anjum, F.M. (2021). Exploring the amino acid profile and microbial properties of locally sweet preserved kachra hareer (*Terminalia chebula*). *Food Science & Nutrition*, 9(2), 909–919.

Im, M., Yoon, Y., Lee, S.K.J., Lee, H.M., Kim, J., Lee, C.U., ... and Shim, H. (2017). Distance and Properties of NGC 4993 as the Host Galaxy of the Gravitational-wave Source GW170817. *The Astrophysical Journal Letters*, 849(1), L16.

Jain, R. and Tiwari, A. (2020). Biological monograph: Haritaki (*Terminalia chebula*). *Matrix Science Pharma*, 4(2), 65.

Jami, M.S.I., Sultana, Z., Ali, M.E., Begum, M.M. and Haque, M.M. (2014). Evaluation of analgesic and antiinflammatory activities on ethanolic extract of *Terminalia chebula* fruits in experimental animal models.

Jeong, H.K., Lee, D., Kim, H.P. and Baek, S.H. (2019). Structure analysis and antioxidant activities of an amylopectin-type polysaccharide isolated from dried fruits of *Terminalia chebula*. *Carbohydrate Polymers*, 211, 100–108.

Jha, A.K. and Sit, N. (2020). Drying characteristics and kinetics of colour change and degradation of phytocomponents and antioxidant activity during convective drying of deseeded *Terminalia chebula* fruit. *Journal of Food Measurement and Characterization*, 14(4), 2067–2077.

Jordaan, M., Van Wyk, A.E. and Maurin, O. (2011). Generic status of Quisqualis (Combretaceae), with notes on the taxonomy and distribution of Q. parviflora. *Bothalia*, 41(1), 161–169.

Jose, M., Sharma, B.B., Shantaram, M. and Ahmed, S.A. (2011). Ethnomedicinal herbs used in oral health and hygiene in coastal Dakshina Kannada. *J Oral Health Comm Dent*, 5(3), 119–123.

Juang, L.J., Sheu, S.J. and Lin, T.C. (2004). Determination of hydrolyzable tannins in the fruit of *Terminalia chebula Retz.* by high-performance liquid chromatography and capillary electrophoresis. *Journal of Separation Science*, 27(9), 718–724.

Juang, L.J., Sheu, S.J. and Lin, T.C. (2004). Determination of hydrolyzable tannins in the fruit of *Terminalia chebula Retz.* by high-performance liquid chromatography and capillary electrophoresis. *Journal of Separation Science*, 27(9), 718–724.

Kadir, M.F., Karmoker, J.R., Alam, M.D., Jahan, S.R., Mahbub, S. and Mia, M.M.K. (2015). Ethnopharmacological survey of medicinal plants used by traditional healers and indigenous people in Chittagong Hill Tracts, Bangladesh, for the treatment of snakebite. Evidence-Based Complementary and Alternative Medicine, 2015.

Kailash, B.R., Charles, B., Ravikanth, G., Setty, S. and Kadirvelu, K. (2022). Identifying the potential global distribution and conservation areas for *Terminalia chebula*, an important medicinal tree species under changing climate scenario. *Tropical Ecology*, 1–12.

Kala, C.P. (2022). Traditional Ecological Knowledge of Tribal Communities and Sustainability of Nature and Natural Resources in Pachmarhi Biosphere Reserve in India. *International Journal of Ecology*, 2022.

Kannan, M. (2018) Open Labeled Randomized Controlled Clinical Trial on Impaired Glucose Tolerance–IGT (Munneerizhivu).

Kapoor, D., Vijayvergiya, R. and Dhawan, V. (2014). Terminalia arjuna in coronary artery disease: ethnopharmacology, pre-clinical, clinical & safety evaluation. *Journal of Ethnopharmacology*, 155(2), 1029–1045.

Kar, S.K., Tripathy, P.K., Mohanty, S.K., Acharya, P.K. and Panda, P.C. (2016). Additions to the flora of Chilika lake and its immediate neighbourhood. *Journal of Economic and Taxonomic Botany*, 40(3), 134–150.

Karel, D.K., Ammar, S., Jari, S., Marja, K., Jyrki, L., Peteri, T., Kalevi, P. (2004). The Structural and Conformational analyses and antioxidant activities of Chebulinic acid and its thrice-hydrolyzed derivative, 2, 4-chebuloylß-d-glucopyranoside, isolated from the fruit of *Terminalia chebula*. *ARKIVOC*. 7:83–105.

Kaur, C.P. (2022). Analeptic Mushrooms. *AGRICULTURE & FOOD E-NEWSLETTER*, 2.

Kausar, F., Amin, K.M.Y., Bashir, S., Parvez, A. and Ahmad, P. (2021). Concept of 'Ihtiraq' in Unani Medicine–A correlation with oxidative stress, and future prospects. *Journal of Ethnopharmacology*, 265, 113269.

Keerthana, J., Priya, A.J. and Devi, R.G. (2018). Physiological and pharmacological activities of Terminalia bellerica. *Drug Invention Today*, 10(10).

Kesharwani, A., Polachira, S.K., Nair R., et al., (2017) Anti-HSV-2 activity of *Terminalia chebula Retz* extract and its constituents, chebulagic and chebulinic acids. *BMC Complement Altern Med* 17:1–11.

Kesharwani, A., Polachira, S.K., Nair, R., Agarwal, A., Mishra, N.N. and Gupta, S.K. (2017). Anti-HSV-2 activity of *Terminalia chebula* Retz extract and its constituents, chebulagic and chebulinic acids. *BMC Complementary and Alternative Medicine*, 17(1), 1–11.

Khan, A., Siddiqui, A., Jafri, M.A. and Asif, M. (2018). Sharbat Ahmad Shahi: A Potent Unani Formulation for Mood Disorders. *Int. J. Sci. Res. in Biological Sciences,* Vol, 5, 6.

Khan, M.U., Khalilullah, H., Akhtar, J. and Elhasan, G. O. (2015). *Terminalia chebula*: an ephemeral glance. *Int J Pharm Pharmac Sci*, 7(2), 40–43.

Khanna, N. and Sharma, S. (2013). Allium cepa root chromosomal aberration assay: a review. *Indian Journal of Pharmaceutical and Biological Research*, 1(03), 105–119.

Kim, T.G., Kang, S.Y., Jung K.K., et al. (2001) Antiviral activities of extracts isolated from Terminalis *chebula Retz.*, Sanguisorba officinalis L., Rubus coreanus Miq. and Rheum palmatum L. against hepatitis B virus. *Phyther Res*, 15:718–720.

Kim, H.J., Song, H.K., Park, S.H., Jang, S., Park, K.S., Song, K.H., ... and Kim, T. (2022). *Terminalia chebula* Retz. extract ameliorates the symptoms of atopic dermatitis by regulating antiinflammatory factors *in vivo* and suppressing STAT1/3 and NF-κB signaling *in vitro. Phytomedicine*, 104, 154318.

Kim, H.L., Choi, B.K. and Yang, S.H. (2022). *Terminalia chebula* Medicinal Uses: A Review of *in vitro* and *in vivo* Studies. *Biotechnology and Bioprocess Engineering*, 1–11.

Koriem, K.M.M. (2014). Antihyperlipidemic activity of the medicinal plants among Kadazan and Dusun communities in Sabah, Malaysia: a review. *Asian Pacific Journal of Tropical Biomedicine*, 4(10), 768–779.

Kumar, A., Kumar, S., Chaudhary, S., Kumar, S., Dwivedi, K.N. and Ram, B. (2017). Encyclopedic Study of Haritaki (*Terminalia Chebula Ritz.*) *In Reference to Prameha. Journal of Medical Science and Clinical Research*, 5(7), 25484–25492.

Kumar, A., Lakshman, K., Jayaveera, K.N., Tripathi, S.M. and Satish, K.V. (2010). Estimation of gallic acid, rutin and quercetin in *Terminalia chebula* by HPTLC. *Jordan J Pharm Sci*, 3(1), 63–68.

Lee, M. and Hwang, Y.S. (2021). Anticaries Effect of Ethanol Extract of *Terminalia chebula. Journal of Dental Hygiene Science*, 21(2), 119–126.

Li, K., Diao, Y., Zhang, H., Wang, S., Zhang, Z., Yu, B., ... and Yang, H. (2011). Tannin extracts from immature fruits of *Terminalia chebula* Fructus Retz. promote cutaneous wound healing in rats. *BMC Complementary and Alternative Medicine*, 11(1), 1–9.

Li, K., Han, X., Li, R., Xu, Z., Pan, T., Liu, J., ... and Liu, X. (2019). Composition, antivirulence activity, and active property distribution of the fruit of *Terminalia chebula Retz. Journal of Food Science*, 84(7), 1721–1729.

Lin L-T, Chen T-Y, Chung C-Y, et al., (2011) Hydrolyzable tannins (chebulagic acid and punicalagin) target viral glycoprotein-glycosaminoglycan interactions to inhibit herpes simplex virus 1 entry and cell-to-cell spread. *J Virol,* 85:4386–4398

Lin, K., Zhou, M., Leng, C., Tao, X., Zhou, R., Li, Y., ... and Liu, W. (2022). Neuroprotective Effect of Polyphenol Extracts from *Terminalia chebula Retz.* against Cerebral Ischemia-Reperfusion Injury. *Molecules*, 27(19), 6449.

Loganathan, M.N. (2019). Harītakī (*Terminalia chebula* Retz.). *The Journal*, 26(4).

Lopez, H.L., Habowski, S.M., Sandrock, J.E., Raub, B., Kedia, A., Bruno, E.J. and Ziegenfuss, T.N. (2017). Effects of dietary supplementation with a standardized aqueous extract of *Terminalia chebula* fruit (AyuFlex®) on joint mobility, comfort, and functional capacity in healthy overweight subjects: a randomized placebo-controlled clinical trial. *BMC Complementary and Alternative Medicine*, 17(1), 1–18.

Mahajan, D.R. and Haswani, N.G. Updated Review: Pharmacological Activities and Bioactive Constituents of *Terminalia Chebula.*

Mahajan, C.P. and Aher, A.N. (2017). A review on ethnobotanical, phytochemical and pharmacological activities of Quisqualis indica Linn. *Research Journal of Pharmacognosy and Phytochemistry*, 9(1), 47–52.

Mahapatra, A.K. and Tewari, D.D. (2005). Importance of non-timber forest products in the economic valuation of dry deciduous forests of India. *Forest Policy and Economics*, 7(3), 455–467.

Mandeville, A. and Cock, I.E. (2018) *Terminalia chebula Retz.* fruit extracts inhibit bacterial triggers of some autoimmune diseases and potentiate the activity of tetracycline. *Indian J Microbiol,* 58:496–506.

Mani, V., Sajid, S., Rabbani, S.I., Alqasir, A.S., Alharbi, H.A. and Alshumaym, A. (2021). Anxiolytic-like and antidepressant-like effects of ethanol extract of *Terminalia chebula* in mice. *Journal of Traditional and Complementary Medicine*, 11(6), 493–502.

Manikandan, R., Balasubramanian, B., Punniyakotti, P., Vijaya Anand, A., Meyyazhagan, A., Velayuthaprabhu, S., ... and Liu, W.C. (2022). Cardio-protective effects of terminalia catappa leaves and *terminalia chebula* fruit extract in doxorubicin-induced cardiomyopathy in rats. *Biomarkers*, 1–8.

Maruyama, T., Abbaskhan, A., Choudhary, M.I., Tsuda, Y., Goda, Y., Farille, M. and Reduron, J.P. (2009). Indusyunic medicine, traditional medicine of herbal. Animal and mineral origin in Pakistan, 104–106, 1997. *Journal of Natural Medicines*, 63(3), 248–253.

Meher, S.K., Panda, P., Das, B., Bhuyan, G.C. and Rath, K.K. (2018). Pharmacological profile of *Terminalia chebula Retz.* and Willd.(Haritaki) in Ayurveda with evidences. *Research Journal of Pharmacology and Pharmacodynamics*, 10(3), 115–124.

Mehra, R., Makhija, R. and Vyas, N. (2012). Role of *Terminalia chebula* on gastrointestinal mucosa. *Research Journal of Pharmacy and Technology*, 5(9), 1183.

Menpara, D., Desai, D. and Chanda, S. (2014). Pharmacognostic, phytochemical, physicochemical and fluorescence analysis of Terminalia bellerica leaf and stem. *World Journal of Pharmaceutical Sciences*, 390–396.

Mishra, V., Agrawal, M., Onasanwo, S.A., Madhur, G., Rastogi, P., Pandey, H.P., ... and Narender, T. (2013). Anti-secretory and cyto-protective effects of chebulinic acid isolated from the fruits of *Terminalia chebula* on gastric ulcers. *Phytomedicine*, 20(6), 506–511.

Mitra, A. and Zaman, S. (2020). Biodiversity and Its Conservation. In Environmental Science–A Ground Zero Observation on the Indian Subcontinent (pp. 143–214). Springer, Cham.

Miyasaki, Y., Rabenstein, J.D., Rhea, J., Crouch, M.L., Mocek, U.M., Kittell, P.E., ... and Liu, G.Y. (2013). Isolation and characterization of antimicrobial compounds in plant extracts against multidrug-resistant Acinetobacter baumannii. *PloS one*, 8(4), e61594.

Mobeen, A. and Moazzam, S.W. (2022). Jawarish Shahi: A special dosage form of herbal formulations for functional gastrointestinal disorders in Unani medicine–A comprehensive review. *Journal of Ethnopharmacology*, 293, 115319.

Myrobalanifera citrina Houtt. in GBIF Secretariat (2022). GBIF Backbone Taxonomy. Checklist dataset https://doi.org/10.15468/39omei accessed via GBIF.org on 2022-12-16.

Myrobalanus gangetica Kostel. in GBIF Secretariat (2022). GBIF Backbone Taxonomy. Checklist dataset https://doi.org/10.15468/39omei accessed via GBIF.org on 2022-12-16.

Nagadesi, P.K., Arya, A. and Albert, S. (2022). Selective and Simultaneous Delignification Capacity of Wood Decay Fungus Trametes pini in Tectona grandis L. f. and Terminalia crenulata (Heyne) Roth. *In Biology, Cultivation and Applications of Mushrooms* (pp. 535-545). Springer, Singapore.

Nigam, M., Mishra, A.P., Adhikari-Devkota A., et al. (2020) Fruits of *Terminalia chebula Retz.*: A review on traditional uses, bioactive chemical constituents and pharmacological activities. *Phyther Res*, 34:2518–2533.

Nigam, M., Mishra, A.P., Adhikari-Devkota, A., Dirar, A.I., Hassan, M.M., Adhikari, A., ... and Devkota, H.P. (2020). Fruits of *Terminalia chebula Retz.*: A review on traditional uses, bioactive chemical constituents and pharmacological activities. *Phytotherapy Research*, 34(10), 2518–2533.

Nyabuto, H.N. (1994). Description and analysis of Goa's Western Ghats' Vegetation with reference to case studies (Doctoral dissertation, Goa University).

Oyuntsetseg, N., Khasnatinov, M.A., Molor-Erdene, P., et al. (2014) Evaluation of direct antiviral activity of the Deva-5 herb formulation and extracts of five Asian plants against influenza A virus H3N8. *BMC Complement Altern Med* 14:1–9.

Pandey, M., Choudhury, H., Abdul-Aziz A., et al. (2020) Advancement on sustained antiviral ocular drug delivery for herpes simplex virus keratitis: Recent update on potential investigation. *Pharmaceutics*, 13:1.

Park, H-J., Kurokawa, M., Shiraki, K., et al. (2005) Antiviral activity of the marine alga Symphyocladia latiuscula against herpes simplex virus (HSV-1) *in vitro* and its therapeutic efficacy against HSV-1 infection in mice. *Biol Pharm Bull,* 28:2258–2262.

Patil, V.S., Harish, D.R., Vetrivel, U., Roy, S., Deshpande, S.H. and Hegde, H.V. (2022). Hepatitis C Virus NS3/4A Inhibition and Host Immunomodulation by Tannins from *Terminalia chebula*: A Structural Perspective. *Molecules*, 27(3), 1076.

Pellati, F., Bruni, R., Righi, D., Grandini, A., Tognolini, M., Prencipe, F.P., ... and Rossi, D. (2013). Metabolite profiling of polyphenols in a *Terminalia chebula* Retzius ayurvedic decoction and evaluation of its chemopreventive activity. *Journal of Ethnopharmacology*, 147(2), 277–285.

Pfundstein, B., El Desouky, S.K., Hull, W.E., Haubner, R., Erben, G. and Owen, R.W. (2010). Polyphenolic compounds in the fruits of Egyptian medicinal plants (Terminalia bellerica, *Terminalia chebula* and Terminalia horrida): characterization, quantitation and determination of antioxidant capacities. *Phytochemistry*, 71(10), 1132–1148.

Philips, C.A., Ahamed, R., Abduljaleel, J.K., Rajesh, S. and Augustine, P. (2021). Critical updates on chronic hepatitis B virus infection in 2021. *Cureus*, 13(10).

Prasad, S.K., Kumar, R., Patel, D.K. and Hemalatha, S. (2010). Wound healing activity of Withania coagulans in streptozotocin-induced diabetic rats. *Pharmaceutical Biology*, 48(12), 1397–1404.

Prasad, S. and Srivastava, S.K. (2020). Oxidative stress and cancer: chemopreventive and therapeutic role of triphala. *Antioxidants*, 9(1), 72.

Rahaman, M.S., Siraj, M.A., Islam, M.A., Shanto, P.C., Islam, O., Islam, M.A. and Simal-Gandara, J. (2022). Crosstalk between xanthine oxidase (XO) inhibiting and cancer chemotherapeutic properties of comestible flavonoids-a comprehensive update. *The Journal of Nutritional Biochemistry*, 109147.

Rai, A.K. and Joshi, R. (2009). Evaluation of antimicrobial properties of fruit extracts of *Terminalia chebula* against dental caries pathogens.

Rajani, S. and Kumar, R.S. (2021). Review of Haritaki (*Terminalia Chebula Retz.*) as an Ekal Dravya Chikitsa in Brihatrayee. *International Journal of Ayurveda and Pharma Research*, 33–40.

Raju, A.S., Lakshmi, P.V. and Ramana, K.V. (2012). Reproductive ecology of Terminalia Pallida Brandis (Combretaceae), an endemic and medicinal tree species of India. *Current Science*, 909–917.

Rana, C.S., Kimothi, G.P., Rai, R., Narayan, S.B. and Sastry, J.L.N. (2018). Triphala churna, myth and microscopic characterization. *Annals of Phytomedicine*, 7(1), 158–164.

Randhawa, M., Meyer, T., Sachdev, M. and Chaudhuri, R.K. (2021). Standardized *Terminalia chebula* Fruit Extract: A Natural Ingredient that Provides Long-Lasting Antioxidant Protection and Reverses Visible Signs of Pollution-Induced Skin Damage. Clinical, *Cosmetic and Investigational Dermatology*, 14, 1257.

Rangari, V.D. (2008). Tannin containing drugs.

Rao, N.K. and Nammi, S. (2006). Antidiabetic and renoprotective effects of the chloroform extract of *Terminalia chebula Retz.* seeds in streptozotocin-induced diabetic rats. *BMC Complementary and Alternative Medicine*, 6(1), 1–6.

Reddy, D.B., Reddy, T.C.M., Jyotsna, G., Sharan, S., Priya, N., Lakshmipathi, V. and Reddanna, P. (2009). Chebulagic acid, a COX–LOX dual inhibitor isolated from the fruits of *Terminalia chebula Retz.*, induces apoptosis in COLO-205 cell line. *Journal of Ethnopharmacology*, 124(3), 506–512.

Riaz, M., Asghar, A., Rashid, R. and Shahzad-ul-Hussan, S. (2015). Treasures hunt in old mines: *Terminalia Chebula*-Based traditional herbal medicinal products. *The Natural Products Journal*, 5(4), 252–267.

Roy, S., Rabinarayan, A., Harisha, C.R. and Switu, J. (2015). Evaluation of pharmacognostical changes in the mature and tender leaves of Panchapallava (group of five leaves), described in Ayurveda. *International Journal of Green Pharmacy,* (IJGP), 9(4).

Ruan, J.J., Nynka M., Haggard, D., et al. (2018) Brightening X-ray emission from GW170817/GRB 170817A: further evidence for an outflow. *Astrophys J Lett,* 853:L4.

Safavi, M., Shams-Ardakani, M. and Foroumadi, A. (2015). Medicinal plants in the treatment of Helicobacter pylori infections. *Pharmaceutical Biology*, 53(7), 939–960.

Saha, M., Kumar, M.B. and Singh, A. (2020). A Review Article on Various Phytochemicals and Different Medicinal Activities of Haritaki. *LANGUAGE*, 12, 14.

Said, M., Barkat, A.K., Naveed, A., Tariq, M., Akhtar, R., Irshad, H., … and Amir, B. (2012). The morphology, extractions, chemical constituents and uses of *Terminalia chebula*: A review. *Journal of Medicinal Plants Research*, 6(33), 4772–4775.

Saleem, A., Husheem, M., Härkönen, P. and Pihlaja, K. (2002). Inhibition of cancer cell growth by crude extract and the phenolics of *Terminalia chebula retz.* fruit. *Journal of Ethnopharmacology*, 81(3), 327–336.

Samuel, O.S., Ramon, B.O. and Johnson, Y.O. (2012). Thermal conductivity of three different wood products of combretaceae family; terminalia superb, Terminalia Ivorensis and Quisqualis indica. *Journal of Natural Sciences Research*, 2(4), 36–43.

Samy, R.P., Pushparaj, P.N. and Gopalakrishnakone, P. (2008). A compilation of bioactive compounds from Ayurveda. *Bioinformation*, 3(3), 100.

Sarabhai, S., Sharma, P. and Capalash, N. (2013). Ellagic acid derivatives from *Terminalia chebula Retz.* downregulate the expression of quorum sensing genes to attenuate Pseudomonas aeruginosa PAO1 virulence. *PLoS one*, 8(1), e53441.

Sarkar, A., Agarwal, R. and Bandyopadhyay, B. (2022). Molecular docking studies of phytochemicals from *Terminalia chebula* for identification of potential multi-target inhibitors of SARS-CoV-2 proteins. *Journal of Ayurveda and Integrative Medicine*, 100557.

Sarkar, R., Hazra, B. and Mandal, N. (2012). Reducing power and iron chelating property of *Terminalia chebula* (Retz.) alleviates iron induced liver toxicity in mice. *BMC Complementary and Alternative Medicine*, 12(1), 1–10.

Schmidt-Przewoźna, K. and Brandys, A. (2016). Utilization of contaminated lands for cultivation of dye producing plants. *In Bioremediation and Bioeconomy* (pp. 329–359). Elsevier.

Shaheen, S., Jaffer, M., Khan, F., Hussain, K., Hanif, U., Younis, S., ... and Ishtiaq, S. (2018). Morpho-palynological assessment of medicinal flora of district Lahore, Pakistan based on LM and SEM. Microscopy Research and Technique, 81(12), 1397–1405.

Shaikh, A.S., Thomas, A.B. and Chitlange, S.S. (2020). Herb–drug interaction studies of herbs used in treatment of cardiovascular disorders—A narrative review of preclinical and clinical studies. *Phytotherapy Research*, 34(5), 1008–1026.

Shao, Y., Dang, M., Lin, Y. and Xue, F. (2019). Evaluation of wound healing activity of plumbagin in diabetic rats. *Life Sciences*, 231, 116422.

Sharifi-Rad J, Iriti M, Setzer W.N., et al. (2018) Antiviral activity of Veronica persica Poir. on herpes virus infection. *Cell Mol Biol*, 64:11–17.

Sharma, S., Singh, B. and Kumar, H. (2019). A critical review of pharmacological actions of haritaki (*Terminalia chebula* Retz) in classical texts. *Journal of Ayurveda and Integrated Medical Sciences,* 4(04), 258–269.

Sheng, Z., Yan, X., Zhang, R., Ni, H., Cui, Y., Ge, J. and Shan, A. (2016). Assessment of the antidiarrhoeal properties of the aqueous extract and its soluble fractions of Chebulae Fructus (*Terminalia chebula* fruits). *Pharmaceutical Biology*, 54(9), 1847–1856.

Shukla, A., Mehrotra, R.C. and Guleria, J.S. (2013). Emergence and extinction of Dipterocarpaceae in western India with reference to climate change: Fossil wood evidences. *Journal of Earth System Science*, 122(5), 1373–1386.

Singamaneni, V., Dokuparthi, S.K., Banerjee, N., Kumar, A. and Chakrabarti, T. (2020). Phytochemical investigation and antimutagenic potential of ethanolic extracts of Emblica officinalis, *Terminalia chebula* and Terminalia bellirica. *The Natural Products Journal*, 10(4), 488–494.

Singh, A.K., Raghubanshi, A.S. and Singh, J.S. (2002). Medical ethnobotany of the tribals of Sonaghati of Sonbhadra district, Uttar Pradesh, India. *Journal of Ethnopharmacology*, 81(1), 31–41.

Singh, M.P. and Sharma, C.S. (2010). Pharmacognostical evaluation of *Terminalia chebula* fruits on different market samples. *Int J ChemTech Res*, 2(1), 57–61.

Sireeratawong, S., Jaijoy, K., Panunto, W. and Soonthornchareonnon, N. (2012). Antiinflammatory activity and toxicity of the water extract of *Terminalia chebula* rezt in rats. *Planta medica*, 78(11), PI112.

Srikantha, K.V., V.K., C.K. and Nagaratna, S.J. (2018). Preparation of haritakyadi eye drops: An ayurvedic formulation for ophthalmia neonatorum.

Suchalatha, S., Srinivasan, P. and Devi, C.S.S. (2007). Effect of *Terminalia chebula* on mitochondrial alterations in experimental myocardial injury. *Chemico-biological interactions*, 169(3), 145–153.

Terminalia acuta Walp. in GBIF Secretariat (2022). GBIF Backbone Taxonomy. Checklist dataset https://doi.org/10.15468/39omei accessed via GBIF.org on 2022-12-16.

Terminalia aruta Buch. ex G.Don in GBIF Secretariat (2022). GBIF Backbone Taxonomy. Checklist dataset https://doi.org/10.15468/39omei accessed via GBIF.org on 2022-12-16.

Terminalia gangetica Roxb. in GBIF Secretariat (2022). GBIF Backbone Taxonomy. Checklist dataset https://doi.org/10.15468/39omei accessed via GBIF.org on 2022-12-16.

Terminalia glandulipetiolata De Wild. in GBIF Secretariat (2022). GBIF Backbone Taxonomy. Checklist dataset https://doi.org/10.15468/39omei accessed via GBIF.org on 2022-12-16.

Terminalia reticulata Roth in GBIF Secretariat (2022). GBIF Backbone Taxonomy. Checklist dataset https://doi.org/10.15468/39omei accessed via GBIF.org on 2022-12-16.

The Ayurvedic Pharmacopoeia of India (API); Government of India Ministry of Health and Family Welfare Department of Ayush, Part-I: Vol-II. 2017.

The Ayurvedic Pharmacopoeia of India (API); Government of India Ministry of Health and Family Welfare Department of Ayush, Part-I: Vol-I. 2013.

The International Plant Names Index and World Checklist of Vascular Plants 2022. Published on the Internet at http://www.ipni.org and https://powo.science.kew.org/

Upadhyay, A., Agrahari, P. and Singh, D.K. (2014) A review on the pharmacological aspects of *Terminalia chebula*. *Int J Pharmacol,* 10:289–298.

Variya, B.C., Bakrania, A.K. and Patel, S.S. (2016). Emblica officinalis (Amla): A review for its phytochemistry, ethnomedicinal uses and medicinal potentials with respect to molecular mechanisms. *Pharmacological Research*, 111, 180–200.

Venkatachalam, P. and Chittibabu, C.V. (2020). Antifungal activity of *Terminalia chebula* fruit extracts. *Curr Bot*, 11, 216–220.

Venkatesan, A., Kathirvel, A., Prakash, S. and Sujatha, V. (2017). Antioxidant, antibacterial activities and identification of bioactive compounds from *Terminalia chebula* bark extracts. *Free Radicals and Antioxidants*, 7(1), 43–49.

Wright M.H., Shalom. J., Matthews B., et al. (2019) Terminalia ferdinandiana Exell: Extracts inhibit Shewanella spp. growth and prevent fish spoilage. *Food Microbiol,* 78:114–122.

Yazarlu, O., Iranshahi, M., Kashani, H.R.K., Reshadat, S., Habtemariam, S., Iranshahy, M. and Hasanpour, M. (2021). Perspective on the application of medicinal plants and natural products in wound healing: A mechanistic review. *Pharmacological Research*, 174, 105841.

17 How Potent is Natural Sugar Destroyer "*Gymnema sylvestre* R. Br. ex Sm." Apropos of Antiviral Activity?

K. Venkata Ratnam[1#], Anu Pandita[2#], Lepakshi Md. Bhakshu[3], Pulala Raghuveer[4] and Deepu Pandita[5*]

1. Taxonomy and Scientific Classification

Gymnema sylvestre (Retz.) R. Br. ex Sm. is usually titled Gurmar because of its decreasing potential (Tiwari et al., 2014; Bindu and Narendhirakannan, 2019). The generic name *Gymnema* has been framed by mingling two words of Greek origin, *Gymnos* (meaning naked) and *nema* (meaning thread), whereas, the species name *sylvestre* has Latin rootage and means *of the forest* (Pramanick, 2016). *Gymnema sylvestre* is botanically also known as *Periploca sylvestre* Retz. *Asclepias geminata*, *Gymnema montanum*, *Asclepias geminata*, or *Gymnema melicida* and in English language as the Periploca of the woods (Kanetkar et al., 2007; Di Fabio et al., 2013). Gurmar is also known as Small Indian ipecac, Ram's horn, Sugar Destroyer, Kalikardori, Vakundi, Gurmarbooti, Adigam, Cherukurinja, Meshashringi, Mardashingi, Merasingi, Kavali, Podapatri, Dhuleti, Sannagerasehambu, Chigengteng or Australian Cowplant (Kanetkar et al., 2007; Di Fabio et al., 2013).

Biological Classification of *Gymnema sylvestre*:

Kingdom	:	Plantae
Class	:	Dicotyledons
Subclass	:	Gamopetale
Series	:	Bicarpellatae
Order	:	Gentianales
Family	:	Apocynaceae
Genus	:	*Gymnema*

[1] Department of Botany, Rayalaseema University, Kurnool, Andhra Pradesh, India.
[2] Vatsalya Clinic, Krishna Nagar, New Delhi, India.
[3] Department of Botany, PVKN Government College (A), Chittoor, Andhra Pradesh, India.
[4] Department of Biotechnology, Indian Institute of Technology Hyderabad, Kandi, Telangana State, India.
[5] Government Department of School Education, Jammu, Jammu and Kashmir, India: ORCID iD-https://orcid.org/0000-0002-8361-1872
[#] *Both these authors have contributed equally*
[*] Corresponding author: deepupandita@gmail.com

DOI: 10.1201/9781003433583-17

Species : *sylvestre*
Binomial name: *Gymnema sylvestre* (Retz.) R.Br. ex Sm

2. Botanical Description of Plant

G. sylvestre known as sugar destroyer for its distinct sugar-lowering property finds use in the Indian Ayurvedic System of Medicine to treat diabetes, and as an anti-diabetic vine in the Indian Pharmacopoeia (Singh et al., 2008; Tiwari et al., 2014). This is a pubescent, vulnerable (VU), perennial woody vine that reaches up to 600 m in height, slow growing straggling shrub that may climb up to tree top, and highly branched and more or less pubescent (Kanetkar et al., 2007; Wu et al., 2012; Kapoor, 2017; Khan et al., 2019; Jamadagni et al., 2021). The shrub has a tap root system and the roots are hairy (Najafi and Deokule, 2011; Rajashekar et al., 2016) whereas the stem is pubescent, cylindrical, branched, hard, twining with terete internodes, 0.7 cm to 17.2 cm in length and 2 mm to 10 mm in the diameter (Najafi and Deokule, 2011; Pramanick, 2016). Aerial parts contain milky latex. The leaves of woody vine are simple, opposite, decussate in pattern with short petioles and green in colour, 2.5 cm to 6 cm in length, 3 mm to 16 mm in diameter, elliptic-ovate shape, tip acute or shortly acuminate, and margins entire with round base (Kanetkar et al., 2007; Gurav et al., 2007; Saneja, 2010; Patel, 2017; Habtemariam, 2019). Leaves are densely velvety pubescent on the lower sides, smooth on the above side, and ciliate on the margins, with trichomes mainly on leaf nerves and have a characteristic smell and a light bitter and astringent taste. Venation of leaves is transverse and reticulate in nature (Kirtikar and Basu, 1975; Saneja, 2010; Pramanick, 2016; Patel, 2017; Habtemariam, 2019). Flowering takes place from June to August with a flowering time of 3 months (Sinha and Mondal, 2017). Flowers are small in small umbels and yellow. The inflorescence is axillary and lateral umbel in cymes (Gurav et al., 2007; Kanetkar et al., 2007). Sepals (calyx) of the flower are 5-lobate, obtuse, ovate, and ciliated whereas the petals (corolla) are campanulated, bell-shaped, and each petal with 5-fleshy lobed scales and the colour is pale yellow (Pramanick, 2016; Sinha and Mondal, 2017). Gurmar flowers have two carpels and five pollinia (Sinha and Mondal, 2017). This woody vine starts to produce fruits in October (Karale and Karale, 2017). The fruit follicles are terete, cylindrical, lanceolate, 3 inches long, and dark green (Kanetkar et al., 2007; Gurav et al., 2007). Seeds of shrubs are flat, oval, 1.3 cm long, and have a thin marginal wing like a white coma (Kirtikar and Basu, 1975; Chopra et al., 2002). Seeds have low viability and alternative propagation approaches include via cuttings of roots in June and/or July and in February and/or March through terminal cuttings (Kirtikar and Basu, 1975).

2.1 Geographical Distribution

This wild *G. sylvestre* is native to India and tropical Asia and prefers 300-700 m elevation. This plant has broad distribution in Eastern Africa to Saudi Arabia,

Fig. 1 (a) Flowering stem (b) Leaves (c) Seed Capsules (d) Seed dispersal from capsules of *Gymnema sylvestre.*
(*Source:* Adapted from https://tropical.theferns.info/imyage.php?id=Gmnema+sylvestre)

tropical rainforests or dry forests of central (Madhya Pradesh and Chhattisgarh) and Peninsular or southern India, Africa, Sri Lanka (Ceylon), Malaysia, Vietnam, Southern China, Japan (Nansei Islands or Ryukyu Arc), Arabian Peninsula, Republic of the Philippines, Republic of Indonesia, Commonwealth of Australia, West and South Africa and Federal Democratic Republic of Ethiopia (Christopoulos et al., 2010; Wu et al., 2012; Vijayakumar and Prabhu, 2014; Kapoor, 2017; Jamadagni et al., 2021) (Fig. 2).

2.2 Phytochemical Constituents and Metabolite Profiling

The *G. sylvestre* popularly known as "Madhu nashini" (Sugar killer) in the Ayurvedic system of medicine shows glucose-lowering activity. Gymnema is a

Fig. 2 Global distribution of *Gymnema sylvestre*
(*Source:* https://powo.science.kew.org/taxon/urn:lsid:ipni.org:names:60454719-2#distributions)

treasure house of several pharmacologically active secondary metabolites and chemical components (Laha and Paul, 2019). The phytochemical components responsible for the sweet suppression activity of this herb include oleanane and dammarane-type triterpenoid saponins and gurmarin polypeptide (Tiwari et al., 2014). The glucose-lowering activity of *G. sylvestre* is attributed to gurmarin, gymnemic acid, and gymnema saponins (Khan et al., 2019). Other chemical constituents of *G. sylvestre* include pentatriacontane, anthraquinones, flavones (flavonoids), hentriacontane (untriacontane), tartaric acid, pcalcium oxalate, hytin (myo-inositol hexakisphosphate), resins, formic acid (methanoic acid), butyric acid (butanoic acid), lupeol (Fagarsterol), β-amyrin related glycosides, and stigmasterol (phytosterol) (Sinsheimer et al., 1970). A series of oleanane-type triterpenoid saponins, called gymnemic acids (gymnemic acid I to VII) were isolated and characterized from dry leaves of *G. sylvestre* (Yoshikawa et al., 1989 a,b). Later, the other five gymnemic acids (VIII, IX, X, XI, and XII) from Gymnema (Figure 2a) were separated and characterized (Yoshikawa et al., 1992b). Dammarane-type triterpenoid saponins known as gymnemosides I to VII and gypenosides XXVIII, XXXVII, LV, LXII, and LXIII (Figure 2b and 2c) were separated and characterized by Yoshikawa et al. (1992a). From above-ground parts of *G. sylvestre* and *Euphorbia ebracteolata*, two new flavonol glycosides *i.e.*, quercetin 3-O-6"-(3-hydroxyl-3-methylglutaryl)-beta-D-galactopyranoside and kaempferol 3-O-beta-glucopyranosyl-(1→4)-alpha-L-rhamnopyranosyl-(1→6)-beta-D-galactopyranoside were isolated (Liu et al. 2004).

A new triterpenoid saponin, 22α-hydroxy-longispinogenin 3-*O*-β-D-glucuronopyranosyl-28-*O*-α-L-rhamnopyranoside was separated along with seven other known components and their structures were elucidated by Liu et al. (2014). The components are conduritol A (Kondurite), stigmasterol (Stigmasterin), lupeol (Fagarasterol), stigmasterol-3-*O*-β-D-glucoside (Stigmasterol glucoside) (Figure

Gymnema sylvestre Phytochemicals

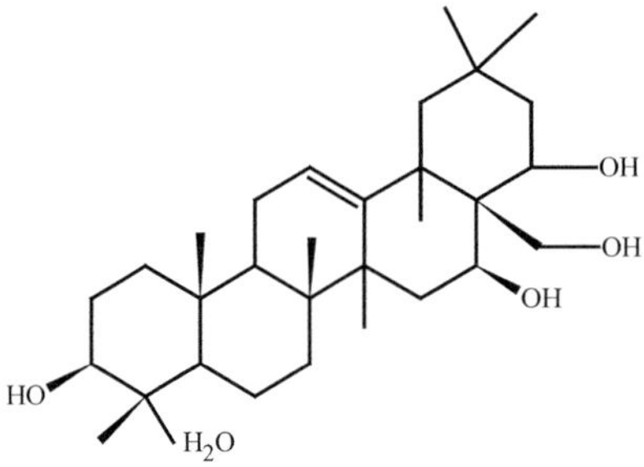

Gymnemanol

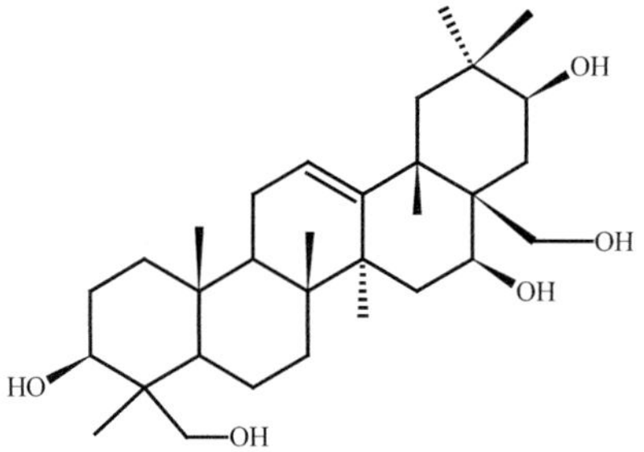

Gymnemagenin

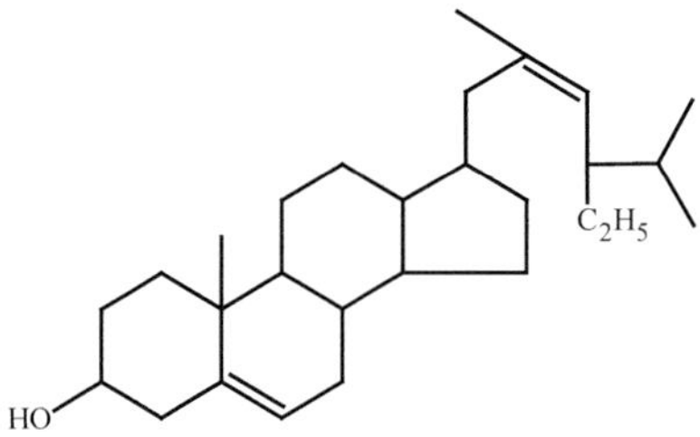

Stigmasterol

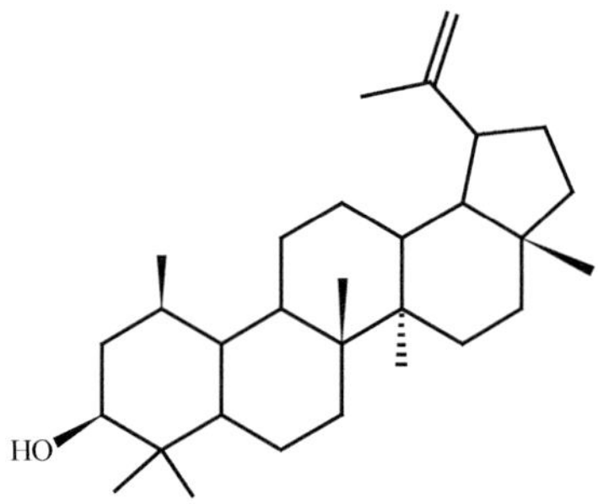

Lupeol

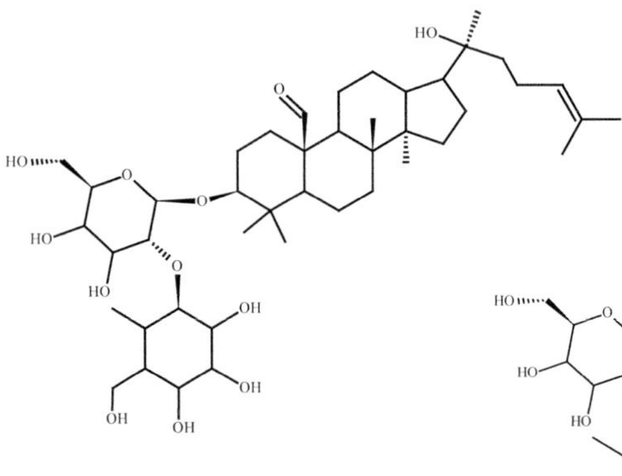

Gypenoside XXVIII

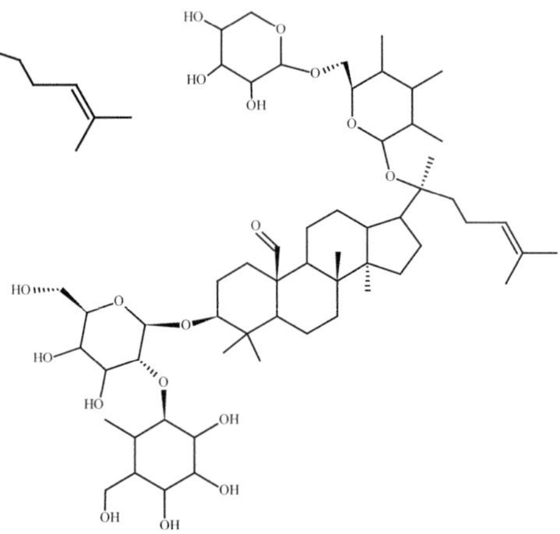

Gypenoside LXIII

Fig. Contd.

Fig. Contd.

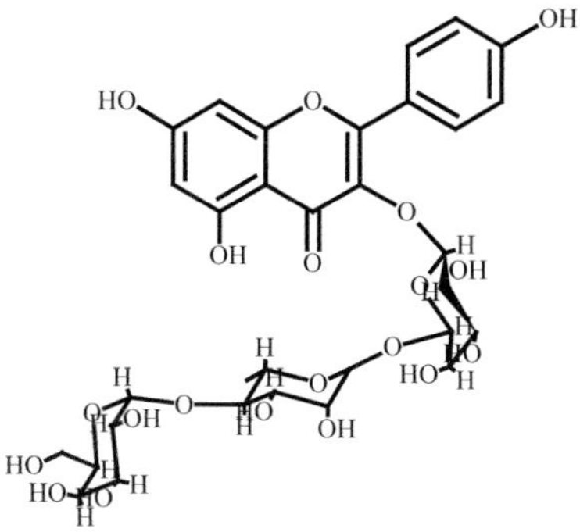

Kaempferol 3-*O*-β
-D-glucopyranosyl-(1-4)-α
-L-rhamnopyranosyl-(1-6)-β
-D-galactopyranoside

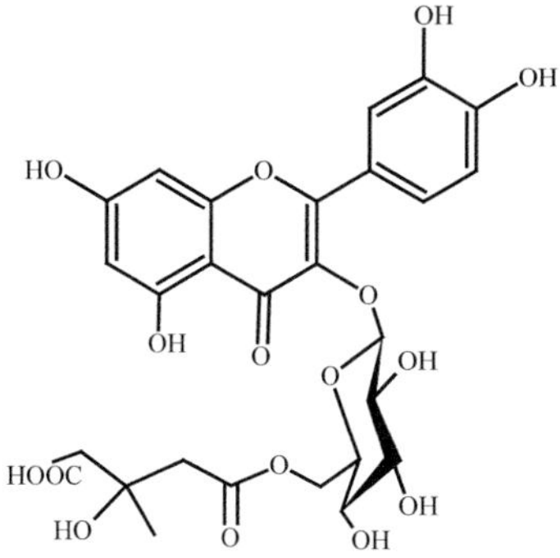

Quercetin 3-*O*-6"-(3-hydroxyl-3-methylglutaryl)-β-D-glucopyranoside

Gymnemasaponins

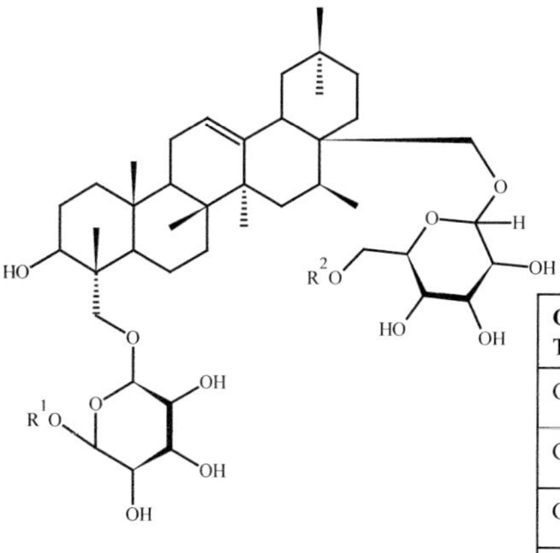

Gymnemasaponin Type	R1	R2
Gymnemasaponin II	H	H
Gymnemasaponin IV	Glu	H
Gymnemasaponin V	Glu	Glu
Glu = β-D-Glucopyranosyl		

Gymnemic acids

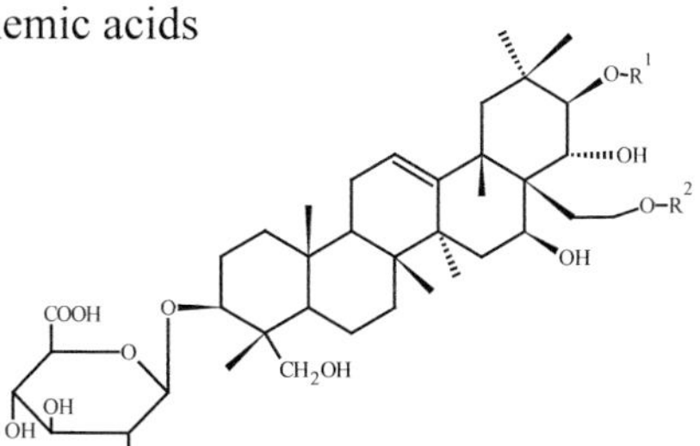

Gymnemic acid

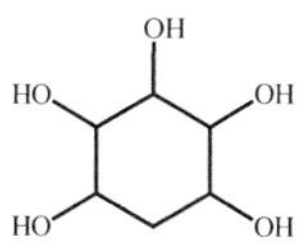

Quecitol

Fig. Contd.

Fig. Contd.

Gymnemic acid Type	**R1**	**R2**
Gymnemic acid I	Tigloyl	Ac
Gymnemic acid II	2-methylbutynyl	Ac
Gymnemic acid III	2-methylbutynyl	H
Gymnemic acid IV	Tigloyl	H

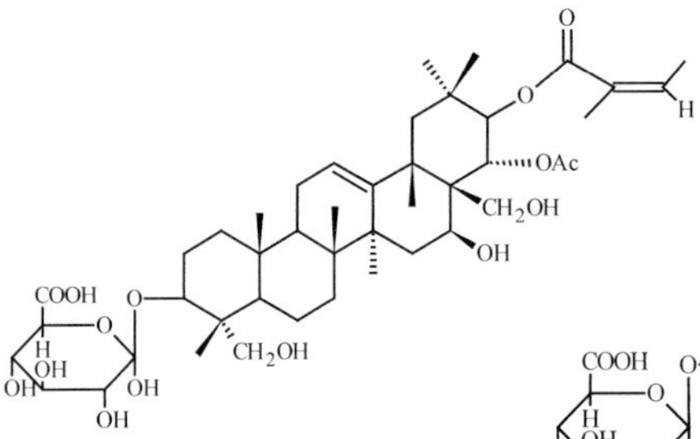

Gymnemoside A

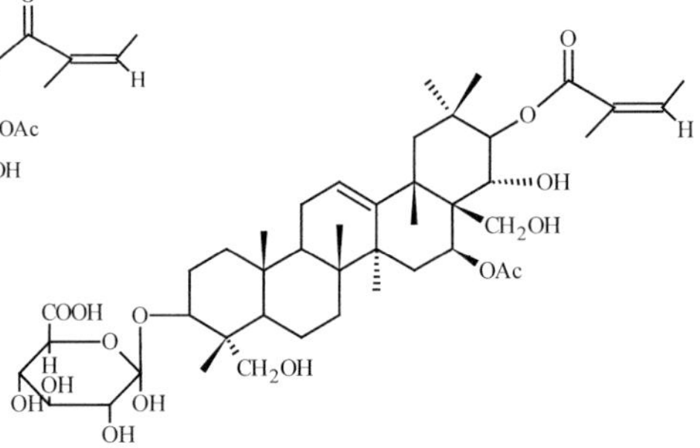

Gymnemoside B

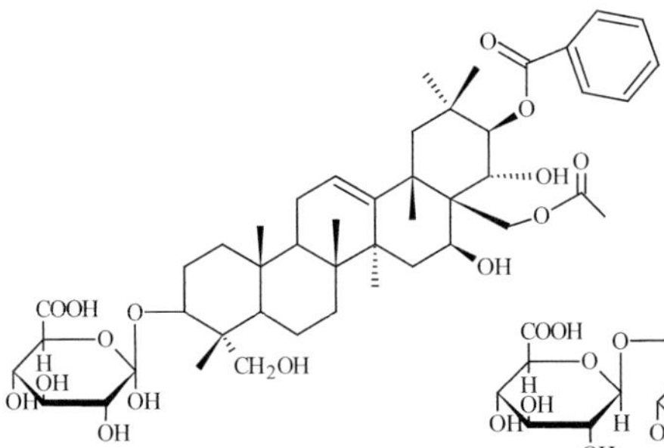

Gymnemoside C

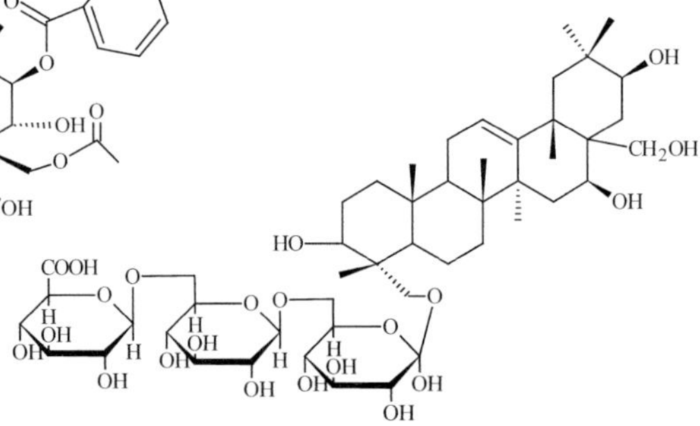

Gymnemoside D

Fig. 3 Structure of some bioactive metabolites of *Gymnema sylvestre* (a) Gymnemic acids (b) Gymnemosides (c) Gypenosides and Flavonoids (d) Chemical constituents. (*Source:* Figure 3a Yoshikawa et al., 1992b), (*Source:* Figure 3b and 3c Yoshikawa et al., 1992a), (*Source:* Figure 3d Imoto et al., 1991)

2d), oleanolic acid-3-*O*-β-D-glucopyranosyl-(1→6)-β-D-glucopyranoside, and sodium salt of 22α-hydroxy-longispinogenin-3-*O*-β-D-glucopyranosyl-(1→3)-β-*D*-glu-curono-pyranosyl-28-*O*-α-*L*-rhamno pyranoside. Gurmarin, a novel polypeptide consisting of 35 amino acids, was separated from the leaves of plants (Imoto et al., 1991).

2.3 Medicinal/Therapeutic Uses

Uses described in Pharmacopoeias and in Traditional Systems of Medicine

Sushrutha Samhitha, an ancient book mentioned *G. sylvestre* is a good remedy for urinary problems and glycosuria (Nadkarni, 1986). In the Indian Ayurvedic System of Medicine, Gymnema is reported as a good medication to treat jaundice, renal calculi, leucoderma, bronchitis, constipation, asthma, hemorrhoids, and with anti-diabetic properties (Sastry, 1994; Nadkarni, 1996; Anis et al., 2000; Mathew, 2004; Jamadagni et al., 2021). In Bhava Prakash Nighantu, *Gymanema* leaves were mentioned as a good remedy for inflammation, roots are useful as an expectorant and emetic, reduce body pains, and for snake bites (Sastry, 1994; Sastry, 2005).

Traditional/Folklore Uses described in Folk Medicine

Indians chew Gymnema leaves early morning to clear urine and decrease glycosuria. Ayurvedic medicinal use of leaves in furunculosis and hyperglycemia is also prevalent in Mumbai and Madras (Vijayakumar and Prabhu, 2014). In the tribes of the Medak district of Telangana, fresh leaf juice is used as eye drops to cure eye diseases. While dry leaf powder is used to treat gastric trouble and diabetes (Reddy et al., 2010). Local people of Chhattisgarh have been using the entire plant of Gymnema as an anti-inflammatory and anti-asth matic drug and leaves as a diuretic (Jain et al., 2006). In Karnataka, the roots of Gymnema find use as a remedy for diabetes (Ghatapanadi et al., 2011). In Kerala, fresh leaf juice is prepared with water and administered orally before meals to treat diabetes (Jayakumar et al., 2010). In Tamil Nadu, the leaf paste of Gymnema along with garlic, pepper, and salt is given orally to animals to treat ephemeral fever, and juice prepared from fresh leaves is used to cure diseases of the eyes (Kiruba et al., 2006). In Jhansi (Uttar Pradesh), fresh leaf juice is applied twice daily on eyelids to treat cataracts (Nigam and Sharma, 2010). In Pondicherry, dry leaf powder or fresh leaves (7 nos) are consumed orally for 15 days as a remedy for diabetes (Dixit and Sudurshan, 2011).

Pharmacological Activities mainly Antiviral

The in-vivo and in-vitro research on the plant of *G. sylvestre* revealed its multifarious pharmacological potentials including anti-oxidant potential against free radicals, antibacterial and anti-microbial, anti-inflammatory, anti-viral, gastro-protective, hepatoprotective, anti-cancer, anti-helmintic, hypoglycemic and anti-hypercholesterolemic, immunostimulatory, and anti-diabetic activities (Saneja, 2010; Di Fabio et al., 2013; Tiwari et al., 2014; Khan et al., 2019; Jamadagni et al., 2021; Karande et al., 2022). The most notable biological effect of this plant is the anti-diabetic action by a decline in blood glucose levels (Shah et al., 2011; Shah et al., 2012; Dholi and Raparla, 2014; Kumar et al., 2015; Kumar et al., 2017; Ahmed et al., 2017; Habtemariam, 2019; Jamadagni et al., 2021). The various pharmacological properties of bioactive metabolites derived from *G. sylvestre* are given in Figure 3 (Khan et al., 2019).

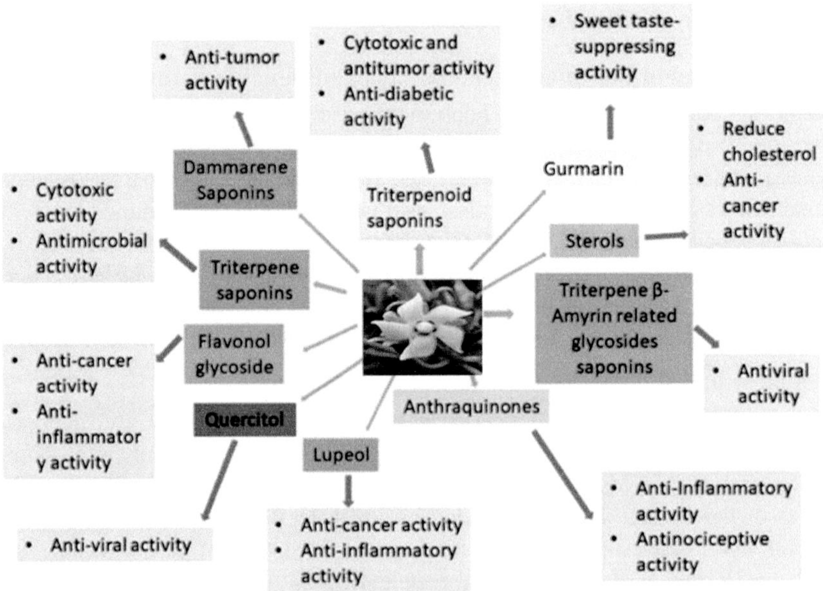

Fig. 4 Various pharmacological activities of bioactive metabolites of *G. sylvestre* (*Source:* Adapted from Khan et al., 2019).

Subashini and Rajendran (2015) investigated the antiviral efficacy of ethanol and methanol extracts of *G. sylvestre* leaves using Hepatitis B surface antigen (HBsAg) binding capacity against the enzyme reverse transcriptase (RT) of Human Immunodeficiency Virus (HIV)-1 and DNA polymerase enzyme of hepatitis B virus. ELISA kit results revealed that methanol extract showed positive results on HBsAg binding capacity. Radiometric assay results indicated that the methanol extract exhibited >50% inhibition of hepatitis B virus DNA polymerase enzyme than ethanol extract. Colorimetric assay results demonstrated that the tested extracts of *G. sylvestre* leaves strongly inhibited the HIV reverse transcriptase enzyme. The antiviral activity of crude extract, fraction, and isolated gymnemic acids A-D was evaluated using *in vitro* method (Sinsheimer et al., 1968). Of the tested samples crude extract exhibited very feeble virus inhibition, ethyl acetate fraction and gymnemic acid B showed moderate activity, gymnemic acid A demonstrated good antiviral activity, whereas, no inhibition was observed by gymnemic acids C and D. Li et al. (2005) evaluated antiviral activity of gymnemic acid against Asian influenza virus. Porchezhian and Dobriyal (2003) reported that the gymnemic acids A and B isolated from Gymnema possess antiviral activity. Gymnemic acid A has the greatest effect on influenza viral infection at 75 mg/kg concentration. The antiviral action of the methanolic extract of *G. sylvestre* was evaluated against mouse coronavirus, Herpes Simplex Virus, and feline calicivirus using *in vitro* methods demonstrating that the extract strongly inhibited herpes simplex virus and mouse corona virus and moderately inhibited feline calicivirus (Vimalanathan et al., 2009). Antiviral efficiency of gymnemagenol isolated from *G. sylvestre* leaves was evaluated against cell lines of Sahul Indian Grouper Eye (SIGE) on infection with nodavirus, which causes nervous necrosis in fish larvae cultures and juvenile marine fishes. The results demonstrated that gymnemagenol at 20 μg/ml concentration significantly suppressed SIGE cell line proliferation (Khanna et al., 2011). The computational analysis including the auto docking or molecular docking on the bio-molecules of *G. sylvestre* such as gymnemic acids showed potential binding on the active sites of viral proteases (3CLpro of Coronavirus disease 2019 (*i.e.*, COVID-19). Subramani et al. (2020) studied the antiviral activity of bioactive gymnemic acids isolated and characterized from *G. sylvestre* against Severe Acute Respiratory Syndrome Coronavirus 2 (SARSCoV2) using *in-silico* molecular docking studies. The energy binding parameters in the molecular docking study interpreted gymnemic acids as the prospective COVID-19 inhibitors by strongly binding with domains (I & II), and domain II-III linker of 3CLpro. The gymnemic acids significantly suppressed the protease enzyme activity by blocking the enzyme linkers, suggesting use as therapeutic against COVID-19. The mechanism of action of various antiviral compounds/extracts is summarized in Table 1.

3. Toxicity and Safety

Acute and sub-acute toxicity of poly-herbo mineral formulation prepared from *G. sylvestre* was evaluated in animal models. It is unconcealed that the formulation was safe and nontoxic to the animals at a dosage of 2000 mg/kg body weight (Vaishali and Chandrashekhar, 2012). Toxicity studies of the homeopathic formulation prepared using *G. sylvestre* were assessed in Sprague Dawley rat models according to OECD -423 guidelines. The results demonstrated that the formulation at a dosage of 500 mg/kg body weight didn't express at all signs and symptoms of toxicity and mortality in the tested animal models (Shukla et al., 2020). The acute and sub-acute toxicity effects of the crude methanol extract and hexane and ethyl acetate fractions from *G. sylvestre* were evaluated in Wistar rats. It demonstrated that crude extract and fractions exhibited acute oral toxicity at a dosage of >5000mg/kg body weight (Raji et al., 2021). The dosage toxicity of extract powder of *G. sylvestre* evaluated in oral repeated dose in mice model for 52 weeks, revealed that no toxicity was observed for males at 504 mg/kg/day and in females at 563 mg/kg/day. The powdered extract didn't express any signs and symptoms of toxicity on body weight, serum biochemical composition, hematological parameters, and in food consumption (Ogawa et al., 2004).

Table 1 Mode of action of Antiviral compounds/extracts.

Extract/Compound	*Virus type*	*Mechanism of Action*	*Reference*
Leaf methanol/ ethanol	Hepatitis B Virus, HIV	Inhibition of DNA polymerase enzyme, Inhibition of HIV-1 Reverse Transcriptase	Subashini and Rajendran (2015)
Leaf methanol extract	Mouse coronavirus, Herpes Simplex Virus (HSV) and Feline calicivirus	Inhibited growth of mouse coronavirus, HSV and feline calicivirus	Vimalanathan et al. (2009)
Gymnemagenol	Grouper Nervous Necrosis Virus (GNNV)	Inhibited proliferation of GNNV	Khanna et al. (2011)
Gymnemagenol	Nodavirus	Inhibited SIGE cell lines on infection with nodavirus	Khanna et al. (2011)
Gymnemic acid	COVID-19/SARS-CoV-2	Blocked protease enzyme linkers	Subramani et al. (2020)

Conclusion

The potential anti-diabetic plant, *G. sylvestre* has been reported for the effective inhibition over a wide range of viruses as well as the key enzyme binder/linkers such as mouse coronavirus, herpes simplex virus, nodavirus and feline calicivirus, protease enzyme activity by blocking the enzyme linkers, HIV-1 (RT enzyme) and hepatitis B virus (DNA polymerase enzyme), clearly indicating its protective role during the viral contagion. Moreover, *G. sylvestre* shows potential for varied biological benefits without severe toxic effects within the recommended dosages, hence may be used for designing the viral drugs after proper *in vivo* investigations.

References

Ahmed, S.K., Cheekavolu, C., Alasyam, N. and Sunil, M. (2017). Evaluation of hypoglycemic property of gurmar (Gymnema sylvestre) leaves methanolic extract (GSLME) in streptozocin (STZ) induced Diabetic Albino Rats. *J. Med. Sci. Clin. Res.* 5(12): 31753–31760. doi: 10.18535/jmscr/v5i12.78.

Anis, M., Sharma, M.P. and Iqbal, M. (2000). Herbal ethnomedicine of the Gwalior Forest division in Madhya Pradesh, India. *Pharm. Biol.* 38(4): 241–253.

Aralelimath, V.R. and Bhise, S.B. (2012). Anti-diabetic effects of *Gymnema sylvestre* extract on streptozotocin induced diabetic rats and possible b-cell protective and regenerative evaluations. *Dig. J. Nanomater. Bios.* 7(1): 135–142.

Bindu Jacob, Narendhirakannan, R.T. (2019). Role of medicinal plants in the management of diabetes mellitus: a review. 3 *Biotech. 2019* Jan; 9(1): 4.

Chopra, R.N., Nayar, S.L., and Chopra, I.C. (2002). Glossary of Indian medicinal plants. National Institute of Science Communication and Information Resources, New Delhi, India.

Christopoulos, M.V., Rouskas, D., Tsantili, E. and Bebeli, P.J. (2010). Germplasm diversity and genetic relationships among walnut (Juglans regia L.) cultivars and Greek local selections revealed by Inter-Simple Sequence Repeat (ISSR) markers. *Sci. Hort.* 125(4): 584–592. doi: 10.1016/j.scienta.2010.05.006.

Dholi, S.K. and Raparla, R.K. (2014). *In vivo* anti-diabetic evaluation of gymnemic acid in streptozotocin induced rats. *J. Pharm. Innov*. 3(7): 82–86.

Di Fabio, G., Romanucci, V., Zarrelli, M., Giordano, M. and Zarrelli, A. (2013). C-4 gemdimethylated oleanes of gymnema sylvestre and their pharmacological activities. *Molecules.* 18(12): 14892–14919. https://doi.org/10.3390/molecules181214892.

Dixit, A.K. and Sudurshan, M. (2011). Review of flora of anti-diabetic plants of Puducherry ut. *Int J Appl Biol Pharm Technol*. 4: 455–462.

Ghatapanadi, S.R., Johnson, N. and Rajasab, A.H. (2011). Documentation of folk knowledge on medicinal plants of Gulbarga district. Karnataka. *Ind. J. Trad. Knowl.*, 10: 349–353.

Gurav, S., Gulkari, V., Duragkar, N. and Patil, A. (2007). Phcog Rev.: Review Article Systemic review : Pharmacognosy, phytochemistry, pharmacology and clinical applications of Gymnema sylvestre R Br. 1(2): 338–343.

Habtemariam, S. (2019). Antidiabetic herbal medicines rebranded as dietary supplements. (Ed. Habtemariam, S.) In Medicinal Foods as Potential Therapies for Type-2 Diabetes and Associated Diseases. 1049–1134. https://doi.org/10.1016/B978-0-08-102922-0.00026-2.

Imoto, T., Miyasaka, A., Ishima, R. and Akasaka, K, (1991)"A novel peptide isolated from the leaves of *Gymnema sylvestre*—I. Characterization and its suppressive effect on the neural responses to sweet taste stimuli in the rat." *Comparative Biochemistry and Physiology*, 100(2): 309–314.

Jain, J., Kumane, S. and Bhattacharya, S. (2006). Medicinal flora of Madhya Pradesh and Chattisgarh –a review. *Indian J Tradit Knowl.* 5(2): 237–342.

Jamadagni, P.S., Pawar, S.D., Jamadagni, S.B., Gautam, M., Gaidhani, S.N., Prasad, G.P., Gurav, A.M. (2021). Recent Updates in Research on Gymnema Sylvestre. *Pharmacogn. Rev.* 15(30): 128–133. doi: 10.5530/phrev.2021.15.15.

Jayakumar, G., Ajithabai, M.D., Sreedevi, S., Viswanathan, P.K. and Remeshkumar, B.(2010). Ethnobotanical survey of the plants used in the treatment of diabetes. *Indian J Tradit Knowl.* 9(1): 100–104.

Kanetkar, P., Singhal, R. and Kamat, M. (2007). Gymnema sylvestre: A Memoir. *J. Clin. Biochem. Nutr.* 41(2): 77–81. doi: 10.3164/jcbn.2007010.

Kang, M.H., Lee, M.S., Choi, M.K., Min, K.S. and Shibamoto, T. (2012) "Hypoglycemic activity of Gymnema sylvestre extracts on oxidative stress and antioxidant status in diabetic rats. *J. Agri. and Food Chem.* 60(10): 2517–2524.

Kapoor, L.D. (2017). CRC Handbook of Ayurvedic Medicinal Plants. S.l. (Boca Raton, Florida, USA: CRC Press). doi: 10.1201/9780203719473.

Karale, P.A. and Karale, M.A. (2017). A review on phytochemistry and pharmacological properties of milkweed family herbs (asclepiadaceae). *Asian J. Pharm. Clin. Res.*, 10(11): 27–34. https://doi.org/10.22159/ajpcr.2017.v10i11.21215.

Karande, S.S., Vikhe, D.N. and Jadhav, R.S. (2022). Gynmema Sylvestre-A Potentional Medicinal Herb. *Res. J. Sci. Technol.* 14:193–196. doi: 10.52711/2349-2988.2022.00032.

Khan, F., Sarker, M.M.R., Ming, L.C., Mohamed, I.N., Zhao, C., Sheikh, B.Y., Tsong, H.F. and Rashid, M.A. (2019). Comprehensive Review on Phytochemicals, Pharmacological and Clinical Potentials of Gymnema Sylvestre. *Front. Pharmacol.* 10:1223. doi: 10.3389/fphar.2019.01223.

Khanna, V.G., Kannabiran, K., Babu, V.S. and Hameed, A.S.S. (2011). Inhibition of fish nodavirus by gymnemagenol extracted from Gymnema sylvestre. *J. Ocean Univ. China* 10: 402–408. https://doi.org/10.1007/s11802-011-1841-2.

Kirtikar, K.R., Basu, B.D. (1975). "Indian medicinal plants, vol. III," in Periodical Experts D-42 (Vivek Vihar Delhi).

Kiruba, S., Jeeva, S. and Dhas, S. (2006). Enumeration of ethnoveterinary plants of Cape Comorin, Tamil nadu. *Indian J Trad. Knowl.* 5(4): 576–578.

Kumar, P., Rani, S., Arunjyothi, B., Chakrapani, P. and Rojarani, A. (2017). Evaluation of Antidiabetic Activity of Gymnema sylvestre and Andrographis paniculata in Streptozotocin Induced Diabetic Rats. *Int. J. Pharmacogn. Phytoch. Res.* 9(1): 22–25.

Kumar, P., Rani, S., Arunjyothi, B., Chakrapani, P. and Rojarani, A. (2017). Evaluation of Antidiabetic Activity of Gymnema sylvestre and Andrographis paniculata in Streptozotocin Induced Diabetic Rats. *Int. J. Pharmacogn. Phytoch. Res.* 9 (1): 22–25. doi: 10.25258/ijpapr.v9i1.8034.

Kumar, S., Paul, S., Walia, Y., Kumar, A. and Singhal, P. (2015). Therapeutic Potential of Medicinal Plants: A Review. *J. Biol. Chem. Chron.* 1: 46–54.

Kumar, V.H., Nayak, I. N., Huilgol, S.V., Yendigeri, S.M. and Narendar, K. (2015). Antidiabetic and hypolipidemic activity of *Gymnema sylvestre* in dexamethasone induced insulin resistance in Albino rats. *Int. J. Med. Res. Health Sci.* 4(3): 639–645.

Laha, S. and Paul, S. (2019). Gymnema sylvestre (Gurmar): A potent herb with anti-diabetic and antioxidant potential. *Pharmacognosy J.*, 11(2): 201–206. https://doi.org/10.5530/pj.2019.11.33.

Li, S.Y., Chen, C., Zhang, H.Q., Guo, H.Y., Wang, H., Wang, L., Zhang, X., Hua, S.N., Yu, J., Xiao, P.G., Li, R.S. and Tan, X. (2005) Identification of natural compounds with antiviral activities against SARSassociated coronavirus. *Antiviral Res.* 67(1): 18–23.

Liu, X., Ye, W., Yu, B., Zhao, S., Wu, H. and Che, C. (2004) "Two new flavonol glycosides from *Gymnema sylvestre* and *Euphorbia ebracteolata. Carbohydrate Research*, 339(4): 891–895.

Liu, Y., Tun-Hai, X.U., Man-QiZHANG, Xue, L.I., Ya-Juan, X.U., Hong-YuJIANG, Tong-HuaLIU, and Dong-Ming, X.U. (2014) Chemical constituents from the stems of Gymnema sylvestre. *Chinese Journal of Natural Medicines*, 12(4): 300–304.

Mall, G.K., Mishra, P.K. and Prakash, V. (2009). Antidiabetic and hypolipidemic activity of Gymnema sylvestre in alloxan induced diabetic rats. *Glob. J. Biotech. Biochem.* 4(1): 37–42.

Mathew, M. (2004). Aromatic and Medicinal Plants Research Station, Odakkali-A Centre for Promoting Medicinal and Aromatic Plants. *Indian Coconut J. Cochin.* 34(10): 10–15.

Nadkarni, K.M. (1986). "*Gymnema sylvestre*," in Indian Materia Medica with Ayurvedic Unani, vol.I (Bombay, India: Popular Prakashan), 596–599.

Nadkarni, K.M. editor. (1996). "Indian materia medica," in Dr. KM Nadkarni's Indian materia medica: with Ayurvedic, Unani-Tibbi, Siddha, allopathic, homeopathic, naturopathic & home remedies, appendices & indexes, vol. 1 (Popular Prakashan).

Najafi, S. and Deokule, S.S. (2011). Studies on *Gymnema sylvestre*–a medicinally important plant of the family Asclepiadaceae. *Trakia. J. Sci.* 9(2): 26–32.

Nigam, G. and Sharma, N.K. (2010). Ethnoveterinary plants of Jhansi district, Uttar Pradesh. *Indian J Tradit Knowl*. 9(4): 664–667

Ogawa, Y., Sekita, K., Umemura, T., Saito, M., Ono, A., Kawasaki, Y., Uchida, O., Matsushima, Y., Inoue, T. and Kanno, J. (2004). *Gymnema sylvestre* leaf extract: a 52-week dietary toxicity study in Wistar rats. *Shokuhin Eiseigaku Zasshi*. 45(1): 8–18.

Patel, M.R. (2017). Pharmacognostic and Phytochemical Evaluation of Gymnema sylvestre Leaf. *World J. Pharm. Pharm. Sci.* 6(7): 1532–1538.

Porchezhian, E. and Dobriyal, R.M. (2003). An overview on the advances of Gymnema sylvestre: chemistry, pharmacology and patents. *Die Pharmazie*, 58(1): 5–12.

Pramanick, D.D. (2016). Anatomical studies on the leaf of Gymnema sylvestre (Retz.) R. Br. ex Schult. (Apocynaceae): A magical herbal medicine for diabetes. *Int. J. Herb. Med.*, 4(1): 70–72.

Rajashekar, J., Kumar, V., Veerashree, V., Poornima, D.V, Sannabommaji, T., Gajula, H. and Giridhara, B. (2016). Protocols for *In Vitro* Cultures and Secondary Metabolite Analysis of Aromatic and Medicinal Plants, Second Edition. 1391: 427–443. https://doi.org/10.1007/978-1-4939-3332-7.

Raji, R.O., Muhammad, H.L. and Abubakar, A. (2021). Acute and sub-acute toxicity profile of crude extract and fractions of *Gymnema sylvestre*. *Clin Phytosci.* 7: 56.

Reddy, K.N., Trimurthulu, G. and Reddy, C.S. (2010). Medicinal plants used by ethnic people of Medak district, Andhra Pradesh. *Indian J Tradit Knowl*. 9(1): 184–190.

Saneja A., Sharma C., Aneja K.R., Pahwa R. (2010). Gymnema Sylvestre (Gurmar): a review. Der Pharmacia Lettre. 2(1): 275–284.

Sastry, B.S. (1994). *Gymnema sylvestre*. Varanasi, India: Bhav Prakash Nighantu, Chaukhambha. p. 443–444.

Sastry, J.L. (2005). *Gymnema sylvestre* R.Br. Dravyaguna Vijnana. Vol. 2. Varanasi, India: Chaukhamba Orientalia; 844–845.

Sathya, S., Kokilavani, R. and Gurusamy, K. (2008). Hypoglycemic effect of *Gymnema sylvestre* (retz.), R. Br leaf in normal and alloxan induced diabetic rats. *Anc. Sci. Life,* 28(2): 12–14.

Shafey, A.A., El-Ezabi, M.M., Seliem, M.M., Ouda, H.H. and Ibrahim, D.S. (2013). Effect of *Gymnema sylvestre* R. Br. leaves extract on certain physiological parameters of diabetic rats. *J. King Saud. Univ. Sci.* 25(2): 135–141.

Shah, K.K., Shiradkar, M.R. and Bindu, V.H. (2012). *In vitro* permeation of aceclofenac through the shed skin of two different species. *Der. Pharmacia Sinic*. 3: 11–19.

Shah, K.K., Shiradkar, M.R. and Hima Bindu, V. (2011). Transdermal delivery of aceclofenac: Effect of Gymnema sylvestre and Caralluma adscendens with its mechanism of action. *Res. J. Pharm. Biol. Chem. Sci*. 2: 762–772.

Shukla, A., Muhammed, I.K., Sheshala, R., Mhaisker, I.U. and Rampal, K. (2020). Acute toxicity evaluation of homeopathic preparation of Gymnema sylvestre and analysis of its chemical constituents. *J Appl Biol Biotech.* 8(04): 033–037. https://dx.doi.org/10.7324/JABB.2020.80405.

Singh, V.K., Umar, S., Ansari, S.A. and Iqbal, M. (2008). Gymnema sylvestre for diabetics. *Journal of Herbs. J. Herbs Spices Med. Plants* 14(1–2): 88–106. doi: 10.1080/10496470802341508.

Sinha, S. and Mondal, A.K. (2017). A phylogenetic study of floral morphology of some members of Asclepiadaceae R. Br. *Annals of Plant Sci.*, 6(2): 1546. https://doi.org/10.21746/aps.2017.02.004.

Sinsheimer, J., Rao, G.S., McIlhenny, H.M., Smith, R.V., Maassab, H.F. and Cochran, K.W. (1968). Isolation and antiviral activity of the gymnemic acids. *Experientia*, 24(3): 302–303.

Sinsheimer, J.E., Rao, G.S. and McIlhenny, H.M. (1970) "Constuents from *Gymnema sylvestre* leaves. V: isolation and preliminary characterization of the gymnemic acids. *Journal of Pharmaceutical Sciences*, 59(5): 622–628.

Subashini, M.S. and Rajendran, P. (2015). *In vitro* screening of anti HBV and anti HIV properties of *Gymnema sylvestre* R. Br. leaves from Kolli Hills, Tamilnadu, India. *Int. J. Curr. Microbiol. App. Sci.* 4(3): 542–547.

Subramani, S.K., Gupta, Y., Manish, M. and Prasad, G.B.K.S. (2020). Gymnema Sylvestre A–Potential Inhibitor of COVID-19 Main Protease by MD Simulation Study. ChemRxiv. Cambridge: Cambridge Open Engage; 2020.

Subramani, S.K., Gupta, Y., Manish, M., Prasad, G. (2020). Gymnema Sylvestre A–Potential Inhibitor of COVID-19 Main Protease by MD Simulation Study. ChemRxiv. 2020; https://doi.org/10.26434/chemrxiv.12333251.v1.

Sugihara, Y.H., Nojima, H., Matsuda, T., Murakami, M., Yoshikawa. and I. Kimura, (2000). Antihyperglycemic effects of gymnemic acid IV, a compound derived from *Gymnema sylvestre* leaves in streptozotocin-diabetic mice. *Journal of Asian Natural Products Research*, 2(4): 321–327.

Tiwari, P., Mishra, B.N. and Sangwan, N.S. (2014). Phytochemical and Pharmacological Properties of Gymnema sylvestre: An Important Medicinal Plant. BioMed Research International, 1–18. https://doi.org/10.1155/2014/830285.

Vaishali, R.U. and Chandrashekhar, D.U. (2012). Acute and sub-acute toxicity studies of the antidiabetic polyherbomineral formulation BSL-150 in experimental animal models. *World J. Pharma. Res.* 1(4): 1204–1211.

Vijayakumar, S. and Prabhu, S. (2014). Gymnema sylvestre-A Key for Diabetes Management–A Review. *Pharm. & Toxicol. Res.*1 (1); 1–10.

Vimalanathan, S., Ignacimuthu, S. and Hudson, J.B. (2009). Medicinal plants of Tamil Nadu (Southern India) are a rich source of antiviral activities. *Pharmaceutical Biology*, 47(5): 422–429. 10.1080/13880200902800196.

Wu, X., Mao, G., Fan, Q., Zhao, T., Zhao, J., Li, F., et al. (2012). Isolation, purification, immunological and anti-tumor activities of polysaccharides from Gymnema sylvestre. *Food. Res. Int.* 48(2): 935–939. doi: 10.1016/j.foodres.2012.02.006.

Yoshikawa, K., Amimoto, K., Arihara, S. and Matsuura, K. (1989a) Gymnemic acid V, VI and VII from gur-ma, the leaves of Gymnema sylvestre R.Br. *Chemical and Pharmaceutical Bulletin*, 37(3): 852–854.

Yoshikawa, K., Amimoto, K., Arihara, S. and Matsuura, K. (1989b) Structure studies of new antisweet constituents from Gymnema sylvestre. *Tetrahedron Letters*, 30(9): 1103–1106.

Yoshikawa, K., Arihara, S., Matsuura, K. and Miyaset, T. (1992a) Dammarane saponins from Gymnema sylvestre. *Phytochemistry*. 31(1): 237–241.

Yoshikawa, K., Nakagawa, M., Yamamoto, R., Arihara, S. and Matsuura, K. (1992b). Antisweet natural products. V. Structures of gymnemic acids VIII-XII from Gymnema sylvestre R.Br. *Chemical and Pharmaceutical Bulletin*, 40(7): 1779–1782.

18 | Exploring the Anti-Viral Potential of Medicinal Gem *Hypericum perforatum* L.

Abhijit Kumar[1], Saurabh Gupta[2] and Pooja Bhadrecha[1*]

1. Taxonomy and Scientific Classification

Hypericum perforatum, commonly known as SJW, klamath weed, tipton weed, goat weed, and enola weed, is a perennial flowering herb belonging to the family *Clusiaceae* (also classified under Hypericaceae and Guttiferae). About 400 species of plants and shrubs belong to the genus Hypericum. Its flowers range in colour from yellow to copper, have 4–5 petals, and multiple stamens in multiple numbers with a single pistil. Herbs of the genus *H. perforatum* normally stand between 40 and 80 centimetres in height and have open, sprawling branches (Gleason and Cronquist 1991). Ovate-shaped leaves with smooth margins are 1–3 cm in length and 0.3–1.0 cm in width and cover the stems and branches densely. When you hold the leaves up to the light, you will see tiny translucent spots scattered throughout. Mature plants' upper regions can bear dozens of yellow blooms with five petals, each measuring one and two centimetres in diameter. Typically, black specks can be seen all around the petal borders. If you crush flowers, you can get a deep red colour, almost like blood. They bloom in the spring, and by late summer, the flowers release capsules that contain hundreds of small, dark brown seeds. While originally from Europe, this species is now found in several temperate regions, including Asia, Africa, Australia, and the Americas (Gleason and Cronquist, 1991). It is a typical sight in wastelands, abandoned mines and quarries, and other places with poor soils. Seven western states in the United States have designated *H. perforatum* as a noxious weed because of concerns about phototoxicity to animals. Canada, California, and Australia are all home to anti-infestation initiatives. This chapter presents the vast body of literature published over the past 30 years on SJW, focusing on its ingredients, mechanisms of action, and potential interactions with other medications. In addition, some suggestions about where future research should go are included.

[1] University Institute of Biotechnology, Chandigarh University, Gharuan, Punjab, India.
[2] Department of Microbiology, Mata Gujri College, Fatehgarh Sahib 140406, Punjab, India.
* Corresponding author: pbhadrecha.pb@gmail.com

DOI: 10.1201/9781003433583-18

2. Synonyms

Hypericum perforatum L. subsp. veronense *(Schrank) H. Lindb.*
Hypericum perforatum L. var. angustifolium *DC.*
Hypericum veronense Schrank"

3. Botanical Description of the Plant

For protection against evil on St. John's Day, Hypericum plants were traditionally hung over a religious icon in the home; the name of the species, Hypericum, comes from the Greek words hyper (above) and eikon (image) (24th June). When held up to the light, the tiny oil glands on the leaves look like little windows (Meruelo, 1993); this is how the species got its name, perforatum. Hypericum is a genus of flowering plants, including annuals and perennials, that range in height from 40 to 80 cm (Gleason and Cronquist, 1991) and is characterised by their free branching and yellow or coppery flowers with four to five petals, numerous stamens, and a single pistil (Degar et al., 1992). Leaves on the stems and branches are rectangular with smooth margins and measure between 1 and 3 centimetres in length and 0.3 and 1.0 centimetres in width. When held up to the light, the leaves reveal tiny translucent dots. There may be dozens of these bright yellow blooms, each with five sepals and a typical width of 1 to 2 centimetres, blooming in the adult plant's top regions. Along the petal's perimeter, you will notice a sprinkling of black dots. The crimson colour is extracted from the blossoms when they are crushed. Seven western states in the United States have designated *H. perforatum* as a noxious weed because of concerns about phototoxicity to animals. Canada, California, and Australia are all home to anti-infestation initiatives.

4. Geographical Distribution

H. perforatum, also called St. John's wort (SJW), is a spreading herb with broad, leafy stems that grows in temperate parts of the world. Since ancient Greece, this species has been used as a herbal cure for both inside and outside problems. Since then, it has been used successfully to treat a wide range of problems, such as stress, depression, wounds, and burns. New research suggests that this herb could be used to treat cancer, reduce inflammation, kill bacteria and viruses, and protect nerve cells. Standardized versions of this herb are prepared by pharmaceutical companies, especially in Europe, and are consumed by millions of individuals every day. The current annual sales of SJW-made products reach several billion dollars around the world. Additionally, SJW produces a diverse range of bioactive compounds, with hypericin and hyperforin being the most promising for therapeutic applications. Flavonoids such as kaempferol, rutin, and quercetin also holds potential as therapeutic agents. Extensive research, including studies on isolated tissue samples, animal models, and human clinical trials has been conducted to deepen the more understanding of *H. perforatum*. In particular, the efficacy of SJW as an antidepressant has been investigated extensively, and the processes involved are well established. When taken alone at the suggested doses, SJW formulations have little unwanted effects. Several documented drug interactions, however, should be taken into consideration. Current studies demonstrate that these interactions occur because SJW components stimulate the production of enzymes in the intestines and liver that helps in the elimination of medicines from the body or convert them to inactive forms. It is majorly found in countries like America, Australia, Canada, China, Chile, Europe, India, Kazakhstan, Kyrgyzstan, Mongolia, North Africa, Russia, Siberia and South Africa.

5. Phytochemical Constituents and Metabolic Profiling

Current medical opinion places a premium on SJW's ability to alleviate depression and other mood disorders. To the tune of USD 6 billion in the late 1990s, Europe was home to a thriving market for SJW-containing pills, capsules, teas, and salves (Ernst, 1999; Greeson, et al., 2001). There was a high in yearly sales of $315 million in the US in 1998, and by 2006, that number had dropped to under $60 million (Tilburt et al., 2008). Numerous pharmacopoeias and monographs have documented SJW including the British Herbal Pharmacopoeia (1996), the European Scientific Cooperative on Phytotherapy (ESCOP) (1996), the American Herbal Pharmacopoeia (1997), and the European Pharmacopoeia (European Directorate for the Quality of Medicines (EDQM)) (2000) (Barnes et al., 2001)".

SJW is commercially accessible in several pharmaceutically-effective forms, most of which are derived from dried plant material. Lichtwer Pharma's LI 160 is a standardised hypericin derivative preparation available in 300 mg capsules with a guaranteed concentration of 0.3%. (Bloomfield et al., 1996). Zeller AG, Switzerland, produces another product called Ze 117, an ethanolic extract with a ratio of 4:1 to 7:1 (herb to extract) compared with LI 160, where hyperforin level is in the range of 1% and 4%, product Ze 117's 0.2% hyperforin content is significantly lower. Ze 117 is administered at a daily dose of 500 milligrams. WS 5570 is an 80% ethanolic extract of Saint John's Wort produced by Dr. Willmar Schwabe Pharmaceuticals in Karlsruhe, Germany. The plant-to-extract ratio varies from 3:1 to 3:1–7:1 containing approximately 5% hyperforin and 0.12%. hypericin (Lecrubier et al., 2002; Szegedi et al., 2005). It is available in 300 mg or 600 mg tablets. Similarly, WS 5572 is formulated with a plant-to-extract ratio of 2.5:1 and 5:1, offering a comparable hyperforin and hypericin profile to WS 5570 (Barrett, 2004). Steiner Arzneimittel's Berlin-made STEI 300 comprises 0.2% to 0.3% hypericin and pseudohypericin and 2% to 3% hyperforin (Philipp et al., 1999). The amount of extract in each capsule is 350 mg. Standardized extract dosages of 900 mg per day are suggested by the German Commission and the European Scientific Cooperation on Phytotherapy (ESCOP) publications (ESCOP, 1996; Blumenthal et al., 1998). Dosages between 300 and 1800 mg daily are safe and effective in clinical trials utilising different *H. perforatum* formulations (Barnes et al., 2001).

In the United States, the Food and Drug Administration classifies SJW as a dietary supplement, as it does for all other herbal medications (FDA). Hence, it is not required to undergo the rigorous safety and efficacy studies that normal pharmaceutical medications must (Clement et al., 2006). The FDA requires that all herbal medications include a disclaimer stating that the agency has not evaluated the effectiveness of the product.

5.1 Active Ingredients

Seven classes of potentially therapeutic chemicals were identified by chemical analysis of *H. perforatum*'s components (Nahrstedt and Butterwick, 1997). Naphthodianthrones, phloroglucinols, flavonoids, and other essential oils are some of the most frequent types (Bombardelli and Morazzoni, 1995; Reuter, 1998; Barnes et al., 2001). Hypericin (a naphtodianthrone) and hyperforin are two primary active components that have been isolated (a phloroglucinol). In contrast, only around 20% of the chemicals that may be extracted are thought to have any biological significance (Staffeldt et al., 1994; Nahrstedt and Butterwick, 1997; Erdelmeier, 1998).

5.1.1 Naphtho Dianthrones

Naphthodianthrones, a group of chemicals isolated from *H. perforatum*, have seen the greatest study and have been standardised in concentrations ranging from 0.1% to 0.3%. (Grainger-Bisset and Wichtl, 2001; Greeson et al., 2001). Most people are familiar with hypericin, the anthraquinone-derived pigment that gives SJW oils their distinctive red hue. The black specks on the petals of these flowers are hypericin.

Due to its chemical composition, hypericin exhibits high photo reactivity. Biochemically, the presence of four hydroxyl groups next to two carbonyl groups sets apart hypericin, a member of the polycyclic quinone family. In a study conducted in (Petrich, 2000). Because of this, while fluorescent light is present, the hydrogen is constantly moving back and forth between the two oxygen atoms (Smirnov and Fulton, 1999). The H-atom transition process is supported by studies of the fluorescence spectra of hypericin and related compounds, which show the presence of a "protonated" carbonyl group (Petrich, 2000). Because of this hydrogen exchange, the surrounding area becomes more acidic (Fehr et al., 1995; Sureau et al., 1996).

5.1.2 Flavonoids

Flavonoids are a diverse class of phytonutrients present in a wide variety of fruits, vegetables, and herbs. They are responsible for the bright colours of these plants, and they also serve as pigments and protectants for the plants. Flavonoids have been shown to have antioxidant and anti-inflammatory properties, and they may also have a positive impact on cardiovascular health and brain function (DerMarderosian and Beutler, 2014). There are six main subclasses of flavonoids: flavones, flavonols, flavanones, flavan-3-ols, anthocyanidins, and isoflavones.

Each subclass has its unique structure and functions. Flavonoids are commonly found in foods such as citrus fruits, berries, apples, onions, parsley, tea, cocoa, and red wine. The health benefits of flavonoids have been studied extensively, and some research suggests that they help reduce the risk of heart disease, cancer, and cognitive decline. However, more research is needed to fully understand the mechanisms behind these benefits and to determine optimal intake levels for different populations (Barnes et al., 2001).

5.1.3 Lipophilic Compoundsompounds

"Many kinds of lipophilic substances, including phloroglucinol derivatives and oils, have been shown to have therapeutic efficacy in SJW extracts. Isolated at concentrations range from 2% to 4.5% (Chatterjee et al., 1998; Greeson et al., 2001), hyperforin is a prenylated phloroglucinol that has been extended into a bicyclo nonaendionol (2,1) and then substituted with numerous lipophilic isoprene chains (Nicolaou et al., 2005). A combination of light and oxygen renders hyperforin unstable (Liu et al., 2005). Total synthesis of hyperforin has not been achieved despite many attempts by many researchers (Nicolaou et al., 2005). Phloroglucinols such adhyperforin (0.2%–1.9%), furohyperforin, and other hyperforin analogues are also found (Hahn, 1992; Barnes et al., 2001; Greeson et al., 2001). Concentrations of essential oils can be anything from 0.05% to 0.9%. (Greeson et al., 2001). Mono, sesquiterpenes, and traces of myrecene, limonene, and caryophyllene, make up the bulk of these compounds (Hahn, 1992; Reuter, 1998)".

5.2 Presence of Other Compounds

The theory behind lipophilic compounds is based on their ability to interact with and dissolve in lipids (fats and oils), rather than in water. This is because lipids are non-polar molecules, meaning they lack an electric charge and do not readily mix with polar molecules such as water. Lipophilic compounds are required to carry out various physiological processes such as the maintenance of the cell membrane structure and its function, hormone production, and energy storage.

However, lipophilic compounds can also pose risks to health if they accumulate in the body in excessive amounts. This can happen when a person consumes high amounts of lipids in their diet, or is exposed to lipophilic toxins such as pesticides or industrial chemicals. In general, lipophilic compounds tend to be more difficult for the body to eliminate than water-soluble compounds, as they are more likely to be stored in fatty tissues and can remain in the body for longer periods . This is why it is important to maintain a balanced diet that includes a variety of nutrients and to avoid exposure to toxins whenever possible (DerMarderosian and Beutler, 2014). Belwal et al. (2019) have precisely enlisted various bioactive compounds from this wonderous plants, which include adhyperforin, amentoflavone, astilbin, avicurarin, I3, II8-biapigenin, cathechin, cyanidin 3-O- α-rhamnoside, cyclopseudohypericin, flavonoids, guaijaverin, hypericin, hyperfirin, hyperforin, hyperoside, isoorientin, isoquercetrin, kaempferol, miquelianin, naphthodianthrones, nicotiflorin, phloroglucinols, procyanidin B2, protohypericin, protopseudohypericin, pseudohypericin, rutin, tannins, quercetin, quercitrin and quercetin 3-O-(2"-acetyl)-β-d-galactopyranoside.

6. Medicinal/Therapeutic Uses

6.1 In Pharmacopoeias and Traditional Systems

St. John's wort (*H. perforatum* L.) is a herbal medicine that has been used for centuries throughout the world. It is even listed in the European Pharmacopoeia, version 6. It's widely used all around the globe to make medicines from plants (Benea et al., 2013; Ravisankar et al., 2016). It has a wide variety of components, including hypericins and hyperforins, which have significant pharmacological effects, and thus a high score (Azizi, 2008). "Diabetes, bruises, rheumatism, burns, malaria, skin wounds, biliary disorders, migraines, eczema, the common cold, gastric ulcer, menorrhagia, diarrhoea, bedwetting, sprains, bronchitis and urogenital diseases, indigestion, haemorrhoids, snake bite, sprains, hysteria, neuralgia, and psychiatric disorders" (Aleksandra, 2015; Asgarpanah, 2012; Božin et al., 2013; Moroydor Derun et al., 2013; Vattikuti and Ciddi, 2005; Bombardelli and Morazzoni,1995; WHO, 2002; Blumenthal et, al., 1998) The polyphenols found in St. John's wort extracts are responsible for the plant's antioxidant action and are therefore being studied as possible cancer preventative. As stated in (Orcic et al., 2011; Bozin et al., 2013) consumers choose natural antioxidants over synthetic ones due to their presumed lesser potential harm.

6.2 Traditional/Folklore uses Described in Folk Medicine

H. perforatum has been used a lot in traditional medical systems like Chinese, Islamic, and Greek medicines, as discussed below.

6.2.1 Traditional Chinese Medicine (TCM)

Traditional Chinese Medicine (TCM) is a healthcare system rooted in thousands of years of practice in China. It operates on the principle that the body is a dynamic system in constant interaction with its environment and that health depends on maintaining balance and harmony within this system. TCM encompasses a range of practices, including acupuncture, herbal medicine, massage therapy, dietary therapy, and mind-body exercises such as tai chi and qigong. These practices are often used to address a wide range of health conditions, from chronic pain to respiratory disorders to digestive issues (Zhang et al., 2020). One of the core principles of TCM is the concept of qi (pronounced "chee"), which is the vital energy that flows through the body's meridian system. According to TCM theory, qi must be balanced and flowing smoothly for the body to be healthy. When the flow of qi is disrupted or blocked, illness and disease can occur. TCM practitioners use a variety of techniques to restore balance and promote the smooth flow of qi, such as acupuncture, which involves the insertion of fine needles into specific points on the body to stimulate the flow of qi, and herbal medicine, which uses plant-based remedies to nourish and strengthen the body's systems.

While TCM is not widely accepted in Western medicine, many people around the world use TCM to manage their health and well-being. The practice is becoming more popular in the West, and research is ongoing to explore the effectiveness and safety of TCM treatments. Despite its lengthy history of usage as an antidepressant in Europe, SJW has not been incorporated into TCM. The plant is neutral and has a bitter and astringent flavour, as described in old texts (Clarke, 2010).

6.2.2 Greek Medicine

Greek medicine refers to the system of healthcare that was practiced in ancient Greece from around 800 BCE to the end of the Roman Empire in the 5th century CE. It was based on the teachings of famous Greek physicians such as Hippocrates, Galen, and Dioscorides (Gunther, 1968) and was influenced by the medical traditions of ancient Egypt, Mesopotamia, and India. Greek medicine emphasized maintaining a healthy lifestyle, including a balanced diet, exercise, and hygiene. It also emphasized the importance of observing the natural world and using natural remedies to promote health and healing (Solati et al.,2021).

One of the key principles of Greek medicine was the concept of the four humors, which were thought to be the fundamental substances that made up the body. These humors were blood, phlegm, yellow bile, and black bile, and they were believed to influence a person's temperament and health. Greek physicians sought to restore balance among the humors through treatments such as herbal medicine, massage, and diet.

6.2.3 Islamic Traditional Medicine

In addition to its purportedly laxative properties, SJW seeds have been noted in several sources. In addition to helping with sciatica, this plant has many other uses (Ibn Beyṭâr et al., 2001). Using a decoction of the plant in wine, Ibn Sinâ (Avicenna) recommended a regimen of treatment lasting forty days (Ibn Sinâ, 2015). Greek medicine also made significant contributions to the field of anatomy, with physicians such as Galen conducting dissections and making detailed observations of the human body. They also developed surgical techniques and instruments and were skilled at treating injuries and wounds. Although Greek medicine is no longer practiced in its original form, its legacy can still be seen in modern medicine. The principles of observing the natural world, using natural remedies, and maintaining a healthy lifestyle are still relevant today, and many medical terms and concepts are derived from Greek terminology.

6.3 Applications Supported by Clinical Data

Studies that appear to validate SJW's usefulness in treating a range of ailments, including depression, have contributed to its growing acceptance as a herbal treatment. Scientists are keen to study the herb and provide a definitive verdict on its efficacy, which has led to increased interest in its use. Clinical studies and meta-analyses of SJW administration to people, as well as studies involving isolated tissue samples and animal models, all fall under this category of research.

6.3.1 Relatinship St. John's Wort with ND Depression

Treatment-resistant depression: Some clinical studies have suggested that St. John's Wort may be effective in treating depression that has not responded to traditional antidepressant medications. However, more research is needed to confirm this finding. Anxiety and sleep disturbances: In addition to its potential antidepressant effects, St. John's Wort may also help treat anxiety and sleep disturbances, which are often associated with depression. It is important to note that while St. John's Wort is a natural remedy, it can interact with certain medications, including antidepressants, birth control pills, and blood thinners. Therefore, it is important to talk to a healthcare provider before using St. John's Wort to treat depression or any other health condition. Lack of amine neurotransmitters has been linked to the disruption of normal brain neurochemistry that underpins depression (5-hydroxytryptamine [5-HT]). Certain neurotransmitters, particularly those found in synapses between nerve cells, are often increased by antidepressant medications (Remick, 2002).

6.3.2 Active Principals: In Vitro Studies

Active principals refer to the specific active compounds or ingredients in natural substances, such as plants or herbs, responsible for their biological activity. *In vitro* studies are laboratory studies conducted using cells or tissues outside of the body. These studies are used to investigate the effects of active principals on cells or tissues, and can provide valuable insights into their potential mechanisms of action. The use of *in vitro* studies to investigate active principals is important because it allows researchers to examine the effects of these compounds on cells or tissues in a controlled environment. *In vitro* studies can provide information on the dose-response relationship of active principals, as well as their potential effects on various cellular pathways. One example of an active principal that has been extensively studied *in vitro* is curcumin, which is the active compound in turmeric. Curcumin has been shown to have anti-inflammatory, antioxidant, and anti-cancer properties in numerous *in vitro* studies. For example, a study published in the journal Antioxidants found that curcumin could protect human endothelial cells from oxidative stress by increasing the expression of antioxidant enzymes. Another study published in the journal Molecular Carcinogenesis found that curcumin could induce cell death in several different types of cancer cells (Chavez and Chavez, 1997; Jellin et al., 2002). Resveratrol is another active principal that has been extensively studied *in vitro*. Resveratrol is a polyphenol found in grapes, red wine, and other foods, and has been shown to have antioxidant, anti-inflammatory, and anti-cancer properties in numerous *in vitro* studies. For example, a study published in the journal Molecular Nutrition and Food Research found that resveratrol could inhibit the growth of breast cancer cells by inducing cell death. Epigallocatechin gallate (EGCG) is a polyphenol found in green tea that has also been extensively studied *in vitro*. EGCG has been shown to have antioxidant, anti-inflammatory, and anti-cancer properties. For example, a study published in the journal Cancer Research found that EGCG was able to inhibit the growth of lung cancer cells by inducing cell death. Quercetin is a flavonoid found in many fruits and vegetables, including onions, apples, and berries, and has been shown to have anti-inflammatory, antioxidant, and anti-cancer properties in numerous *in vitro* studies. For example, a study published in the journal Nutrition and Cancer found that quercetin was able to inhibit the growth of colon cancer cells by inducing cell death.

While *in vitro* studies provide valuable insights into the potential activity of active principals, it is important to note that the results of these studies may not always translate to humans. *In vitro* studies are conducted in a controlled environment, and may not reflect the complexity of human physiology or the effects of active principals on other parts of the body. Furthermore, the concentrations of active principals used in *in vitro* studies may not reflect the concentrations that are achievable in the body through normal dietary or supplemental intake (Jones et al., 2002).

Another limitation of *in vitro* studies is that they do not take into account the potential interactions between active principals and other compounds that may be present in natural substances. For example, the anti-cancer effects of a natural substance may be due to a combination of several different compounds, rather than a single active principal. Despite these limitations, *in vitro* studies are an important tool for investigating the potential activity of active principals in natural substances. By providing information on the mechanisms of action of active principals, *in vitro* studies can help guide further research and development of natural products for use in healthcare (Treiber et al., 2005).

This is in contrast to synthetic antidepressants, which block 5-HT receptors. Nonselective cation channels were shown to mediate $Na+$ inflow by Treiber et al. (2005). This behavior is undesirable because the absence of a gradient inhibits the reuptake of monoamine neurotransmitters. Since hyperforin has been proven to exert its effects via a mechanism distinct from that of traditional antidepressants, it may point the way towards a new category of such medications (Leuner et al., 2007).

6.3.3 Clinical Studies

Computerized searches of multiple databases were used to compile this information from published and unpublished studies. Every review had its own set of inclusion standards, that change over time. Comparison of trials for SJW formulations with placebos or other antidepressants were required for inclusion in their 1998 evaluation. The total number of patients involved in the analysed studies was 2291, from 27 separate trials. Double-blind studies, no combining of SJW with other plant extracts, a minimum 4-week SJW intervention period, and the exclusive use of conventional antidepressants were all requirements of the 2005 review (Linde et al., 2005b). There were 37 studies that fulfilled their criteria, including 3405 patients. In contrast, trials were limited to those involving individuals with serious depression in their 2009 analysis.

In 1998 Linde and Mulrow showed that SJW extract formulations were more efficacious than placebo and on par with conventional antidepressants. The authors concluded that Hypericum extracts are superior to placebo for the short-term therapy of mild to moderately severe depressive disorders based on this evidence. A comparison of SJW's efficacy to that of other antidepressants was not found to be statistically significant.

Barnes et al., in 2001 outlined the results of seven more clinical trials looking at SJW mono formulations. SJW extracts were associated with a higher improvement in mood than the placebo group and were on par with synthetic antidepressants in terms of their effects on patients' lives. They pointed out that critics of studies comparing SJW to synthetic antidepressants pointed out that the latter were given excessively low amounts. According to Spira (2001), the study was flawed due to its use of older tricyclics and its short length of analysis (6 weeks).

Two U.S. studies found that SJW was ineffective in treating moderate to major depression, contrasting sharply with the results of most German trials conducted in 1990s. Two hundred adult outpatients with a diagnosis of major depression were included in an evaluation by Shelton et al. (2001) that took place between 1998 and 2000 at eleven different academic medical facilities in the United States. They found a rather low rate of remission overall, between 4.9% and 14.3% for the treatment and placebo groups

SJW (St. John's Wort) is a herbal supplement derived from the *H. perforatum* plant. It has been used for centuries as a traditional remedy for various ailments, including depression, anxiety, and sleep disorders. SJW contains several biologically active compounds, including hypericin and hyperforin, which are believed to be responsible for its therapeutic effects. Hypericin has been shown to have antiviral and antidepressant properties, while hyperforin has been shown to inhibit the reuptake of serotonin, dopamine, and norepinephrine, similar to the mechanism of action of some antidepressant medications. Clinical studies have suggested that SJW may be effective in treating mild to moderate depression and anxiety, as well as some other conditions such as menopausal symptoms and neuropathic pain. However, the evidence is not as strong as that for conventional antidepressant medications, and the quality of some of the studies has been criticized. One advantage of their study was that it contained comparisons to two groups: one receiving a placebo and the other receiving the SSRI sertraline. Those treated with SJW were less likely to report a full or partial response than those given placebo (55.9%) or sertraline (61.1%).

Previous German research had found a statistically significant difference between Hypericum and placebo, therefore the current study's results were strikingly at odds with those. They worried that controlled research on natural and synthetic antidepressants could be misinterpreted due to the heterogeneity in placebo response between trials. German studies have focused on individuals with mild to moderately severe depression, which may explain why these results differ from those found in American trials (Linde et al., 2002). They suggest that further research is needed to understand why SJW appears highly effective in Germany but not in other regions.

Linde et al. (2005a) define a Cochrane Review as a systematic examination of the existing evidence on a specific healthcare intervention to provide high-quality data to guide clinical practice. A Cochrane Review on St. John's Wort for the treatment of significant depression was published in 2016. The review included 35 randomized controlled trials (RCTs) involving a total of 6,993 participants. The review concluded that SJW may be an effective treatment option for mild to moderate depression, but may not be suitable for more severe cases of depression. The review also noted that SJW was generally well-tolerated, with fewer side effects than conventional antidepressants. The RCTs compared SJW with placebo or conventional antidepressants for the treatment of significant depression. The review found that SJW was more effective than placebo for the treatment of depression, but was not as effective as conventional antidepressants. Synthetic antidepressants like tricyclics and SSRIs were compared to SJW and found to be as effective. Those who were given SJW had fewer side effects and lasted in the studies for longer than those who were given synthetics. Most recently, a Cochrane Review that focused solely on trials including people who had major depressive disorder symptoms (Linde et al., 2009) found that SJW treatment resulted in greater heterogeneity than placebo treatment. Despite this, the results showed that Hypericum was helpful for individuals who took it. Earlier Cochrane Reviews, indicated that the difference was most significant between smaller trials conducted in German-speaking countries and larger trials in non-German-speaking countries. Those who were administered synthetic SSRIs or SJW extracts showed no discernible difference. Little negative side effects were reported by Linde et al., 2009 when SJW was administered; however, the authors did warn about drug interactions when SJW was used with other medications.

Kasper et al. (2008a), Brattstrom (2009), Rahimi et al., (2009), and other studies reporting clinical trials which were published after the July 8, 2008 and cut off for the Linde, Berner, and Kriston (2009) review all demonstrated an advantage over placebo or similar benefits compared to synthetic antidepressants".

According to the results of clinical trials cited in the literature, Hypericum is better than synthetic antidepressants, as it produces fewer side effects and shows similar adverse reactions to those observed in placebo-controlled groups. As a result, SJW is considered a promising option for alleviating depression.

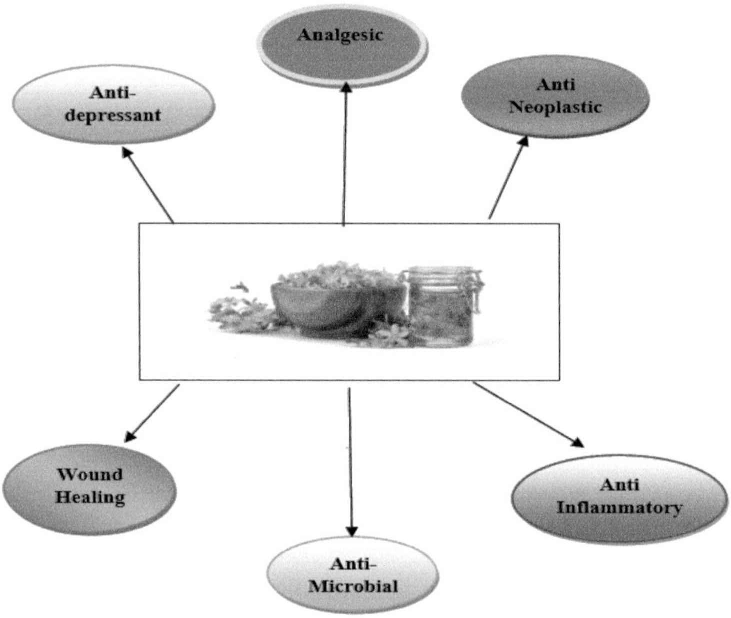

Fig. 1 The pharmacological effects of *Hypericum perforatum*.

6.4 Pharmacological Activities-mainly Antiviral

SJW extracts have been extensively used to treat heal cuts, abrasions, and other wounds from thousands of years. The fact that it also acts as an antibacterial agent may help explain why it is so effective in reducing inflammation. New studies point to its potential application in viral prevention and treatment as well. Hypericin is a compound found in St. John's Wort, a plant that has been used for centuries as a natural remedy for a variety of health conditions. Hypericin has been shown to have a number of potential therapeutic benefits, including aiding in the treatment of certain skin conditions. Hypericin has been studied for its potential antiviral and antifungal properties (Degar et al., 1992). *In vitro* studies have shown that hypericin can inhibit the replication of certain viruses, including HIV and herpes simplex virus. It has also been shown to have antifungal properties and can be effective against certain fungal infections. Hypericin has also been studied for its potential use in the treatment of certain skin conditions, including psoriasis and eczema. It has been shown to have anti-inflammatory properties, which can help to reduce inflammation and redness associated with these skin conditions. Hypericin has also been shown to have photodynamic properties, which means that it can be activated by exposure to light. This makes it a potential treatment for skin conditions that can be treated with photodynamic therapy, such as actinic keratosis and some types of skin cancer (Maury et al., 2009).

In addition, hypericin has been studied for its potential use in the treatment of depression. While more research is needed to confirm its effectiveness, some studies have shown that hypericin can be as effective as conventional antidepressant medications in the treatment of mild to moderate depression. It is important to note that hypericin can interact with certain medications, including antidepressants, birth control pills, and blood thinners. Therefore, it is important to talk to a healthcare provider before using hypericin to treat any health condition. Additionally, hypericin can increase the sensitivity of the skin to sunlight, which can cause sunburn and other skin damage. Therefore, it is important to avoid excessive exposure to sunlight while using hypericin.

7. Toxicity and Safety

Although SJW-based treatments are typically safe, they are not without the potential for side effects, particularly when used with other drugs.

7.1 Adverse Effects

Recent *in vitro* studies suggest that the use of SJW may result in phototoxicity. Both hypercin and light were found to harm lens-crystallin proteins (Schey et al., 2000). Yet, there have been no instances of SJW having any negative effects on humans' visual abilities. SJW users should still take the usual precautions to protect their eyes from harsh sunlight.

7.2 Drug Interactions: Overview

Those taking SJW as a standalone supplement are unlikely to experience adverse pharmacological responses; however, when combined with other medications that may interact with its ingredients, the daily amount becomes a worry. Several outstanding evaluations investigate the potential for interactions between the species' preparations and a wide variety of medications (Barnes et al., 2001). In several studies, using the plant in conjunction with another component increased its efficacy. This boost could be beneficial for the person, or it could cause the compound's effects to become toxically high. On the other hand, it's possible that the herb could dampen or even nullify the impact of certain medicine ingredients (Parker et al., 2001).

7.2.1 P-glycoprotein

P-glycoprotein (P-gp), is also termed as multidrug resistance protein 1 (MDR1), is a transmembrane protein that is found in many different tissues in the body, including the liver, kidneys, intestines, and blood-brain barrier. It plays an important role in the elimination of many drugs and toxins from the body by pumping them out of cells. P-gp is involved in the efflux of a wide range of substrates, including chemotherapy drugs, antiviral agents, and immunosuppressive agents. Genetic differences, pharmacological combinations, and illness conditions can all have an impact on its expression and activity. The MDR/TAP subfamily, of which it is a part, is engaged in the transport of different molecules across both the extracellular and intracellular membranes, and it plays a role in multidrug resistance (Dean et al., 2002).

P-gp expression was found to be induced by SJW extracts in numerous studies. Laboratory investigations, animal studies (mostly rats and mice), and human studies have all been performed. One of the key ingredients that increased P-gp activity was hypericin (Mannel, 2004). Because of its role in drug elimination, P-gp can affect the efficacy of many drugs, especially those that are substrates of P-gp. In addition, P-gp can also affect the absorption and distribution of drugs, particularly those that are lipophilic or have low water solubility. There are many ongoing efforts to develop drugs that can modulate P-gp activity, either to enhance the efficacy of drugs that are substrates of P-gp or to inhibit the efflux of drugs that are toxic to cancer cells. While hypericin had no effect on LS 180 intestinal cancer cells, hyperforin was observed to boost P-gp expression. Within 48 hours of having SJW removed, the P-gp level was back to normal (Tian et al., 2005).

7.2.2 Cytochrome P450

Cytochrome P450 (CYP450) enzymes are a family of enzymes that play a crucial role in the metabolism of drugs, toxins, and other foreign substances in the body. They are primarily found in the liver, but can also be found in other tissues, such as the intestines and lungs. There are over 50 different CYP450 enzymes that have been identified in humans, each with a specific role in the metabolism of different compounds. The most important CYP450 enzymes in drug metabolism are CYP1A2, CYP2C9, CYP2C19, CYP2D6, CYP3A4, and CYP3A5.

CYP450 enzymes work by adding an oxygen atom to a substrate molecule, which helps to make it more water-soluble and easier for the body to eliminate. This process is known as oxidation, and it is the primary way that drugs and other foreign substances are metabolized in the body. The Cytochrome P450 (CYP) superfamily is an extensive and diversified group of haemoprotein enzymes found in virtually all known species. The first digit represents the family, the next one the subfamily, and the third the individual gene within that family (Nelson et al., 1993).

Several different CYP isotypes have been shown to be induced by the SJW extract. It is well established that CYP3A activity increases, and CYP3A4 in particular (Madabushi et al., 2006). Wenk et al. (2004) discovered that extracts successfully induced CYP1A2 in females but not in men. By contrast, neither CYP1A nor CYP2D6 expression was induced nor inhibited by the extracts which discovered that hyperforin reduced CYP3A4 activity when given acutely at 5 and 10 M.

Hyperforin is a naturally occurring compound found in the herb St. John's Wort (*Hypericum perforatum*). St. John's Wort is a popular herbal supplement used to treat mild to moderate depression, anxiety, and sleep disorders. Hyperforin is believed to be one of the main active constituents of St. John's Wort, along with other compounds such as hypericin. It is thought to work by inhibiting the reuptake of several neurotransmitters, including serotonin, dopamine, norepinephrine, and GABA, which can help to elevate mood and reduce anxiety. In addition to its potential antidepressant effects, hyperforin has also been studied for its potential benefits. It has also been shown to have antibacterial and antiviral properties. However, it is important to note that St. John's Wort can interact with a number of medications, including antidepressants, anticoagulants, and certain HIV medications. This is because hyperforin can induce the activity of several drug-metabolizing enzymes in the liver, including CYP3A4 and P-glycoprotein, can lead to faster metabolism of other drugs and potentially reduce their effectiveness. Similar observations were reported by Mueller et al. (2009), after the administration of SJW product with less hyperforin content and showed clinically little CYP3A induction. Further research is needed to confirm whether reduced CYP3A induction occurs following the administration of low-dose hyperforin extracts, as suggested by Whitten et al. (2006) in their review.

It has been hypothesized that hyperforin, an extract of SJW, is responsible for the induction of some cytochromes.

Hepatocytes are the main type of liver cells and are responsible for performing many important functions in the liver, including detoxification, metabolism of nutrients, and production of bile. There are numerous genes that are expressed in hepatocytes, and they play a crucial role in the functioning of the liver. The complex then enters the nucleus, where it binds to the peptide precursor of the pregnancy.

7.2.3 Specific Drug Interactions

Here we take a quick look at the reported drug interactions between SJW and a few different medications. Fexofenadine is an antihistamine medication that is used to treat symptoms of seasonal allergies, such as hay fever, as well as chronic idiopathic urticaria (hives). It works by blocking the effects of histamine, a substance that is released by the body during an allergic reaction and can cause symptoms such as itching, sneezing, runny nose, and watery eyes. Fexofenadine is classified as a second-generation antihistamine, which means that it is less likely to cause sedation than first-generation antihistamines, such as diphenhydramine. This makes it a preferred option for people who need to remain alert and focused during the day. Fexofenadine is available in both prescription and over-the-counter forms, and is typically taken orally in tablet or liquid form. The usual recommended dose is 60 mg twice daily for adults and children over 12 years of age, or 30 mg twice daily for children aged 6 to 11 years. It is generally well-tolerated, but common side effects can include headache, nausea, dizziness, and dry mouth (Dresser et al., 2003).

Digoxin is a medication that belongs to a group of drugs known as cardiac glycosides. It is derived from the leaves of the Digitalis plant and is used to treat various heart conditions, such as heart failure and certain arrhythmias. Other cardioactive drugs from the Digitalis plant include digitoxin and ouabain. These drugs work by inhibiting the sodium-potassium ATPase pump in cardiac cells, leading to an increase in intracellular calcium concentrations and enhancing the contractility of the heart muscle. It is important to note that all of these drugs can have significant toxic effects if taken in excess or if used in patients with certain medical conditions or drug interactions. Close monitoring of blood levels and clinical symptoms is necessary when using these drugs in patients with heart conditions.

Amitriptyline is a medication that is used to treat depression, neuropathic pain, and other conditions. It is metabolized in the liver by two cytochrome P450 enzymes, CYP2C19 and CYP2D6. CYP2C19 is responsible for the metabolism of approximately 10% of amitriptyline, while CYP2D6 is responsible for the metabolism of the majority of amitriptyline. These enzymes play an important role in determining how quickly amitriptyline is metabolized and eliminated from the body. CYP2C19 is a relatively minor pathway for amitriptyline metabolism, and genetic variations in the CYP2C19 gene are not thought to significantly affect amitriptyline metabolism. However, CYP2D6 is a major pathway for amitriptyline metabolism, and genetic variations in the CYP2D6 gene can affect how quickly or slowly amitriptyline is metabolized (Johne et al., 2002).

Ivabradine is a selective inhibitor of the sinus node channels, which reduces the activity of the heart's pacemaker, resulting in more relaxed heart muscle and a more relaxed heart rate. Ivabradine is metabolized by CYP3A4 in the liver and the intestines (Zhou and Lai, 2008). After 14 days of treatment with SJW, the AUC of ivabradine was reduced by 61.7% in 12 healthy volunteers, and the maximum plasma concentration was reduced by 52.9%. (Portoles et al., 2006).

Anti-rejection drugs like cyclosporin and tacrolimus are frequently used to prevent organ rejection in transplant patients. Both cyclosporin and tacrolimus undergo metabolism via cytochrome P450 3A4 and P-glycoprotein Clinical studies from transplant patients raised initial concerns about SJW's potential to interfere with cyclosporin (Barone et al., 2000). As SJW treatment was discontinued, patients showed considerable improvement in immunosuppression. The plasma levels of cyclosporin were shown to decrease during SJW administration and to increase quickly after SJW was discontinued, as was the case in larger investigations involving 30 anti-rejection drugs are medications that are used to suppress the immune system and prevent rejection of transplanted organs. Some examples of anti-rejection drugs include calcineurin inhibitors (such as cyclosporine and tacrolimus), corticosteroids, and antimetabolites. The theory behind the interaction between SJW and anti-rejection drugs is that SJW can increase the activity of drug-metabolizing enzymes in the liver, leading to faster metabolism of anti-rejection drugs. This can result in decreased effectiveness of the anti-rejection drugs, potentially leading to rejection of the transplanted organ. Several case reports and studies have suggested that there may be an interaction between SJW and anti-rejection drugs. For example, a case report published in the Journal of Heart and Lung Transplantation described a patient who experienced acute rejection of a transplanted heart after taking SJW in addition to her prescribed anti-rejection medications (Mai et al., 2003). Verapamil, a calcium channel blocker, is primarily metabolized by CYP3A4 (Tannergren et al., 2004). When compared to the control group, the AUC and Cmax of verapamil were reduced by 78-80% and 76-78%, respectively, when given to 8 males for 14 days through SJW (Tannergren et al., 2004).

St. John's Wort (SJW) is an herbal supplement commonly used to treat mild to moderate depression, anxiety, and sleep disorders. Benzodiazepine sedatives, such as alprazolam and diazepam, are medications used to treat anxiety and sleep disorders by enhancing the activity of the neurotransmitter GABA in the brain.

Theoretically, SJW may interact with benzodiazepine sedatives through various mechanisms, including: Induction of CYP3A4: SJW has been shown to induce the activity of the liver enzyme CYP3A4, which is responsible for metabolizing many drugs, including benzodiazepines. This can lead to a decrease in the plasma concentration of benzodiazepines, reducing their efficacy. Interaction with GABA receptors: SJW has been shown to bind to GABA receptors in the brain, potentially enhancing the activity of benzodiazepines and leading to an increased risk of adverse effects.

Serotonin syndrome: Serotonin syndrome is a potentially life-threatening condition that occurs when there is an excess of serotonin in the body. The syndrome can be caused by various factors, including the use of certain medications, such as selective serotonin reuptake inhibitors (SSRIs), monoamine oxidase inhibitors (MAOIs), and some opioids. The theory behind serotonin syndrome is that it is caused by an excess of serotonin in the brain and body. Serotonin is a neurotransmitter involved in many important physiological functions, including mood regulation, appetite, and sleep. In normal circumstances, serotonin is released in response to certain stimuli, and then reabsorbed by the nerve cells that release it. However, when there is an excess of serotonin in the body, it can lead to overstimulation of the nervous system, resulting in a range of symptoms that characterize serotonin syndrome. Symptoms of serotonin syndrome can include agitation, confusion, rapid heart rate, high blood pressure, dilated pupils, muscle rigidity, tremors, and seizures. In severe cases, serotonin syndrome can lead to coma or death.

Steroid contraceptives which are the substrates used for and inducers of CYP3A4 include certain types of oral contraceptives that contain synthetic hormones such as ethinyl estradiol and norethindrone. CYP3A4 is an enzyme that is responsible for the metabolism of many drugs, including steroid contraceptives. When a drug is a substrate for CYP3A4, it is broken down by the enzyme. However, when a drug is also an inducer of CYP3A4, it can cause an increase in the activity of the enzyme, leading to faster metabolism of both the inducing drug and any other drugs that are substrates for CYP3A4. For examples the combination of ethinyl estradiol and norethindrone, which is used in many oral contraceptive pills. Other drugs that are substrates for CYP3A4 include certain antifungal medications, antibiotics, and HIV medications. If taken together these drugs can lead to decreased effectiveness of the contraceptives (Guengerich, 1988). While administering SJW, the authors and others saw an increase in breakthrough bleeding When women experience breakthrough bleeding, they frequently stop taking their oral contraceptives, which can lead to an unwanted pregnancy (Rosenberg and Waugh, 1998). Resuming regular cycles after stopping SJW use, although other methods of contraception should be considered if SJW use is maintained (Schwarz et al., 2003).

When combined with other anti-HIV treatments, SJW has been shown to impair the efficacy of two of the most prevalent classes of these drugs: In both cases, the induction of cytochrome P450 3A4 is mostly responsible for the impact (Zhou, 2008). The plasma levels of the protease inhibitor indinavir fell by 57% in eight healthy persons (Piscitelli et al., 2000). Ho et al. (2009) observed that hepatic and intestinal CYP3A were better able to lower plasma levels of indinavir after administering SJW extract to rats. We also discovered that when SJW was administered concurrently, the concentrations of NNRTIs such as nevirapine and lamivudine decreased. In contrast, in 2006 L'Homme et al. discovered that administration of SJW to a limited group of healthy women did not diminish the concentration of nevirapine.

Anticoagulants like warfarin are produced from coumarin, a benzopyrone present in nature in a number of plants (Runciman et al., 2002; Bovill et al., 2004; Yarnell and Abascal, 2009). First reported in Sweden, incidents of SJW interacting with coumarin-derived anticoagulants have now been reported elsewhere. Anticoagulants are medications that are commonly used to prevent blood clots and reduce the risk of stroke or heart attack. In a clinical trial, the treated group is the group of participants who receive the experimental intervention or treatment being studied. If anticoagulants are used as the experimental intervention in a clinical trial, the treated group would receive anticoagulant medication while the control group would receive either a placebo or standard of care treatment. The primary endpoint of the trial would be to compare the efficacy and safety of the anticoagulant treatment in preventing blood clots or reducing the risk of stroke or heart attack compared to the control group.

Voriconazole, an antifungal medication, is mostly metabolized by CYP2C19, while it is also processed by CYP3A4 and CYP2C9 (Hyland et al., 2003). The AUC of voriconazole was reduced by 59% and oral clearance was increased when it was given alongside SJW therapy to 16 males for 15 days (Rengelshausen et al., 2005). Antidepressants and anxiolytics are two classes of medications used to treat mental health conditions such as depression and anxiety disorders. Voriconazole is an antifungal medication that is used to treat serious fungal infections. It belongs to a class of drugs called triazoles, which work by inhibiting the growth of fungi. Voriconazole is used to treat a variety of fungal infections, including aspergillosis, candidiasis, and infections caused by Scedosporium or Fusarium species. It can be administered orally or intravenously, depending on the severity of the infection and the patient's condition. Like any medication, Voriconazole has potential side effects, including nausea, vomiting, headache, and changes in vision. Patients taking Voriconazole should be monitored closely for any signs of liver or kidney problems, as well as for any allergic reactions. It is important to follow the dosing instructions provided by the prescribing doctor and to complete the full course of treatment, even if symptoms improve. Failure to do so could result in the infection returning or becoming more difficult to treat.

Anxiolytics, on the other hand, are medications used to treat anxiety disorders by reducing anxiety symptoms, such as excessive worry or fear. Anxiolytics include benzodiazepines and non-benzodiazepine medications, such as buspirone. Based on the available evidence, upregulation of cytochrome P450 (CYP) and phosphoglucoside phosphohydrolase (P-gp) is not the mechanism of action. However, the reason for the cumulative action remains unclear because CYP2D6 is the primary enzyme responsible for metabolizing SSRIs (Zhou and Lai, 2008). However, until the safety profile can be characterized, it is best to avoid combining SSRIs and SJW. It is important to discuss the use of SJW and other dietary supplements with a healthcare provider before starting any new treatment regimen. In addition, patients should inform their healthcare provider of any prescription medications they are taking, as SJW may interact with certain medications, such as antidepressants, birth control pills, and blood thinners.

Conclusion

SJW has been used traditionally as a herbal remedy for various conditions. Over the past 20 years, it has shown promise in treating cancer, inflammation, bacterial, and viral infections, among other illnesses, and is now widely recognised as an alternative treatment for depression. Since the mid-1980s, the efficacy of SJW has been extensively studied due to its popularity. These analyses have primarily focused on the pharmacology of its components and the outcomes of clinical trials. Pharmacological studies have demonstrated that SJW extracts possess neuroactive properties, with Hyperforin rather than hypericin, which has been studied for a longer duration appears to be responsible for these notable effects.

German research from the 1990s has shown that people with mild to moderate depression who take SJW experience improvement in their mood at rates higher than those taking a placebo. The improvement rates were considered comparable to those observed by people using synthetic antidepressants. It has been argued, however, that the clinical studies comparing SJW to synthetics are flawed as they were conducted over too short a period and used synthetic medications at doses that were too low. No significant improvement in depression symptoms was observed with SJW compared to a synthetic antidepressant or a placebo when studies were conducted in the United States in 2001 and 2002. Since then, however, research and meta-analyses have accumulated that support SJW's utility in the treatment of many mood disorders.

SJW does appear to be useful in treating some problems, like skin ailments, inflammation, and different types of cancer. It is effective against bacteria and viruses, although its use as an HIV treatment is controversial. Its protective effects on the brain and nervous system have only recently come to light. Clinical trials have shown that, when used in the customary doses, SJW poses little risk to the user. However, large doses may cause phototoxicity in those who are already vulnerable. There is evidence that SJW extracts interact with other pharmaceuticals, particularly those that affect enzyme function in the liver and intestines. Those who are taking SJW in addition to other medications should be aware of the possibility of adverse drug reactions and inform their doctors of this. SJW, like any other herbal treatment, should only be used if medically recommended by a qualified practitioner.

References

Azizi, M. (2008). Change in content and chemical composition of *Hypericum perforatum* L. oil at three harvest time. *J. Herbs, Spices Med. Plants*, 13: 79–85.

Aleksandra, D. Chemical composition of *Hypericum perforatum* L. essential oil. (2015). *Adv. Technol*, 4: 64-68.

Asgarpanah, J. (2012). Phytochemistry, pharmacology and medicinal properties of *Hypericum perforatum* L. *Afr. J. Pharm. Pharmacol*, 6: 1387–1394.

Barnes, J., Anderson, L.A. and Phillipson, J.D. (2001). St. John's wort (*Hypericum perforatum* L.): A review of its chemistry, pharmacology, and clinical properties. *J Pharm Pharmacol*, 53:583–600.

Barone, G.W, Gurley, B.J., Ketel, B.L, Lightfoot, M.L., Abul-Ezz, S.R. Drug interaction between St. John's wort and cyclosporine. Ann Pharmacother. 2000;34(9):1013–1016. [PubMed]

Barrett, M. (2004). The Handbook of Clinically Tested Herbal Remedies. Vol. 2. Binghamton, NY: Haworth Press; 2004.

Belwal, T., Devkota, H.P., Singh, M.K., Sharma, R., Upadhayay, S., Joshi, C., Bisht, K., Gour, J.K., Bhatt, I.D., Rawal, R.S. and Pande, V. (2019). St. John's Wort (*Hypericum perforatum*). In: Nabavi, S.M., Silva, A.S. (eds.). Nonvitamin and Nonmineral Nutritional Supplements. 415–432. https://doi.org/10.1016/B978-0-12-812491-8.00056-4, ISBN 9780128124918

Benea, A., Gonceariuc, M., Kulcitki, V., Dragalin, I. and Nistreanu, A. (2013). Essential Oil Chemical Composition Biodiversity in the *Hypericum* L. Species from the Spontaneous Flora of the Republic of Moldova. *Oltenia J. Stud. Nat. Sci. Museum of Oltenia Craiova*, 29: 47–52.

Bloomfield, H.H., Nordfors, M. and McWilliams, P. (1996). *Hypericum* and Depression. Los Angeles, CA: Prelude Press; 1996.

Blumenthal, M., Busse, W.R. and Goldberg, A. (1998). The Complete German Commission E Monographs: Therapeutic Guide to Herbal Medicines; American Botanical Council: Boston, 1998.

Bovill, E.G, Fung, M. and Cushman, M. (2004). Vitamin K and oral anticoagulation: Thought for food. *Am J Med*, 116(10): 711–713. [PubMed]

Bozin, B., Nebojsa, K., Nevena, G., Goran, A., Isidora, S., Neda, G. and Branislava Srdenovic, C. (2013). Impact of origin and biological source on chemical composition, anticholinesterase and antioxidant properties of some St. John's wort species (*Hypericum* spp., *Hypericaceae*) from the Central Balkans. *Molecules*, 18: 11733-11750.

Bombardelli, E. and Morazzoni, P. (1995). *Hypericum perforatum*. *Fitoterapia* (Milano), 66: 43–68.

Brattström, A. (2009). Long-term effects of St. John's wort (*Hypericum perforatum*) treatment: A 1-year safety study in mild to moderate depression. *Phytomedicine*, 16(4): 277–283. [PubMed]

Chatterjee, S., Bhattacharya, S.K., Wonnemann, M., Singer, A. and Muller, W.E. (1998). Hyperforin as possible antidepressant component of *Hypericum* extracts. *Life Sci*, 63: 499–510. [PubMed]

Chavez, M.L. and Chavez, P.I. (1997). Saint John's wort. *Hosp Pharm*. 32(12), 1621–1632.

Clarke, R. (2010). Are modern preparation methods compromising the efficacy of Chinese herbal medicines? An investigation using Guan Ye Lian Qiao (*Hypericum perforatum* L.). *Journal of Chinese Medicine*, 35–38.

Clement, K., Covertson, C., Johnson, M.J. and Dearing, K. (2006). St. John's wort and the treatment of mild to moderate depression: A systematic review. *Holist Nurs Pract*, 20(4): 197–203. [PubMed]

Dean, M., Hamon, Y. and Chimini, G. (2002). The human ATP-binding cassette (ABC) transporter superfamily. *J Lipid Res*, 42: 1007–17. [PubMed]

Degar, S., Prince, A.M., Pascual, D. et al. (1992). Inactivation of the human immunodeficiency virus by hypericin: Evidence for photochemical alterations of p24 and a block in uncoating. *AIDS Res Hum Retroviruses*, 8: 1929–1936.

DerMarderosian, A. and Beutler, J. (2014). Review of Natural Products: The Most Complete Source of Natural Product Information (Facts and Comparisons). 8th Edition. Lippincott Williams & Wilkins. 1574393685.

Dresser, G.K., Schwarz, U.I., Wilkinson, G.R. and Kim, R.B. (2003). Coordinate induction of both cytochrome P4503A and MDR1 by St John's wort in healthy subjects. *Clin Pharmacol Ther*, 73(1): 41–50. [PubMed]

Erdelmeier, C. (1998). Hyperforin, possibly the major non-nitrogenous secondary metabolite of *Hypericum perforatum* L. *Pharmacopsychiatry*, 31: 2–6. [PubMed]

Ernst, E. (1999). Second thoughts about the safety of St. John's wort. *Lancet*. 354: 2014–15. [PubMed]

European Scientific Cooperative on Phytotherapy. 1996. St. John's wort. Monographs on the medicinal use of plant drugs. Fascicule 1. Hyperici herba.

Fehr, M., McCloskey, M.A. and Petrich, J.W. (1995). Light-induced acidification by the antiviral agent hypericin. *J Am Chem Soc*, 117: 1833–1837.

Gleason, H.A. and Cronquist, A. (1991). Manual of Vascular Plants of Northeastern United States and Adjacent Canada. 2nd ed. Bronx, NY: The New York Botanical Garden.

Grainger-Bisset, N. and Wichtl, M. (2001). Herbal Drugs and Phytopharmaceuticals. 2nd ed. Stuttgart, Germany: *Medpharm GmbH Scientific Publishers*; 2001.

Greeson, J.M., Sanford, B. and Monti, D.A. (2001). St. John's wort (*Hypericum perforatum*): A review of the current pharmacological, toxicological, and clinical literature. *Psychopharmacology*, 153: 402–414. [PubMed]

Guengerich, F.P. (1988). Oxidation of 17 alpha-ethynylestradiol by human liver cytochrome P-450. *Mol Pharmacol*, 33(5): 500–508. [PubMed]

Gunther, R. (1968). Dioscorides,'De materia medica,'The Greek herbal of Dioscordes. New York: Hafner Publishing Co; 1968.

Hahn, G. (1992). *Hypericum perforatum* (St. John's wort): A medicinal herb used in antiquity and still of interest today. *J Naturopathic Med*, 3: 94–96.

Ho, Y.F., Huang, D.K., Hsueh, W.C., Lai, M.Y., Yu, H.Y. and Tsai, T.H. (2009). Effects of St. John's wort extract on indinavir pharmacokinetics in rats: Differentiation of intestinal and hepatic impacts. *Life Sci*, 85(7-8): 296–302. [PubMed]

Hyland, R., Jones, B.C. and Smith, D.A. (2003). Identification of the cytochrome P450 enzymes involved in the N-oxidation of voriconazole. *Drug Metab Dispos*, 31(5): 540–547. [PubMed]

Ibn Sinâ, H.A. (2015). Al-Qânun fi aṭ-Ṭibbe (Canon of Medicine). Tehran Iran Almaʿee Publication.

Ibn Beyṭâr, A.A. Al-Jâmʿee le Mofradât al-Adwyah wa al-Aghḏiah (Comprehensive Book in Simple Drugs and Foods). Beirut: Dâr al-Kotob al-Ilamiyah; 2001.

Jellin, J.M., Gregory, P.J., Batz, F. and Hitchen, K. (2002). Pharmacist's Letter/Prescriber's Letter Natural Medicines Comprehensive Database. 4th ed. Stockton, CA: Therapeutic Research Faculty.

Johne, A., Schmider, J. and Brockmöller, J. et al. (2002). Decreased plasma levels of amitriptyline and its metabolites on comedication with an extract from St. John's wort (*Hypericum perforatum*). *J Clin Psychopharmacol*, 22(1): 46–54. [PubMed]

Jones, B.J. and Blackburn, T.P. (2002). The medical benefit of 5-HT research. *Pharmacol Biochem Behav*, 71: 555–681. [PubMed]

Kasper, S., Gastpar, M., Müller, W.E. et al. (2008a). Efficacy of St. John's wort extract WS 5570 in acute treatmentof mild depression: A reanalysis of data from controlled clinical trials. *Eur Arch Psychiatry Clin Neurosci*, 258(1): 59–63. [PubMed]

Lecrubier, Y., Clerc, G., Didi, R. and Kieser, M. (2002). Efficacy of St. John's wort extract WS 5570 in major depression: A double-blind, placebo-controlled trial. *Am J Psychiatry*, 159: 1361–1366. [PubMed]

Leuner, K., Kazanski, V., Müller, M. et al. (2007). Hyperforin-a key constituent of St. John's wort specifically activates TRPC6 channels. *FASEB J*, 21(14): 4101–4111. [PubMed]

L'Homme, R.F., Dijkema, T., van der Ven, A.J. and Burger, D.M. (2006). Brief report: Enzyme inducers reduce elimination half-life after a single dose of nevirapine in healthy women. *J Acquir Immune Defic Syndr*, 43(2): 193–196. [PubMed]

Liu, F., Pan, C., Drumm, P., Ang, C.Y. Liquid chromatography-mass spectrometry studies of St. John's wort methanol extraction: Active constituents and their transformation. J Pharm BiomedAnal. 2005;37:303–312. [PubMed]

Linde, K., Melchart, D., Mulrow, C., Berner, M. St John's wort and depression. JAMA. 2002;288:446. [PubMed]

Linde, K., Mulrow, C.D., Berner, M. and Egger M. (2005a). St John's wort for depression. Cochrane Database Syst Rev, CD:000448. [PubMed]

Linde, K., Berner, M., Egger, M. and Mulrow, C. (2005b). St. John's wort for depression: Meta-analysis of randomized, controlled trials. *Br J Psychiatry*, 184: 99-107. [PubMed]

Linde, K., Berner, M. and Kriston, L. (2009). St John's wort for major depression. *Cochrane Database Syst Rev*, Art. No.: CD000448. DOI: 10.1002/14651858.CD000448.pub3.

Linde, K. and Mulrow, C.D. (1998). The Cochrane Library. Vol. 4. Oxford. St. John's wort for depression (Cochrane Review) pp. 1–15.

Madabushi, R., Frank, B., Drewelow, B., Derendorf, H. and Butterweck, V. 920060. Hyperforin in St. John's wort drug interactions. *Eur J Clin Pharmacol*, 62(3): 225–233. [PubMed]

Mai, I., Stormer, E., Bauer, S., Kruger, H., Budde, K. and Roots, I. (2003). Impact of St. John's wort treatment on the pharmacokinetics of tacrolimus and mycophenolic acid in renal transplant patients. *Nephrol Dial Transplant*, 18(4): 819–822. [PubMed]

Mannel, M. (2004). Drug interactions with St. John's wort: Mechanisms and clinical implications. *Drug Saf*, 27(11): 773–797. [PubMed]

Maury, W., Price, J.P. and Brindley, M.A. (2009). Identification of light-independent inhibition of human immune-deficiency virus-1 infection through bio-guided fractionation of *Hypericum perforatum*. *Virol J*, 6: 101. [PMC free article] [PubMed]

Mueller, S.C., Majcher-Peszynska, J., Mundkowski, R.G. et al. (2009). No clinically relevant CYP3A induction after St. John's wort with low hyperforin content in healthy volunteers. *Eur J Clin Pharmacol*, 65(1):81–87. [PubMed]

Meruelo, D. (1993). The potential use of hypericin as inactivator of retroviruses and other viruses in blood products. *Blood*, 82: 205A.

Moroydor Derun, E.; Eslek, Z. and Piskin, S. (2013). Extraction and Analysis of Hypericum perforatum L. from Turkey. *Int. J. Chem. Mol. Nucl. Mater. Metall. Eng.*, 7: 495–499.

Nahrstedt, A. and Butterwick, V. (1997). Biologically active and other chemical constituents of the herb *Hypericum perforatum* L. *Pharmacopsychiatry*, 30: 129-134. [PubMed]

Nelson, D.R., Kamataki, T., Waxman, D.J. et al. (1993). The P450 superfamily: Update on new sequences, gene mapping, accession numbers, early trivial names of enzymes, and nomenclature. *DNA Cell Biol*, 12(1) :1–51. [PubMed]

Nicolaou, K.C., Carenzi, G.E. and Jeso, V. (2005). Construction of highly functionalized medium-sized rings: Synthesis of hyperforin and perforatumone model systems. *Angew Chem*, 44: 3895–3899. [PubMed]

Orcic, D.Z., Neda M.M-D., Marina, M.F., Slobodan, S.P. and Emilija, D.J. (2011). Antioxidant activity relationship of phenolic compounds in *Hypericum Perforatum* L. *Chem. Cent. J.*, 5, No. 34.

Parker, V., Wong, A., Boon, H. and Seeman, M. (2001). Adverse reactions to St John's wort. *Can J Psychiatry*, 46: 77–79. [PubMed]

Petrich, J.W. (2000). Excited-state intramolecular H-atom transfer in nearly symmetrical perylene quinones: Hypericin, hypocrellin, and their analogues. *Int Rev Phys Chem.*, 19: 479– 500.

Philipp, M., Kohnen, R. and Hiller, K.O. (1999). Hypericum extract versus imipramine or placebo in patients with moderate depression: Randomized multicentre study of treatment for eight weeks. *BMJ*, 319: 1534–1539. [PMC free article] [PubMed]

Piscitelli, S.C., Burstein, A.H., Chaitt, D., Alfaro, R.M. and Falloon, J. (2000). Indinavir concentrations and St. John's wort. *Lancet*, 355(9203): 547–548. [PubMed]

Portoles, A., Terleira, A., Calvo, A., Martinez, I. and Resplandy, G. (2006). Effects of Hypericum perforatum on ivabradine pharmacokinetics in healthy volunteers: An open-label, pharmacokinetic interaction clinical trial. *J Clin Pharmacol*, 46(10): 1188–1194. [PubMed]

Ravisankar, N., Jerrine, J., Manikkan, R. and Thirunavukarasu, R. (2016). *In vitro* cytotoxicity of methanol extracts of Hypericum wightianum and *Hypericum hookerianuim* against 3T3L1 cell lines. *Bangladesh J. Pharmacol*. 11: 328–329.

Rahimi, R., Nikfar, S. and Abdollahi, M. (2009). Efficacy and tolerability of Hypericum perforatum in major depressive disorder in comparison with selective serotonin reuptake inhibitors: A meta-analysis. *Prog Neuropsychopharmacol Biol Psychiatry*, 33(1): 118–127. [PubMed]

Remick, R.A. (2002). Diagnosis and management of depression in primary care: A clinical update and review. *CMAJ*, 167: 1253–1260. [PMC free article] [PubMed]

Rengelshausen, J., Banfield, M. and Riedel, K.D. (2005). Opposite effects of short-term and long-term St. John's wort intake on voriconazole pharmacokinetics. *Clin Pharmacol Ther*, 78(1): 25–33. [PubMed]

Reuter, H. (1998). Chemistry and biology of *Hypericum perforatum* (St. John's wort). *ACS Symp Ser*, 691: 287–298.

Rosenberg, M.J. and Waugh, M.S. (1998). Oral contraceptive discontinuation: A prospective evaluation of frequency and reasons. *Am J Obstet Gynecol*, 179: 577–582. [PubMed]

Runciman, D.J., Lee, A.M., Reed, K.F. and Walsh, J.R. (2002). Dicoumarol toxicity in cattle associated with ingestion of silage containing sweet vernal grass (*Anthoxanthum odoratum*). *Aust Vet J*, 80(1–2): 28–32. [PubMed]

Schey, K.L., Patat, S., Chignell, C.F., Datillo, M., Wang, R.H. and Roberts, J.E. (2000). Photooxidation of lens alpha-crystallin by hypericin (active ingredient in St. John's wort). *Photochem Photobiol*, 72(2): 200-203. [PubMed]

Schwarz, U.I., Buschel, B. and Kirch, W. (2003). Unwanted pregnancy on self-medication with St John's wort despite hormonal contraception. *Br J Clin Pharmacol*, 55(1): 112–113. [PMC free article] [PubMed]

Shelton, R.C., Keller, M.B., Gelenberg, A. et al. (2001). Effectiveness of St John's wort in major depression: A randomized controlled trial. *JAMA*, 285: 1978-86. [PubMed]

Smirnov, A. and Fulton, D.B. (1999). Exploring ground-state heterogeneity of hypericin and hypocrellin A and B. *JAmChem Soc*, 121: 7979–7989.

Solati, K., Karimi, M., Rafieian-Kopaei, M., Abbasi, N., Abbaszadeh, S. and Bahmani, M.J. (2021). Phytotherapy for wound healing: The most important herbal plants in wound healing based on Iranian ethnobotanical documents. *Mini Rev Med Chem*, 21: 500–519.

Spira, J.L. (2001). Study design casts doubt on value of St John's wort in treating depression. *BMJ*, 322: 493. [PubMed]

Staffeldt, B., Kerb, R., Brockmoller, J., Ploch, M. and Roots, I. (1994). Pharmacokinetics of hypericin and pseudohypericin after oral intake of Hypericum extract LI 160 in healthy volunteers. *J Geriatr Neurol*, 7: S47–53. [PubMed]

Sureau, F., Miskovsky, P., Chinsky, L. and Turpin, P. (1996). Hypericin-induced cell photosensitization involves an intracellular pH decrease. *JAmChem Soc*, 118: 9484–9488.

Szegedi, A., Kohnen, R., Dienel, A. and Kieser, M. (2005). Acute treatment of moderate to severe depression with Hypericum extract WS 5570 (St John's wort): Randomised controlled double blind non-inferiority trial versus paroxetine. *BMJ*, 330(7490): 503. [PMC free article] [PubMed]

Tannergren, C., Engman, H., Knutson, L., Hedeland, M., Bondesson, U. and Lennernas, H. (2004). St John's wort decreases the bioavailability of R- and S-verapamil through induction of the first-pass metabolism. *Clin Pharmacol Ther*. 75(4): 298–309. [PubMed]

Tian, R., Koyabu, N., Morimoto, S., Shoyama, Y., Ohtani, H. and Sawada, Y. (2005). Functional induction and de-induction of P-glycoprotein by St. John's wort and its ingredients in a human colon adenocarcinoma cell line. *Drug Metab Dispos*, 33(4): 547–554. [PubMed]

Tilburt, J.C., Emanuel, E.J. and Miller, F.M. (2008). Does the evidence make a difference in consumer behavior? Sales of supplements before and after publication of negative research results. *J Gen Intern Med*, 23(9): 1495–1498. [PMC free article] [PubMed]

Treiber, K., Singer, A., Henke, B. and Müller, W.E. (2005). Hyperforin activates nonselective cation channels (NSCCs). *Br J Pharmacol*, 145(1): 75–83. [PMC free article] [PubMed]

Vattikuti, U.M.R. and Ciddi, V. (2005). An overview on *Hypericum perforatum* Linn. *Nat. Prod. Radiance*, 4: 368–381.

Wenk, M., Todesco, L. and Krähenbühl, S. (2004). Effect of St. John's wort on the activities of CYP1A2, CYP3A4, CYP2D6, N-acetyltransferase 2, and xanthine oxidase in healthy males and females. *Br J Clin Pharmacol*, 57(4): 495–499. [PMC free article] [PubMed]

World Health Organization, G., Switzerland., WHO Monograph on Selected Medicinal Plants. 2002; Vol. 2, pp 149–171.

Whitten, D.L., Myers, S.P., Hawrelak, J.A. and Wohlmuth, H. (2006). The effect of St John's wort extracts on CYP3A: A systematic review of prospective clinical trials. *Br J Clin Pharmacol*, 62(5): 512–526. [PMC free article] [PubMed]

Yarnell, E. and Abascal, K. (2009). Plant coumarins: Myths and realities. *Altern Complement Ther*, 15(1): 24–30.

Zhang, R., Ji, Y., Zhang, X., Kennelly, E.J. and Long, C. (2020). Ethnopharmacology of *Hypericum* species in China: A comprehensive review on ethnobotany, phytochemistry and pharmacology. *J Ethnopharmacol*, 254: 112686.

Zhou, S.F. (2008). Drugs behave as substrates, inhibitors and inducers of human cytochrome P450 3A4. *Curr Drug Metab*, 9(4): 310–322. [PubMed]

Zhou, S.F. and Lai, X. (2008). An update on clinical drug interactions with the herbal antidepressant St. John's wort. *Curr Drug Metab*, 9(5): 394–409. [PubMed]

19 Ocimum spp. (Tulsi): The Queen of Herbs with Antiviral Potential

Mahantesha B.N. Naika[1*], Mithraa T[2] and Shashidhar M. Dodamani[3]

1. Introduction

Between 3500 and 1600 B.C., Rigveda documented the utilization of plants for medicinal purposes. The Indian Himalayas are home to approximately 8000 species of vascular plants, with 1748 being medicinal. In Ayurveda, Tulsi is renowned for its therapeutic effects on the mind, body, and spirit, and is commonly referred to as the "Queen of Herbs." Tulsi, also known as *Ocimum* sp., belongs to the Lamiaceae family. It is divided into two groups: holy basil and M editerranean basil. Holy basil, which includes *O. sanctum* (Rama-tulsi), *O. tenuiflorum* (Amrita-tulsi), and *O. gratissum* (Vana-tulsi), is the most revered and is associated with Hinduism in India. The second group, Mediterranean basil, consists of various species such as *O. basilicum* (Sweet basil) and *O. thyrsiflora* (Thai basil) (Kaur et al., 2020). *O. sanctum*, also known as Rama-tulsi, is India's most used type of tulsi species (Fig. 1). Traditional Indian medical systems like Unani, Ayurveda, and Siddha utilize different parts of the tulsi plant for medicinal purposes. Additionally, ancient Greek and Roman medicinal systems also acknowledge the healing properties of tulsi.

2. Chemical Constituents

Tulsi has a complex chemical composition of several nutrients and biochemical compounds utilised in pharmacological activities. The essential oil extracted from *Ocimum* sp. contain the compounds including methyl chavicol, methyl-eugenol, trans-anethole, eugenol, b-eudesmol, β-caryophyl-lenesoxide, β-caryophyllene, γ-terpinen, farnesol, 4-allylanisole, dihydrocarvone, cuminol, cuminyl aldehyde, isoborneol, camphor, d-limonene, 2-decanol, camphene, bornylacetate, 2-heptanol, isopulegol, methylheptane, nerol, and citral (Table 1) (Raghav and Saini, 2018).

[1] Centre for Biotechnology Research, Department of Biotechnology and Crop Improvement, College of Horticulture, Bagalkot, University of Horticultural Sciences, Bagalkot, Karnataka, India.
[2] Division of Plant Genetic Resources, ICAR- Indian Agricultural Research Institute, New Delhi.
[3] Department of Plantation, Spices, Medicinal and Aromatic crops, KRCCH, Arabhavi, University of Horticultural Sciences, Bagalkot, Karnataka, India.
* Corresponding author: mahantesha.bnn@uhsbagalkot.edu.in

DOI: 10.1201/9781003433583-19

Fig. 1 Morphological presentation and field-grown Tulsi plants
(A) Krishna tulsi (*Ocimum tenuiflorum)* (B) Shri tulsi (*Ocimum tenuiflorum)*
(C) Camphor Basil (*Ocimum kilimandscharicum*) (D) Clove Basil (*Ocimum gratissimum)*.

3. Antiviral Effects of Tulsi

The extracts derived from different parts of tulsi plants have been utilized for their antioxidant, antidiabetic, anti-stress, anti-inflammatory, anti-helminthic,

Table 1 Major compounds in essential oils extracted from *Ocimum* species.

Ocimum sp.	*Major compounds*	*References*
O. sanctum	Eugenol (1.50%), methyl eugenol (72.41%), beta-elemene (2.48%), isoeugenol (6.31%) and beta-caryophyllene (12.00%)	Bhattacharya et al., 1996
O. gratissimum	Limonene (1.98%), eugenol (87.68%), beta-caryophyllene (1.84%) and beta-selinene (2.82%)	Bhattacharya et al., 1996
O. *tenuiflorum*	Eugenol (79.0–82.7%) and β-caryophyllene (7.9–9.8%)	Machado et al., 1999
O. *kilimandscharicum*	Eugenol (74.28%), α-pinene (8.20%), and germacrene-D (5.13%).	Ayoubi et al., 2023
O. basilicum	Linalool (4-47.39 %), methyl cinnamate (35.16-84.53 %), methyl chavicol (8.92 %)	Gangrade et al., 2000
O. thysiflora	Linalool (68 %), methyl chavicol (20%)	Avetisyan et al., 2017
O. citriodorum	Citral (21%), nerol (23%)	Avetisyan et al., 2017
O. cinnamon	Estragole (87.869%), α-Bergamotene (2.922%), τ-Cadinol (2.770%), and Linalool (1.347%)	Tran et al., 2018

antifungal, antimicrobial, analgesic, anti-hypolipidemic, and radioprotective properties. Basil extract contains secondary metabolites like eugenol, apigenin, urosolic acid, and linalool, that exhibit antiviral effects against various viruses. The antiviral properties of Ocimum have been observed against deadly viral agents such as Adenovirus, Para influenza virus Type 3, Hepatitis C virus, Herpes Simplex Virus, Human Immunodeficiency Virus, Enterovirus 71, and Infectious Hematopoietic Necrosis Virus (Kumar et al., 2022). *O. basilicum* has shown efficacy against DNA and RNA viruses, including human adenovirus, by inhibiting DNA synthesis (Amber et al., 2017). *O. gratissimum* has demonstrated activity against Herpes simplex virus 1 (HSV–1), HSV-2, Murine CMV, and Influenza – A virus by suppressing viral replication (Raja et al., 2015).

4. Tulsi Against COVID-19

Since December 2019, the Corona Virus Disease COVID-19 pandemic has been spread by the novel viral strain SARS–CoV–2 (Severe Acute Respiratory Syndrome Coronavirus-2). Several studies have recommended herbal therapy as a viable COVID-19 treatment and preventive strategies. Countries like India and China combined herbal therapy with conventional systems to enhance the immune systems of COVID-19 patients (Bhattacharya and Bose, 2023). The extracts of *O. sanctum* act against the SARS-CoV-2 by inhibiting RNA-Binding Domain (Muthumanickam et al., 2021), protease (Shrivastava et al., 2020), SARS-CoV-2 main protease, spike glycoproteins, RNA (Kumar, 2020), SARS-CoV-2 spike and Mpro proteins and SARS-CoV-2 RNA dependent RNA (Kar et al., 2022). Likewise, *O. basilicum* plays a role in inhibiting the binding of the N protein of SARS-CoV-2 (Nallusamy et al., 2020).

5. Genomic Resources on Tulsi for the Discovery of Potential Antiviral Compounds

Genomic resources are essential for comprehending the genetic composition of plants like Tulsi and discovering compounds that possess antiviral properties. Currently, there is a dedicated genomic database called TulsiDbDatabase of *Ocimum tenuiflorum* (accessible at http://caps.ncbs.res.in/Ote/) solely for Tulsi. However, studies on Tulsi's transcriptome and genome have been carried out to learn more about its genetic composition and the processes that lead to the synthesis of bioactive chemicals (Upadhyay et al., 2015). Numerous studies have focused on sequencing the genome of *Ocimum sanctum* to identify crucial genes responsible for the synthesis of bioactive compounds, including those with potential antiviral properties. Transcriptomic studies have analyzed the gene expression profiles of Tulsi, aiding in the identification of genes accountable for producing compounds that may exhibit antiviral activity (Rastogi et al., 2014). Metabolomics, on the other hand, concentrates on studying the complete range of small molecules (metabolites) produced by the plant. This approach assists in the identification of potential antiviral compounds present in Tulsi.

6. Antiviral Compounds

Research has suggested that Tulsi contains various bioactive compounds that may exhibit antiviral properties, although more extensive studies are often necessary to confirm these effects. Some of the compounds found in Tulsi that have shown antiviral potential include:

- Eugenol: This compound possesses antimicrobial properties and has been studied for its antiviral effects.
- Ocimumosides A and B: These compounds have demonstrated antiviral activity in some studies.
- Ursolic Acid: Found in Tulsi, this compound exhibits antiviral effects in certain experimental setups.

7. Future Directions

- Bioinformatics: Further exploration of Tulsi's genome through bioinformatics tools can identify potential gene clusters responsible for synthesizing antiviral compounds.
- Drug Discovery: Computational approaches and high-throughput screening can aid in identifying and validating potential antiviral compounds derived from Tulsi.

While genomic resources for Tulsi might not be as comprehensive as those for extensively studied plants, ongoing research continues to delve into its genetic makeup and bioactive compounds. These studies aim to uncover the potential antiviral properties of Tulsi and its application in drug discovery and herbal medicine.

References

Amber, R., Adnan, M., Tariq, A. and Mussarat, S. (2017). A review on antiviral activity of the Himalayan medicinal plants traditionally used to treat bronchitis and related symptoms. *Journal of Pharmacy and Pharmacology*, *69*(2), 109–122.

Avetisyan, A., Markosian, A., Petrosyan, M., Sahakyan, N., Babayan, A., Aloyan, S. and Trchounian, A. (2017). Chemical composition and some biological activities of the essential oils from basil Ocimum different cultivars. *BMC Complementary and Alternative Medicine*, *17*(1), 1–8.

Ayoubi, R., Singh, G. and Pandey, D.K. (2023). Chemical Composition of *Ocimum kilimandscharicum* Guerke Essential Oil Cultivated in Phagwara, Punjab, India. *Hamdard Medicus*, *66*(1).

Bhattacharya, A.K., Kaul, P.N. and Rajeswara Rao, B.R. (1996). Essential oils of *Ocimum gratissimum* L. and *Ocimum tenuiflorum* L. (syn. *Ocimum sanctum* L.) grown in Andhra Pradesh. *Indian Perfumer*, *40*(3), 73–75.

Bhattacharya, R. and Bose, D. (2023). A review of the sustainable development goals to make headways through the COVID-19 pandemic era. *Environmental Progress and Sustainable Energy*, e14093.

Gangrade, S. K., Mishra, P. K. and Sharma, R. K. (2000). Variability in essential oil constituents of ocimum species. *J Med Aromat Plant Sci*, *22*(23), 13–16.

Kar, P., Kumar, V., Vellingiri, B., Sen, A., Jaishee, N., Anandraj, A. and Subramaniam, M. D. (2022). Anisotine and amarogentin as promising inhibitory candidates against SARS-CoV-2 proteins: a computational investigation. *Journal of Biomolecular Structure and Dynamics*, *40*(10), 4532–4542.

Kaur, S., Sabharwal, S., Anand, N., Singh, S., Baghel, D.S. and Mittal, A. (2020). An overview of Tulsi (Holy basil). *European Journal of Molecular & Clinical Medicine*, *7*(7), 2833–2839.

Kumar, A. (2020). Molecular docking of natural compounds from tulsi (Ocimum sanctum) and neem (*Azadirachta indica*) against SARS-CoV-2 protein targets.

Kumar, R., Saha, P., Lokare, P., Datta, K., Selvakumar, P. and Chourasia, A. (2022). A Systemic Review of *Ocimum sanctum* (Tulsi): Morphological Characteristics, Phytoconstituents and Therapeutic Applications. *International Journal for Research in Applied Sciences and Biotechnology*, *9*(2), 221–226.

Machado, M.I.L., de Vasconcelos Silva, M.G., Matos, F. J.A., Craveiro, A.A. and Alencar, J.W. (1999). Volatile constituents from leaves and inflorescence oil of *Ocimum tenuiflorum* L. (syn. *O. sanctum* L.) grown in Northeastern Brazil. *Journal of Essential Oil Research*, *11*(3), 324–326.

Muthumanickam, S., Kamaladevi, A., Boomi, P., Gowrishankar, S. and Pandian, S.K. (2021). Indian ethnomedicinal phytochemicals as promising inhibitors of RNA-binding domain of SARS-CoV-2 nucleocapsid phosphoprotein: an *in-silico* study. *Frontiers in Molecular Biosciences*, *8*, 637329.

Nallusamy, S., Mannu, J., Ravikumar, C., Angamuthu, K., Nathan, B., Nachimuthu, K. and Neelakandan, K. (2020). Shortlisting phytochemicals exhibiting inhibitory activity against major proteins of SARS-CoV-2 through virtual screening.

Raghav, P.K. and Saini, M. (2018). Antimicrobial properties of Tulsi (*Ocimum sanctum*) in relation to shelf-life enhancement of fruits and vegetables. *Int J Green Herbal Chem*, *7*, 20–32.

Raja, M.C., Srinivasan, V.A.R.S.H.A., Selvaraj, S.H.A.R.M.I.L.A. and Mahapatra, S.K. (2015). Eugenol: A versatile phytomedicine. *Int. J. Pharm. Pharm. Sci*, *7*, 35–40.

Rastogi, S., Meena, S., Bhattacharya, A., Ghosh, S., Shukla, R.K., Sangwan, N.S. and Shasany, A.K. (2014). De novo sequencing and comparative analysis of holy and sweet basil transcriptomes. *BMC Genomics*, *15*, 1–18.

Shrivastava, G., Pagare, S., Khanna, A. and Aglawe, V. (2020). Antimitotic Effect of Artesunate and Tulsi Extract on Onion Root Tip System.

Tran, T.H., Nguyen, H.H.H., Nguyen, D.C., Nguyen, T.Q., Tan, H., Nhan, L.T.H. and Nguyen, T.D. (2018). Optimization of microwave-assisted extraction of essential oil from vietnamese basil (*Ocimum basilicum* L.) using response surface methodology. *Processes*, *6*(11), 206.

Upadhyay, A.K., Chacko, A.R., Gandhimathi, A., Ghosh, P., Harini, K., Joseph, A.P., ... and Sowdhamini, R. (2015). Genome sequencing of herb Tulsi (Ocimum tenuiflorum) unravels key genes behind its strong medicinal properties. BMC plant biology, 15, 1–20.

20 Antiviral Potential of *Vitex negundo* L.

Karishma Rajbhar Khan[1*], Himanshu Dawda[1] and Usha Mukundan[1*]

1. Introduction

A new coronavirus called SARS (severe acute respiratory) syndrome coronavirus surfaced in late 2019. This virus, the causative agent of the global pandemic COVID-19, has since reshaped our world. In terms of health care, the economy, and cultures globally, the COVID-19 pandemic has presented humanity with extraordinary challenges. According to surveys, approximately 400 million instances were reported at the beginning of 2022. The COVID-19 pandemic was a deadly virus outbreak that significantly impacted the world economy and humanity (Ng *et al.*, 2022; Mieres-Castro *et al.*, 2023).

Control of COVID-19 infections worldwide with the help of COVID-19 vaccination was very challenging to handle as it faced hurdles. The global battle against COVID-19 faces several challenges: unequal vaccine access, distribution logistics, special storage needs, two-dose requirements, waning immunity within six months, and variants like Omicron that evade vaccine-induced immunity. These issues highlight the limitations of vaccines and the necessity for complementary treatments, such as broad-spectrum, orally administered antiviral drugs. Developing these antivirals involves two main approaches: targeting viral proteins or host factors essential for the virus's life cycle and addressing the pathogenesis of SARS-CoV-2 infections. These strategies are crucial to bolstering our defences against this evolving adversary (Ng *et al.*, 2022).

2. Plant Displaying Antiviral Therapeutic Activity

Research and development related to antiviral therapeutic activities must mainly focus on the innovation of new therapeutic drugs and efficiently manage the challenges of viral diseases like COVID-19. Many updated documents reveal that numerous natural compounds and their complex moiety derivatives are effective against viral infection. Several herbal extracts mostly belonging to the families like *Myrtales, Verbenaceae, Dicksoniaceae, Polygonaceae,* and *Ganodermataceae* (fungi) have a mechanism to block the interaction between the viral S protein and

---

[1] Department of Botany, HVPS Ramniranjan Jhunjhunwala College (Empowered Autonomous), Ghatkopar (W), Mumbai, India.

* Corresponding author: usha.rjc@gmail.com

DOI: 10.1201/9781003433583-20

the host. Emodin metabolites are well known for such activity (Ng *et al.*, 2022; Chavda *et al.*, 2022).

Flavonoids such as myricetin and scutellarin have been used to validate the inhibition mechanism of SARS-CoV-2 helicase NSP13, which plays a key role in RNA viral replication, transcription, and translation. Additionally, flavonoids and triterpenes like herbacetin, rhoifolin, pectolinarin, betulinic acid, and savinin have demonstrated a substantial reduction in viral protease activity due to their anti-inflammatory properties. Cysteine protease like papain has a vital role in RNA viral genomic replication. Hence, they are being targeted to develop anti-viral drugs more intricately. *Ganodermataceae* (fungi) species extracts were effective against COVID-19 as they mainly targeted RNA polymerase dependent on RNA of SARS-CoV-2. The synthesis of both viral and non-viral RNA requires the essential enzyme RNA polymerase (Chavda *et al.*, 2022).

Thus, a need for more alternatives in therapeutic medicinal plants can be explored to find alternative medicines, drugs, and active constituents having antiviral activity. Our current review data emphasize the value of Verbenaceae family *Vitex negundo* L. active constituents revealing antiviral activity (Nallusamy *et al.*, 2021; Alagarasu *et al.*, 2022).

Databases utilised have been critically reviewed by applying Google Scholar, PubMed Central mainly for viral diseases, ScienceDirect, and Scopus. The data literature on active compounds exhibiting antiviral action was finalised by reviewing recent articles. In provision, the literature review was carried out according to the keywords used for the search: "active constituents + *Vitex negundo*", "active constituents + antiviral activity", "antiviral activity + clinical trial", "antiviral + pharmacological effects". For the current review experimental and review articles related to plant derivatives and active constituents revealing antiviral activity published in peer-reviewed journals to date were selected. At the start, the surface analysis was carried out by just reading abstracts and conclusion of the screened articles. The data was compiled from all of the screened publications and was thoroughly analysed based on application, including a detailed examination of the active constituent content for antiviral content. After reviewing the full-text articles an eligibility criterion was utilised in the section so that only reviews with active constituents revealing antiviral activity and pharmacological effects of those constituents against viral symptoms and clinical complications of viral disease were studied to assess the properties of active constituents associated with inflammation were included in the discussion.

Herbal medicine seeks not only to cure specific diseases but also to restore the body's natural state of health. The phytochemicals in medicinal plants often work individually, additively, and synergistically to enhance overall well-being. For *Vitex negundo*, extensive analysis of its chemical components highlights its therapeutic potential. This plant is widely used to treat various ailments and its medicinal properties can be categorised into three main areas: traditional medicine, folk medicine, and pharmacological evidence. Traditional medicine documents its long-standing use, folk medicine reflects its cultural applications, and pharmacological evidence supports its efficacy through scientific research. This comprehensive approach emphasizes the significant role of *Vitex negundo* in promoting health (Vishwanathan *et al.*, 2010).

Traditional medicine, including systems like Indian Ayurveda, Arabic, Unani, and traditional Chinese medicine, remains crucial in regions such as Asia and Latin America, where cultural and historical influences keep it relevant. Notably, it accounts for about 40% of healthcare in China and serves up to 80% of Africa's population. *Vitex negundo*, a key plant in these traditions, is well-documented for its therapeutic properties. Its anthelmintic qualities and effectiveness in treating syphilitic skin illnesses, sinusitis, gland sores, headaches, and sexual diseases are mentioned in the Charaka Samhita of Ayurveda. Traditional Chinese medicine recommends its fruit for eye conditions, headaches, and arthritic joints. In rural and tribal areas, folk medicine extensively utilises *Vitex negundo*, highlighting its ethnobotanical diversity and enduring significance. The plant's broad medicinal applications across various cultural practices underscore its vital role in natural medicine and its comprehensive therapeutic potential (Vishwanathan *et al.*, 2010).

2.1 Introduction of Vitex negundo

A renowned medicinal plant from the *Verbenaceace* family for respiratory disorders is *Vitex negundo*. *Verbenaceae* comprises 75 genera consisting of 2500 species of which *Vitex negundo* Linn. Synonyms used for *Vitex negundo* Linn. are Nirgundi, Sambhalu, Sinduvara, and in English, it is also referred to as the five-leaved chaste tree, it is renowned for its medicinal properties (Singh *et al.*, 2020).

Fig. 1 *Vitex negundo* Linn. Twig. (ICMR 2005 and ICMR 2013)

2.2 Plant Description and Geographical Distribution of Selected Plant

Vitex negundo is a significant aromatic shrub growing 4 to 5 meters. The plant is a shrub or under tree, leaves digitate with three or five leaflets, distributed throughout

India, and common in wastelands, and on roadsides; it is also cultivated in gardens as shrubby borders and as a hedge plant. In India, it is spread in the greater part of the outer Himalayas with an altitude of 1500 meters. *Vitex negundo*, also prevalent in Sri Lanka, Burma, China, Pakistan, Malaysia, tropical Africa, and the Philippines, is a slender, erect tree reaching 2 to 5 meters, with quadrangular branches. Its distinctive features include palmately arranged lanceolate leaflets, typically hairy underneath, ranging from 4 to 10 cm in length. The tree boasts numerous bluish-purple flowers and succulent, black fruits about 4 mm in diameter when ripe. Vegetative propagation via cutting is feasible. Commonly utilized in traditional medicine, *Vitex negundo* (Linn.) exhibits a wide array of pharmacological activities (Bano *et al.*, 2015).

Fig. 2 *Vitex negundo* Linn. Flowering twig and Dried roots. (ICMR 2005 and ICMR 2013)

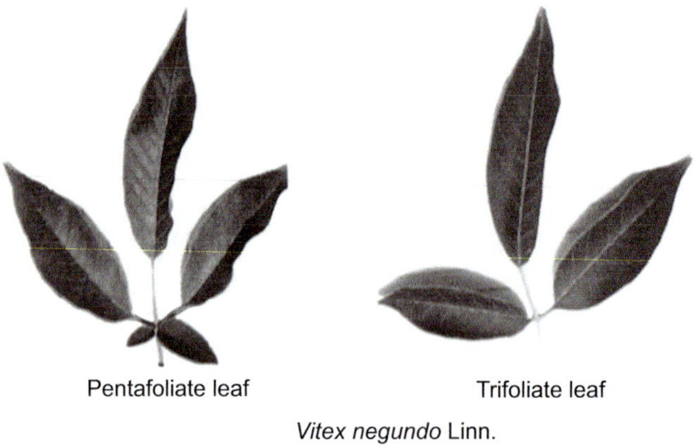

Fig. 3 *Vitex negundo* Linn. Leaves. (ICMR 2005 and ICMR 2013)

Vitex negundo (Linn.) fresh leaves are leathery and arranged in palmately trifoliate to pentafoliate patterns. The middle leaflet is the largest, measuring a length of 6–12 cm long and the width is 1.5–3 cm, attached to a 1 to 3 cm long petiole at the base. The two lower lateral leaflets of pentafoliate leaves are small or nearly sessile, with a long rachis. The lanceolate leaflets are acute, mostly entire, occasionally crenate, and have a pubescent, dark green upper surface, while the lower surface is tomentose and whitish. The venation is unicostate reticulate, with lateral veins arising at a 45° angle and running almost parallel. The leaf is characterised by an aromatic odour and is bitter. The upper epidermis of the leaflet shows cluster crystals of calcium oxalate and oil globules (ICMR 2005 and ICMR 2013).

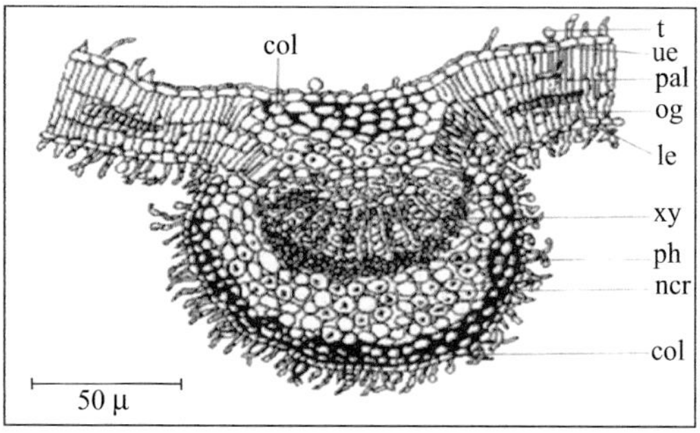

Fig. 4 Microscopy of transverse section of *Vitex negundo* leaf (col-collenchyma; mcr-microrosette crystals of calcium oxalate; ph-phloem; xy-xylem; le-lower epidermis; og-oil globules; pal-palisade; ue-upper epidermis; t-trichome). (ICMR 2005 and ICMR 2013)

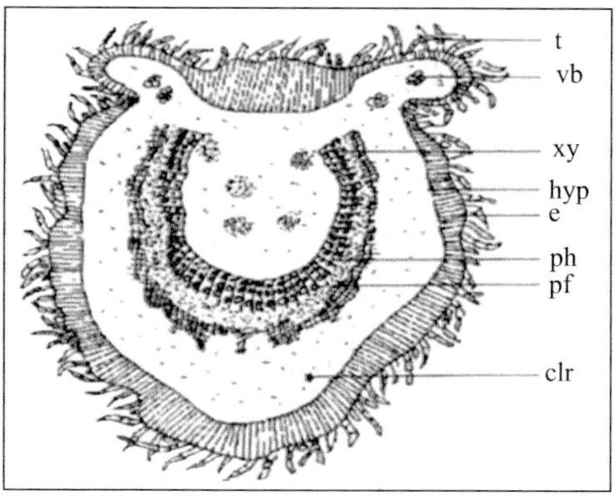

Fig. 5 Microscopy of the transverse section of *Vitex negundo* petiole (clr-cluster crystals of calcium oxalate; pf-pericycle fibres; ph-phloem; e-epidermis; hyp-hypodermis; xy-xylem; vb-vascular bundle; t-trichome). (ICMR 2005 and ICMR 2013).

Microscopic analysis of a transverse section of the *Vitex negundo* leaflet reveals a unique structure, with the leaflet strongly convex on the lower epidermis side and slightly flattened on the upper side. An arc of meristele is centrally located,

thickened cells, with a vertical bar-like thickened lignified outermost wall. The endocarp is stony, composed of many rows of sclereids and few stone cells of various sizes and shapes, the majority of sclereids being wide lumened and having beaded walls, occasionally filled with yellowish brown contents and prismatic crystals of calcium oxalate. Powder microscopy analysis of *Vitex negundo* fruit has revealed the occurrence of a fragment of epicarp with a trichome and groups of stone cells, some trichomes from the sepals, some septate and non-septate fibres, and fragments of longitudinally cut pitted and spiral vessels, pitted parenchyma from the pith and glandular and non-glandular trichomes, sclereids and fibres from the pericarp, reticulated parenchyma of testa (ICMR 2005 and ICMR 2013).

Table 1 Therapeutic applications of different plant parts of *Vitex negundo* (Vishwanathan *et al.*, 2010; Ullah *et al.*, 2012; Basri *et al.*, 2014)

Plant parts used	*Used for Diseases*
Whole Plant	Anti-Flatulent, Malaria
Powdered Roots/ Tincture Roots	Dysentery, Flatulence/Irritable Bladder, Dyspepsia, Colic, Respiratory Disorders Cough, Bronchitis, Asthma, Rheumatism, Malaria, Wounds and Ulcers, Diuretic, Tonic, Antitoxin
Stem and Bark	Burns, Rheumatism, Cancer
Leaves	Headache, Improves Eyesight, Respiratory Disorders Cough, Common Cold, Flu, Sore Throat, Whooping Cough, Sinusitis, Post-Partum Bath And Recovery, Rheumatism, Cancer, Gout, Catarrhal Fever, Wounds and Ulcers, Dentistry, Diuretic, Astringent, Tonic, Anti Allergic, Analgesic, Antifeedant, Fumigant, Insecticide, Reducing Swellings, Backache and Chest Pain
Flowers	Gastrointestinal Disorders, Diarrhoea, Cholera, Cancer, Jaundice/ Liver Disorders, Astringent, Tonic
Fruits	Eye Disease/Cataract/Watery Eyes, Galactogogue, Analgesic, Backache, Chest Pain, Bodyache

3.1 Nutritive value of Vitex negundo

Proximate analysis of *Vitex negundo* bark revealed its chemical composition, fat, ash, moisture, crude fibre, and protein. In-depth phytochemical analysis of *Vitex negundo* leaves and bark reveals the occurrence of carbohydrates, terpenoids, steroids, saponins, phenolic compounds, and flavonoids, while tannins and alkaloids were notably absent. Interestingly, the alcoholic extract of *Vitex negundo* exhibits a different chemical profile, containing phenolic compounds, alkaloids, carbohydrates, saponins, tannins, and flavonoids, with terpenoids and steroids absent. Further investigations have identified volatile oils, lignans, flavonoids, flavonoid glycosides, triterpenes, diterpenes, sesquiterpenes, and steroids in *Vitex negundo*. Notably, the bark alcoholic extract contains negundoside, agundoside, vitegnoside, hydroxybenzoic acid, and β-sitosterol, adding to the plant's pharmacological diversity and potential therapeutic applications (Kumar *et al.*, 2013; Koirala *et al.*, 2020).

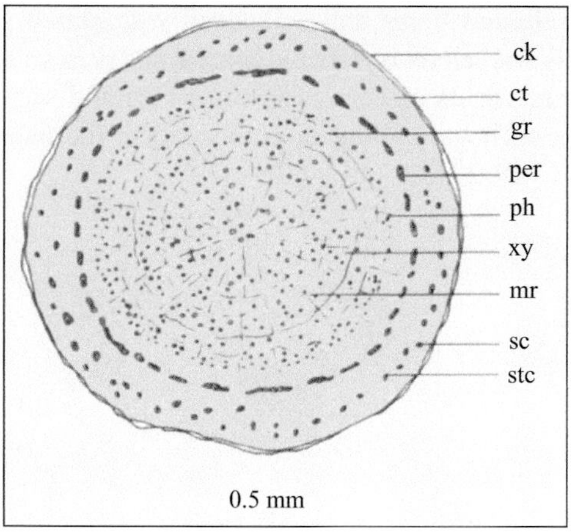

Fig. 8 Transverse section of *Vitex negundo* Linn. Root (ck-cork; ct-cortex; gr-growth ring; mr,-medullary rays; per- pericycle; ph-phloem; sc-sclereids; stc-stone cells; xy,-xylem). (ICMR 2005 and ICMR 2013)

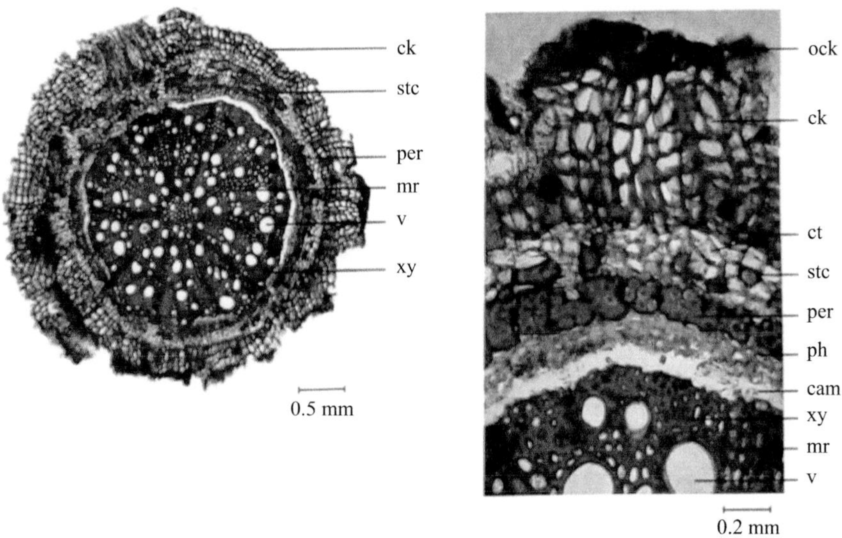

Fig. 9 Microscopy of *Vitex negundo* Linn. Root (cam-cambium; ck,-cork (lignified); ct-cortex; mr-medullary rays; ock-outer cork (suberised); per-pericycle; ph-phloem; stc-stone cells; v,-vessel; xy-xylem). (ICMR 2005 and ICMR 2013)

Powder microscopy of *Vitex negundo* root

Powder microscopy of *Vitex negundo* root shows abundant fragments of thick-walled xylem and pericyclic fibres of various sizes and shapes, pitted vessels, tracheids, and parenchyma; radially and tangentially, longitudinally cut medullary rays crossing vessels and fibres and embedded with starch grains; isolated or groups of stone cells and sclereids from the cortical and pericyclic regions; lignified and non-lignified fragments of cork cells in surface and sectional view; starch grains scattered as such throughout or embedded in the parenchymatous cells (ICMR 2005 and ICMR 2013).

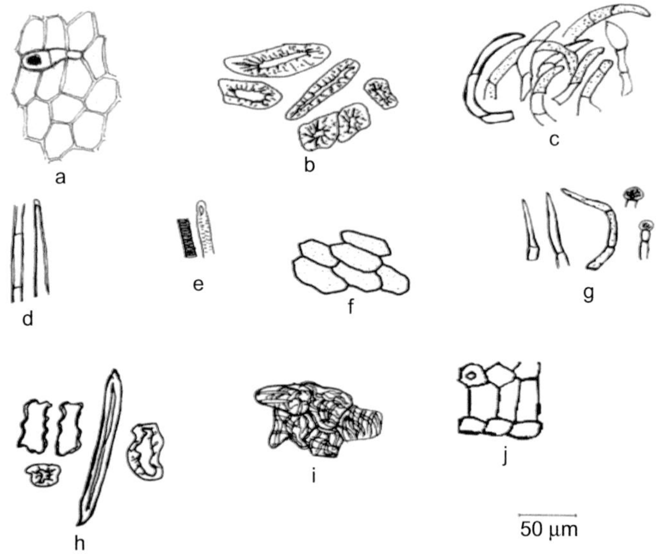

Fig. 10 Powder microscopy of *Vitex negundo* root (a) fragment of cork in surface view; (b) fragment of cork in sectional view; (c) stone cells; (d) sclereids; (e) fragment of pitted vessel; (f) fragments of fibre; (g) fragment of xylem showing longitudinally cut vessel and fibres crossing medullary ray cells; (h) starch grains; (i) wood parenchyma; (j) tracheids). (ICMR 2005 and ICMR 2013).

Table 2 Nutrient analysis of *Vitex negundo.* (Kumar *et al.*, 2013 and Koirala *et al.*, 2020)

Element compounds	*Nutrients/100gm*
Ash	7.86 g
Calcium	5267.13 mg
Crude Fibre	6.52 g
Crude Fibre	0.9 g
Crude protein	9.687 g
Fat	6.19 g
Iron	9.4 mg
Moisture	18.35 g
Phosphorus	680.3 mg
Vitamin C	180.9 mg

Table 3 Compounds detected in methanolic extract of *Vitex negundo* (Kumar *et al.*, 2013 and Koirala *et al.*, 2020).

Name of Compounds	*Area%*
Caryophyllene (E)	0.03
Caryophyllene oxide	0.01
Virdiflorol	0.10
Methyl palmitate	0.14
Beyerene isomer	0.04
Longifolene	0.02
Kolavenol	0.03
Mannol	0.18
Phytol	0.08
Methyl octadecanaote	0.12
Monoolein	0.57
Ledol	0.01

3.2 Phytochemical Constituent and Metabolite Profiling and Pharmacological Activities

Vitex negundo chemical components profiling has revealed the existence of essential oil isolates from its leaves, including P-caryophyllene, caryophyllene oxide, camphene, camphor, carene, cineole, and benzaldehyde. Some exotic varieties of *Vitex negundo* leaves also consisted of pinene, camphene, citral, linalool, terpinyl acetate, cineole, sabinene, bornyl acetate, camphene, *P*-caryophyllene, and caryophyllene oxide as principal constituents of essential oil. Thus, when enlisted and explored out of thirty-five compounds which make up around 74.96% of oil, compounds like viridiflorol, caryophyllene, sabinene, 4-terpineol, terpinene, caryophyllene oxide, 1-oceten-3-ol and globulol are mainly seen. Viridiflorol is specific in the oil of *Vitex* species, it can be used as a marker of the species (Vishwanathan *et al.*, 2010; Ullah *et al.*, 2012 ; Basri *et al.*, 2014).

While various parts of *Vitex negundo* are utilised in indigenous medicine, its leaves are particularly potent for medicinal purposes. Decoctions of the leaves find application in treating a range of conditions, including eye diseases, toothaches, sinus issues, and bronchitis, and as tonics for inflammation, leucoderma, rheumatoid arthritis, and spleen enlargement. Moreover, the leaves are documented for their effectiveness against ulcers, cancers, catarrhal fever, gonorrhoea, and scrofulous sores. Their antibacterial, analgesic, antihistaminic, and antipyretic qualities, as well as their antigenotoxic and antifertility effects, have been the subject of extensive research. Overall, the leaves of *Vitex negundo* emerge as a versatile and invaluable resource in traditional medicine (Vishwanathan *et al.*, 2010; Ullah *et al.*, 2012; Basri *et al.*, 2014).

Chemical constituents majorly present are negundoside and nishindaside. Besides this chemical constituents are agnuside, mussaenoside tetraacetate, mussaenoside pentaacetate., 6’-p-hydroxybenzoyl mussaenosidic acid, 2’-p-hydroxybenzoyl mussaenosidic acid, sabinene, p-cymene, phellandrene, y-terpinene, terpinen-4-ol, P-caryophyllene, viridiflorol, P-eudesmo1, a-pinene, 6-carene, limonene, camphene, citral, carophyllene, methyl heptanone, linalool, camphor, 1,8-cineole, cc-terpineol, geraniol, carophyllene oxide, terpenyl acetate, geranyl acetate, benzaldehyde, cinnamaldehydea, 5-hydroxy-pentamethoxy flavone’, 5,3’-dihydroxy-trimethoxy flavanone, 5,3’-dihydroxy-trimethoxy flavanone, dimethoxy-trans-stilbene and heptamethoxy flavone (Vishwanathan *et al.*, 2010; Ullah *et al.*, 2012; Basri *et al.*, 2014).

Past investigations into *Vitex negundo* phytochemistry have unveiled a rich array of chemical components, including essential and volatile oils, flavonoids, terpenes (such as triterpenes, diterpenes, and sesquiterpenes), iridoids, lignans, and steroids. The chemical constituents have unique medicinal and therapeutic value confirming the shrub’s high biological value. Considering all these ethnobotanical significance, the chemical contents of *Vitex negundo* leaves must be investigated for detailed therapeutic value data (Vishwanathan *et al.*, 2010; Ullah *et al.*, 2012; Basri *et al.*, 2014).

Vitex negundo (Linn.) fresh leaf extract shows anti-inflammatory activities and was used in arthritis and rheumatic disorders. Because *Vitex negundo* leaf extract has analgesic properties, it is also used to treat central nervous system disorders. Solvent extract of leaves has potential activity in the sedative-hypnotic effect of pentobarbitone and anticonvulsant activity when verified with an extremely high dose. *Vitex negundo* has an antiulcerogenic effect so inhibits vascular permeability. The leaf extracts are proven for the prevention of genotoxicity, and antibacterial· hepatoprotective activity, and antifungal activities. *Vitex negundo* major therapeutic claims are for its anti-inflammatory activity. *Vitex negundo* leaf extract has shown much therapeutic activity as discussed by Rizwan *et al.*, 2012; mainly focusing on antibacterial, and anticancer, and applied to sinuses and scrofulous sores. However, the safety doses and aspects of LD50 of the leaf alcoholic extract are documented for 1.25 g/kg in mice (Zahoor *et al.*, 2012).

3.3 Phytochemical constituents of different plant parts of Vitex negundo (Vishwanathan et al., 2010; Ullah et al., 2012; Basri et al., 2014)

3.3.1 Chemical Constituents in Leaves: *p*-cymene, *cis*-ocimene, citronellal, β-curcumene, β-caryophyllene, α-guaiene, guaia-3,7-diene, δ-guaiene, valencene, caryophyllene epoxide, ethyl-9-hexadecenoate, palmitic acid, (E)-nerolidol, humulene epoxide 1, globulol, humulene epoxide 2, epi-α-cadinol, α-muurolol, α-cadinol and α-bisabolol acetate; hydroxy-3,6,7,3′,4′-pentamethoxyflavone; 6′-p-hydroxybenzoyl mussaenosidic acid; 2′-p-hydroxybenzoyl mussaenosidic acid; protocatechuic acid; oleanolic acid; flavonoids; 5, 3′-dihydroxy-7,8,4′-

trimethoxyflavanone; 5,3′-dihydroxy-6,7,4′- trimethoxyflavanone; viridiflorol; β-caryophyllene; sabinene; 4-terpineol; gamma-terpinene; caryophyllene oxide; 1-oceten-3-ol; globulol; angusid; casticin; vitamin-C; nishindine; gluco-nonitol; p-hydroxybenzoic acid; sitosterol; betulinic acid [3β-hydroxylup-20-(29)-en-28-oic acid]; ursolic acid [2β-hydroxyurs-12-en-28-oic acid]; n-hentriacontanol; β-sitosterol; p-hydroxybenzoic acid, protocatechuic acid; oleanolic acid; flavonoids, angusid; casticin; vitamin-C; nishindine; gluco-nonitol; p-hydroxybenzoic acid; sitosterol; δ-guaiene; guaia-3,7-dienecaryophyllene epoxide; ethyl-hexadecenoate; α-selinene; germacren-4-ol; caryophyllene epoxide; (E)-nerolidol; β-selinene; α-cedrene; germacrene D; hexadecanoic acid; and valencene.

3.3.2 Chemical Constituents in Seeds: 3β -acetoxyolean-12-en-27-oic acid; 2α, 3α-dihydroxyoleana-5,12-dien-28-oic acid; 2β,3α diacetoxyoleana-5,12-dien-28-oic acid; 2α, 3β-diacetoxy-18-hydroxyoleana-5,12-dien-28-oic acid; vitedoin-A; vitedoin-B; a phenylnaphthalene-type lignan alkaloid, vitedoamine-A; β-sitosterol; p-hydroxybenzoic acid; 5-oxyisophthalic acid; n-tritriacontane, n-hentriacontane; n-pentatriacontane; n-nonacosane; 6-hydroxy-4-(4-hydroxy-3-methoxy-phenyl)-3-hydroxymethyl-7-methoxy-3, 4-dihydro-2-naphthaldehyde.

3.3.3 Chemical constituents in flowers: formic acid, n-heptane, *p*- cymene, β-caryophyllene, trans-α-bergamotene, valencene, α-selinene, β-selinene, germacren-4-ol, caryophyllene epoxide, (E)-nerolidol and *P*-(1,1-dimethylethyl) toluene; sabinene, linalool, terpinen-4-ol, β-caryophyllene, α-guaiene and globulol; sesquiterpenes, monoterpenes, terpenoids and sterols; guaia-3,7-dienecaryophyllene epoxide; ethyl- hexadecenoate; α-cedrene; germacrene D; hexadecanoic acid; and valencene.

3.3.4 Chemical constituents in fruits: α-copaene, β-caryophyllene, α-cedrene, α-guaiene, guaia-3, 7-diene, α-humulene, aristolene, germacrene D, β-selinene, caryophyllene oxide, n- hexadecanoic acid, palmitolic acid and traces of acetyl lactyl glycerate; δ-guaiene; guaia-3,7-dienecaryophyllene epoxide; ethyl-hexadecenoate; α-selinene; germacren-4-ol; caryophyllene epoxide; (E)-nerolidol; β-selinene; p-cymene and valencene.

3.3.5 Chemical constituents in Roots: 2β, 3α-diacetoxyoleana-5,12-dien-28-oic acid; 2α,3α-dihydroxyoleana-5,12-dien-28-oic acid; 2α,3β-diacetoxy-18-hydroxyoleana-5,12-dien-28-oic acid; vitexin and isovitexin, negundin-A; negundin-B; (+)- diasyringaresinol; (+)-lyoniresinol; vitrofolal-E and vitrofolal-F, acetyl oleanolic acid; sitosterol; 3-formyl-4.5-dimethyl-8-oxo-5H-6,7-dihydronaphtho (2,3-b)furan.

4.1 Antiviral Potency of Vitex Negundo: Bioactive Insights

Numerous bioactive substances, including flavonoids, terpenoids, glycosides, and phenolic compounds, are found in *Vitex negundo*. It is particularly rich in luteolin derivatives, which are essential to its antiviral qualities. These compounds are pivotal in hindering various viruses by disrupting viral replication and impeding viral protein functions (Singh *et al.*, 2020; Nallusamy *et al.*, 2021).

Vitex negundo exhibits remarkable antiviral and as well as anti-flu activity. The extracts display substantial antiviral efficacy, particularly against the Dengue virus and potentially influenza viruses. These extracts effectively inhibit viral replication and diminish viral RNA levels, underscoring their promise as natural antiviral agents. Moreover, traditional uses and pharmacological investigations corroborate its efficacy in alleviating flu-like symptoms, possibly attributed to its anti-inflammatory and immune-boosting properties (Singh *et al.*, 2020). Research indicates that an ethanolic extract of *Vitex negundo* displays effectiveness against the Dengue virus, though its activity against the Chikungunya virus is comparatively limited. This suggests that the antiviral properties of *Vitex negundo* may vary depending on the specific virus being targeted. These studies have shown significant reductions in viral RNA levels and infectious virus particles when treated with *Vitex negundo* extracts, highlighting its potential as a potent antiviral agent against these viral infections. Compounds isolated from *Vitex negundo* exhibit potential activity against HIV-1 reverse transcriptase, indicating broader antiviral capabilities. The overall antiviral efficacy of *Vitex negundo* underscores its potential as a source of bioactive compounds for developing antiviral treatments, particularly given its traditional use in diverse medicinal preparations (Nallusamy *et al.*, 2021; Alagarasu *et al.*, 2022). Some of the key highlights of the studies are listed below:

- **Antiviral Activity Against Chikungunya and Dengue Viruses:** *Vitex negundo* has demonstrated significant in vitro antiviral activity against the Dengue virus (DENV-2), effectively inhibiting its replication. However, its activity against the Chikungunya virus (CHIKV) is limited (Alagarasu *et al.*, 2022).
- **Activity Against SARS-CoV-2 and HIV-1:** Phytochemicals derived from *Vitex negundo* exhibit the ability to inhibit crucial proteins of SARS-CoV-2, including the main protease and RNA-dependent RNA polymerase. Additionally, the plant demonstrates activity against HIV-1 Reverse Transcriptase, suggesting broad-spectrum antiviral properties (Nallusamy *et al.,* 2021).
- **General Antiviral Efficacy:** Studies have underscored *Vitex negundo* broad-spectrum antiviral capabilities, showcasing its effectiveness against various human pathogenic viruses such as Chikungunya and HIV-1. These findings reinforce the plant's potential in traditional and modern medicinal applications (Singh *et al.*, 2020; Alagarasu *et al.*, 2022; Nallusamy *et al.*, 2021).
- **Potential Against Influenza:** Although specific studies on influenza are not extensively highlighted, the general antiviral properties of *Vitex negundo* suggest its potential effectiveness against flu viruses. The plant's ability to inhibit RNA virus replication, as demonstrated with DENV-2 and CHIKV, supports this hypothesis (Alagarasu *et al.*, 2022; Nallusamy *et al.*, 2021).

- **Traditional and Modern Usage:** *Vitex negundo* has been utilised in traditional medicine for its wide range of therapeutic properties, including antiviral and antiflu activities. Modern pharmacological studies have supported these traditional uses, offered scientific validation, and explored new therapeutic potentials (Singh *et al*., 2020).
- **Safety and Efficacy:** Studies have demonstrated that the ethanolic extract of *Vitex negundo* leaves is safe, with non-toxic doses observed in oral administration. This supports its potential for development into antiviral therapies. Supplementary research is necessary to fully establish its efficacy and safety in clinical settings (Singh *et al.,* 2020).

4.2 Commercial Products Market with Vitex Negundo Active Constituents

Vitex negundo and its derivatives have undergone extensive clinical testing, leading to the global commercialisation of numerous products. Manufacturers integrate *Vitex negundo* into various formulations either as a primary ingredient or as an adjunct in products like V-Gel, Liv.52, Nigundi ghrita, Nirgundi Churna, Vitex women health supplements, Relief cream, Pilex, Acne-n-Pimple Cream, Muscle and Joint Rub, massage oils, syrups, and tablets. Essential oil extracted from its leaves finds application in pain relief. A synergistic combination of *Vitex negundo* water extract and matra basti in 500 mg tablets has confirmed efficacy in alleviating sciatica pain, numbness, and discomfort associated with conditions such as low back pain, weakness, herniated disk, piriformis syndrome, and spinal stenosis across a diverse age group ranging from 20 to 60 years (Tandon 2005).

Table 4 Various companies and Commercial products market with *Vitex negundo* active constituents (Tandon 2005).

Product	*Usage*
Leaf Syrup - Phlemex	For cough and asthma
Acne-n-pimple cream; Joint care B cream, V Gel, Liv-52, Pilex tablet - Himalayan D rug Corp.	For acne and skin disorders rheumatic disorder, cervicitis, detoxify liver, haemorrhoids
Jigrine - Hamdard lab.	For increasing liver capacity
Massage oil - IndSwift Ltd	For joint pain, gout, frozen shoulders
Nigundi ghrita - VHCA herbal	For traumatic weakness
Nirgundi Churna - Vindhya herbal	For kidney pain
Vitex women health - NOW food	To maintain hormone balance
Relief cream - Surya Herbal Ltd	For decreasing muscle pain and joints
Syrup and tablet - Pascual Labs	For cough and asthma

4.3 Vitex Negundo Active Constituent's Toxicity Study

Vitex negundo is a ubiquitous herb, that finds its way into various pharmaceutical products, yet its journey from nature to industry isn't without intrigue. Initial studies on its ethanolic leaf extract revealed remarkable non-toxicity up to a recorded LD50 dose of 7.58g/kg in albino rats, highlighting its safety profile. However, deeper investigations uncovered intriguing dose-dependent histomorphological changes in vital organs like the heart, liver, and lungs, shedding light on potential toxicity concerns. Dyspnea, observed post-administration, hinted at cardiac involvement, emphasising the need for cautious exploration. Enter OECD guidelines, providing a structured approach to studying acute and sub-chronic toxicity, and guiding researchers through meticulous evaluations.

Meanwhile, investigations into its essential oil on Wistar rats uncovered an unexpected tale of safety, with no adverse effects observed even after prolonged exposure. Classified as Category 5 under the globally harmonised system, its LD50 value over 2000 mg/kg solidifies its benign nature. Yet, caution flags wave for those with hormone-dependent conditions, urging restraint in consumption. Despite warnings, Vitex's *negundo* allure persists, its bioactive compounds tempting researchers to uncover new realms of therapeutic potential. While documented side effects exist, the lure of its benefits eclipses these concerns. The quest for isolated bioactive compounds propels drug discovery, promising novel therapies yet unseen. As the future beckons, *Vitex negundo* stands as a beacon of exploration, its mysteries waiting to be unveiled, its potential as a remedial agent against diverse maladies ensuring its place in the spotlight. With each discovery, the narrative of this botanical marvel evolves, promising a tapestry of innovation and healing in the years to come (Tandon 2005, Gill *et al.*, 2018).

4.4 Vitex Negundo Active Constituent's Recommended Dosage

Various parts of *Vitex negundo*, including leaves, roots, bark, fruits, flowers, and seeds, are harnessed for therapeutic purposes. They are administered as juice, decoction, powder, oil, tincture, or dry extract with sugar, water, or honey paste. For adults, recommended dosages range from 10–20 ml for juice, 50–100 ml for decoction, 1.5–3 g for leaf powder, and 300–600 mg for dry leaf extract, ensuring optimal utilisation of its active constituents for medicinal benefits (Tandon 2005; Gill *et al.,* 2018).

Conclusion

Vitex negundo stands out as a rich source of biologically active constituents, supported by numerous empirical studies; it has a strong foundation in both traditional knowledge and experimental research. The findings collectively support the antiviral and antiflu activities of *Vitex negundo*, positioning it as a valuable candidate for further investigation and potential therapeutic use against viral infections. *Vitex negundo* stands out as a potent medicinal plant with significant antiviral and antiflu activities, underpinned by its rich phytochemical composition, including luteolin derivatives. While more research, particularly clinical trials, is necessary, existing evidence supports its use in traditional and modern antiviral therapies. *Vitex negundo* is also a promising candidate in the pharmaceutical industry. Emerging data highlight the significance of nutritional supplementation, particularly in bolstering immune function and potentially minimising viral load when administered in higher-than-usual dietary doses. Deficiencies in vitamins and micronutrients can compromise immune performance, emphasising the significance of addressing nutritional gaps. Rigorous clinical investigations suggest that active constituents and nutritional supplements offer a straightforward and effective approach, providing broad-ranging health benefits. Importantly, the risk-to-benefit ratio of vitamins and micronutrients appears favourable, making them viable options for enhancing immune resilience. Hence, in these findings, nutrient supplementation emerges as a proficient strategy in combating viral infections, offering a simple yet impactful means to support overall health and well-being.

References

Alagarasu, K., Patil, P., Kaushik, M., Chowdhury, D., Joshi, R.K., Hegde, H.V.,... and Parashar, D. (2022). *In vitro* antiviral activity of potential medicinal plant extracts against dengue and chikungunya viruses. *Frontiers in Cellular and Infection Microbiology;* 12: 1–17.

Anonymous, 2013. ICMR, New Delhi; *Quality standards of Indian Medicinal Plants Edition*, Volume 11.

Anonymous, 2005. ICMR, New Delhi; *Quality standards of Indian Medicinal Plants Edition*, Volume 03.

Bano, U., Jabeen, A., Ahmed, A. and Siddiqui, M.A. (2015). Therapeutic uses of *Vitex negundo. World Journal of Pharmaceutical Research*; 4(12): 589–606.

Basri, F., Sharma, H.P., Firdaus, S., Jain, P. and Ranjan, A. (2014). A review of ethnomedicinal plant-*Vitex negundo* Linn. *Int. J. Adv. Res*; 2(3): 882–894.

Chavda, V.P., Kapadia, C., Soni, S., Prajapati, R., Chauhan, S.C., Yallapu, M.M. and Apostolopoulos, V. 2022. A global picture: therapeutic perspectives for COVID-19. *Immunotherapy*; 14(5): 351–371.

Gill, B. S., Mehra, R. and Kumar, S. 2018. *Vitex negundo* and its medicinal value. *Molecular Biology Reports*; 45(6): 2925–2934.

Koirala, N., Dhakal, C., Munankarmi, N.N., Ali, S.W., Hameed, A., Martins, N.,... and Salehi, B. (2020). *Vitex negundo* Linn.: phytochemical composition, nutritional analysis, and antioxidant and antimicrobial activity. *Cellular and Molecular Biology*; 66(4): 1–7.

Kumar, M., Dandapat, S., Kumar, A. and Sinha, M.P. (2013). Determination of nutritive value and mineral elements of five-leaf chaste tree (*Vitex negundo* L.) and Malabar nut (Adhatoda vasica Nees). *Academic Journal of Plant Sciences*; 6(3): 103–108.

Mieres-Castro, D. and Mora-Poblete, F. (2023). Saponins: research progress and their potential role in the post-COVID-19 pandemic era. Pharmaceutics, 15(2): 348.

Nallusamy, S., Mannu, J., Ravikumar, C., Angamuthu, K., Nathan, B., Nachimuthu, K.,... and Neelakandan, K. 2021. Exploring phytochemicals of traditional medicinal plants exhibiting inhibitory activity against main protease, spike glycoprotein, RNA-dependent RNA polymerase and non-structural proteins of SARS-CoV-2 through virtual screening. *Frontiers in Pharmacology*; 12: 1–29.

Ng, T.I., Correia, I., Seagal, J., DeGoey, D.A., Schrimpf, M.R., Hardee, D.J.,... and Kati, W.M. (2022). Antiviral drug discovery for the treatment of COVID-19 infections. *Viruses* 14(5): 961.

Rizwan, U.H., Azhar-ul-Haq, A.S., Zahoor, U., Habib, U. and Rafeeq, A.K. (2012). Antitussive and toxicological evaluation of Vitex negundo Linn. *Natural product resesrch*, 26(5), 484–488.

Singh, Y., Mishra, P. and Kannojia, P. (2020). Morphology, phytochemistry and pharmacological activity of Vitex negundo: an overview. *Journal of Drug Delivery and Therapeutics*; 10(3s): 280–285.

Tandon, V.R. (2005). Medicinal uses and biological activities of *Vitex negundo*. *Natural product radiance*; 4(3): 162–165.

Ullah, Z., Ullah, R., Ahmad, I. and Haider, S. 2012. Phytochemical and biological evaluation of *Vitex negundo* Linn: A Review. *International Journal of Pharmaceutical Sciences and Research*; 3(8): 2421.

Vishwanathan, A.S. and Basavaraju, R. 2010. A review on *Vitex negundo* L.: A medicinally important plant. *Eur J Biol Sci*; 3(1): 30–42.

21 *Azadirachta indica* A. Juss. with Antiviral Potential

Nagesh Nilawar[1] and Tabassum Khan[2*]

1. Introduction

Azadirachta indica A. Juss. (*A. indica*), an ancient tree belonging to the family Meliaceae is found abundantly in Australia and other South Asian countries and is a native plant of India. A. indica was named the "Tree of the twenty-first century" by the United Nations. The US National Academy of Science published a report in 1992 entitled "Neem: A tree for solving a problem", which shows the importance of this tree in science. For decades *A. indica* tree has been used as a traditional medicine to treat various diseases, across civilizations. The usage of *A. indica is* mentioned in ancient history like Siddha, Palm leaf manuscript, and Indus Valley civilization. The Sanskrit mythology has a specific name for *A. indica* "Arista" meaning "perfect in all aspects" (Jagannathan, 2015).

All the parts of A. *indica* tree are significant , have diverse properties, and are used for curing skin infections, fever, ulcers, cancers, respiratory problems, and digestive problems. The pharmacologically important secondary metabolites in *A. indica* have been used in the development of medicine (Girish et al., 2008; Sultana et al., 2011; Pankaj et al., 2011; Kumar et al., 2013; Gupta et al., 2017) .

Taxonomy and Scientific Classification

Kingdom	: Plantae
Sub kingdom	: Tracheobionta
Domain	: Eukaryota
Class	: Magnoliopsida
Order	: Sapindales
Suborder	: Rutinal
Phylum	: Spermatophyta
Subphylum	: Angiospermae
Family	: Meliaceae
Subfamily	: Melioideae
Genus	: Azadirachta
Species	: *indica* (Hashmat et al., 2012; More et al., 2015, Singh et al., 2020)

[1] Department of Quality Assurance, SVKM's Dr. Bhanuben Nanavati College of Pharmacy, Mumbai, Maharashtra, India.
[2] Department of Pharmaceutical Chemistry & Quality Assurance, SVKM's Dr. Bhanuben Nanavati College of Pharmacy, Mumbai, Maharashtra, India.
* Corresponding author: tabassum.khan@bncp.ac.in

DOI: 10.1201/9781003433583-21

Synonyms

A. indica is known by different names in various regions and languages; English; pride of China, China berry, Hindi; bakan, nim, nimb, Marathi; nimbay, Sanskrit; arista, nimba, nimbha, pakvakrita, Tamil; veppi, sengumaru, malaivempu, Kannada; turakabevu, bemu, bivu, kaybevu, Malayalam; malaveppu, hutchubevu, karinnvembu, Panjabi; bakam, drek, nim chein, jek, Gujrati; limbado, limbra,danujhada, Bengali; mahnim, ghoramin, niumgachh, Urdu; neem (Hashmat et al., 2012; Singh et al., 2020; Mehta et al., 2021).

2. Botanical Description of the Plant

A. indica tree can grow fast up to 20–25 feet within a few years. The tree has a thick straight trunk with broader stems and branches growing outward in the shape of a round crown. It requires a small percentage of water to survive (17–18 inches) and hence can thrive in regions of low rainfall. The anticipated life of a tree is between 150 to 200 years (Ahmed et al., 1989; Yerima et al., 2012; Kumar et al., 2013; Raj, 2015; Gupta et al., 2017). Flowering starts nearly at the end of winter and beginning of summer. Fruits are produced 2–5 years after plantation. Flowers bloom in clusters and are white to yellowish. Fruits are greenish on maturation they turn yellow, emitting a garlic-like smell. Leaves are fully-grown with many leaflets as shown in Fig. 1. The bark is hard, scaly, rough, and brown (Van der Nat et al., 1991; Hashmat et al., 2012).

Fig. 1 (a) Leaf (b) Leaflet.
(*Source:* Botanical Garden SVKM's Dr. Bhanuben Nanavati College of Pharmacy)

3. Geographical Distribution

There are two main species of *A. indica*; *A. Juss* local to India *and Azadirachta excelsa Kack* local to Indonesia and Philippines. India, Sri Lanka, Bangladesh, Pakistan, Burma, Indonesia, Malaysia, and Thailand are the major producers of *A. indica*. The spread of these trees is now seen over several continents. This plant has been successfully grown in over 72 countries, including Asia, South America, North America, Central America, Africa, and Australia. (Sujarwo et al., 2016; Queensland, 2023).

The estimated value of cultivation of *A. indica* is 20 to 25 million trees cultivated every successive year. In India, the highest cultivation is found in Uttar Pradesh (55–56%) followed by Tamil Nadu (17–18%) and Karnataka (5–6%). Other states like Maharashtra, Kerala, Gujrat, Delhi, Bihar, Haryana, Panjab, Rajasthan, Madhya Pradesh, Odisha, West Bengal, and Himachal Pradesh also produce *A. indica* in small quantities. India stands first in the civilization of *A. indica* trees (Ogbeuwu et al., 2011; Chhabra et al., 2021).

4. Phytochemical Constituents and Metabolite Profiling

There are about 300 pharmacologically active phytoconstituents present in *A. indica*. They are further divided into two major types, isoprenoids and non-isoprenoids. The isoprenoid class comprises diterpenoids and triterpenoids and the non-isoprenoid class comprises proteins, carbohydrates, sulfur-containing compounds, tannins, flavonoids, coumarins, aliphatic compounds, and dihydrochalcones. The major phytoconstituents reported are nimbin, nimbinin, nimbidin, nimbidol, sodium nimbinate, quercetin, limonoids, azadirone, marganolone, margalone, and azadirachtin (Maithani et al., 2011; Tesfay et al., 2020). Some others include gedunin, salannin, 6-desacetylnimbinene, nimbolide, and *n*-hexacosanol, 7-desacetyl-7-benzoylgedunin, 17-hydroxyazadiradione, 7-desacetyl-7-benzoylazadiradione, ascorbic acid, and ß-sitosterol (Govindachari et al., 1998; Hwisa et al., 2014; Alzhhairy et al., 2016; Kraus et al., 1981). Table 1 gives the phytoconstituents reported in various parts of *A. indica* and Fig. 1. Gives the structures of some of the phytoconstituents reported in *A. indica* (Saleem et al., 2018).

Table 1 Phytoconstituents in *A. indica.*

Parts of A. indica	*Extract*	*Phytoconstituents*	*References*
Seeds	Methanol	Azadirachtin M, azadirachtin N, 11-epi-azadirachtin H, triterpenoid (1α,7α-diacetoxyapotirucall-14-ene-3α,21, 22,24,25-pentaol), odoratone, 2β,3β,4β-trihydroxypregnan-16-one 11-Hydroxyazadirachtin-B, 1-tigloyl-3-acetylazadirachtinin, 1,2-diacetyl-7-tigloyl-12 hydroxyvilasinin, 23-desmethyllimocin-B, 1α,2α-Epoxy-17β-hydroxyazadiradione, 1α,2α-epoxynimolicinol and 7-deacetylnimolicinol along with epoxyazadiradione, 17β-hydroxyazadiradione, gedunin, nimbin, and nimolicinol	(Kumar et al., 1996; Luo et al., 1999; Luo et. al., 2000)

Contd.

Table 1 *Contd.*

Parts of A. indica	*Extract*	*Phytoconstituents*	*References*
	n-Hexane	Genistein 7-O-glucoside (–) epi-catechin, and salannin	(Yamasaki et al., 1988; Kanwal et al., 2011)
	Dichloromethane	Azadirachtin	(Jarvis et al., 2000)
	Petroleum ether extract dissolved in methanol	1α-Methoxy-1,2 dihydroepoxyazadiradione, 1β,2β diepoxyazardiradione, 7 acetylneotrichilenone, desacetyl-7-benzoylazadiradione 7-desacetyl-7 benzoylepoxyazadiradione, and 7-desacetyl-7-benzoyl-gedunin	(Siddiqui et al., 1988)
Leaves	Ethanol	Zafaral, meliacinanhydride, nimocinol, isomeldenin nimonol nimbothalin, n-tridecyl benzene Isoprenoid, flavanone (8,3′-di-isoprenyl-5,7 dihydroxy-4′ methoxyflavanone), nonisoprenoids, meliacin, and Tetracyclic triterpenes	(Siddiqui et al., 1988; Gopalakrishnan et al., 2002; Siddiqui et al., 2004)
	Methanol	22,23-Dihydronimocinol, desfurano-6-α hydroxyazadiradione Meliatetraolenone, odoratone, and nimonol	(Suresh et al.,1997; Siddiqui et al., 2000; Siddiqui et al., 2003)
	n-Hexane	alpha-Linolenic acid	(Nair et al., 1997)
	Chloroform	Steroid, glycoside, flavonoids, triterpenoid, carbohydrate, alkaloids, and antiquinone	(Raphael et al., 2012; Prashanth et al., 2014)
	Dichloromethane	Stigmasterol, terpinen-4-ol, sugiol, 4-cymene, nimbiol, α-terpinene, and vitamin E	(Gupta et al., 2012)
Roots	Dichloromethane and methanol	Nimolinin and nimbilin	(Ara et al., 1989)
Flowers	Methanol	Prenylated flavonoids, Flowerine, flowerone, o-methylazadironolide, Diepoxyazadirol, Azharone, azadirone, and isoazadironolide	(Nakhara et al., 2003; Siddiqui et al., 2009)
	N-Hexane	Sesquiterpenes, aromatic compounds, fatty acids, fatty acid esters, steroids, and few hydrocarbons	(Siddiqui et al., 2006)

Contd.

Table 1 *Contd.*

Parts of A. indica	*Extract*	*Phytoconstituents*	*References*
Fruits	Ethanol	Mahmoodin, azadirachtol, naheedin, salimuzzalin, azadirolic acid, azadiradinol, azadironol limocinone, limocin A, limocin B, limocinol, and limocinin	(Siddiqui et al., 1991)

5. Medicinal / Therapeutic Uses

5.1 Uses Described in Pharmacopoeias and Traditional Systems of Medicine

A. indica is reported to possess diverse utility based on the parts used, the leaves are reported for leprosy, skin diseases, intestinal worms, chicken pox, respiratory disorders, and eye problems. Because of its analgesic properties, the bark can treat malaria, and fever, and is used in cosmetic items. Neem oil has been used in intestinal worms, constipation, soaps, toothpaste, and leprosy. The seeds are used in intestinal worms, gastro intestinal tract infections, and leprosy. lowers used in lowering bile levels. The fruits are used in reducing pain, and eye problems, in leprosy, and have wound-healing properties. The gum obtained from trees is used for several skin problems; ringworm, and ulcers, and as a tonic, and stimulant. The entire tree is used in several conditions like skin conditions, leprosy, dental plaque, insect repellent, ulcers, head lice, diabetes, heart diseases, malaria, psoriasis, contraceptives, and stomach upset (Biswas et al., 2002; Drabu et al., 2012; Saleem et al., 2018).

5.2 Traditional/Folklore Uses described in Folk Medicine

The use of *A. indica* tree is deeply rooted in traditional medicine systems, especially in India. Many medical systems, including Ayurveda, Homeopathy, Siddha, and Unani, provide evidence of the traditional use of A. indica. "Nimba" means "a good health condition" in Sanskrit, while "Sarv aroga nivarini" means "curing of all disease conditions." Also, the ayurvedic system says that neem can help relieve pain and hence it is tagged as "Arishtha". The tree is also named as "Divine tree" or "Village pharmacy" because of its varied local uses. Ayurveda medicine system primarily uses the *A. indica* plant parts to resolve many problems such as *Pitta, Kapha,* and *Vatic*. Plant parts; leaves, bark extracts, and oil are used as folk medicine. Neem oil has a significant effect in curing skin disease. Decoction of the leaf generally called "Kadha" is reported as antiseptic and used in ulcers. The small branches are used for cleaning teeth and have antibacterial properties. The bark tonic is used to control nausea and vomiting in several conditions. The leaf and bark extracts are used in the treatment of snake and scorpion poisoning. The bark

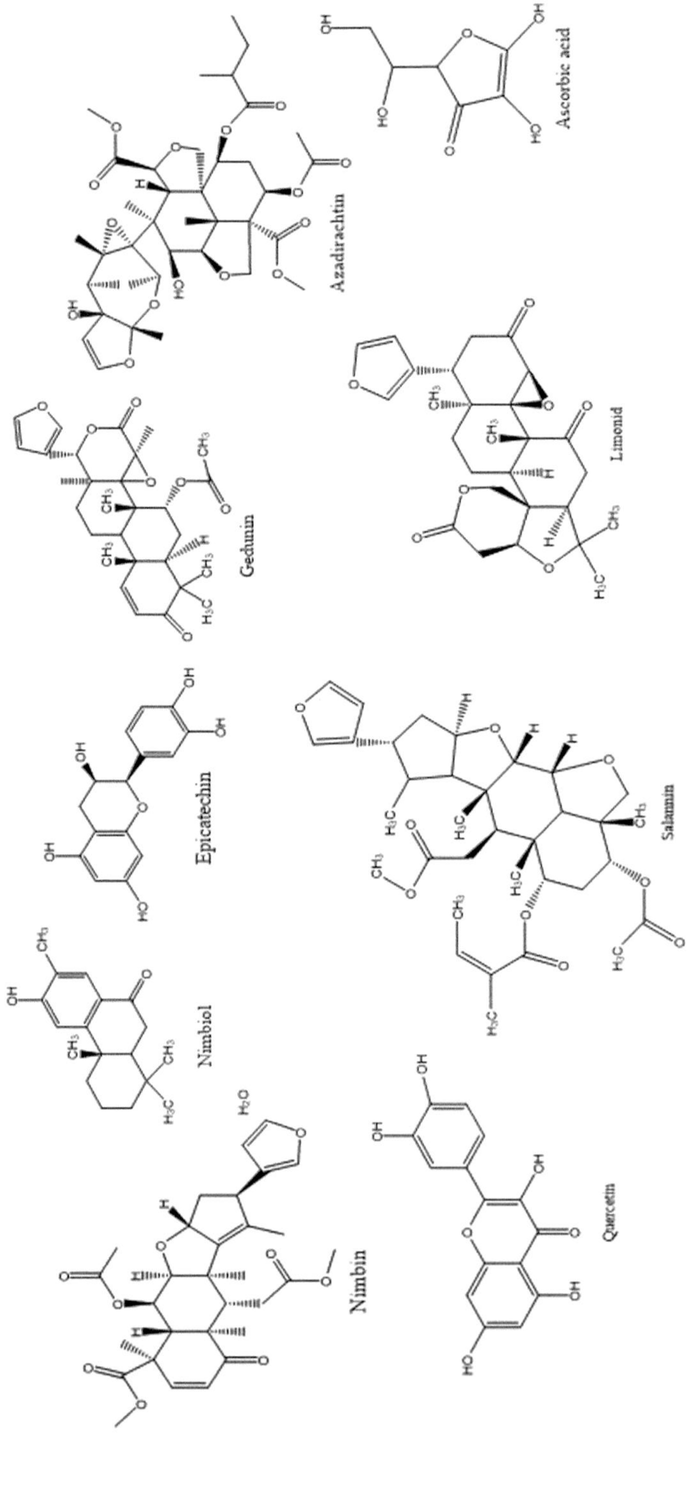

Fig. 2 Structures of few phytoconstituents reported in *A. indica.*

powder, leaves, and seeds are used to purify blood and have anti-acne properties. In homeopathic medicine, TA. indica is used to treat scabies-like conditions, joint pain, and eczema. A. indica is used in the homeopathic medical system to treat eczema, scabies-like disorders, and joint pain. The Unani medicine system uses *A. indica* as a blood purifier, for ulcers, clearing urinary passages, bronchitis, and several skin conditions. Neem oil obtained is directly applied to the body parts in skin diseases (Reddy et al., 2022).

5.3 Applications Supported by Clinical data

The clinical trial data (Table 2) and the patents (Table 3) on *A. indica* demonstrate the benefits of this tree in medicine thus providing proof of the translational potential with viable benefits.

Table 2 Clinical trials of *A. indica.* (Clinical trials, 2023)

S. No.	*Parts Used*	*Clinical trial No.*	*Title*	*Disease*	*Phase*	*Status*
1	Neem oil	NCT02974088	Neem lotion with combing for lice	Pediculoses head lice	Phase 2	Completed
2	Neem extract	NCT03166163	Evaluation of bleeding on brushing using neem extract mouthwash versus chlorhexidine mouthwash (mouthwash)	Gingival bleeding	–	–
3	Neem leaf extract	NCT02943759	Resolution of pain and swelling after the use of neem leaf extract compared to chlorhexidine gluconate as an intracanal irrigant during pulpectomy	Periapical abscess	Phase 2 Phase 3	Completed
4	Neem leaf extract	NCT02352987	Effect of neem extract, propylene glycol and salicylic acid combination in the treatment of arsenical palmar keratosis	Chronic arsenic poisoning	Phase 2	Completed
5	Neem leaf extract	NCT04686474	Effect of the compound extracted from *Azadirachta indica* leaves in the treatment of arsenical keratosis	Keratotic nodular size	–	Recruiting
6	Neem	NCT00985335	Effectiveness of theraneem lip therapy for herpes simplex labialis (neem)	Herpes Simplex	Phase 1	Completed

Contd.

Table 2 *Contd.*

S. No.	*Parts Used*	*Clinical trial No.*	*Title*	*Disease*	*Phase*	*Status*
7	Neem extract	NCT04341688	A clinical trial of gargling agents in reducing intraoral viral load among COVID-19 patients (Gargles)	COVID-19	–	Not yet recruiting
8	Neem leaf extract	NCT01251250	Azadirachta indica in treating patients with chronic lymphocytic leukemia	Chronic lymphocytic leukemia	Phase 1	Withdrawn

Table 3 Patents on *A. indica.*

Application No./Publication No.	*Title*
202241068954	*Azadirachta indica* is an effective photosensitizers for dye-sensitized solar cells
202221066848	Apparatus and method for production of methyl esters of commercial grade *Azadirachta indica* oil
202221051288	A novel formulation of herbal medicated monolithic matrices using extracts of *Azadirachta indica*
202241036462	Anti-fungal cream formulation and evaluation with *Nulembo nucifera* and *Azadirachta indica* leaves extract
202221051288	A novel formulation of herbal medicated monolithic matrices using extracts of *Azadirachta indica*
48440/93	Therapeutic compounds derived from the neem tree
1173/MAS/1995	A process for extracting a therapeutically useful fraction designated as IRDNB from neem leaves

5.4 Pharmacological Activity

The phytoconstituents reported in various parts and extracts of *A. indica* play a very important role in curing the disease. Some of the reported mechanisms of action of *A. indica* are activating various enzymes, rupturing cell walls, and inhibiting cellular pathways. Reported pharmacological activities are; antimicrobial, antidiabetic, antiviral, antibacterial, antioxidant, hepatoprotective, wound healing, immunomodulatory, antimalarial, contraceptives, antiplasmodial, insecticidal, insect repellent, molluscidal, antifeedant, cardioprotective, nephroprotective (Atawodi et al., 2009; Alzhhairy et al., 2016). The main focus of this chapter is a review of the antiviral activity of *A. indica.*

Antiviral activity of *A. indica*

Several researchers have discovered that A. indica has exceptional antiviral activity when compared to other antiviral medications. Phytoconstituents of

A. indica have been known to inhibit several viruses like herpes simplex virus type 1 (HSV-1), Newcastle disease virus, Dengue virus type-2, Duck plague virus, Pox viruses, Smallpox, chicken pox, Warts, Poliovirus, Influenza virus, HIV, and Coxsackie B group virus.

Tiwari et al. (2010) reported the antiviral activity of aqueous neem bark extracts against HSV-1. Labial herpes, ocular keratitis, encephalitis, and genital diseases are common diseases caused by this virus. In their study, they used neem bark powder extract. Neem bark extract inhibits the HSV-1 virus entry in the human host. The polysaccharides present in the neem bark extract possess antitumor and anti-inflammatory activity. Wild-type Chinese hamster ovary cells were used for the study, cell model was used to demonstrate the inhibition of the virus, by blocking the entry of the HSV-1 virus and glycoprotein-mediated cell-cell fusion. A blocking action was observed when the HSV-1 virus was previously incubated with neem bark extract, indicating the virucidal effect of the extract. (Tiwari et al., 2010).

Parida et. al. reported the effect of an aqueous extract of neem leaves and *azadirachtin* on the replication of Dengue virus type-2. Dengue is one of the deadliest diseases reported. The dengue fever occurrence in rural and urban areas via mosquito vectors (*Aedes aegpyti*) is more common. The modification of these viruses is responsible for the occurrence of more deaths, forming some conditions like Dengue haemorrhagic fever (DHF) and Dengue shock syndrome (DSS). They carried out both in vitro and in vivo investigations; the former was carried out using C6/36 cells, while the latter was carried out with mice (an animal model). Conformation of these tests was done using RT-PCR. The use of neem leaf extract and *azadirachtin* showed inhibition of the replication of dengue virus type 2 while the phytoconstituents *azadirachtin* showed promising results compared with neem leaf extract (Upadhyay et al., 2002).

Khurshid and coworkers reported the antiviral activity of neem leaf extract. Hepatitis C virus (HCV) is a chronic infection, that leads to liver failure and is common in all age groups. HCV replicates in the liver and increases concentrations of liver enzymes; aspartate transaminase (AST) and alanine transaminase (ALT). Phytoconstituent *3-Deacetyl-3-cinnamoyl-azadirachtin* present in the leaves of neem can bind with the NS3 protease of HCV leading to inhibition of protease. HCV seropositivity was higher (78.5IU/ml) in the patients before taking the neem leaves extract and after taking neem leaves extract it was lower (26.5 IU/ml). After consuming neem leaf extract, the liver enzyme concentrations of AST and ALT were 7.5 U/L and 8.5 U/L, respectively, compared to 13.8 U/L and 15.84 U/L. Their study found that neem leaf extract showed effective antiviral activity against HCV and significantly decreased AST and ALT levels (Khurshid et al., 2022).

The use of extracts obtained from various parts of *A. indica* plant and phytoconstituents obtained from its various scientists reported. Neem leaf extracts possess effects on the inhibition of viral DNA polymerase of the hepatitis B, C virus, also the extract obtained from neem kernels is effective in inhibiting the herpes virus. Neem leaf methanolic extracts have anti-Coxsackie B group viral properties. It acts by inhibiting the development of plaque in six distinct Coxsackie B group virus antigenic variants. Neem leaf extract is most effective against Coxsackie B-4 virus by inactivating and reducing their formation speed. The reported MIC was not toxic to African green monkey kidneys (Vero). A trypan blue dye exclusion test was used to confirm the subtoxic concentration of 8000 μg/ml and the cytotoxic concentration of 10000 μg/ml (Tembe-Fokunang et al., 2019; Uzzaman et al., 2019).

Neem leaves have been used in numerous investigations to suppress the poliovirus, a finding later validated by Galhardi et al. (2012). His team isolated polysaccharides from neem leaves and evaluated their activity against poliovirus. The leaves were extracted with water, dialyzed, ethanol precipitated, and freeze-dried to produce the native polysaccharide-containing fractions (P1 and P2) and their sulphated forms (PS1 and PS2). These polysaccharides were tested for antiviral activity using a plaque reduction assay. The compounds showed a significant reduction in polio virus load with P1 and PS1 showing better inhibition when added concomitantly with the viruses. The virucidal assay also confirmed that polysaccharides demonstrated a 27–80% reduction in viral titre with native compounds showing better activity than sulphated forms. The study thus concluded that isolated polysaccharides from neem leaves could be developed as potential molecules for the suppression of poliovirus growth. The same polysaccharides were also evaluated for activity against herpesvirus. Examination of P1, P2, PS1, and PS2 on HSV-1 replication and cell viability showed that P1, PS1, and PS2 inhibited the HSV-1 infection, in cell, with IC50 of 105.2 μg/mL, 31.1 μg/mL, and 80.5 μg/mL. The sulphated forms showed better inhibition of viral load as compared to native polysaccharides. Compounds showed the best activity when added during the inoculation of viruses. Adsorption inhibition studies indicated that P1 and PS2 were responsible for better adsorption of HSV-1 in HEp-2 cells. Overall, the study concluded that polysaccharides inhibited virus adsorption more effectively with no effect on viral penetration. With a better understanding of the mechanism of action and potentiality, *A. indica* polysaccharides can be used for suppressing HSV infection (Galhardi et al., 2019).

Many researchers have focussed on establishing the activity of neem in controlling avian diseases. Ong et al. attempted to prove the antiviral effect of neem bark and leaves aqueous extract against avian viruses of the Velogenic Newcastle disease strain. Prepared extracts with different concentrations (0.1 mg/ml to 15 mg/ml) were used to determine a non-toxic dose later used for the antiviral assay. An *in-vitro* virus inhibition assay revealed that neem bark extract at a dosage of 5.6 mg/ml effectively inhibited the development of the Velogenic Newcastle disease virus. Neem extracts were found to be non-toxic at all concentrations. Neem leaves however did not support this activity and hence the study concluded that chickens can be supplementation with neem bark extract for protection against Newcastle disease (Ong et al., 2014).

COVID-19 was one of the most recent outbreaks that created a worldwide crisis because of mutations in the SARS-CoV-2 genome. To understand the role of *A. indica* and its nimbin isomers in restriction of β-coronaviral infection and replication; Sarkar et al. conducted a study on anti-SARS-CoV-2 activity of methanolic extract of neem bark. The extract showed significant inhibition of CoV-2 (SS)-induced cell toxicity and susceptibility in a concentration-dependent manner against cells infected with the viral SARS CoV-2 strain. The extract also inhibits SARS-induced cytopathy and viral replication. The *in-vivo* studies showed that extract significantly restricted m-CoV-RSA59 replication and proinflammatory chemokine expression. Treatment with extract helped in the reduction of meningitis, encephalitis, myelitis, and hepatitis. Molecular docking confirmed that nimbin was one of the major phytoconstituents contributing to the antiviral activity. The study concluded that methanolic neem bark extract and isolated compounds have promising potential to provide therapeutic benefits in SARS-CoV-2. (Sarkar et al., 2022).

5.5 Future of Research and Drug Delivery

As discussed above, many studies have already established the potential of *A. indica* in the treatment of viral infections. The antiviral potential of trees is attributed to the presence of active phytoconstituents; however; the isolation process of these active compounds is difficult and cost-intensive . Thus, there is an unprecedented need to find newer techniques for the separation of therapeutically active phytoconstituents from plant extracts.

Extracts have a lower quantity of actives; thus, necessitating larger doses of extract to be administered for achieving therapeutic activity. The requirement of larger doses and non-specific distribution of compounds contribute to adverse effects like; allergic reactions, liver and kidney toxicities, hypoglycaemia, and irritation to the gastric mucosa. These effects can be alleviated by using suitable drug delivery platforms.

6. Toxicity and Safety

The use of herbal compounds in medicines is important because of their visible advantages and ease of application but there are some disadvantages associated with them. Several toxicities studies are reported to check their toxicity and safety. Using reported data and safety profiles drugs can be formulated. The use of *A. indica* in a short period like our ancestors used it in their daily routine from using it as mouthwash, drinking its "kadha", chewing its leaf, using its small branches (Twigs) as toothbrush, leaf extract and bark extract as antiseptic and using its flower as an ingredient in various Indian recipes. However, long-term use may cause some side effects and might cause harm to our body mainly the liver and kidneys. Reported toxicities studies are listed in Table 4 below.

Table 4 Toxicity and safety of *A. indica.*

Plant part used	*Extracts*	*Duration of study*	*Type of study*	*Animals model*	*LC50/ LD50*	*References*
Leaves	Aqueous	24 hrs	Acute	Fish (*Cirrhinus mrigala*)	1.035 g/L	(Saravanan et al., 2011)
	Ethanol	14 days		Wistar rats	–	(Kanagasanthosh et al., 2015)
	Methanol	6 days	Toxicity effects	*P. xylostella*	0.61% of extract	(Sharma et al., 2014)
	Hexane				3.95% of extract	
Neem oil	Neem oil	6 days	Acute	*P. nigrispinus*	41.92 g/ml	(Zanuncio et al., 2016)
		96 hrs	Acute	*D. rerio*	0.22 ml/L	(Maranho et al., 2014)
		21 days	Chronic	*D. magna*	0.17 ml/L	
		14 days	Acute and subacute	Kunming mice	31.95/kg	(Deng et al., 2013)
		28 days	Subacute		–	
Seed	Aqueous	48 hrs	Acute	Rats	9.4 ml/kg	(Saleem et al., 2018)
		72 hrs	Contact toxicity	*T. castaneum*	74.27 g per insect	(Islam et al., 2005)
Seed kernel	Seed kernel	35 days	Toxicity implications	Cockerel chicks	–	(Uko et al., 2006)
Stem bark	Ethanol	21 days	Toxicity implications	Wistar rats	–	(Ashafa et al., 2012)
Flowers	Methanol	90 days	Acute and subacute	Rats	>12 g/kg	(Saleem et al., 2018)
Seed	Neem seed oil	–	Acute	Rats and rabbits	14 ml/kg and 24 ml/kg	(Tembe-Fokunang et al., 2019)

Conclusion

Since time immemorial, plant-based drugs have been used in the treatment of various ailments. *A. indica* is one of the most versatile plants with numerous medicinal benefits owing to the presence of diverse phytoconstituents. The entire plant, including roots, stems, leaves, and extracts, are reported in a variety of applications including; anti-inflammatory, antifungal, antioxidant, antiviral, antibacterial, anticancer, antiulcer, and immunomodulatory. Isoprenoids and nonisoprenoids are two major classes of phytoconstituents present in neem. Many studies have shown that phytoconstituents isolated from various extracts and fractions of *A. indica* have shown potent antiviral activity. It inhibits one or more steps involved in viral replication thus suppressing the proliferation of viruses. It is reported to inhibit many viruses including polio virus, dengue virus, hepatitis C virus, and herpes simplex virus. Many preclinical and clinical studies also support the uses of this plant but more *in vivo* and molecular biology mechanistic studies are required to confirm the safety and efficacy of this plant.

The activity of plants sometimes cannot be attributed to a single phytoconstituent owing to synergism between different compounds in the extract. Plant activity can be verified by a variety of *in vitro* investigations, but *in vivo* research is required to comprehend the behavior, accumulation, and safety of plant parts *in vivo*. There also exists a need for the design of suitable carrier vectors for the delivery of isolated compounds to the target tissues.

References

Ahmed, S., Bamofleh, S. and Munshi, M. (1989). Cultivation of neem (*Azadirachta indica,* Meliaceae) in Saudi Arabia. *Economy Botany*, 43: 35–38.

Alzhhairy, M.A. (2016). Therapeutic role of *Azadirachta indica* (Neem) and their active constituents in diseases prevention and treatment. *Evidence Based Complementary and Alternative Medicine*, 1: 1–11.

Ara, I., Siddiqui B.S., Faizi, S. and Siddiqui, S. (1989). Two new terpenoids from root bark of *Azadirachta indica. Journal of Natural Products*, 52(6): 1209–1213.

Ashafa, A.O., Orekoya, L.O., and Yakubu, M.T. (2012). Toxicity profile of ethanolic extract of *Azadirachta indica* stem bark in male Wistar rats. *Asian Pacific Journal of Tropical Biomedicine*, 2(10): 811–817.

Atawodi, S.E. and Atawodi, J.C. (2009). *Azadirachta indica* (neem): A plant of multiple biological and pharmacological activities. *Phytochemistry Reviews*, 8: 601–620.

Biswas, K., Chattopadhyay, I., Banerjee, R.K and Bandyopadhyay, U. (2002). Biological activities and medicinal properties of neem (*Azadirachta indica*). *Current Science*, 82: 1336–1345.

Chhabra, M., Saini, B.S. and Dwivedi, G. (2021). Impact assessment of biofuel from waste neem oil. *Energy Sources A: Recovery utilization, and environmental effects*, 43(24): 3381–3392.

Clinical trials. National library of medicine. Accessed on (September 23, 2023). Available at: https://clinicaltrials.gov/

Deng, Y., Cao, M., Shi, D., Yin, Z., Jia, R., Xu, J., Wang, C., Cheng, L., Liang, X., He, C.L., Yang, Z. and Zhao, J. (2013). Toxicological evaluation of neem (*Azadirachta indica*) oil: Acute and subacute toxicity. *Environmental Toxicology and Pharmacology*, 35(2): 240–246.

Drabu, S., Khatri, S. and Babu, S. (2012). Neem: Healer of all ailments. *Research Journal of Pharmaceutical, Biological and Chemical Sciences*, 3 (1): 120–126.

Galhardi, L., Ray, S., Lopes, N., Ali, I., Espada, S.F., Santos, J.P., Ray, B., Linhares, R.E.C. and Nozawa, C. (2019). Assessment of antiherpetic activity of nonsulfated and sulfated polysaccharides from *Azadirachta indica. International Journal of Biological Macromolecules*, 137: 54–61.

Galhardi, L., Yamamoto, K.A., Ray, S., Ray, B., Linhares, R.E.C. and Nozawa, C. (2012). The *in vitro* antiviral property of *Azadirachta indica* polysaccharides for poliovirus. *Journal of Ethnopharmacology*, 142(1): 86–90.

Girish, K. and Shankara, B.S. (2008). Neem – a green treasure. *Electronic Journal of Biology*, 4(3): 102–111.

Gopalakrishnan, G., Singh, N.D.P. and Kasinath, V. (2002). Photooxygenation of nimonol, a tetranortriterpenoid from *Azadirachta indica*. A. Juss. *Molecules*, 7: 112–118.

Govindachari, T.R., Suresh, G., Gopalakrishnan, G., Banumathy, B. and Masilamani, S. (1998). Identification of antifungal compounds from the seed oil of *Azadirachta indica*. *Phytoparasitica*, 26 (2): 109–116.

Gupta, A., Verma, U.P., Lal, N. and Ojha, S.K. (2017). Evolution and exploration of *Azadirachta indica* in dentistry: an update. *British Journal of Medicine and Medical Research*, 21(8): 1–15.

Gupta, R., Nand, P. and Drabu, S. (2012). Insignificant anti-acne activity of *Azadirachta indica* leaves and bark. *Journal of Negative Results*, 3(1): 29–33.

Hashmat, I., Azad, H. and Ahmed, A. (2012). Neem (*Azadirachta indica* A. Juss)-A nature's drugstore: an overview. *International Research Journal of Biological Sciences*, 1(6): 76–79.

Hwisa, N.T., Assaleh, F.H., Gindi, S., Melad, F., Chandu B.R. and Katakam, P. (2014). A study on antiurolithiatic activity of *Melia azadirachta* L. aqueous extract in rats. *American Journal of Pharmacological Sciences*, 2(1): 27–31.

Islam, M.S. and Talukder, F.A. (2005). Toxic and residual effects of *Azadirachta indica, Tagetes erecta,* and *Cynodon dactylon* seed extracts and leaf powders towards *Tribolium castaneum*. *Journal of Plant Diseases and Protection*, 112(6): 594–601.

Jagannathan, R., Ramesh, R.V. and Kalyanakumar, V. (2015). A review of neem derivatives and their agricultural applications. *International Journal of Pharmacology and Pharmaceutical Technology*, 6: 3010–3016.

Jarvis, A.P., Morgan, E.D. (2000). Analysis of small samples of limonoids of neem (*Azadirachta indica*) using solid phase extraction from tissue culture. *Phytochemical Analysis*, 11(3): 184–189.

Kanagasanthosh, K., Shanmugapriyan, S. and Kavirajan, V. (2015). Evaluation of acute toxicity, anti-inflammatory activity and phytochemical screening of ethanolic extract of *Azardirachta indica* leaves. *International Journal of Research and Development in Pharmacy and Life Sciences*, 4(5): 1737–1742.

Kanwal, S., Ullah, N., Afzal, I. and Mirza, B. (2011). Antioxidant, antitumor activities and phytochemical investigation of *Hedera nepalensis* K. Koch, an important medicinal plant from Pakistan. *Pakistan Journal of Botany*, 43: 85–89.

Khurshid, R., Majeed, S., Saghir, S., Saad, M., Ashraf, H. and Fayyaz, I. (2022). Antiviral activity of extract of neem (*Azadirachta indica*) leaves: an *in vivo* study. *Pakistan Journal of Medical and Health Sciences*, 16 (4): 10–11.

Kraus, W., Cramer, R. and Sawitzki, G. (1981). Tetranortriterpenoids from the seeds of *Azadirachta indica*. *Phytochemistry*, 20: 117–120.

Kumar, C.S.S.R, Srinivas, M. and Yakkundi, S. (1996). Limonoids from the seeds of *Azadirachta indica*. *Phytochemistry*, 43(2): 451–455.

Kumar, V.S. and Navaratnam, V. (2013). Neem (*Azadirachta indica*): Prehistory to contemporary medicinal uses to humankind. *Asian Pacific Journal of Tropical Biomedicine*, 3(7): 505–514.

Luo, X., Ma, Y., Wu, S. and Wu, D. (1999). Two novel azadirachtin derivatives from *Azadirachta indica*. *Journal of Natural Products*, 62(7): 1022–1024.

Luo, X.D., Wu, S.H., Ma, Y.B. and Wu, D.G. (2000). A new triterpenoid from *Azadirachta indica*. *Fitoterapia*, 71: 668–672.

Maithani, A., Parcha, V., Dhulia, I. and Kumar, D. (2011). *Azadirachta indica* (Neem) leaf: a review. *Journal of Research in Pharmacy*, 4(6): 1824–1827.

Maranho, L.A., Botelho, R.G., Inafuku, M.L., Nogueira, R., Alves de Olinda, B.A. Inácio de Sousa, and Tornisielo V. L. (2014). Testing the neem biopesticide (*Azadirachta indica* A. Juss) for acute toxicity with Danio rerio and for chronic toxicity with Daphnia magna. *Journal of Agricultural Science*, 16(1): 105–111.

Mehta, N., Chandra, A. and Thakur, A. (2021). Neem and its derivatives-a tropical trove. *International Journal of Current Microbiology and Applied Sciences*, 10(02): 3228–3235.

More, Y.W., Raut, R.R., and Taware, A.S. (2015). Review on biological activities of *Azadirachta indica* (Neem) and its medicinal uses. *International Journal of Informative Futuristic Research*, 2: 1327–1334.

Nair, M.S., Gopal, S. and Issac, D. (1997). Optimised isolation procedure for biologically active compounds nimbolide and 28-deoxonimbolide from *Azadirachta indica* leaves. *Phytochemistry*, 46 (3): 1177–1178.

Nakahara, K.M., Roy, M.K., R., Ono, H., Maeda, I., Ohnishi-kameyama, M., Yoshida, M. and Trakoonntivakorn, G. (2003). Prenylated flavanones isolated from flowers of *Azadirachta indica* (the neem tree) as antimutagenic constituents against heterocyclic amines. *Journal of Agricultural and Food Chemistry*, 51(22): 6456–6460.

Neem Tree Risk Assessment. Queensland department of agriculture and fisheries. Available at: https://www.daf.qld.gov.au/__data/assets/pdf_file/0006/63168/IPA-Neem-Tree-Risk-Assessment.pdf

Ogbuewu, I.P., Odoemenam, V.U., Obikaonu, H.O., Opara, M.N., Emenalom, O.O., Uchegbu, M.C., Okoli, I.C., Esonu B.O. and Iloeje, M.U. (2011). The growing importance of neem *Azadirachta indica* A. Juss) in agriculture, industry, medicine and environment a review. *Journal of Medicinal Plants Research*, 5(3): 230–245.

Ong, G.H., Syamsiah, A.S., Hasrul, A.H., Zunaida, B., Maizatul, Z., M. Jihan, R.M., Redzean, S., Leow, B.L., Faizul, F.M.Y., Chandrawathani, P. and Ramlan, M. (2014). Antiviral effect of aqueous neem extract from branches of neem tree on Newcastle disease virus. *Malaysian Journal of Veterinary Research*, 5: 5–9.

Pankaj, S., Lokeshwar, T., Mukesh, B. and Vishnu, B. (2011). Review on neem (*Azadirachta indica*): thousand problems one solution. *International Research Journal of Pharmacy*, 2(12): 97–102.

Prashanth, G.K. and Krishnaiah, G.M. (2014). Chemical composition of the leaves of *Azadirachta indica* Linn (Neem). *International Journal of Advancement in Engineering Technology, Management and Applied Science*, 1(5): 21–31.

Raj, A. (2015). Beneficial and harmful effects of *Azadirachta indica*: a review environmental science. *International Journal of Research*, 4(3): 6–7.

Raphael, E. (2012). Phytochemical constituents of some leaves extract of aloe vera and *Azadirachta indica* plant species. *Global Advanced Research Journals*, 1(2): 14–17.

Reddy, I.V.S. and Neelima, P. (2022). Neem (*Azadirachta indica*): A review on medicinal kalpavriksha. *International Journal of Economic Plants*, 9(1): 059–063.

Saleem, S., Muhammad, G., Hussain, M.A. and Bukhari, S.N.A. (2018). A comprehensive review of phytochemical profile, bioactives for pharmaceuticals, and pharmacological attributes of *Azadirachta indica*. *Phytotherapy Research*, 32(7): 1241–1272.

Saravanan, M., Ramesh, M., Malarvizhi, A. and Petkam, R. (2011). Toxicity of neem leaf extracts (*Azadirachta indica* A. Juss) on some haematological, ionoregulatory, biochemical and enzymological parameters of Indian major Carp, *Cirrhinus mrigala*. *Journal of Tropical Forest Science*, 1(1): 14–26.

Sarkar, L., Oko L., Gupta, S., Bubak, A., Das, B., Gupta, P., Safiriyu, A., Singhal, C., Neogi, U., Bloo, D., Banerjee, A., Mahalingam, R., Cohrs, R., Koval, M., Shindler, K., Pal, D., Nagel, M. and Sarma, J. (2022, Feb). *Azadirachta indica* A. Juss bark extract and its nimbin isomers restrict β-coronaviral infection and replication. *Virology*, 569: 13–28.

Sharma, S., Senrung, A. and Singh, A.K. (2014). Toxic effect of neem, *Azadirachta indica* (A. Juss) foliage extracts against diamondback moth (DBM), *Plutella xylostella* (L.) (Lepidoptera, Plutellidae). *Journal of Biopesticides*, 7: 99–105.

Siddiqui, B. S., Ali, S.T., Rajput, M.T., Gulzar, T., Rasheed, M. and Mehmood, R. (2009). GC-based analysis of insecticidal constituents of the flowers of *Azadirachta indica* A. Juss. *Natural Product Research*, 23(3): 271–283.

Siddiqui, B.S., Ghiasuddin, G., Faizi, S. and Siddiqui, S. (1991). Terpenoids from fruits coatings of *Azadirachta indica*. *Phytochemistry*, 30: 1615–1619.

Siddiqui, B.S., Afshan, F., Faizi, S., Naqvi, S., Tariq, H.E. and Karachi, H. (2000). Two insecticidal tetranortriterpenoids from *Azadirachta indica*. *Phytochemistry*, 53: 371–376.

Siddiqui, B.S., Afshan, F., Gulzar, T. and Hanif, M. (2004). Tetracyclic triterpenoids from the leaves of *Azadirachta indica*. *Phytochemistry*, 65(16): 2363–2367.

Siddiqui, B.S., Ali, T.S. and Ali, S.K. (2006). A new flavonoid from the flowers of *Azadirachta indica*. *Natural Product Research*, 20(3): 241–245.

Siddiqui, B.S., Tariq Ali, S., Rasheed, M. and Nadeem Kardar, H.M. (2003). Chemical constituents of the flowers of *Azadirachta indica*. *Helvetica Chimica Acta*, 86: 2791–2796.

Siddiqui, S., Siddiqui, B.S., Faizi, S. and Mahmood, T. (1988). Tetracyclic triterpenoids and their derivatives from *Azadirachta indica*. *Journal of Natural Products*, 5(1): 30–43.

Singh, B., Pandya, D. and Mankad, A. (2020). A review on different pharmacological and biological activities of *Azadirachta indica* A. Jusm. and *Melia azedarach* L. *Journal of Plant Science and Research*, 36(1-2): 53–59.

Sujarwo W., Keim A.P., Caneva G., Toniolo C. and Nicoletti M. (2016). Ethnobotanical uses of neem (Azadirachta indica A. Juss.; Meliaceae) leaves in Bali (Indonesia) and the Indian subcontinent in relation with historical background and phytochemical properties. *Journal of Ethnopharmacology*, 189:186–193.

Sultana, S., Khan, M.A., Ahmad, M., Bano, A., Zafar, M. and Shinwari, Z.K. (2011). Authentication of herbal medicine neem (*Azadirachta indica* A. Juss.) by using taxonomic and pharmacognostic techniques. *Pakistan Journal of Botany*, 43: 141–150.

Suresh, G., Narasimhan, N.S. and Palani, N. (1997). Structure of nimonol from fresh whole green leaves of *Azadirachta indica*. *Phytochemistry*, 45(4): 807–810.

Suresh, G., Narasimhan, N.S., Masilamani, S. and Gopalakrishnan, G. (1997). Antifungal fractions and compounds from uncrushed green leaves of *Azadirachta indica*. *Phytoparasitica*, 5(1): 33–39.

Tembe-Fokunang, E.A., Charles, F., Kaba, N., Donatien, G., Michael, A. and Bonaventure, N. (2019). The potential pharmacological and medicinal properties of neem (*Azadirachta indica* A. Juss) in the drug development of phytomedicine. *Journal of Alternative and Complementary Medicine*, 7(1): 1–18.

Tesfay, T., Chaithanya, Karri, K., Rao, K.K., Gebremedhin, G., Krishna C.K. and Kamalakararao, K. (2020). Phytochemical screening and *in vitro* anti-dandruff activities of bark extracts of neem (*Azadirachta indica*). *Drug Invention*, 13: 707–713.

Tiwari, V., Darmani, N.A., Yue, B. and Shukla, D. (2010). *In vitro* antiviral activity of neem (*Azardirachta indica* L.) bark extract against herpes simplex virus type-1 infection. *Phytotherapy Research*, 24(8): 1132–1140.

Uko, O.J. and Kamalu, T.N. (2006). Protein quality and toxicity of full-fat neem (*Azadirachta indica* A. Juss) seed kernel. *Archivos de Zootecnia*, 55(209): 51–62.

Upadhyay, P., Pandya, G. and Jana, A.M. (2002). Inhibitory potential of neem (*Azadirachta indica* Juss) leaves on dengue virus type-2 replication. *Journal of Ethnopharmacology*,79: 273–278.

Uzzaman, S. (2019). Pharmacological activities of neem (*Azadirachta indica*): A review. *International Journal of Pharmacognosy and Life Science*, 1(1): 38–41.

Van der Nat, J.M., Van der Sluis, W.G., De Silva, K.T.D. and Labadie, R.P. (1991). Ethnopharmacognostical survey of *Azadirachta indica* A. Juss (Meliaceae). *Journal of Ethnopharmacology*, 35(1): 1–24.

Yamasaki, R.B., Ritland, T.G., Barnby, M.A. and Klocke, J.A. (1988). Isolation and purification of salannin from neem seeds and its quantification in neem and chinaberry seeds and leaves. *Journal of Chromatography A*, 447: 277–283.

Yerima, M.B., Jodi, S.M. and Oyinbo, K. (2012). Effect of neem extracts (*Azadirachta indica*) on bacteria isolated from adult mouth. *Nigerian Journal of Basic and Applied Sciences*, 20(1): 64–67.

Zanuncio, J.C., Mourão, S.A., Martínez, L.C., Wilcken, C.F., Ramalho, F.S., Plata-Rueda, A., Soares, M.A. and Serrão, J.E. (2016). Toxic effects of the neem oil (*Azadirachta indica*) formulation on the stink bug predator, *Podisus nigrispinus* (Heteroptera: Pentatomidae). *Scientific Reports*, 6: 30261.

22 Deciphering Promising Antiviral Effects of Medicinal Marvel *Xanthoceras sorbifolium* Bunge

Swaati Sharma[1], Ishita Pandey[1] and Pooja Bhadrecha[1*]

1. Introduction

Xanthoceras sorbifolium Bunge is a medicinal shrub that is commonly used to treat a variety of diseases such as Alzheimer's disease, arteriosclerosis, hyperlipidemia, hypertension, chronic hepatitis etc. It is also involved in treating cancer and has anti-inflammatory effects. Xanthoceras sorbifolium is an ancient plant from China with fantastic uses as a biofuel source and a useful method for making unusual medications. The usage of Yellowhorn and its anatomical and morphological makeup in the treatment of various ailments has been documented by chronology and numerous studies by different academics. This has led to the analysis of the central role of Xanthoceras in ancient Chinese and Mongolian medicine. It comes under the family name *Sapindaceae* and genus name *Xanthoceras* which is known to be a monotypic species (a type of species in which no further subspecies is present) majorly disseminated in every part of China.

It is commonly known as the Yellow horn or golden horn, it is local to the northern part of China in the districts of Gansu, Hebei, Henan, Shaanxi, etc. This plant additionally is grown in some parts of Russia and has been shipped in that region past the nineteenth Centennial. Yellowhorn has shiny greenish compound leaves which look like the leaves of mountain ash (*Sorbus*), consequently signifying a specific designation for sorbifolium. The compound leaves promote moderately glossy and tremendously sharp leaflets. This flora has a smooth appearance in terms of topography and the fall hue ranges from a fragile gold to a glass like yellow. Although it is rather sensitive to uncurbed mugginess or soils that stay damp, this plant does not resist mild clamminess. Yellowhorn has become used to soil and is not overly picky about it, although a loamy soil would be ideal, it is not essential for the success of this hardy tree; soils that are slightly acidic to alkaline are also beneficial. A pH limit amidst 5.5 to 8.5 is taken into consideration in terms of soil. Yellowhorn would not thrive or blossom well in areas of heavy shadow. The best location for growth is in the sun. Although coral spot fungus has been observed in some recent studies, this plant does not have any significant arthropod or disease

[1] University Institute of Biotechnology, Chandigarh University, Punjab, India.
* Corresponding author: pbhadrecha.pb@gmail.com

DOI: 10.1201/9781003433583-22

problems. (Harrington et al., 2005). In this chapter, we will be focusing on various aspects of *X. sorbifolium* in detail.

Xanthoceras sorbifolium comes under the kingdom *Plantae*, subdivided into subkingdoms and infra-kingdoms respectively as *Viridiplantae* and *Streptophyta*. As we move downwards further taxa such as super division include *Embryophyta*, division include *Tracheophyta*, subdivision include *Spermatophyte*, class includes *Magnoliopsida*, superorder as *Rosanae*, order as *Sapindales*, family as *Sapindaceae* genus, as*XanthocerasBunge* and species as *Sorbifolium*. Talking about the parts of the plant we have leaves, fruits, peduncles etc. In this chapter we will focus on the foliage, fruits, and peduncle characteristics of Yellowhorn.

1. Popular Name: Yellowhorn
2. Genre: Deciduous Shrub
3. Family: *Sapindaceae*
4. Native Range: Northern China
5. Range of Zone: 4 To 7 Feet
6. Height of The Plant: 8.00 to 25.00 Feet
7. Distribution: 10.00 to 25.00 Feet
8. Flowering Span: May
9. Blooming Depiction: White with Red or Yellow Center
10. Sun: Adequate Sunlight
11. Water Content: Medium
12. Maintenance Required: Low
13. Use: Flowering Tree
14. Flower: Showy, Good Cut
15. Berry: Showy
16. Tolerance: Drought

1.1 Leaves

The leaves of the Yellowhorn are characteristically deciduous and glossy green in color and appearance. These leaves are long-lasting and appear gold/yellow often. The leaves are compound (pinnately, bipinnately, palmately, alternate). The compound leaves are of two types pinnate and bipinnate; pinnate leaves are attached along a rachis which is an extension of petiole. They consist of an odd number of leaves and in bipinnate leaflets themselves are pinnately veined. The length of these leaves' ranges from 1–3 inches.

1.2 Fruits

Fruits of *Xanthoceras* tend to appear quite greenish, these arequite showy to the gardeners. The type of fruit that exists is a capsule which ranges 1–3 inches in length and width. The green fruits look like buckeyes. They can be either round or pear-shaped and have capsules that contain large pea-sized seeds. When roasted, they are said to have a taste like macadamia nuts. These seeds are purplish brown. The peduncle is also known as the stalk of the plant. The peduncle of *Xanthoceras* appears to be brown, copper red or red burgundy in appearance and texture. These peduncle are nonaromatic and non-suckering.

2. Taxonomy and Scientific Classification

Therapeutic frondescence is used extensively in developing countries since it is more accessible and less expensive than contemporary drugs. One such medicinal plant is Xanthoceras sorbifolium, also called Chinese flowering chestnut, Yellowhorn, shiny leaf Yellowhorn, or golden horn. It is a type of angiosperm shrub that is frequently found in the Sapindaceae family, which is essentially known as the family soapberry. Xanthoceras is the only variety of the genus. The most prominent elementary portion of the Sapindaceae family is the genus Xanthoceras, which is usually referred to as "Yellowhorn."

The specific epithet (specific epithet resembles the species within that genus). *Sorbifolium* resembles the foliage, which look like those of the obscure interconnected rowans commonly called (Sorbs). Yellowhorn was firstly recognized in the Chinese Material Medical by "*Jib Huang Ben Cao*" in 1406 Alzheimer disease under the title "Went Guan Husain China". Additionally, this plant has been used to treat rheumatoid arthritis, acute hepatitis, hypercholesterolemia, hypertension, and atherosclerotic cardiovascular disease. In addition to 1868, this member of the Sapindaceae family was gradually polished in Europe (Paris) after being originally compiled and classified in 1833. This shrub was described in 1887 by Joseph Hooker, the Director of the Royal Botanic Gardens at Kew (1865–1885), as "one of the extremely charismatic and fascinating woody garden shrubs that has been oriented in many years." The Royal Horticultural Society Award of Garden Merit, which should hint at its capacitance worth and topographic ramifications, has also been acknowledged by Xanthoceras (Harrington et al., 2005).

2.1 Synonyms

The Yellowhorn (*Xanthoceras sorbifolium*) is known by different names commonly referred to as Golden horn, white rain tree and Chinese flowering chestnut. Yellowhorn also acknowledged the Royal Horticultural Society Award of Garden Merit that clues at its plausible use along with slam in the topography. Another familiar title for the plant is "popcorn shrub". The span of flowering lasts around two weeks. Additionally, Yellowhorn was first introduced as a small urban tree in Beijing, China, and is now often seen in many other localities. There is usually a huge orchard of *Xanthoceras sorbifolium* in China as the seeds are manifesting eminent assurance as a largely applicable biofuel (Harrington et al., 2005).

2.2 Botanical Description of Plant

Yellowhorn is an ephemeral shrub which is huge or short tree of approximately 8 meters in length. The foliage is typically 12 to 30 centimeters long and is arranged in an alternate way. Leaves are basically compound pinnate (lamina is divided into multiple leaflet), with 9 to 17 leaflets, the leaflets are 3 to 6 centimeters lengthy, with a piercingly jagged margin. The flowers are 2 to 3 centimeter in broadness, possessing pentamerous white petals, and are generate in upright raceme (much branched inflorescence) which are 10 to 20 centimeter in length in middle of the spring season. The berry is an egg-shaped rugged capsule which is 5 to 6 centimeter in broadness and broken into tri sections at blossoming to generate 6 to 18 seeds; these seeds are usually blackish in texture measuring 1.5 centimeter in broadness, like a small horse chestnut seed. As the aging proceeds, the type becomes more erect along with an inflexible, boorish branching habit whereas the beautiful flowers might duck down the branches to a specific degree when blooming is at its peak. Embodiments are generally numerous trunks whereby few are monotonous trunks and branched reasonably insignificant to the ground. *Xanthoceras* may gradually and steadily cover up an area along with few roots suckering which isn't a property of each sample. The corpulent, hairy roots build up this shrub demanding to transplant which is complementary. The usually noticed high transplant death rate appears to relate to a conscious root structure. Soft and cautious handling throughout all stages of planting is recommended. The various parts of the plant include – plant, berries, seeds, flowers, wood, leaves, husks. The plant has unusually fragrant flowers which bloom in the month of May and appear on terminal racemes. These are whitish in color with a very dim greenish smear and a midpoint which vary from being yellowish on the fresh flowers to a beautiful red orange on inflorescence that are older. This abundant flower bunch can be approximately 10 inches in length, also independent flowers are slight 1 inch in broadness and pentamerous. *Xanthocerassorbifolium* grows up to 2 to 5 meters in height and has sturdy baby branches which are brown reddish in color along with captivating with plate-like floweret scales arranged on upper and edge buds. The foliole branches are approximately about 15 to 30 centimeters stretched out. There are 4–8 doublets of baby leaves, lamina ceousor tacky, lanceolate. The flanking veins are delicate and a little hitched up on twain sides. The flowers are usually synoicous, and the bloom is at its extremity. The alary of the male bloom is 12 to 20 centimeters in size and upright with a little wholesome stalk and a remains of bud scale at the bottom. The stalk is usually about 1.2 to 2 centimeters in size. The folioles are 0.5 to 1 centimeter in length. The leaves are 6 to 7 mm in length with hairs of grey texture on the dual sides. The corolla is usually white in texture, but the bottom part is purple-reddish or yellowish in texture. The berry possesses a capsule which is usually 5 to 6 centimeter in broadness along a trio of the seed chamber which are 1 to 1.5 centimeter in broadness. The count of seeds per chamber can vary from one to six (Zang et al., 2021).

2.3 Geographical Distribution

Temperate zones usually are warm temperature areas located in the middle of the equatorial and the arctic regions. The two regions in the middle of the southern Tropic and the polar Circle or the northern Tropic and the polar circle are called the moderate or temperate zones. Yellowhorn develops in moderate and mild temperate region, where the height usually varies from 300 directed towards 2000 meters and the procumbent ranch is 28°34′–47°20′ East and 73°20′–120°25′ north. Yellowhorn overall are scattered in eighteen districts in China, including Inner Mongolia, Shanxi, Hebei, and Henan.

In accordance with a recent analysis, Chifeng a small province in Inner Mongolia has the largest diverse and dense dwellers and at present have the largest swamps and mangrove mangals in China. The species is having a good life rate (approximately up to 1,000 years), and it could develop in soil of pH range from 7 directed towards 8.5. It could abide scarcity, little temperature, and soils that consist of clay, sand, or loam, counting those which are basic in nature, also of less fertile rates. It develops well in mountains of desert, desolate valleys, sandy lands, and steep hillsides. The cultural conditions include gleam range which ranges to full sun. Soil texture appears to be loamy. The pH range of the soil ranges from acidic to neutral (Zang et al., 2021).

3. Phytochemical Constituents and Metabolite Profiling

The chemistry of the plant and its processes along with products is known as phytochemistry. Among the two hundred seventy-eight compounds which have been segregated and recognized from Yellowhorn, triterpenes and flavonoids are considered as an important and main bioactive substance because of their diversity, therapeutically activities, and content.Triterpenoids illustrate a huge component of the chemical constituents in the Yellowhorn with of approximately about one hundred twenty-four triterpenoid compounds which have been recognized from the arils, foliage, seeds along with stem extracted the seed oil dregs of Yellowhorn. Flavonoids are basically a cluster of essentially developing compounds that consist of a benzopyran heterocycle connected to a hexene ring. At present, forty-eight flavonoids are isolated from various parts of the plant such as trunks, branches, foliage, aril, and flowers of Yellowhorn. Amidst above mentioned compounds, quercetin along with myricetin are the major aglycons which are organic compounds in combination with the sugar part of a glycoside. Zhang and Bao acclimatized nylon and gel column chromatography to obtain the chemical components of lignum *Xanthoceras*.

Cinnamic acids are essential; compounds with hexene rings. These usually possess a carbolic acid framework and are a class of phenolic substances. Fourteen uncomplicated cinnamic acids have been isolated deriving out of Yellowhorn, with their major components being coumarins as well as lignans phenylpropanoids are essential compounds with hexene rings. Zhu et al. 2019 obtained and refined the chemical make-up of seed oil remains of Yellowhorn by WD40TM, megaporous, sephadex lh-20, and octadecyl silica, paper chromatography. Nine alkaloids came into possession from the methanol extract of the seed and aril. Dong et al. acclimatized silica gel column, pharmadex lh-20 paper chromatography and preparative thin layer chromatography (TLC) to obtain compounds via berries of Yellowhorn furthermore recognized its framework viaspectral mechanism. Conclusion drawn was that four quinone compounds were obtained; physicone, 5-dimethoxy-p-benzoquinone, emodin, 8-dihydroxy-3-methylanthraquinone (Amorati and Valgimigli, 2015).

3.1 Phenolic Compounds obtained from Xanthoceras sorbifolium

Yellowhorn is used as a conventional drug in north China to assess the worth of the anti-cancerous furthermore radical-scrounge activities of Yellowhorn; thirty-seven polyphenols were extracted from the aril of the plant. First few compounds were extracted by bioassay-mediated fractionation. some of the compounds didn't show high degree of toxicity to living cells in opposition to the quadruplet checked cancerous cell lines (a549, hepg2, mgc-803) in contrast to paclitaxel and beneath the circumstances checked in the anti- assay, but some of them showed greater radical-scrounging activity than vitamin c in a peroxidase substrate namely synthetic method for anti-oxidative potential like 2,2-Azino bis-3-ethylbenzothiazoline-6-sulphonic acid assay (ABTS).

3.2 Sterols Obtained from Xanthoceras sorbifolium

Steroids are found in nearly all plants and display a wide range of significant biological activity. At present, there Text been 17 steroids which are found in the wood extract, husks, stem, stalk, seed oil remains and kernel oii of Xanthoceras sorbifolia. Various scientists used Thin Layer Chromatography technique to separate kernel oil obtained from the plant with petroleum ether as the production assistant. The analeptic compounds or sterols such as β-sitosterol campesterol, acetate, cholesterol and campesterolacetate were recognized by Thin Layer Chromatography, Mass Spectroscopy, and Fourier transform infrared and Gas Chromatography-Mass spectroscopy (Chen et al., 1984).

Fourteen unornamented phenylpropanoids were isolated from the various parts of the plant, with their important constituents being coumarins and lignans. Moving ahead with the other class of biologically active compounds that are steroids which are found in the majority of plants and express important biological activity. Plant sterol is one such steroid from the C_{17} side chain with eight to ten carbon atoms in the side chain. Two phenolic compounds, hydroquinone along with 4-hydroxybenzylcyanide were recognized based on physicochemical characteristics. Approximately about nine alkaloids were obtained from the methanol composition of the seeds of yellowhorn and arils. Examples of alkaloids included, allantoin, indole-3-acetylaspartic acid and indole-3-carboxaldehyde (Zang et al., 2021; Yang et al., 2016).

3.3 Quinones Obtained from Xanthoceras sorbifolium

Four quinones were moreover found in the berries and wood of yellowhorn scientists acclimatized silica gel column, preparative Thin Layer Chromatography, and pharmadex LH-20 paper chromatography to obtain certain compounds from the berries of Yellowhorn and recognized their structural framework by a spectral mechanism. Conclusion is that quadruplet of quinone compounds were obtained: chrysophanolphysicone, emodin and 2,5-dimethoxy-p-benzoquinone (Chen et al., 1984; Zang et al., 2021; Yang et al., 2016).

4. Medicinal and Therapeutic uses of *Xanthoceras Sorbifolium*

4.1 Pharma Copoiesis and Traditional System of Medicine

Medicinal plants are certain herbs and shrubs that have certain medicinal properties and contain chemical extracts which are used to treat various disorders. Therapeutic plants basically supply three important types of assistance: health assistance to those who ingest them as drugs; financial assistance to those who gather crops, process, and disperse them for sale; and society-wide benefits. *Xanthoceras sorbifolium* is one such medicinal plant that provides us various medicinal benefits (Zang et al., 2021).

4.2 Traditional Uses Described in Folklore Medicine

The seeds of yellowhorn are prosperous enough in unsaturated sterols and are generally used in preparing vegetable oil. Morsel could be inserted inside seasonal milk products or prepared protein drinks. The foliage is also acclimated to tea. The plant could also be acclimated in cosmetics and to make biodiesel. The arils of the plant (by-products) are acclimatized to develop chemical substantial, including enforced carbon, birch sugar and furfuraldehyde.The composition processed from its arils has anti-infectious and anti-tumor characteristics; it furthermore helps in prohibiting the action acquired immunodeficiency syndrome (HIV) and boosts learning and consciousness, of the patients (Aderogba et al., 2013; Zang et al., 2021).

4.3 Pharmacological Activities-Antiviral

In recent research it has been proven that almost 278 constituents from diverse sections of Yellowhorn, including triterpenoids, flavonoids, fatty acids, and other secondary metabolites have been isolated. These plentiful bioactive elements have a vast range of therapeutic activities, including better learning potential and consciousness, anti-infectious, anti-tumor, antioxidant, anti-HIV, and vascular relaxation effects, as well as discourages activity of steapsin. The plant is also used to cure hyperlipidemia, high blood pressure, chronic hepatitis, and rheumatism (Chan, 2007; Chan et al., 2008; Bai et al., 2010; Bao et al., 2012).

The most interest-bearing fact about the plant is that each division of it, has an impending therapeutical and fitness values and is accustomed to cure certain disorders. The flower and calyx part of the plant possess a chemical called baicalin, which has febrifuge activity, somniferous, anti-convulsive, and anti-cancerous effects. Their seeds are accustomed to precluding and treat arterial sclerosis. In Inner Mongolia region, various parts of the tree such as trunks along with branches were acclimatized to treat bone related disorder, for example-arthritis. Xanthoceraside is a triterpene saponin isolated out of the arils of the tree. It has a great ability to improve learning and memory. Xanthocerasidecan also be used a big asset in the prevention and treatment ofAlzheimer disease which is a neurodegenerative disease that display persistent progress in cognition disabilities and consciousness disorder. A sequence of oleanane type of glycosides were taken from *Xanthoceras sorbifolium,* based on which an atom-based model was constructed (Chen et al., 2020 a; Chen et al., 2020 b).

5. Applications supported by Clinical Data

1. *Xanthoceras sorbifolium* is used in various therapeutic activities which include antioxidant, anti-neuro infectious, anti-cancerous, anti-adipogenesis, anti-obesity, gastroprotective, immunoregulatory, and anti-inflammatory activities.
2. Yellowhorn is an ancient medicinal plant which is used as a classical herbal medicine and originated in China to deal with autoclaving, destroying male gametes of homo sapiens, stabilizing capillary, to maintain constant body temperature commonly known as homeostasis, reducing cholesterol, rheumatism, and pediatric enuresis (Yang et al., 2016).
3. The reason behind AD, commonly referred to as dementia at present, is still very unclear, and despite ongoing research, there are still no medicines that work effectively. Therefore, the research for a new medicine to cure this disease is a pressing priority. Experiments were performed on a no targeted metabolomics survey of the urine isolated from various clusters of mice using gas chromatography-mass spectrometry (GS-MS). Interestingly there were fourteen potential biomarkers which were recognized in the mice's urine, highlighting five metabolic areas of keen interest. This research opens new pathways about the potential pathogen production of this disease and the achievable medicinal procedure of Yellowhorn, giving us a clinical application of *Xanthocerassorbifolia* (Li et al., 2012; Yu et al., 2012; Wan et al., 2015).
4. Seeds of this plant contain seed kernel and a hard-outer tegument which contains oil. The oil content inside the shell is approximately 30–50%, and within the seed kernel it was about 55–65% which is higher than the previous seeds. It is a great source of double or triple bonded fatty acids (85–93%) and also is enriched in linoleic and oleic acids. It could be acclimatized in consumable oils, along with oil yielding benefits. The plant has a great dietetic value and diverse therapeutic applications and the oil obtained from Yellowhorn is an edible vegetable oil (Zang et al., 2021).

6. Toxicity and Safety

Xanthoceras sorbifolium Bungeis known for its therapeutic acclimatization as an ancient herbal drug in China. A dditionally, it is attributed as an oil-bearing plant for the manufacturing of vegetable oil due to the fitness-promoting effects. In current times, due to its high-salary potential and minimal investment requirements, Yellowhorn has become more and more popular worldwide. Speaking about safety, it plays a fantastic part in improving learning and conscience problems, which are primarily identified by the worsening symptoms of AD. Cell death, glial cell transformation to start infectious cascades, oxidative stress, increased intracellular calcium ion generation, and a decrease in cell membrane fluidity are some of the mechanisms mentioned. Previous research depicted that barrigenol–a type of triterpenoids which exhibits commendable protective results against spatial memory disabilities. Although, they have the capacity to be acclimated in the treatment of Alzheimer disease and various neurodegenerative diseases (Chirva and Kintya, 1971; Zang et al., 2021).

6.1 Inflammatory Activity

Current research shows that extracts and compounds obtained out of Yellowhorn chiefly affect neural inflammation, vascular infection along with rheumatism. The flavonoids and phenylpropanoids obtained from the foliage of Yellowhorn are involved in reduced nitric oxide (NO) generation in the lipopolysaccharide induced BV2 microglial cells. Currently, reviews only acclimatized the crude extract to authenticate the anti-infectious activity of Yellowhorn. Therefore, more comprehensive therapeutic studies should be taken into consideration to refine the procedure under the anti-infectious impact of Yellowhorn and its active compounds to supply decent documents to acknowledge the growth and uptake of Yellowhorn (Chen et al., 2022).

6.2 Anti-Tumor Activity

Commonly, the anti-cancerous tasks of essential products are many times analyzed by their capacity to put a break on the multiplication of oncogenic cells and they prevail upon immune cells to release cytokines that proceed on cancerous cells. The working components of natural flora utilities as anti-tumor associates have become famous to an increasing extent, with much of them targeting barrigenol triterpenes (Han et al., 2022).

6.3 Therapeutics

Yellowhorn has several diverse therapeutic activities, including anti-AIDS, and has pivotal and protective functions in cardiovascular and cerebrovascular disorders. Moreover, it generally blocks the effects of pancreatic lipase and tyrosinase. Scientists also figured out that Xanthoceraside could notably advance cerebral artery ischemia-reperfusion injury in rats. Its procedure may encourage synaptic remodeling and decrease synaptic structural and functional harm. Interestingly, flavonoids and saponins obtained from the arils of Yellowhorn have been known to show blocking repercussions on tyrosinase. The important parts or active constituents with anti-Alzheimer's effects can be obtained out of Yellowhorn which could be acclimatized to get ready the working foods or medicine to advance learning and consciousness and have the capacity to embellish the topmost anti-Alzheimer's medicines through more exploration and development (Ji et al., 2014).

Xanthoceras sorbifolium Bungea special woody plant commonly known as Yellowhorn which was also known to be a bioenergy plant in the 21st century in China. The seeds of this plant act as a raw material to produce biodiesel. The therapeutic benefits have been in records since ancient times which include treatment of pediatric enuresis. Xanthoceraside is a triterpene extracted from the husks of the plant and research showed that this triterpene has very efficient effects on learning and memory disorder in the several animal models of Alzheimer's, for example – A β25-35 induced Alzheimer disease mice model. Bladder cancer has a malignancy in the urinary system, each year, more than 400,000 new bladder cancer cases are reported around the world. The statistical data of National Cancer Institute of China depicted that new cases of bladder cancer were 80,500, and deaths caused due to bladder cancer were 32,900 in 2015 in China. Bladder cancer is derived from the transitional epithelium commonly known as urothelium. The recent observation is made that xanthoceraside could induce apoptosis of cancerous bladder cells (Chai et al., 2021).

Conclusion

In summary, Xanthoceras sorbifolium, the scientific name for Yellowhorn, is a local plant in China that has medicinal and sustainable uses and is rich in resources. A total of 278 chemicals have been found to address the photochemical features of Yellowhorn, including terpenoids, phenylpropanoids, flavonoids, phenols, steroids, fatty acids, quinones, alkaloids, and other substances. The improvement of learning and memory impairments, as well as anti-infectious and antioxidant properties, have been the main targets of the therapeutic effects. There are still several research obstacles that need to be overcome despite the fact that there are so many established uses and benefits. This plant is an excellent resource for research and development because of its large secondary metabolite content and numerous medical benefits, including anti-inflammatory and anti-cancer effects. In addition, several compounds, including baicalin, are extracted from the plant and contribute to its antipyretic qualities. In terms of look and flowering, they thrive in loamy soil and yield blooming blooms. However, the plant has significant drawbacks, such as oil production, which produces a lot of trash, including teguments and arils, among other things. As the adage goes, "a double-edged sword cuts both ways." According to recent studies, specific compounds found in the residues of different plant sections might be further examined to make better use of the waste products that are extracted from the plant.

References

Aderogba, M.A., Ndhlala, A.R., Rengasamy, K.R. and Van Staden, J. 2013. Antimicrobial and selected *In Vitro* enzyme inhibitory effects of leaf extracts, flavonols and indole alkaloids isolated from croton Menyharthii. *Molecules*. 18 (10), 12633–12644.

Amorati, R. and Valgimigli, L. 2015. Advantages and limitations of common testing methods for antioxidants. Free Radic Res. 49(5):633–49. doi: 10.3109/10715762.2014.996146.

Bai, D., Liu, R.X., Hu, C.Y., Li, Y.H. and Yin, S.N. 2010. Research on seedlings transplanted technique of *Xanthoceras sorbifolia bunge* in Shandong mining area. *J. Inner Mongolia Agr. Univ.* 31, 72–76.

Bao, X., Go, Y., Yu, R., Li, W.M., Yang, N. and Zhu, H.N. 2012. Determination of fatty acids in the seed kernel oil of *Xanthoceras sorbifolia bunge* by precolumn-HPLC derivatization. *North. Pharma.* 9 (1), 3–4.

Chai, X., Zhang, J.W., Li, S.H., Cheng, Q.S., Qin, M.M., Yang, C.Y. et al. 2021. Xanthoceraside induces cell apoptosis through downregulation of the PI3K/Akt/Bcl-2/Bax signaling pathway in cell lines of human bladder cancer. *Indian Journal of Pathology and Microbiology* (294–301).

Chan, P.K. 2007. Acylation with diangeloyl groups at C21-22 positions in triterpenoid saponins is essential for cytotoxicity towards tumor cells. *Biochem. Pharmacol.* 73(3), 341–350.

Chan, P.K., Zhao, M., Che, C.T. and Mak, E. 2008. Cytotoxic acylated triterpene saponins from the husks of *Xanthoceras sorbifolia* . *J. Nat. Prod.* 71, 1247–1250.

Chen, G., Xie, Y., Zhou, D., Yang, L., Zou, L., Cheng, M., et al. 2020 (a). *In vivo* evaluation and atom-based 3d-qsar studies on saponins from shells of *Xanthoceras sorbifolium bunge* as anti-AD agents. *Bioorg. Chem.* 94.

Chen, G., Xie, Y., Zhou, D., Yang, Y., Liu, J. and Hou, Y. et al. 2020(b). Chemical constituents from shells of *Xanthoceras sorbifolium*. *Phytochemistry*. 172.

Chen, X., Lei, Z., Cao, J., Zhang, W.,Wu, R., Cao, F., Guo, X. and Wang, J. 2022. Traditional uses, phytochemistry, pharmacology and current uses of underutilized *Xanthoceras sorbifolium bunge*: A review. *Journal of Ethanopharmacology*. 283, 114747.

Chen, Y.J., Takeda T., Ogihara Y. and Iitaka Y. 1984. Studies on the constituents of Xanthoceras sorbifolia Bunge. II. Major sapogenol and a prosapogenin from the fruits of Xanthoceras sorbifolia Bunge. Chem Pharm Bull (Tokyo). 32(9):3378–2283. doi: 10.1248/cpb.32.3378.

Chirva, V.Y. and Kintya, P.K. 1971. The structure of *Xanthoceras Saponin*. *Khim. Prir. Soedin.* 7(4).

Han, R., Wang, M., Wang, L., Zhang, Y., Li, X., Hou, Y., Yan, J. and Maulidiani, X.P.M. 2022. GC/MS-Based Urine Metabolomics Study on the Ameliorative Effect ofXanthoceras sorbifolia Extract on Alzheimer's Disease in Mice. *Hindawi Evidence Based Complementary And Alternative Medicine*, Article Id –3390034 (1–13).

Harrington, M.G., Edwards, K.J., Johnson, S.A., Mark, W., Chase, M.W. and Gadek, P.A. 2005. Phylogenetic inference in *Sapindaceae sensulato* using plastid matK and rbcL DNA sequences. *Systematic Botany* 30 (2): 366–382. https://doi.org/10.1600/0363644054223549 for cross-check

Ji, X. F., Chi, T.Y., Xu, Q., He, X.L., Zhou, X.Y., Zhang, R. and Zou, L.B. 2014. Xanthoceraside ameliorates mitochondrial dysfunction contributing to the improvement of learning and memory impairment in mice with intracerebroventricular injection of Aβ1-42. *Evid Based Complement Alternate Med 969342* (1-11). PMCID-PMC 4058193.

Liu, Y., Han, Y., Liu, Y., Huang, C., Feng, W., Cui, H., Li, M. 2024. Xanthoceras sorbifolium leaves alleviate hyperuricemic nephropathy by inhibiting the PI3K/AKT signaling pathway to regulate uric acid transport, Journal of Ethnopharmacology, 327: 117946, ISSN 0378-8741, https://doi.org/10.1016/j.jep.2024.117946.

Li, Y.Y., Xiang, Z., Cui, H., Xiao, H., Kang, T.G., Dou, D.Q. and Kuang, H.X. 2012. Two new oleanane-type saponins from the husks of *Xanthoceras sorbifolia* Bunge. *Nat. Prod. Res.* 208–214.

Wan, G.S., Ren, Y.H., Gao, H.Y., Bai, S., Xi, R.G. and Wang, X.B. 2015. Isolation and identification of chemical constituents from the husks of *Xanthoceras sorbifolium* Bunge. *J. Shenyang Pharm. Univ.* 8–21.

Yang, C.Y., Ha, W., Lin, Y., Jiang, K., Yang, J.L. and Shi, Y.P. 2016. Polyphenols Isolated from Xanthoceras sorbifolia Husks and Their Anti-Tumor and Radical-Scavenging Activities. *Molecules* (Basel, Switzerland). 21(12), 1694. https://doi.org/10.3390/molecules21121694.

Yu, L.L., Wang, X.B., Wei, X.C., Wang, M.M., Chen, L.X., Cao, S.J., Kang, N. and Qiu, F. 2012. Triterpenoid saponins from Xanthoceras sorbifolia Bunge. and their inhibitory activity on human cancer cell lines. *Bioorg. Med. Chem.* Lett. 5232–5238.

Zang, E., Qiu, B., Chen, N., Li, C., Liu, Q., Zhang, M., Liu, Y., Li, M. (2021). Xanthoceras sorbifolium Bunge: a review on botany, phytochemistry, pharmacology, and applications. Frontiers in Pharmacology 12:12633 DOI 10.3389/fphar.2021.708549.

Zang, R., Qiu, L., Chen, X., Li, X., Liu, Y., Zhang, Y., Liu, X. and Li, J. (2021). Xanthoceras sorbifolium Bunge: A Review on Botany, Phytochemistry, Pharmacology, and Applications. Frontiers in Pharmacology, 12, 708549. https://doi.org/10.3389/fphar.2021.708549

Zhu R-B, Wang Q, Guan W-B, et al. (2019). Conservation of genetic diversity hotspots of the high-valued relic yellowhorn (Xanthoceras sorbifolium) considering climate change predictions. Ecol Evol. 2019; 9: 3251–3263. https://doi.org/10.1002/ece3.4944

23 *Glycyrrhiza glabra* L. (Liquorice): Unveiling Biological Marvels and Therapeutic Promise, and Emphasizing Antiviral Potency

K. Venkata Ratnam[1#], Anu Pandita[2#], Lepakshi Md. Bhakshu[3], Pulala Raghuveer Yadav[4] and Deepu Pandita[5]

1. Taxonomy and Scientific Classification

Glycyrrhiza glabra L. is a perennial pea family Fabaceae herbaceous plant known as Liquorice or licorice in English. The genus name Glycyrrhiza is derived from the Greek word glykos which means sweet and rhiza which means root and thus Glycyrrhiza means sweet root (Fiore et al., 2005; Dastagir and Rizvi, 2016). *G. glabra* has 3 varieties, viz., *G. glabra* var. violacea is known as Persian and Turkish licorice, *G. glabra* var. gladulifera which is known as Russian licorice and *G. glabra* var. typica L. is known as Spanish and Italian licorice (Nomura et al., 2002).

Below we mention Bentham and Hooker's classification of *Glycyrrhiza glabra* L.

Scientific Classification

Kingdom	:	Plantae
Class	:	Dicotyledons
Subclass	:	Polypetalae
Series	:	Calyciflorae
Order	:	Rosales
Family	:	Leguminosae or Fabaceae
Genus	:	*Glycyrrhiza*
Species	:	*glabra*

[1] Department of Botany, Rayalaseema University, Kurnool, Andhra Pradesh, India.
[2] Vatsalya Clinic, Krishna Nagar, New Delhi, India.
[3] Department of Botany, PVKN Government College (A), Chittoor, Andhra Pradesh, India.
[4] Department of Biotechnology, Indian Institute of Technology Hyderabad, Kandi, 502 285, Telangana State, India.
[5] Government Department of School Education, Jammu, Jammu and Kashmir, India: ORCID iD-https://orcid.org/0000-0002-8361-1872.
[#] *Both these authors have contributed equally.*
* Corresponding author: deepupandita@gmail.com

DOI: 10.1201/9781003433583-23

Synonyms

Glycyrrhiza glandulifera (Waldst. & Kit.)
Glycyrrhiza brachycarpa (Boiss.)
Glycyrrhiza hirsuta (Pall.)
Glycyrrhiza pallida (Boiss.)
Glycyrrhiza violacea (Boiss.)

2. Botantical Description of *GLYCYRRHIZA GLABRA* L.

Glycyrrhiza glabra L. is a perennial herb of the pea family and grows on clay or sandy soils near streams or river banks. It grows up to 1 meter in height. The leaves are compound, alternate, imparipinnate, of about 7 to 15 cm length and have 9–17 leaflets each (Fig. 1a, b). Inflorescence is raceme in axillary position. Several flowers 8 to 12 mm in size are loosely arranged in inflorescence, and possess a purple or pale whitish blue color (Fig. 1c). The fruit is a legume or oblong pod of 2–3 centimeters long and contains 2 to 5 seeds (Fig. 1d) Sharma et al., (2018). Roots are about 1 m long and 1cm in diameter, stoloniferous, soft, flexible, and bright yellow on the inner side with a distinctive sweetness (Fig. 1e).

Fig. 1 (a) Drawing of *G. glabra* (Source: *G. glabra* from Koehler's *Medicinal-Plants*), (b, e) Foliage and Roots (*Source:* Adapted from https://en.wikipedia.org) (c, d) Inflorescence and legume fruits (pod) of *G. glabra* (Source: Adapted from https://www.gbif.org)

3. Geographical Distribution

Glycyrrhiza glabra L is a perennial herbaceous species, and is commercially the most important species in the Glycyrrhiza genus. *G. glabra* is distributed globally in India, China, Spain, Russia, Italy, Turkey, Caucasus, Central Asia, and Iran (Sharma et al., 2018; Hayashi et al., 2019). This plant is native to Syria, Afghanistan, Persia, Southern Europe, and North Africa (Chin and Keng, 1992;

Ghazanfar, 1994). Countries like Iran, India, the United States, China, Uzbekistan, Syria, Azerbaijan, Spain, France, Greece, England, Afghanistan, Turkmenistan, and Iraq are cultivating this plant as a commercial crop (Sokolov and Zamotayev, 1985; Chevallier, 1997).

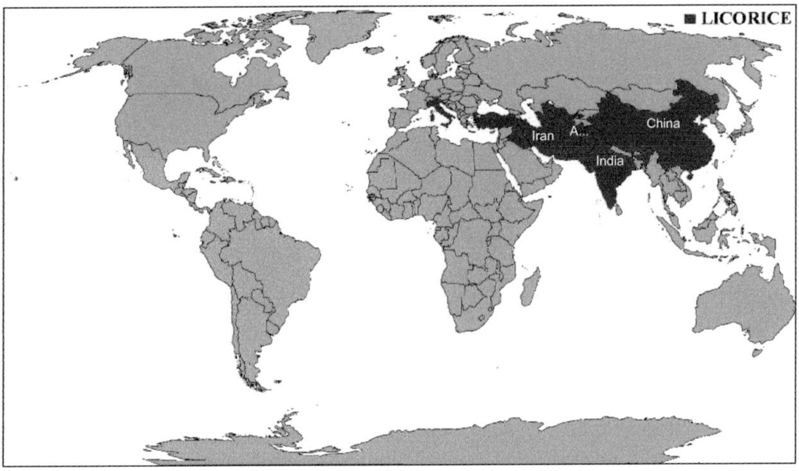

Fig. 2 Licorice producing countries in the world. (*Source:* Wahab et al., 2021)

4. Phytochemical Constituents and Metabolite Profiling

The rhizomes/roots of liquorice are rich in biochemical components of nutritional and therapeutic importance. The nutritional components isolated from liquorice are sugars, polysaccharides, proteins, amino acids, resins, starch, mineral salts such as sodium, iron, calcium, copper, manganese, zinc, phosphorus, selenium, silicon, and vitamins like B complex, coumarins and sterols (Wang et al., 2015a, 2015b). Bioactive chemical constituents isolated from liquorice are triterpenoids, flavonoids, and saponins (Wang et al., 2015b; Rizzato et al., 2017). Glycyrrzine, triterpenoid saponins is the main chemical constituent of roots of liquorice, responsible for its sweet taste, and is the primary active principle of it. Glycyrrzine is a concoction of salts like magnesium, potassium, and calcium of Glycyrrhizin or glycyrrhizic acid or glycyrrhizinic acid (Fig. 3). It presents as 10% of the quantity in dry roots of liquorice and it is 50 times sweeter than sucrose (Yu et al., 2015; Rizzato et al., 2017).

Various groups of flavonoids were isolated from liquorice roots are flavonols, flavones, isoflavones, chalcones, flavanones, isoflavenes, and isoflavanones (Fig. 3). The major flavonoid glycosides isolated from roots of liquorice are isoliquiritigenin (2′,4,4′-trihydroxychalcone, liquiritigenin (4′,7-dihydroxyflavanone), licuraside, liquiritin, liquiritin apioside and isoliquiritin (Rizzato et al., 2017). Five novel flavonoids isolated from dried roots of liquorice are shinflavanone,

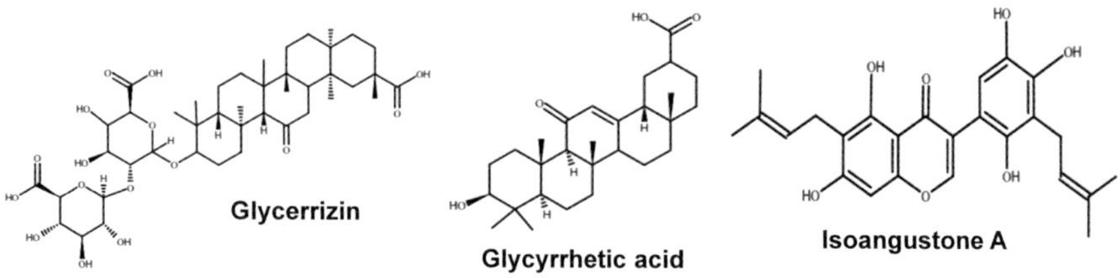

Fig. 3 Terpenoid saponins and terpenoids of licorice (*Source:* Rizzato et al., 2017)

prenyllicoflavone A, licopyranocoumarin, 1-methoxyphaseolin, rhamnoliquirilin, liquiritigenin, licoarylcoumarin, shinpterocarpin, glisoflavone, and coumarin-GU-12 and glucoliquiritin apioside (El-Saber Batiha et al., 2020). Two chemical constituents isolated from liquorice leaves are licoflavanone and pinocembrin (Fukui et al., 1988). Glabrene, hispaglabridin A, glabridine, hispaglabridin B (Fig. 4), 4′-O-Methylglabridin, Isoliquiritigenin, and Formononetin were reported from the roots of liquorice (Vaya et al., 1997; Simmler et al., 2013). Glycyrrhizic acid is a natural saponin that consists of two molecules i.e. glucuronic acid (hydrophilic part) and glycyrrhetic acid (hydrophobic part) (Obolentseva et al., 1999).

The volatile components reported from essential oil isolated from liquorice are hexanol, pentanol, α-terpineol, terpinen-4-ol, geraniol, maltol, propionic acid, furfuraldehyde, trimethylpyrazine, benzoic acid, furfuryl formate, 2,3-butanediol, and 1-methyl-2-formylpyrrole (Chouitah et al., 2011). Coumarins isolated from licorice are glycycoumarin, licopyranocoumarin, licocoumarin, glabrocoumarone A&B (Sharma et al., 2018b). Isoprenoid substituted phenol components isolated from licorice are isoangustone A, 1-methoxyficifolinol, Semilicoisoflavone B, licoriphenone (Fenwick et al., 1990).

5. Medicinal/Therapeutic Uses

5.1 Uses described in Pharmacopoeias and in Traditional Systems of Medicine

The traditional uses of liquorice can be traced back to ancient cultures of Indian, Chinese, Egyptian, and Assyrian societies. In the Chinese traditional medicinal system, liqu rice is considered a vital natural medication. It has been used as an enhancing herbal recipe for the spleen and heart since 25 AD. It is also an effective medication for respiratory conditions, coughs, and analgesia (Wahab et al., 2021). In Indian traditional medicinal systems such as Ayurveda and Siddha, licorice has been used to treat ulcers, as purgative, and expectorant. L icorice based pharmaceutical formulations are Gut Guard as antioxidant and for healthy support to gastro-intestinal tract, Health Aid Licorice for the dry cough and free breathing. The Himalaya Company prepared a whitening face wash to remove dark spots and clarify impurities using licorice. The Solaray Licorice capsules find use as nutraceuticals (Thakur and Raj, 2017). In German traditional medicine

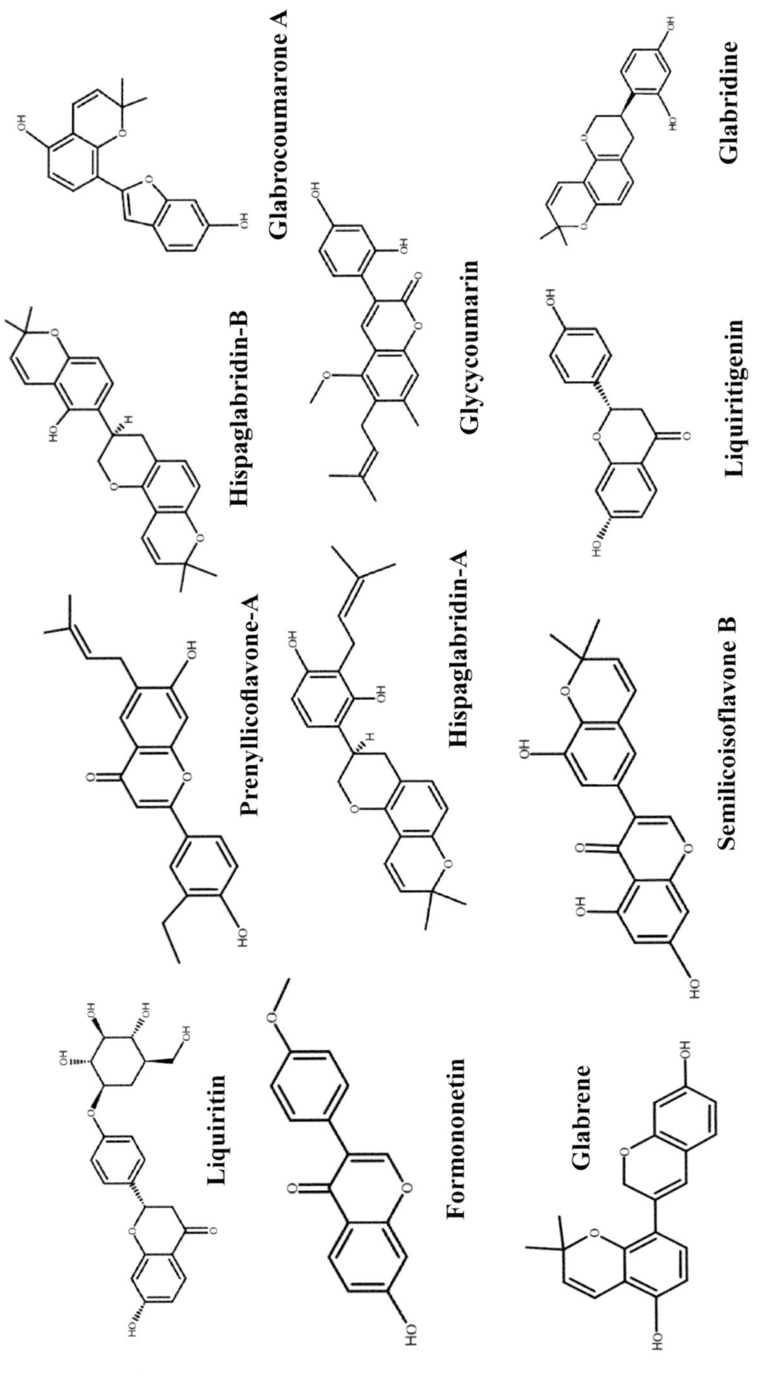

Fig. 4 Phenolic Components of Licorice
(*Sources*: Rizzato et al., 2017; El-Saber et al., 2020; Fukui et al., 1988).

licorice is used as an oral drug to promote gastric function. In France two licorice formulations are used to treat throat infections and for digestion problems (EMA/HMPC Report, 2011).

5.2 Traditional/Folklore Uses described in Folk Medicine

In ethno-medicine licorice has been used to treat different human ailments. Paste prepared by mixing licorice powder with butter and applied externally is used on wounds or burns. Licorice powder is mixed with cow's milk and given orally to improve lactation in feeding mothers. A paste prepared with licorice powder, milk, and gingelly seeds mixed with butter is applied externally on skin swellings/inflammation. It is also used as a root extract applied as eye drops to cure eye diseases. Poultices are prepared using licorice dry root powder and honey and given orally to cure anemia. *Glycyrriza glabra* along with *Picirrhiza kurroa* is made into paste with sugar water, and given orally to cure cardiovascular diseases. Licorice powder is mixed with sandal wood powder and used to treat haematemesis (Kumar and Dora, 2012). In Pakistan, licorice root powder is mixed with flour and made into a paste with oil, and given orally to cattle to increase lactation and fertility (Aziz et al., 2018). In Turkey, the root sap of licorice is used to prepare wine (Kargioglu et al., 2010). In Italy, decoction is prepared from licorice root as a mild laxative (Motti, 2021). In Nepal, juice is prepared from the stem and root of licorice as a tonic, stimulant and purgative (Luitel et al., 2014).

5.3 Applications supported by Clinical Data

Antiviral effect of glycyrrhizic acid against Kaposi's Sarcoma-Associated Herpesvirus (KSHV) in chronic hepatitis B patients of 211 members of both males and females of the 31–40 age group in China was investigated. It was observed that the patients treated with GA demonstrated reduced levels of viral DNA indicating its antiviral effects (Xie et al., 2011). Antiviral efficiency of glycyrrhizine (GL) and lamivudine was assessed in patients with chronic HBV Carriers with NonHodgkin Lymphoma. It revealed that both the drugs significantly inhibited replication of virus (Matsuo et al., 2001). A clinical investigation conducted by Mori et al. (1989) demonstrated that glycyrrhizin substantially inhibited HIV in patients with hemophilia. GL maintained normal levels of lipids, serum electrolytes, protein, and renal function, and without any serious side effects during the treatment period.

Antiviral activity of GL along with four drugs such as mycophenolic acid, 6-azauridine, ribavirin and pyrazofurin was tested against FFM-1 and FFM-2 isolated from SARS patients, revealed that among the tested drugs GL exhibited very strong antiviral activity that other drugs (Cinatl et al., 2003). The SARS inhibitory effect of GL injection CGI in 26 patients revealed that the CGI significantly inhibited viral replication, reduced liver damage and enzyme activities with minimal side effects (Lu et al., 2003). The effect of GL on chest X-ray manifestations in 60 cases of SARS patients revealed that the group treated with GL significantly reduced SARS induced lesions in the lungs of patients than the control group (Wu et al., 2004). Combined therapy of diammonium glycyrrhizinate (DG), with vitamin C indicated that the patient with severe COVID-19 successively recovered with COVID-19 symptoms. This shows its potential therapeutic effects as an alternative medicine to treat COVID -19 symptoms (Ding et al., 2020). An antiviral activity of a Persian herbal medicine consisting of licorice, *Rheum palmatum,* and pomegranate was assessed in hospitalized COVID-19 patients demonstrating that the formula significantly increased the respiratory rate in COVID-19 patients (Setayesh et al., 2022).

5.4 Pharmacological Activities mainly Antiviral Studies

The antiviral efficiency of phyto-constituents glycyrrhizin and 18β-glycyrrhetinic acid isolated from liquorice was investigated against Hepatitis C Virus (HCV), Human Immunodeficiency Virus (HIV), Herpes Simplex Virus (HSV), coxsackievirus B3 (CVB3), Duck Hepatitis Virus (DHV), Influenza A virus (subtype H5N1), Influenza virus, Enterovirus 71 (EV71), Herpes Simplex Virus (HSV), rota virus and Human Respiratory Syncytial Virus (HRSV) (Wang et al., 2015a). The antiviral efficiency of glycyrrhizic acid (GA) isolated from *Glycyrrhiza* species was evaluated against the herpes simplex virus (HSV), demonstrating that the component exhibited good antiviral activity (Laconi et al., 2014). Matsumoto et al. (2013) studied the antiviral activity of glycyrrhizine (GL) against the hepatitis C Virus-infected Huh7 cells and revealed that the constituent significantly inhibited the attachment of HCV to the Huh7 cells.

The antiviral effects of GA liver cells infected with HCV at non-toxic dose revealed that GL significantly inhibited HCV in concentration-dependently. In a combined study with interferon, it showed synergistic effects to inhibit the viral particles (Ashfaq et al., 2011). Glycyrrhizic acids have strong antiviral properties and were studied against HSV *in vitro* (Pompei et al., 1979). Glycyrrhizine derivatives i.e. carbenoxolone sodium (CBX) and Cicloxolone sodium (CCX) exhibited potent antiviral activity against HSV 1 and 2. CCX at 300 μM and CBX 500 μM inhibited virus replication 10,000 to 1, 00, 000 folds (Dargan et al., 1986). I n vitro study results revealed that GL and its derivatives exhibited potent antiviral activity against HIV-1 and HSV-1 (Hirabayashi et al., 1991). *In vivo* study results demonstrated that GL significantly enhanced the survival capacity of mice with herpes simplex encephalitis (Sekizawa et al., 2001). Ikeda et al., (2005) identified that glycyrrhizic acid is 10 times more potent than glycyrrhizin to inhibit HSV-1. The antiviral mechanism of GA in lymphocytes infected with Kaposi's sarcoma-associated herpesvirus showed that GA significantly altered the enhancement of RNA polymerase II complex, and transcription factors at the cellular level and reduced sister chromatid cohesion (Kang and Lieberman, 2011).

Huang et al. (2012) studied the inhibitory effect of GL on intercellular adhesion molecules on HSV using polymorphonuclear leukocytes (PMN) and cerebral capillary vessel endothelial cells (CCECs) isolated from rat models.

Micropipette aspiration technique results indicated that GL significantly reduced inflammatory responses of HSV between polymorphonuclear leukocytes and cerebral capillary vessel endothelial cells. The therapeutic efficiency of GL against Coxsackievirus B3-induced inflammatory cardiomyopathy in mice models was evaluated. GL significantly attenuated Coxsackievirus B3-induced inflammatory cardiomyopathy. It was evidenced by a reduction in serum cardiac enzymes, myocardial inflammation, and enhanced survival rate (Zhang et al., 2012). The a ntiviral activity of GL against Duck Hepatitis Virus (DHV) was assessed using *in vitro* and *in vivo* models. *In vitro* investigation revealed that GL significantly inhibited the DHV. After four days of inoculation, the results of an *in vivo* experiment showed that GL, both alone and in combination, greatly increased the proliferation of lymphocytes (Soufy et al., 2012).

The inhibitory effects of GL on the H5N1 influenza A virus in human macrophages were assessed using in vitro methods. It revealed that GL significantly inhibited the expression of pro-inflammatory cytokines in human macrophages (Michaelis et al., 2010). The modulatory effect of GL in combination with glutamyl-tryptophan against influenza virus-induced inflammation in mice was evaluated. It demonstrated that both showed protective effects against viral infection and reduced lung inflammation and death of viral-infected animals (Smirnov et al., 2012). Anti-viral efficiency of glycyrrizic acid (GA) was assessed using epithelial cells infected with rotavirus. Immunoblot analysis revealed that GA significantly altered viral protein synthesis (Hardy et al., 2012). The a ntiviral effect of water extracts of licorice i.e. Raxix glycyrrhizae and Radix glycyrrhizae preparata against Human Respiratory Syncytial Virus (HRSV) and two active principles of licorice *i.e.*, glycyrrhizine and gycyrrhizic acid on human upper (HEp-2) and low (A549) respiratory tract cell lines. Both water extract preparations of licorice significantly inhibited HRSV in a concentration dependent manner. Gycyrrhizic acid strongly inhibited HRSV (Yeh et al., 2013).

Glycyrrhizic acid (GA) reported inhibitory activity of B-lymphocytes infected with kaposi sarcoma-associated HSV (Curreli et al., 2005). An in vitro investigation conducted by (Lin, 2003) on inhibitory capacity of GA against Epstein-Barr-Virus infected Raji Cells demonstrated concentration dependent antiviral activity. Antiviral activity of glycyrrhizic acid and its derivatives was assessed against Epstein-Barr-Virus, revealing that GA exhibited dose-dependent antiviral activity. It was observed that 18β-glycyrrhetinic acid, a derivative of GA, was found to be 7.5 folds more potent than GA (Lin et al., 2008). Bentz et al. (2019) investigated that GA significantly inhibited the processes of sumoylation, reduced cell growth by inducing programmed cell death in EBV-associated lymphoid cells. Glycyrrhizic acid isolated from herbal drugs used in Chinese traditional medicine was evaluated for antiviral effects against hepatitis B virus, demonstrating moderate antiviral activity without any toxicity to host cells (Romero et al., 2019). Biochemical assays were carried out to understand the antiviral mode of the action of glycyrrhizine (GL) against hepatitis B virus using PLC/PRF/5 cells. It was noted that the GL significantly reduced secretion levels of hepatitis B surface antigen (HBsAg) and sialylation in a concentration-dependent manner (Takahara et al., 1994; Sato et al., 1996).

A comparative antiviral effect of GL alone and in combination with interferon was assessed in HCV-infected liver cells, demonstrating that GL alone exhibited a 50% growth reduction of liver cells in concentration concentration-dependent manner and transcription of HCV 3a core gene in live cells. GL increased its therapeutic potential after adding interferon (Ashfaq et al., 2011). In combined therapy with GL and interferon results demonstrated that the components significantly altered hepatitis C virus-induced liver damage (Fujisawa, 1991). An open pilot study conducted with the oral administration of lamivudine and the intravenous administration of GL significantly attenuated sub-acute hepatitis caused by hepatitis B and E viruses (Tandon et al., 2001). The a ntioxidant capacity of stronger neo-minophagen C preparation made from glycyrrhizin against CCl4 induced oxidative stress in transgenic mice expressing HCV poly-protein. It revealed that the preparation significantly altered CCl4-induced oxidative stress and mitochondrial injury in hepatocytes of mice (Hidaka et al., 2007; Korenaga et al., 2011).

Recently Shi et al. (2020) investigated the antiviral capacity of GA against liver inflammation induced by the mouse virus 1 in mouse models indicating that GA significantly altered the expression levels of cytokines associated with the HMGB1-TLR4 signaling pathway. An *in vitro* study, Pompei et al. (1983) evaluated the antiviral activity of GA against chicken eggs infected with influenza and Newcastle Disease virus. The outcomes displayed that GA significantly inhibited the retrieval of hemagglutinins in both viruses. The antiviral efficiency of GL was studied against influenza A virus-infected lung cells *in vitro* conditions. It demonstrated that GL significantly reduced cytopathic effect and hem agglutination (Wolkerstorfer et al., 2009). The inhibitory effects of GL against the H5N 1 virus-induced pro-inflammatory cytokines in human macrophages, revealed that GL strongly inhibited H5N1-induced cytokines such as COX, CCL and IL 6 (Michaelis et al., 2010).

The mechanism of the antiviral activity of GL against human para-influenza virus type 2 (hPIV-2) indicated that GL significantly inhibited viral cells by inhibiting actin filament formation. RT-PCR results showed GL significantly inhibited protein synthesis and mRNA expression (Sakai-Sugino et al., 2017). Glycyrrhizin significantly and concentration-dependently suppressed HIV type 1 replication in MOLT-4 cells (Ito et al., 1988). The antiviral effect of GL against the peripheral blood mononuclear cells of patients infected with the non-syncytium-inducing HIV variant and syncytium-inducing HIV variant. The results demonstrated that GL strongly inhibited HIV replication in patients (Sasaki et al., 2002). The pre-treatment of MDM cultures supplemented with recombinant CCL2 and IL-10 and polymorphonuclear neutrophils infected with R5 HIV with GL significantly inhibited production of CCL2 and IL-10 in both cell lines (Takei et al., 2005; Yoshida et al., 2009).

The anti-SARS-CoV activity of GL and its derivatives demonstrated that the GL conjugated with amides amino acid residues and 30-COOH function showed potent antiviral activity rather than the individual components (Hoever et al., 2005). GL and GA significantly reduced HIV infection by inhibiting the expression levels of transmembrane serine protease (Murck, 2020). GL strongly inhibited perturbations on lung cells expressing Severe Acute Respiratory Syndrome Coronavirus 2 (SARS-CoV-2) viral proteins and pro-inflammatory proteins associated with lung cells and it inhibited SARS-CoV-2 replication in Vero cells at higher doses without cytotoxicity dose-dependently (Gowda et al., 2021). The inhibitory activity of GL against SARS-CoV-2 was evaluated in computer-aided drug design models (Sinha et al., 2021; van de Sand et al., 2021; Yu et al., 2021). Similarly, GA has been found to have binding capacity with multiple essential SARS-CoV-2 proteins through virtual screening and molecular dynamics simulations (Sinha et al., 2020; Dharmendra Kumar, 2021). The antiviral efficiency of *G. glabra* extract against COVID-19-induced lung infection in the hamster model revealed that licorice significantly enhanced body weight in hamsters and viral load in lungs (Rizvi et al., 2022). GA bioactive molecule proves effective against SARS-CoV-2 virus by directly targeting vital protein structures, inhibiting of virus invasion and replication process (Sun et al., 2021).

6.5 Mode of Action of Antiviral Compounds/Extracts

Liqu orice and its active principles strongly inhibited the following virus types as Hepatitis C Virus (HCV), Herpes Simplex Virus (HSV), Coxsackie Virus B3 (CVB3), Duck Hepatitis Virus (DHV), Influenza A virus (subtype H5N1), Influenza virus, Enterovirus 71 (EV71), Herpes Simplex Virus (HSV), etc by suppressing the replication, adhesion capacity and gene expression (Wang et al., 2009; Michaelis et al., 2010; Ashfaq et al., 2011; Huang et al., 2012; Zhang et al., 2012; Soufy et al., 2012; Matsumoto et al., 2013). The mechanism of antiviral activity of licorice and its active principles is summarized in Table 1.

Table 1 Mechanism of Antiviral Action of Licorice Extracts/Active Principles

Compound/ Extract	*Virus type*	*Mode of action*	*Reference*
Glycyrrhizin (GL)	CVB3	Inhibited inflammatory markers for instance TNF-α, IL-1β, IL6 and NFκB	Zhang et al. (2012)
Glycyrrhizin	HSV CCECs and PMN	Inhibited intercellular adhesion molecules, reduced adhesion forces, and stress in CCECs & PMN cells	Huang et al. (2012)
Glycyrrhizin	Hepatitis C virus, Huh7 cells	Inhibited phospholipase A2 and phospholipase A2 group IB, inhibited the attachment of HCV to the Huh7 cells.	Matsumoto et al. (2013)

Contd.

Table 1 *Contd.*

Compound/ Extract	*Virus type*	*Mode of action*	*Reference*
Glycyrrhizin	Hepatitis C virus infected liver cells	Inhibited HCV 3a core gene expression at mRNA and protein level	Ashfaq et al. (2011)
Glycyrrhizin	Duck hepatitis virus	Inhibited viral replication in in vitro conditions In vivo studies it inhibited proliferation of lymphocytes	Soufy et al. (2012)
Glycyrrhizin	H5N1 influenza A virus	It significantly altered H5N1-induced production of CCL5, IL6 and CXCL10, and inhibited H5N1-induced apoptosis	Michaelis et al. (2010)
Glycyrrhizin	Hepatitis B Virus	Reduced the secretion levels of hepatitis B surface antigen (HBsAg), intracellular transport , and its sialylation	Takahara et al. (1994)
Glycyrrhizin	Influenza A Virus	Reduced viral RNA within cells and in supernatants of cells	Wolkerstorfer et al. (2009)
Glycyrrhizin	H5N1 virus	Attenuated pro-inflammatory cytokines COX, CCL and IL6	Michaelis et al. (2010)
Glycyrrhizin	Human Parainfluenza virus type 2	Inhibited actin filament formation, mRNA, and protein synthesis	Sakai-Sugino et al. (2017)
Glycyrrhizin	HIV -1 MOLT cells	Inhibited protein kinase C activity	Ito et al. (1988)
Glycyrrhizin	R5 HIV-infected PMN cells and MDM cultures	Inhibited production of CCL2 and IL-10 markers	Takei et al. (2005); Yoshida et al. (2009)
Glycyrrhizin	SARS-CoV-2	Inhibited viral replication and pro-inflammatory cytokines	Gowda et al. (2021)
Glycyrrhizin	SARS-CoV-2	Targeting of viral structural proteins and replication	Sun et al. (2021)
Stronger Neo-Minophagen C (SNMC)–GL preparation	Hepatitis C Virus	Attenuated mitochondrial injury, enhanced m-RNA expression levels of gamma-GCS, and restored GSH and ALT levels in mice	Hidaka et al. (2007); Korenaga et al. (2011)
Glycyrrhizine & Interferon	Hepatitis C Virus	Synergetic effects on inhibition of expression levels of HCV 3a core gene at gene level and protein level	Ashfaq et al. (2011)

Contd.

Table 1 *Contd.*

Compound/ Extract	*Virus type*	*Mode of action*	*Reference*
Glycyrrhizic acid (GA)	Herpes Simplex Virus-1 (HSV-1)/ Epithelial cells	Induced autophagy activator Beclin 1 in Epithelial cells	Laconi et al. (2014)
Glycyrrhizic acid	Rotavirus	Inhibited the synthesis of viral proteins VP6, NSP2 and VP2	Hardy et al. (2012)
Radix glycirrhizae/ Glycyrrhizic acid	Human respiratory syncytial virus	Inhibited viral attachment, penetration of viral cells into host cells and enhanced up-regulation of IFN secretion	Yeh et al. (2013)
Carbenoxolone sodium (CBX), Cicloxolone sodium (CCX)	Herpes Simplex Virus-1 (HSV-1 & 2)	Inhibited virus replication, reduced HSV induced the synthesis of nuclear and cytosolic polypeptides	Dargan et al. (1986)
GA	Herpes Simplex Virus-1 (HSV-1	P53-mediated apoptosis, upregulated viral cyclin expression levels and downregulated latency-associated nuclear antigen	Curreli et al. (2005)
GA	Kaposi's Sarcoma-Associated Herpesvirus	Reduced enhancement of RNA polymerase II complex, inhibited c-myc genes and reduced sister chromatid cohesion	Kang and Lieberman (2011)
GA	Epstein-Barr Virus	Inhibited viral replication at cell cycle and synthesis of viral DNA	Lin. (2003)
GA	Mouse hepatitis virus	Inhibited the expression levels of IL 22 and IL17 associated with the HMGB1-TLR4 signaling pathway	Shi et al. (2020)
Licorice extract	COVID-19 in Hamster model	Inhibited pro-inflammatory markers like IL17A and IL 4, differentiation of Th2 and Th17	Rizvi et al. 2022

6. Toxity and Safety

Scientific studies on the adverse effects or toxicity of *Glycyrriza glabra* and glycyrrhizin were compiled (Wahab et al., 2021). Excess consumption of liqu orice causes primary hyperaldosteronism, hyper mineralocorticoidism, a condition characterized by the retention of salt, hypoaldosteronism, metabolic alkalosis, hypertension, hypokalemia, and low renin activity (Omar et al., 2012). Isbrucker et al. (2006) reported that consumption of a high amount of glycyrrhizin leads to hyper mineralocorticoidism and use of a high dose of glycyrrhizic acid inhibits the enzyme 11-β-hydroxysteroid dehydrogenase. An excess intake of licorice induced hypertension, metabolic alkalosis, hypokalemia, hypoaldosteronism, and low-renin activity (Celik et al., 2012). Feasting on candy cigars containing glycyrrhizic acid (GA) is present in licorice-induced hypertension, peripheral edema, and weight gain in a 49-year-old female (Johns, 2009). Nazari et al. (2017) disclosed the toxicological effects of liqu orice and its active principal GL which has moderate toxicity in humans. These showed selective toxicity to cancer cells. The central harmful effects of licorice and GL are hypokalemic-induced secondary disorders and hypertension.

Conclusion

The well-known medicinal plant licorice is utilised in many traditional remedies and is said to provide a wide range of therapeutic benefits. The inhibitory properties of viral replication have been highly regarded by researchers worldwide, which may offer rational support for its efficacy and safe use. Higher doses of it, however, may have unintended consequences and raise blood pressure. Therefore, the verified practitioners may give detailed instructions on how to use it. Additionally, cultivated forms of licorice are being improved to become nontoxic without compromising their original therapeutic potential or efficacy.

References

Ashfaq, U.A., Masoud, M.S., Nawaz Z. and Riazuddin S. (2011). Glycyrrhizin as antiviral agent against Hepatitis C Virus. *J Transl Med.* 9: 112. doi:10.1186/ 1479-5876-9-112.

Aziz, M.A., Khan, A.H., Adnan, M. and Ullah, H. (2018). Traditional uses of medicinal plants used by Indigenous communities for veterinary practices at Bajaur Agency. *Pakistan. J. Ethnobiol. Ethnomed.* 14 (1): 11.

Bentz, G.L., Lowrey, A.J., Horne, D.C., Nguyen, V., Satterfield, A.R. and Ross, T.D. (2019). Using Glycyrrhizic Acid to Target Sumoylation Processes during Epstein-Barr Virus Latency. *PLoS One*. 14 (5): e0217578.

Celik, M.M., Karakus, A., Zeren, C., Demir, M., Bayarogullari, H., Duru, M. and Al, M. (2012). Licorice induced hypokalemia, edema, and thrombocytopenia. *Hum. Exp. Toxicol.* 31: 1295–1298.

Chevallier, A. (1997). The encyclopedia of medicinal plants. *Choice Rev.* Online. 34: 34.

Chin W.Y. and Keng H. (1992). An Illustrated Dictionary of Chinese Medical Herbs. CRCS Publications, Singapore.

Chouitah, O., Meddah, B., Aoues, A. and Sonnet, P. (2011). Chemical composition and antimicrobial activities of the essential oil from *Glycyrrhiza glabra* leaves. *Journal of Essential Oil-Bearing Plants*, 14(3): 284–288.

Cinatl, J., Morgenstern, B., Bauer, G., Chandra, P., Rabenau, H. and Doerr, H. (2003). Glycyrrhizin, an Active Component of Liquorice Roots, and Replication of SARS-Associated Coronavirus. *The Lancet* 361 (9374): 2045–2046.

Curreli, F., Friedman-Kien, A.E. and Flore, O. (2005). Glycyrrhizic Acid Alters Kaposi Sarcoma-Associated Herpesvirus Latency, Triggering P53-Mediated Apoptosis in Transformed B Lymphocytes. *J. Clin. Invest.* 115 (3): 642–652.

Dargan, D.J. and Subak-Sharpe, J.H. (1986). The Antiviral Activity against Herpes Simplex Virus of the Triterpenoid Compounds Carbenoxolone Sodium and Cicloxolone Sodium. *J. Antimicrob. Chemother Encephalitis.* 18 (Suppl. B): 185–200.

Dastagir, G., and Rizvi, M.A. (2016). Review - Glycyrrhiza glabra L. (Liquorice). *Pakistan Journal of Pharmaceutical Sciences*, *29*(5): 1727–1733.

Dharmendra K.M. (2021). Evaluation of Yashtimadhu (Glycyrrhiza glabra) Active Phytochemicals Against Novel Coronavirus (SARS-CoV-2). Res. Square. doi:10.21203/rs.3.rs-26480/v1

Ding, H., Deng, W., Ding, L., Ye, X., Yin, S. and Huang, W. (2020). Glycyrrhetinic Acid and its Derivatives as Potential Alternative Medicine to Relieve Symptoms in Nonhospitalized COVID-19 Patients. *J. Med. Virol.* 92: 2200–2204.

El-Saber Batiha, G., Magdy Beshbishy, A., El-Mleeh, A., Abdel-Daim, M.M. and Prasad Devkota, H. (2020). Traditional Uses, Bioactive Chemical Constituents, and Pharmacological and Toxicological Activities of *Glycyrrhiza glabra* L. (Fabaceae). *Biomolecules*. 10(3):352.

EMA/HMPC. Assessment report on *Glycyrrhiza glabra* L. and/or *Glycyrrhiza inflata* Bat. and/or *Glycyrrhiza uralensis* Fisch ex D.C., radix. Volume 44 European Medicines Agency; Amsterdam, The Netherlands: 2011.

Fenwick, G.R., Lutomski, J. and Nieman, C. (1990). Liquorice, *Glycyrrhiza glabra* L. composition, uses and analysis. *Food Chem.* 38 (2): 476–477.

Fiore, C., Eisenhut, M., Ragazzi, E., Zanchin, G. and Armanini, D. (2005). A history of the therapeutic use of liquorice in Europe. *Journal of Ethnopharmacology*, *99*(3): 317–324. https://doi.org/10.1016/j.jep.2005.04.015.

Fujisawa, K. (1991). Interferon Therapy in Hepatitis C Virus (HCV) Induced Chronic Hepatitis: Clinical Significance of Pretreatment with Glycyrhizine. *Trop. Gastroenterol.* 12 (4): 176–179.

Fukui, H., Goto, K. and Tabata, M. (1988). Two antimicrobial flavanones from the leaves of *Glycyrrhiza glabra*. *Chemical & Pharmaceutical Bulletin*, 36(10): 4174–4176.

Ghazanfar, S.A. (1994). Handbook of Arabian Medicinal Plants, CRC Press, Boca Raton, Florida, 1994.

Gowda, P., Patrick, S., Joshi, S.D., Kumawat, R.K., and Sen, E. (2021). Glycyrrhizin Prevents SARS-CoV-2 S1 and Orf3a Induced High Mobility Group Box 1 (HMGB1) Release and Inhibits Viral Replication. *Cytokine*, 142: 155496.

Hardy, M.E., Hendricks, J.M., Paulson, J.M. and Faunce, N.R. (2012).18β-glycyrrhetinic acid inhibits rotavirus replication in culture. *Virol J.* 9: 96. doi: 10.1186/1743-422X-9-96.

Hayashi, H., Yokoshima, K., Chiba, R., Fujii, I., Fattokhov, I. and Saidov, M. (2019). Field survey of *Glycyrrhiza* plants in Central Asia (5). Chemical characterization of *G. bucharica* Collected in Tajikistan. *Chem. Pharm. Bull.* 67: 534–539.

Hidaka, I., Hino, K., Korenaga, M., Gondo, T., Nishina, S. and Ando, M. et al. (2007). Stronger Neo-Minophagen C?, a Glycyrrhizin-Containing Preparation, Protects Liver against Carbon Tetrachloride-Induced Oxidative Stress in Transgenic Mice Expressing the Hepatitis C Virus Polyprotein. *Liver Int.* 27 (6): 845–853.

Hirabayashi, K., Iwata, S., Matsumoto, H., Mori, T., Shibata, S. and Baba, M., et al. (1991). Antiviral Activities of Glycyrrhizin and its Modified Compounds against Human Immunodeficiency Virus Type 1(HIV-1) and Herpes Simplex Virus Type 1(HSV-1) *In Vitro*. *Chem. Pharm. Bull.* 39 (1): 112–115.

Hoever, G., Baltina, L., Michaelis, M., Kondratenko, R., Baltina, L. and Tolstikov, G.A., et al. (2005). Antiviral Activity of Glycyrrhizic Acid Derivatives against SARS–Coronavirus. *J. Med. Chem.* 48 (4): 1256–1259.

Huan, C., Xu, Y., Zhang, W., Guo, T., Pan, H. and Gao, S. (2021). Research Progress on the Antiviral Activity of Glycyrrhizin and its Derivatives in Liquorice. *Frontiers in Pharmacology*, 12: 680674.

Huang, W., Chen, X., Li, Q., Li, P., Zhao, G.N. and Xu, M.M. (2012). Inhibition of intercellular adhesion in herpex simplex virus infection by glycyrrhizin. *Cell Biochem Biophys*. 62:137–140.

Ikeda, T., Yokomizo, K., Okawa, M., Tsuchihashi, R., Kinjo, J. and Nohara, T., et al. (2005). Anti-herpes Virus Type 1 Activity of Oleanane-type Triterpenoids. *Biol. Pharm. Bull.* 28 (9): 1779–1781.

Isbrucker, R.A. and Burdock, G.A. (2006). Risk and safety assessment on the consumption of licorice root (*Glycyrrhiza* sp.), its extract and powder as a food ingredient, with emphasis on the pharmacology and toxicology of glycyrrhizin. *Regul. Toxicol. Pharmacol.* 46: 167–192.

Ito, M., Sato, A., Hirabayashi, K., Tanabe, F., Shigeta, S. and Baba, M., et al. (1988). Mechanism of Inhibitory Effect of Glycyrrhizin on Replication of Human Immunodeficiency Virus (HIV). *Antiviral Res.* 10 (6): 289–298.

Johns, C. (2009). Glycyrrhizic acid toxicity caused by consumption of licorice candy cigars. *Can. J. Emerg. Med.* 11: 94–96.

Kang, H. and Lieberman, P.M. (2011). Mechanism of Glycyrrhizic Acid Inhibition of Kaposi's Sarcoma-Associated Herpesvirus: Disruption of CTCF-CohesinMediated RNA Polymerase II Pausing and Sister Chromatid Cohesion. *J. Virol.* 85 (21): 11159–11169.

Kargioglu, M., Cenkci, S., Serteser, A., Konuk, M. and Vural, G. (2010). Traditional uses of wild plants in the middle Aegean region of Turkey, *Hum. Ecol.* 38 (3): 429–450.

Korenaga, M., Hidaka, I., Nishina, S., Sakai, A., Shinozaki, A. and Gondo, T., et al. (2011). A Glycyrrhizin-Containing Preparation Reduces Hepatic Steatosis Induced by Hepatitis C Virus Protein and Iron in Mice. *Liver Int.* 31 (4): 552–560.

Kumar, A. and Dora, J. (2012). Review on *Glycyrrhiza glabra* (Liquorice). *Journal of Pharmaceutical and Scientific Innovation (JPSI)*, 1(2): 1-4.

Laconi, S., Madeddu, M.A. and Pompei, R. (2014). Autophagy activation and antiviral activity by a licorice triterpene. *Phytother Res.* 28:1890–1892.

Lin, J.C. (2003). Mechanism of Action of Glycyrrhizic Acid in Inhibition of Epstein-Barr Virus Replication *In Vitro*. *Antiviral Res.* 59 (1): 41–47.

Lin, J.C., Cherng, J.M., Hung, M.S., Baltina, L. A., Baltina, L. and Kondratenko, R. (2008). Inhibitory Effects of Some Derivatives of Glycyrrhizic Acid against Epstein-Barr Virus Infection: Structure-Activity Relationships. *Antiviral Res.* 79 (1): 6–11.

Lu, H. Y., Huo, N., Wang, G. F., Li, H. C., Nie, L. G., and Xu, X. Y. (2003). Clinical Observation of Therapeutic Effect of Compound Glycyrrhizin on SARS. *China Pharm.* (10): 34–36.

Luitel, D.R., Rokaya, M.B., Timsina, B. and Munzbergova, Z. (2014). Medicinal plants used by the Tamang community in the Makawanpur district of central Nepal. *J. Ethnobiol. Ethnomed.* 10 (1): 5.

Matsumoto, Y., Matsuura, T., Aoyagi, H., Matsuda, M., Hmwe, S.S. and Date, T. (2013). Antiviral activity of glycyrrhizin against hepatitis C virus *in vitro*. *PLoS One.* 8: e68992.

Matsuo, K., Takenaka, K., Shimomura, H., Fujii, N., Shinagawa, K. and Kiura, K., et al. (2001). Lamivudine and Glycyrrhizin for Treatment of Chemotherapy-Induced Hepatitis B Virus (HBV) Hepatitis in a Chronic HBV Carrier with NonHodgkin Lymphoma. *Leuk. Lymphoma,* 41 (1-2): 191–195.

Michaelis, M., Geiler, J., Naczk, P., Sithisarn, P., Ogbomo, H. and Altenbrandt, B. (2010). Glycyrrhizin inhibits highly pathogenic H5N1 influenza A virus–induced pro-inflammatory cytokine and chemokine expression in human macrophages. *Med Microbiol Immunol.* 199: 291–297.

Mori, K., Sakai, H., Suzuki, S., Sugai, K., Akutsu, Y. and Ishikawa, M., et al. (1989). Effects of Glycyrrhizin (SNMC: Stronger Neo-Minophagen C) in Hemophilia Patients with HIV Infection. *Tohoku J. Exp. Med.* 158 (1): 25–35.

Motti, R. (2021). Wild plants used as herbs and spices in Italy: an ethnobotanical review. *Plants (Basel)*, 10 (3): 563.

Murck, H. (2020). Symptomatic Protective Action of Glycyrrhizin (Licorice) in COVID-19 Infection?. *Front. Immunol.* 11: 1239.

Nazari, S., Rameshrad, M. and Hosseinzadeh, H. (2017). Toxicological effects of *Glycyrrhiza glabra* (Licorice): A Review. *Phytother. Res.* 31: 1635–1650.

Nomura, T., Fukai, T. and Akiyama, T. (2002). Chemistry of phenolic compounds of licorice (Glycyrrhiza species) and their estrogenic and cytotoxic activities. Pure and Applied Chemistry, 74(7): 1199–1206.

Obolentseva, G.V., Litvinenko, V.I., Ammosov, A.S., Popova, T.P. and Sampiev, A.M. (1999). Pharmacological and therapeutic properties of licorice preparations (a review). *Pharm Chem J,* 33:24–31.

Omar, H.R., Komarova, I., El-Ghonemi, M., Fathy, A., Rashad, R., Abdelmalak, H.D., Yerramadha, M.R., Ali, Y., Helal, E., Camporesi, E.M. (2012). Licorice abuse: Time to send a warning message. *Ther. Adv. Endocrinol. Metab*. 3: 125–138.

Pompei, R., Flore, O., Marccialis, M.A., Pani, A. and Loddo, B. (1979). Glycyrrhizic Acid Inhibits Virus Growth and Inactivates Virus Particles. *Nature*, 281 (5733): 689–690. doi:10.1038/281689a0.

Pompei, R., Paghi, L., Ingianni, A. and Uccheddu, P. (1983). Glycyrrhizic Acid Inhibits Influenza Virus Growth in Embryonated Eggs. *Microbiologica,* 6 (3): 247–250.

Rizvi, Z.A., Babele, P., Sadhu, S., Madan, U., Tripathy, M.R., Goswami, S. and Dikshit, M. (2022). Prophylactic treatment of *Glycyrrhiza glabra* mitigates COVID-19 pathology through inhibition of pro-inflammatory cytokines in the hamster model and NETosis. bioRxiv. *Front. Immunol*, 15. DOI 10.3389/fimmu.2022.945583.

Rizzato, G., Scalabrin, E., Radaelli, M., Capodaglio, G. and Piccolo, O. (2017). A new exploration of licorice metabolome. *Food Chemistry*, 221: 959–968.

Romero, M.R., Efferth, T., Serrano, M.A., Castaño, B., Macias, R.I.R. and Briz, O., et al. (2005). Effect of Artemisinin/artesunate as Inhibitors of Hepatitis B Virus Production in an "In Vitro" Replicative System. *Antiviral Res*. 68 (2): 75–83.

Sakai-Sugino, K., Uematsu, J., Kamada, M., Taniguchi, H., Suzuki, S. and Yoshimi, Y., et al. (2017). Glycyrrhizin Inhibits Human Parainfluenza Virus Type 2 Replication by the Inhibition of Genome RNA, mRNA and Protein Syntheses. *Drug Discov. Ther.* 11 (5): 246–252.

Sasaki, H., Takei, M., Kobayashi, M., Pollard, R.B., and Suzuki, F. (2002). Effect of Glycyrrhizin, an Active Component of Licorice Roots, on HIV Replication in Cultures of Peripheral Blood Mononuclear Cells from HIV-Seropositive Patients. *Pathobiology* 70 (4): 229–236.

Sato, H., Goto, W., Yamamura, J., Kurokawa, M., Kageyama, S. and Takahara, T., et al. (1996). Therapeutic Basis of Glycyrrhizin on Chronic Hepatitis B. *Antivir. Res*. 30 (2-3): 171–177.

Sekizawa, T., Yanagi, K. and Itoyama, Y. (2001). Glycyrrhizin Increases Survival of Mice with Herpes Simplex. *Acta Virol*. 45 (1): 51–54.

Setayesh, M., Karimi, M., Zargaran, A., Abousaidi, H., Shahesmaeili, A., Amiri, F. and Hasheminasab, F. S. (2022). Efficacy of a Persian herbal medicine compound on coronavirus disease 2019 (COVID-19): a randomized clinical trial. *Integrative Medicine Research*, 11(3):100869.

Sharma, V., Katiyar, A. and Agrawal, R.C. (2018). Glycyrrhiza glabra: Chemistry and Pharmacological Activity. In: Mérillon, JM., Ramawat, K. (eds) Sweeteners. Reference Series in Phytochemistry. Springer, Cham. https://doi.org/10.1007/978-3-319-27027-2_21.

Shi, X., Yu, L., Zhang, Y., Liu, Z., Zhang, H. and Zhang, Y., et al. (2020). Glycyrrhetinic Acid Alleviates Hepatic Inflammation Injury in Viral Hepatitis Disease via a HMGB1-TLR4 Signaling Pathway. *Int. Immunopharmacol*. 84: 106578.

Simmler, C., Pauli, G.F. and Chen, S.N. (2013). Phytochemistry and biological properties of glabridin. *Fitoterapia*, 90: 160–184.

Sinha, S.K., Prasad, S.K., Islam, M.A., Chaudhary, S.K., Singh, S. and Shakya, A. (2021). Potential Leads from Liquorice Against SARS-CoV-2 Main Protease Using Molecular Docking Simulation Studies. *Comb. Chem. High Throughput. Screen* 24 (4): 591–597.

Sinha, S.K., Prasad, S.K. and Islam, M.A. (2020). Identification of Bioactive Compounds from glycyrrhiza Glabra as Possible Inhibitor of Sars-Cov-2 Spike Glycoprotein and Non-structural Protein-15: A Pharmacoinformatics Study. *J. Biomol. Struct. Dyn.* 1–15. doi:10.1080/07391102.2020.1779132.

Smirnov, V.S., Zarubaev, V.V., Anfimov, P.M. and Shtro, A.A. (2012). Effect of a combination of glutamyl-tryptophan and glycyrrhizic acid on the course of acute infection caused by influenza (H3H2) virus in mice. *Vopr Virusol.* 57: 23–27.

Sun, Z., He, G., Huang, N., Thilakavathy, K., Lim, J.C.W., Kumar, S.S. and Xiong, C. (2021). Glycyrrhizic Acid: A Natural Plant Ingredient as a Drug Candidate to Treat COVID-19. *Front. Pharmacol.* 12:707205. doi: 10.3389/fphar.2021.707205.

Sokolov, S., Zamotayev, I. Directory of Medicinal Plants; Medicina: Moscow, Russia, 1985. (In Russian)

Soufy, H., Yassein, S., Ahmed, A.R., Khodier, M.H., Kutkat, M.A. and Nasr, S.M. (2012). Antiviral and immune stimulant activities of glycyrrhizin against duck hepatitis virus. *Afr J Tradit Complement Altern Med.* 9: 389–395.

Takahara, T., Watanabe, A. and Shiraki, K. (1994). Effects of Glycyrrhizin on Hepatitis B Surface Antigen: a Biochemical and Morphological Study. *J. Hepatol.* 21 (4): 601–609.

Takei, M., Kobayashi, M., Li, X.-D., Pollard, R.B. and Suzuki, F. (2005). Glycyrrhizin Inhibits R5 HIV Replication in Peripheral Blood Monocytes Treated with 1-methyladenosine. *Pathobiology*, 72 (3): 117–123.

Tandon, A., Tandon, B.N., and Bhujwala, R.A. (2001). Treatment of Sub-acute Hepatitis with Lamivudine and Intravenous Glycyrrhizin: a Pilot Study. *Hepatol. Res.* 20 (1): 1–8.

Thakur A.K. and Raj P. (2017). Pharmacological perspective of *Glycyrrhiza glabra* Linn.: A Mini-Review. *J. Anal. Pharm. Res.* 5: 00156. doi: 10.15406/japlr.2017.05.00156.

van de Sand, L., Bormann, M., Alt, M., Schipper, L., Heilingloh, C.S. and Steinmann, E., et al. (2021). Glycyrrhizin Effectively Inhibits SARS-CoV-2 Replication by Inhibiting the Viral Main Protease. *Viruses*, 13 (4): 609.

Vaya, J., Belinky, P.A. and Aviram, M. (1997). Antioxidant Constituents from Licorice Roots: Isolation, Structure Elucidation and Antioxidative Capacity Toward LDL Oxidation. *Free Radical Biology and Medicine*, 23 (2): 302–313.

Wahab, S., Annadurai, S., Abullais, S.S., Das, G., Ahmad, W., Ahmad, M.F., Kandasamy, G., Vasudevan, R., Ali, M.S. and Amir, M. (2021). *Glycyrrhiza glabra* (Licorice): A Comprehensive Review on Its Phytochemistry, Biological Activities, Clinical Evidence and Toxicology. *Plants (Basel).* 10(12): 2751. doi: 10.3390/plants10122751.

Wang, J.H., Kwas, C. and Wu, L. (2009). Intercellular adhesion molecule 1 (ICAM-1), but not ICAM-2 and -3, is important for dendritic cell-mediated human immunodeficiency virus type 1 transmission. *J Virol.* 83: 4195–4204.

Wang, L., Yang, R., Yuan, B., Liu, Y. and Liu, C. (2015a). The antiviral and antimicrobial activities of licorice, a widely-used Chinese herb. *Acta Pharmaceutica Sinica B*, 5(4): 310–315. doi: 10.1016/j.apsb.2015.05.005.

Wang, Q., Qian, Y., Wang, Q., Yang, Y.f., Ji, S., Song, W. and Ye, M. (2015b). Metabolites identification of bioactive licorice compounds in rats. *Journal of Pharmaceutical and Biomedical Analysis*, 115: 515–522.

Wolkerstorfer, A., Kurz, H., Bachhofner, N. and Szolar, O. H. J. (2009). Glycyrrhizin Inhibits Influenza A Virus Uptake into the Cell. *Antiviral Res.* 83 (2): 171–178.

Wu, C.H., Xu, X.Y., Lu, H.Y., ling, X.H., Hou, F.Q., Yu, Y.Y., et al. (2004). Analysis of the Chest X-ray Manifestations in SARS Patients Treated with Compound Glycyrrhizin. *China Pharm.* (01): 40–42.

Xie, Y., Ruan, B., Chen, Y., Wu, N., Hu, M. and Zhu, B. (2011). Kaposi's Sarcoma Associated Herpesvirus Infection in Chinese Patients with Chronic Hepatitis B. *J. Med. Virol.* 83 (5): 879–883.

Yeh, C.F., Wang, K.C., Chiang, L.C., Shieh, D.E., Yen, M.H. and Chang, J.S. (2013). Water extract of licorice had anti-viral activity against human respiratory syncytial virus in human respiratory tract cell lines. *J. Ethnopharmacol.* 148: 466–473.

Yoshida, T., Kobayashi, M., Li, X. D., Pollard, R.B. and Suzuki, F. (2009). Inhibitory Effect of Glycyrrhizin on the Neutrophil-dependent Increase of R5 HIV Replication in Cultures of Macrophages. *Immunol. Cel Biol.* 87 (7): 554–558.

Yu, J.Y., Ha, J.Y., Kim, K.M., Jung, Y.S., Jung, J.C. and Oh, S. (2015). Anti-inflammatory activities of licorice extract and its active compounds, glycyrrhizic acid, liquiritin and liquiritigenin, in BV2 cells and mice liver. *Molecules*, 20(7): 13041–13054.

Yu, S., Zhu, Y., Xu, J., Yao, G., Zhang, P., Wang, M., et al. (2021). Glycyrrhizic Acid Exerts Inhibitory Activity against the Spike Protein of SARS-CoV-2. *Phytomedicine,* 85: 153364.

Zhang, H.C., Song, Y.X. and Zhang, Z.C. (2012). Glycyrrhizin administration ameliorates coxsackievirus B3–induced myocarditis in mice. *Am J Med Sci.* 344: 206–210.

Index

For Product Safety Concerns and Information please contact our EU representative GPSR@taylorandfrancis.com Taylor & Francis Verlag GmbH, Kaufingerstraße 24, 80331 München, Germany